OXFORD SERIES ON ADVANCED MANUFACTURING

SERIES EDITORS

J. R. CROOKALL
MILTON C. SHAW
NAM P. SUH

METAL CUTTING PRINCIPLES

MILTON C. SHAW

Professor of Engineering
Arizona State University

CBS Publishers & Distributors Pvt. Ltd.

New Delhi • Bengaluru • Chennai • Kochi • Kolkata • Mumbai
Hyderabad • Nagpur • Patna • Pune • Vijayawada

ISBN: 81-239-0136-4

First Indian Edition: 1992
Reprint: 1999, 2002, 2003

Published by:
Satish Kumar Jain for CBS Publishers & Distributors Pvt. Ltd.,
4819/XI Prahlad Street, 24 Ansari Road, Daryaganj, New Delhi - 110002
delhi@cbspd.com, cbspubs@airtelmail.in • www.cbspd.com
Ph.: 23289259, 23266861, 23266867 • Fax: 011-23243014

Corporate Office: 204 FIE, Industrial Area, Patparganj, Delhi - 110 092
Ph: 49344934 • Fax: 011-49344935
E-mail: publishing@cbspd.com • publicity@cbspd.com

Branches:
- ***Bengaluru:*** 2975, 17th Cross, K.R. Road, Bansankari 2nd Stage, Bengaluru - 70 • Ph: +91-80-26771678/79 • Fax: +91-80-26771680 E-mail: cbsbng@gmail.com, bangalore@cbspd.com
- ***Chennai:*** No. 7, Subbaraya Street, Shenoy Nagar, Chennai - 600030 Ph: +91-44-26681266, 26680620 • Fax: +91-44-42032115 E-mail: chennai@cbspd.com
- ***Kochi:*** Ashana House, 39/1904, A.M. Thomas Road, Valanjambalam, Ernakulum, Kochi • Ph: +91-484-4059061-65 Fax: +91-484-4059065 • E-mail: cochin@cbspd.com
- ***Kolkata:*** 6-B, Ground Floor, Rameshwar Shaw Road, Kolkata - 700014 Ph: +91-33-22891126/7/8 • E-mail: kolkata@cbspd.com
- ***Mumbai:*** 83-C, Dr. E. Moses Road, Worli, Mumbai - 400018 Ph: +91-9833017933, 022-24902340/41 • E-mail: mumbai@cbspd.com

Representatives:
- Hyderabad: 0-9885175004
- Nagpur: 0-9021734563
- Patna: 0-9334159340
- Pune: 0-9623451994
- Vijayawada: 0-9000660880

Printed at:
Nazia Printers, Delhi

Dedicated to

my wife

Mary Jane

Dedicated to

my wife

Mary Jane

PREFACE

Metal cutting is one of the most important methods of removing unwanted material in the production of mechanical components. This treatment identifies the major problem areas and relates observed performance to fundamentals of physics, chemistry, materials behavior, and the engineering sciences of heat transfer, solid mechanics, and surface science (tribology).

The basic two dimensional (orthogonal) cutting process is first analyzed in detail followed by consideration of representative three dimensional cutting operations. Special attention is directed toward cutting temperatures, tool wear and tool life, and the integrity of finished surfaces. Machining economics and optimization of cutting processes are discussed in terms of representative examples.

Cutting processes are unusually complex largely due to the fact that two basic operations occur simultaneously in close proximity with strong interaction:

1. large strain plastic deformation in a zone of concentrated shear;
2. material transport along a heavily loaded region of relative motion between chip and tool.

In general, several simplified models which emphasize different aspects of the problem such as thermal, material, and surface considerations are operative simultaneously with varying degrees of importance depending on specific machining conditions. Due to complexities of the problem, a general predictive theory is not possible. Instead of seeking the impossible, a more practical approach is adopted in which a wide variety of experiences are explained in fundamental terms. Thus, the aim is to illustrate how fundamental concepts may be used to explain observed results from carefully planned experiments and how solutions to new machining situations may be achieved by application of scientific principles. Where possible, theoretical discussions are kept simple by use of techniques such as dimensional analysis.

The book should serve as a valuable reference to those engaged in research in metal cutting or as a text in a graduate subject concerned with manufacturing engineering. The background material concerning the plastic flow and fracture of solid materials and the special behavior of surfaces contains many new ideas and should be of interest to a wider group of engineers than those engaged in production engineering.

This monograph began as a set of photo-offset lecture notes first published in 1950 by Technology Press (MIT, Cambridge, Ma.) with the title *Metal Cutting*

Principles. Although these notes were issued in three successive editions (the third and last in 1954), they never appeared in complete form. For example, there was no Chapter 8. This was not because of any superstition attached to that number but because Chapter 8 was reserved for what has come to be known as the 'shear angle problem' (also Chapter 8 in the present work). In the early 1950s there was considerable interest in this problem and always the hope that a satisfactory general solution was soon to be found. While there have been countless attempts, the ideal solution has never been found largely for reasons discussed in the present Chapter 8.

I have worked continuously on metal removal problems with many very talented graduate students for over thirty-five years (Massachusetts Institute of Technology (MIT), 1946–61; Carnegie-Mellon University (CMU), 1961–78; and Arizona State University (ASU), 1978–). This book largely records and integrates their collective ideas and carefully performed experiments. Generous support for this research has been received from a number of industrial companies, the US Government and the three institutions with which I have been associated.

Metal cutting is a time honored activity having a rich literature much of which goes back well before the work of F. W. Taylor at the turn of the present century. No attempt has been made to review all that has been written on a given subject. I have been selective in citing literature that seems to support points of view I consider closest to fact in each problem considered. Ideas and experimental results have been considered from a number of sources and in each case I have tried to clearly acknowledge them throughout the book.

While I have done my best to eliminate errors of all types, I do not expect complete success in this endeavor and should appreciate being informed of errors which still persist.

I wish to acknowledge with thanks the efforts of Mr Young Moon Lee of Kyungpook National University, Taegu, Korea, who critically reviewed the entire manuscript and uncovered many errors and inconsistencies. Special thanks are also due to Ms Elinor M. Lindenberger of ASU for carefully typing most of the manuscript. My wife, Mary Jane, spent many hours editing the text, reading proofs and generally improving the manuscript for which I am most grateful.

Tempe, Arizona M. C. S.
August 1983

CONTENTS

SYMBOLS

A	area
	apparent area of contact
	constant in eqn (9.11)
A_m	factor defined in eqn (12.11)
A_R	real area of contact
A_S	area of shear plane
$\bar{A}$	area factor (a function of aspect ratio of slider, m/l in Fig. 12.9)
AB	chordal size of 'ear' type chip (Fig. 18.9)
AB^*	nondimensional chip shape $= AB/R_c$ (Fig. 18.9)
B	volume worn away (Chapter 11)
	Bainite
	constant in eqn 9.11
C	constant in Taylor tool life equation (eqn 11.12)
	stress concentration factor
	constraint factor for an indentor
	volume specific heat
	corner angle of milling cutter
	controlled-contact tool-face length
	concentration of additive in cutting fluid (volume fraction)
	cutting efficiency
C_e	end-cutting edge angle
C_s	side-cutting edge angle
$\bar{C}$	iron carbide (Fe_3C)
CE	chip equivalent (Chapter 17)
D	diameter
E	Young's modulus of elasticity
	output voltage from strain gage
F	force (fundamental dimension in dimensional analysis)
	shear force
	fraction failed in Weibull statistics
F_C	cutting force component parallel to tool face
F_o	probability of failure at given stress level
F_P	cutting force component in power direction (i.e. parallel to V)
F_Q	cutting force component in undeformed chip thickness direction
F_R	cutting force component perpendicular to F_P and F_Q
F_S	cutting force component parallel to shear plane
G	shear modulus of elasticity
H_B	Brinell hardness

H_K	Knoop hardness
H_M	Meyer hardness
H_{RA}	Rockwell A hardness
H_{RB}	Rockwell B hardness
H_{RC}	Rockwell C hardness
H_S	Moh's hardness
H_V	Vickers hardness
I	area moment of inertia
J	polar moment of inertia
	mechanical equivalent of heat
K	bulk modulus
	stress intensity factor in fracture mechanics
	spring constant $[FL^{-1}]$
	probability that a real contact will result in a wear particle
	ratio of (shear stress on shear plane)/(shear stress on tool face)
	incremental speed factor in machine tool drive
	pressure coefficient of ductility in Bridgman fracture equation
L	liquid phase
L	fundamental dimension for length in dimensional analysis
	nondimensional contact length (a'/t)
	helical length of cut in turning
	sliding length (Chapter 11)
	pitch length of helix (drill)
	active length of cutting edge (Fig. 17.32b)
	nondimensional velocity quantity in Chapter 12 = $Vl/2K$
M	drilling torque
	cost machinability ratio = ϕ_1/ϕ_2
M_f	temperature at which martensitic transformation is complete
	fracture torque in drilling
M_s	temperature at which martensitic transformation begins
M_T	twisting moment
$M^{\#}$	cost machinability ratio under cost optimum conditions
N	r.p.m.
	cycles to fracture
N_C	cutting force component perpendicular to tool face
N_F	nondimensional fracture number $=(\sigma^2K)/(Eu^2b)$ (Chapter 11)
N_o	characteristic life in Weibull statistics
N_S	cutting force component perpendicular to shear plane
N_W	wear number = BH/LP (Chapter 11)
P	pearlite
P	force
	principal stress in Griffith crack initiation analysis
	probability

P_f brittle fracture load
P'_f brittle fracture load per unit length
Q energy input per unit time
principal stress in Griffith crack initiation analysis
total heat flux
R resultant force on tool face
profile radius at neck of tensile specimen
heat partition coefficient in Chapter 12 = fraction of heat flowing to extensive member of sliding pair (R_1 for shear plane, R_2 for tool face)
cost ratio $= xT_d + y/x$ in Chapter 19
risk of rupture in Weibull statistics
resistance (ohms)
chip packing ratio = volume of chips/equivalent volume of uncut metal
R' resultant force on shear plane
R_a arithmetic average surface roughness
R_n indenting force due to tool tip radius (Chapter 10)
radius of principal cutting edge (Fig. 10.14)
R_t maximum peak-to-valley surface roughness
R_{nP} power component of Rn (Fig. 10.15)
R_{nQ} feed component of Rn (Fig. 10.15)
S Spherodite
S normal tensile stress (based on original area A_o)
percent surviving in Weibull statistics
total saving with stepless machine tool drive
S_d Steadite
T tempered martensite
T fundamental dimension (time) in dimensional analysis
life in Taylor tool life eqn (11.12)
surface energy $[FL^{-1}]$
absolute temperature
drilling thrust
T_c cutting time
T_d down-time to change and reset tool
T_e component of drill thrust due to 'extrusion' at web
T_f fracture thrust for drill
T_H homologous temperature = absolute temperature/absolute melting temperature (both K)
T_m velocity modified temperature
T_P surface energy associated with plastic crack growth
$T^{\#}$ cost optimum tool life (mins)
$T^{\#\#}$ production rate optimum tool life (mins)

U	total cutting energy per unit time
	elastic energy stored at crack tip per unit length in Griffith analysis (Chapter 6)
V	cutting speed
	fundamental dimension (velocity) in dimensional analysis
	voltage
	volume in Weibull statistics
V_C	chip speed
V_W	velocity of work (Chapter 16)
V'	a volume
$V^{\#}$	cost optimum cutting speed
$V^{\#\#}$	production rate optimum cutting speed
V_{60}	60-min tool-life
V_{240}	240-min tool-life
W	weight
	limiting wattage for strain-gage
Y	flow (yield) stress in uniaxial tension
Z	nondimensional quantity in Chapter 12 = $(4\pi DK)/(Vt^2)$
a	radius of contact area in indentation hardness
	radius of neck of tensile specimen
	acceleration
	imperfection spacing (Chapter 6)
a'	tool face contact length
a_1, a_2	atomic spacings
b	width of cut (depth of cut in turning)
	slider width (Chapter 10)
	width of beam
	offset distance for chip breaker (Fig. 18.8)
b_C	deformed chip width
c	half-length of crack
	length of chisel edge of drill
d	diameter
	maximum depth of crater on tool face
e	nominal strain based on original gage length (l_o)
	e.m.f.
	secondary cutting edge length (Chapter 14)
f	feed per revolution (drilling)
	feed per tooth (milling)
	width of flat in disc test
f_e	exciting frequency
g	acceleration due to gravity
	cost of cutting fluid additive per gallon (Chapter 19)
h	height of beam

	depth of heated layer (Chapter 10)
	height of scallop left on milled surface
	drop height in drop test
h_c	critical drop height for fracture
i	inclination angle
	electrical current
k	coefficient of thermal conductivity
	plane strain flow stress in shear
	Boltzmann's constant
l	length
	undeformed chip length
	axial length of cut in turning
	half-length of slider (Chapter 12)
l_C	deformed chip length
m	mass
	strain rate sensitivity index
	half-width of slider (Chapter 12)
	Weibull slope
n	strain hardening index
	exponent in Taylor tool-life equation
	number of cutter teeth
	number of revolutions
	number of planes per unit distance (Chapter 9)
p	pressure
	point half angle of drill
	normal stress at center of Mohr's circle in slip line field analysis
$\bar{p}$	mean pressure on punch face (Meyer hardness)
q	heat flux in Chapter 12 $[FL^{-2}T^{-2}]$
r	cutting ratio $= t/t_C$
	moment arm length
	nose radius of tool
t	undeformed chip thickness (feed in turning)
t	time
	axial thickness in disc test
t_C	deformed chip thickness
t_m	maximum undeformed chip thickness (milling)
u	total cutting energy per unit volume
u_A	surface energy per unit volume
u_c	specific elastic tensile energy for fracture in impact
	fracture energy
u_F	friction energy per unit volume
u_M	momentum energy per unit volume
u_o	specific energy of thermal origin (Chapter 9)

u_S	shear energy per unit volume
w	extent of wear land on clearance face of tool
	web thickness of drill
	crack width
w_C	weight of chip
x	cost of machine, operator and overhead per unit time
y	mean value of single cutting edge
y_o	initial displacement
Δy	thickness of shear plane
α	rake angle
	ferrite
	slip line coordinate direction
	fraction of A_R strongly bonded to mating surface
	angular deformation
α'	decrease in rake angle due to secondary shear (Fig. 3.17)
α_b	back rake angle
α_e	effective rake angle
α_n	normal rake angle
α_r	radial rake angle
α_s	side rake angle
α_v	velocity rake angle
β	friction angle on tool face = $\tan^{-1}(F_C/N_C)$
	slip line coordinate direction
β'	$\tan^{-1}(\tau/\sigma)$ on shear plane
γ	shear strain
γ'	specific weight of work material
$\dot{\gamma}$	time rate of shear strain
δ	linear deformation
	extent of secondary shear zone (Fig. 12.29)
	wedge angle of principal cutting edge (Fig. 10.14)
	inclination of free surface (Chapter 17)
	helix angle (drill)
ϵ	true (ln) strain
	feed angle of drill
ϵ_e	effective strain
ϵ_f	strain at fracture in chip (Chapter 18)
ϵ_U	strain at point of onset of necking = ultimate strain
ϵ_y	strain at yield point
$\epsilon_1, \epsilon_2, \epsilon_3$	principal strains
ζ	setting angle
η	angle between tool face and plane of maximum shear stress (Fig. 8.3)
η'	angle between shear plane and plane of maximum shear stress
η_C	chip flow angle (Fig. 16.5)

η_S shear flow angle (eqn 16.8)

θ temperature (fundamental dimension in dimensional analysis)
clearance angle
angular extent of BUE (built-up edge)
angle of twist in torsion test
chip breaker face angle (Fig. 18.1a)
cone half angle (for conical indentor or punch θ_e = effective cone half angle)
angle in Fig. 8.4a

θ_f workpiece surface temperature 180° from point of cutting

θ_e end-relief angle

θ_m maximum temperature

θ_o ambient temperature of work (Chapter 12)

θ'_o ambient temperature of tool (Chapter 12)

θ_s side relief angle

$\bar{\theta}$ mean cutting temperature

$\bar{\theta}_S$ mean shear plane temperature (Chapter 12)

$\bar{\theta}_T$ mean temperature rise in cutting (Chapter 12)

$\bar{\theta}_t$ mean tool temperature by chip-tool thermocouple technique

$\Delta\bar{\theta}_F$ mean temperatue rise on tool face (Chapter 12)

λ chip compression ratio (reciprocal of cutting ratio $= 1/r$)

μ coefficient of tool-face friction $= \tan\theta = F_C/N_C$

ν Poisson's ratio

Π nondimensional group in dimensional analysis

ρ radius at tip of abrasive particle (Chapter 11)
radius of curvature of tool tip

ρC volume specific heat

σ normal stress

σ_C nominal uniaxial compressive stress at fracture

σ_c critical tensile stress at fracture
normal stress corrected for nonuniform stress in neck of tensile specimen

σ_D nominal tensile stress in disc test at fracture

σ_e effective normal stress

σ_f normal stress at fracture

σ_{fo} constant in Bridgman fracture equation

σ_H mean principal stress ($\cong$ hydrostatic stress)

σ_o characteristic stress in Weibull statistics

σ_x elastic tensile stress at disc center in disc test

σ_y elastic compressive stress at disc center in disc test

σ_T nominal uniaxial tensile stress at fracture

σ_U stress at onset of necking = ultimate stress

σ_{xD} nominal tensile stress at center of disc at fracture

σ_{xT}	nominal tensile stress in uniaxial tensile test at fracture
$\sigma_1, \sigma_2, \sigma_3$	principal stresses ($\sigma_1 > \sigma_2 > \sigma_3$)
$\sigma_1', \sigma_2', \sigma_3'$	principal deviator stresses
$\bar{\sigma}$	mean normal stress
	effective normal stress
τ_0	theoretical shear strength
ϕ	shear angle
	angle of shear coordinate rotation in slip line field analysis
ϕ_n	normal shear angle (Chapter 16)
ψ	some function (in dimensional analysis)
	angle between shear plane and direction of maximum deformation in chip (Fig. 18.3)
ω	angular velocity (rad s^{-1})
	angle in Fig. 8.5
¢	total cost per part
¢*	optimum cost per part
¢#	optimum cost per cut
w/o	weight percent

1

INTRODUCTION

There are several methods of changing the geometry of bulk material to produce a mechanical part:

1. By putting material together (+).
2. By moving material from one region to another (0).
3. By removing unnecessary material (—).

These operations may be performed on the atomic, micro, or macro scales. For example, electroplating and electroforming are plus operations at the atomic level while the fabrication of a structure by welding is a plus operation at the other end of the spectrum (macro joining). Rolling, forging, and extrusion are examples of (0) operations performed at the macro level while surface burnishing is at the micro level. Removal operations (—) with which this book is concerned are performed at the macro level (cutting).

Another way of classifying operations is in terms of the temperature pertaining. Mechanical properties are related to the amplitude of vibration of adjacent atoms which varies linearly with absolute temperature. The melting point of a metal represents a critical temperature where the amplitude of atomic vibration is sufficient to cause a structural change from that of a solid to that of a liquid. At equal percentages of melting temperature on the absolute temperature scale, metals have similar properties and this suggests an homologous temperature (T_H) scale which corresponds to the fraction of the melting temperature on the absolute (K or R) temperature scale. Metals deformed below $T_H = 0.5$ behave differently than those deformed above $T_H = 0.5$. Deformation below $T_H = 0.5$ is called cold working and takes place primarily within individual crystals and is relatively strain rate insensitive but strongly strain sensitive (strain hardening). Deformation above $T_H = 0.5$ is called hot working, occurs primarily by grain boundary rearrangement, and is relatively strain rate sensitive but insensitive to strain (negligible strain hardening). Machining that occurs at temperatures above $T_H = 0.5$ is called hot machining. For example, steel has a melting temperature of about 1540 °C (2800 °F) or 1540 + 273 ≃ 1810 °K (or 2800 + 460 = 3260 °R). For hot machining, steel should be cut at an homologous temperature of at least $T_H = 0.5$ (or 630 °C or 1160 °F).

There are still other ways of classifying material removal operations that will be developed in subsequent chapters of this book.

Importance of Material Removal

The importance of material removal operations in the scheme of things may be realized by considering the total cost associated with this activity, including expendable tool cost, labor cost, and cost of capital investment. In the USA, the yearly cost associated with material removal has been estimated at about 10 per cent of the gross national product.

The importance of the cutting process may be further appreciated by the observation that nearly every device in use in our complex society has one or more machined surfaces or holes.

There are several reasons for developing a rational approach to material removal:

1. To improve cutting techniques—even minor improvements in productivity are of major importance in high volume production.
2. To produce products of greater precision and of greater useful life.
3. To increase the rate of production and produce a greater number and variety of products with the tools available.

All basic fields of industrial endeavor have taken similar paths in the course of their development. The earliest work has generally been carried out on a purely empirical basis and in many instances such activities have been highly developed by following the case method. While this method presents a clear picture of each specific job, a great many cases must be considered before sufficient examples have been presented to enable all common situations to be covered. This approach has been extensively used in metal cutting as well as in other fields, such as machine design, hydraulics, metallurgy, and even such nonengineering activities as law and medicine. The weakness of the method lies in its failure to provide a direct means for solving problems which lie beyond the range of current experience. Each new case that is established must be arrived at by a costly procedure of trial and error.

Not too many years ago, steam turbines and power machinery were designed largely in accordance with the judgement of the designer, rather than by following the more rational approach involving stress analysis that is in wide use today. Similarly, the design of hydraulic conduits and machinery that once was done by rule-of-thumb procedures is now being accomplished largely with the aid of the principles of fluid mechanics. In the field of metallurgy, steelmaking is being carried out by considering it a special problem in physical chemistry instead of employing the age-old recipe technique. Metal cutting tools and procedures are still largely established by the old case method. This activity has resisted the impact of modern technology and the scientific method, mainly due to the complexity of the operations but also partly due to the attitude held toward metal cutting in engineering schools.

Traditionally, metal cutting has been part of the training of mechanical engineers. However, in the past, a trade-school approach was generally adopted, emphasis being placed entirely upon nomenclature, the mastery of machine manipulation, and the learning of a large number of disconnected empirical rules. In some instances, the major objective has actually been the production of trinkets, thus appealing to the hobby instincts of the student rather than developing the ability to apply fundamental concepts.

Approach to Subject

In this treatment of the subject we will consider the cutting process in fundamental terms. The objective is to explain a number of commonly observed results rather than to present a large mass of empirical constants and a large number of empirical relationships of limited applicability. In this first chapter, we will briefly consider the subject from a qualitative point of view. Basic definitions and concepts will be given and discussed. The wide scope of material removal will be presented and the relation to other fields of science and engineering indicated.

Classification of Material Removal Operations

The entire field of material removal may be divided into the following categories mainly in terms of the size of the individual elements removed:

Cutting
Grinding
Special techniques.

Cutting operations involve the removal of macroscopic chips in the form of ribbons or particles having a thickness of from about 0.025 mm (10^{-3} in) to 2.5 mm (10^{-1} in). A wide range of kinematic arrangements briefly discussed in the next chapter are involved in cutting. Grinding operations usually involve subdivision of the material removed into smaller particles than in cutting. Grinding chips will usually range in thickness from 0.0025 mm (10^{-4} in) to 0.25 mm (10^{-2} in). Other removal techniques such as electrochemical machining (ECM), electrodischarge machining (EDM), ultrasonic machining (UM), or electron beam machining (EBM) involve chips of atomic or submicroscopic size.

Overview of the Cutting Process

The chisel (Fig. 1.1) was probably one of the first cutting tools used by man. The earliest stone implements were undoubtedly blunt as shown in Fig. 1.1(a) but as experience was gained, the importance of three basic angles became apparent—the rake angle (α), the clearance angle (θ), and the setting angle (ζ)

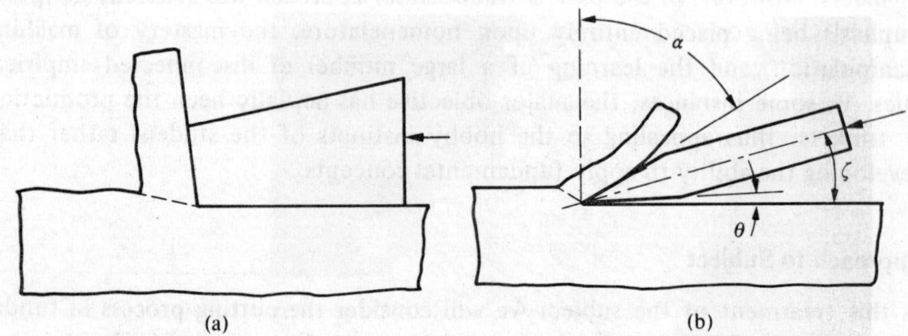

FIG. 1.1 Probable forms of early cutting implements. (a) Blunt implement (b) Implement of improved design requiring less effort. θ = clearance angle, ζ = setting angle, α = rake angle.

(Fig. 1.1(b)). Modern tools have a wide variety of forms and their geometry and kinematics differ considerably from those of the chisel. However, all of our modern tools have effective rake, clearance, and setting angles.

Certain superficial observations may be made by merely observing a metal cutting tool in operation. These include:

1. The basic difference between the cutting of wood and metal. Formerly it was believed that when metal was cut, the material merely split off in front of the tool as the tool advanced—like the chip formed when an axe splits a log. When thickness of a metal chip is measured and compared with the depth of layer removed, it is found that the chip is thicker than the actual depth of layer removed and the chip correspondingly shortened.

2. There is essentially no flow of metal at right angles to the direction of chip flow. For the purpose of simplifying the geometry involved in cutting, it is advantageous to start with a two dimensional process. We then need to consider what happens in but one representative plane. Although most cutting operations involve tools and processes which are not strictly two dimensional, many processes such as planing, sawing, and certain turning operations are essentially two dimensional. Later we will briefly discuss the complications introduced by the three dimensional aspects of cutting tools.

3. Flow lines are evident on the side and back of a chip. These lines suggest that cutting involves a shearing mechanism.

4. Some chips are in the form of a continuous ribbon while others are discontinuous, being composed of individual segments.

5. The chip, tool, and workpiece are hot to the touch. Considerable thermal energy is associated with the cutting process.

The significance of these observations will be discussed in subsequent chapters.

A photomicrograph of a partially formed chip reveals much concerning the

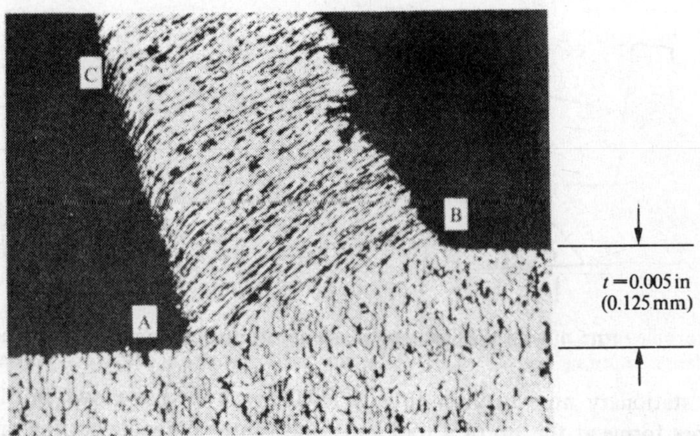

FIG. 1.2. Photomicrograph of partially formed chip. Work material, AISI 1015 steel cut at 24 f.p.m. (0.13 m s^{-1}) using water + 0.1 $^{w}/_{o}$ $NaNO_2$. Undeformed chip thickness (t) = 0.005 in (0.125 mm).

cutting process (Fig. 1.2). Such photomicrographs are obtained in the following manner. During the course of a cutting operation, the tool is brought to a sudden stop. Then the tool is carefully removed leaving a partially formed chip attached to the workpiece. The section of the metal in the vicinity of the partially formed chip is cut from the workpiece and mounted in plastic for convenience in handling. The mounted specimen is ground and polished to produce a very smooth flat surface. Next, the polished surface is etched with a fluid such as a 1 percent mixture of nitric acid in alcohol which reacts at different rates with the different components of the metal. The etched surface of the specimen is photographed through a microscope. Examination of a photomicrograph of a partially formed chip such as that of Fig. 1.2 reveals:

1. There is generally no crack extending in front of the tool point.

2. A line separates the deformed and undeformed regions. Line AB in Fig. 1.2 divides the work from the chip. The material below this line is undeformed. The chip or material above the line has been deformed by a concentrated shearing process. When the line AB is projected perpendicular to the paper and parallel to itself, it describes what is known as the shear plane (Fig. 1.3), making an angle ϕ with the direction of cut.

3. The chip is in intimate contact with the rake face of the tool from A to C and is subject to a substantial shear stress sufficient to cause the secondary subsurface to shear (as evident in Fig. 1.2 along the tool face).

4. The rate at which metal is deformed along shear plane AB is high as a consequence of the thinness of the region in which shear occurs.

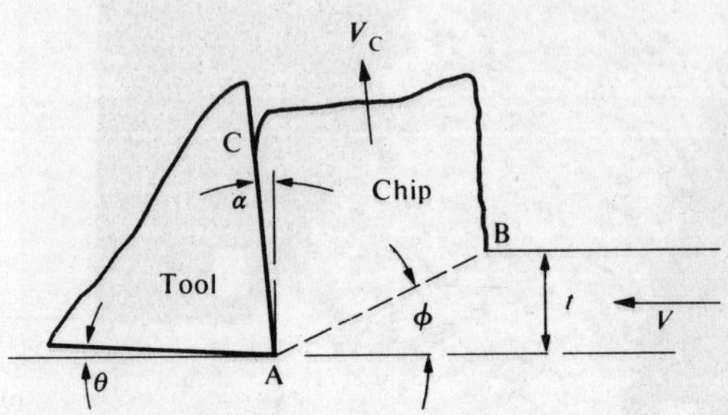

FIG. 1.3. Orthogonal (two dimensional) cutting process.

5. A stationary nose or built-up edge (BUE) such as that shown in Fig. 1.4 sometimes forms at the tip of a tool and significantly alters the cutting process. The BUE is one of the major sources of surface roughness and also plays an important role in tool wear. Its cause and significance will be treated in detail in a later chapter.

The friction between the chip and the tool plays a significant role in the cutting process. This friction may be reduced by:

1. Improved tool finish and sharpness of the cutting edge.
2. Use of low-friction work or tool materials.
3. Increased cutting speed (V).
4. Increased rake angle (α).
5. Use of a cutting fluid.

When tool face friction is decreased there is a corresponding increase in shear angle and an accompanying decrease in the thickness of the chip. As shown in Chapter 3, the plastic strain in the chip decreases as the shear angle increases (Fig. 1.5). Also, the length of the shear plane is seen to be significantly decreased as the shear angle increases. The force along the shear plane will vary as the area of the shear plane, assuming the shear stress on the shear plane remains constant.

The temperature of a cutting tool may reach a high value particularly when a heavy cut is taken at high speed. This is evident when the work or tool is touched, by the presence of temper colors on the chip, work, or tool, or may even be evident due to the loss of hardness of the tool point with an attendant loss of tool geometry and failure by excessive flow at the cutting edge.

The operational characteristics of a cutting tool are generally described by a single word—machinability. There are three main aspects of machinability:

1. Tool life.

FIG. 1.4 Photomicrograph of partially formed chip showing large built-up edge (BUE) and portions of BUE along finished surface and face of chip. Work material, AISI 1020 steel; cutting speed, 90 f.p.m. (4.84 m s^{-1}); fluid, air; $t = 0.008$ in (0.2 mm).

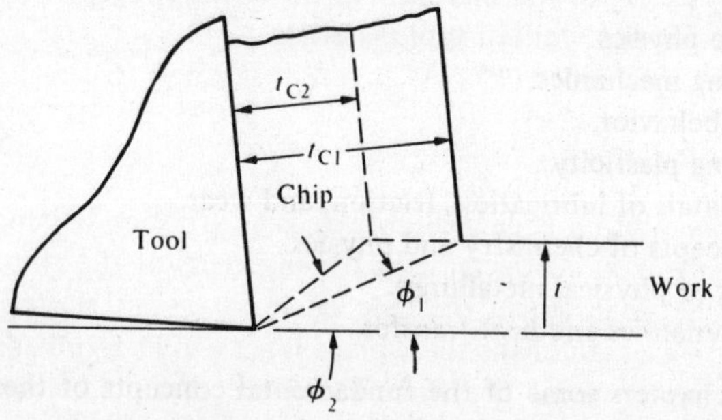

FIG. 1.5. Effect of small (ϕ_1) and large (ϕ_2) shear angle on chip thickness (t_C) and length of shear plane for a given tool and undeformed chip thickness (t).

2. Surface finish.
3. Power required to cut.

These important items will be considered in detail later.

There are three regions of interest in the cutting process. The first area shown in Fig. 1.6 extends along the shear plane and is the boundary between the deformed and undeformed material or the chip and the work. The second

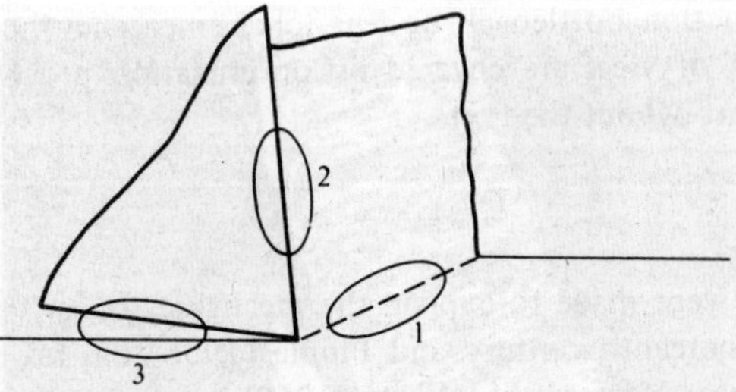

FIG. 1.6. Principal areas of interest in machining.

area includes the interface between the chip and the tool face, while the third area includes the finished or machined surface and the material adjacent to that surface. We are primarily interested in the plastic deformation characteristics of the material cut in the first area, the friction and wear characteristics of the tool-work combination in the second area, and the surface roughness produced and surface integrity in the finished surface constituting the third area.

The understanding of what happens in each of the three major regions of the cutting process involves a knowledge of several fields of engineering and science including:

1. Solid state physics.
2. Engineering mechanics.
3. Materials behavior.
4. Engineering plasticity.
5. Fundamentals of lubrication, friction, and wear.
6. Basic concepts of chemistry and physics.
7. Principles of physical metallurgy.
8. Thermodynamics and heat transfer.

In subsequent chapters some of the fundamental concepts of these many fields of endeavor will be reviewed and applied to the cutting process.

Process Optimization

Since material removal is a workshop related activity where very strong economic or production rate constraints pertain, it is important that the selection of tools, fluids, operating conditions, etc., be based upon a rational optimization procedure. Therefore, the basic principles of engineering economics will be presented and applied to optimize typical removal operations at appropriate points in the text. While the most important optimization variable is usually total cost per part (including tool, operator, and overhead costs), there are occasions when maximum rate of production must take precedence (as in the case of a production bottleneck or a national emergency); consequently, both of these points of view are covered by optimization procedures discussed at several points throughout the text.

History

Early attempts were made to explain the mechanics of cutting in fundamental terms in the nineteenth century and Finnie (1956) has reviewed some of this early work as has Zorev (1965). Shaw (1968) has reviewed some of the major metal cutting studies of the first half of the twentieth century.

References

Finnie, I. (1956). *Mech. Engng.* 78, 715.

Shaw, M. C. (1968). In *Metal transformations.* Gordon and Breach, New York, pp. 211–60.

Zorev, N. N. (1965). *Metal cutting mechanics.* Pergamon Press, Oxford.

2

TYPICAL CUTTING OPERATIONS

Principal Cutting Operations

The three most widely used cutting operations are:

Turning
Milling
Drilling

Turning (Fig. 2.1) is a process using a single point tool that removes unwanted material to produce a surface of revolution. The machine tool on which this is accomplished is a lathe (Fig. 2.2). The variables adjusted by the operator are the cutting speed V (f.p.m. or m s^{-1}), the feed t (i.p.r. or mm rev^{-1}), and the depth of cut b (in or mm). Since the depth of cut (b) is usually at least five times the feed (t), the chip is produced in plane strain and hence the width of chip is equal to the undeformed chip width (i.e. there is no spread of material parallel to the cutting edge during chip formation). The cutting edge is usually straight and may be perpendicular to the cutting speed vector or have an inclination angle (i) other than 0° (Fig. 2.1b). When $i = 0°$ and b is at least $5t$, the chip

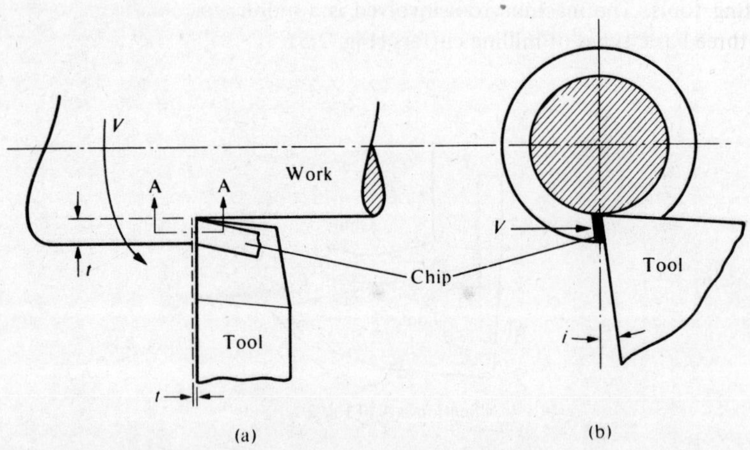

FIG. 2.1 Single point turning operation. (a) Plan view (b) Side view.

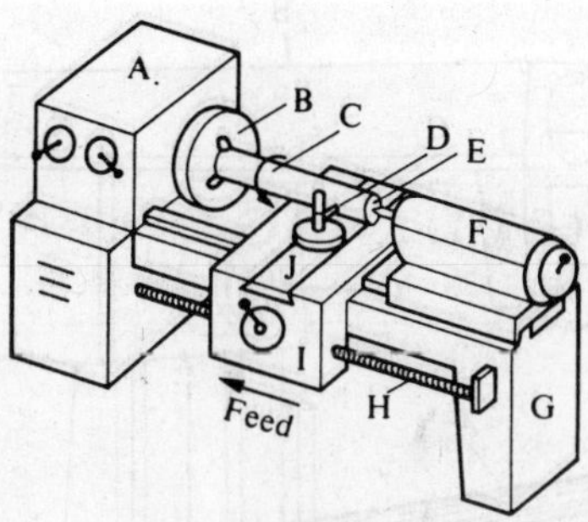

FIG. 2.2. Engine lathe.

forming process is two dimensional and the chip flows up the tool face in a direction perpendicular to the cutting edge. This is referred to as *orthogonal cutting* (Fig. 2.3). The effective rake angle (α) is then that ground on the tool measured in a plane perpendicular to the cutting edge. The undeformed chip thickness corresponds to the feed rate and the depth of cut to the chip width. When the inclination angle (i) is not 0°, the chip flow up the tool face is not perpendicular to the cutting edge and the effective rake angle should then be measured in a plane containing the chip velocity vector (V_C). As a first approximation, orthogonal cutting is assumed for all turning operations whether the inclination angle is 0° or not. In Chapter 16, the implications of inclination angles other than 0° will be considered.

Milling is a process for producing flat and curved surfaces using multipoint cutting tools. The machine tool involved is a milling machine (Fig. 2.4). There are three basic types of milling cutters (Fig. 2.5).

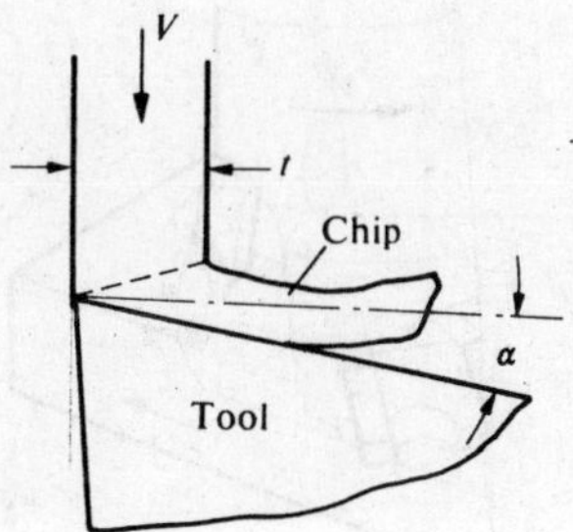

Section AA of Fig. 2.1 (i=0)

FIG. 2.3. Orthogonal turning operation.

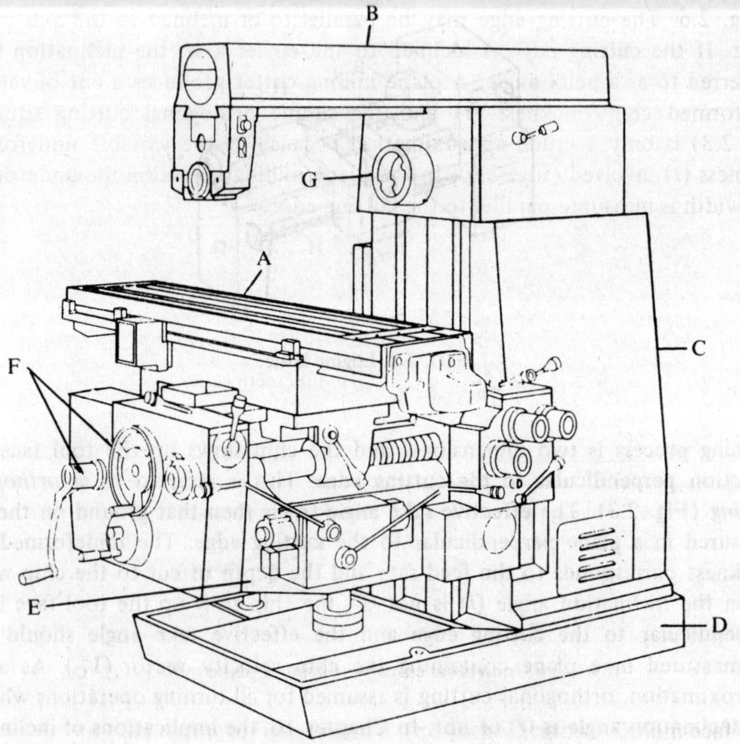

FIG. 2.4. Horizontal milling machine.

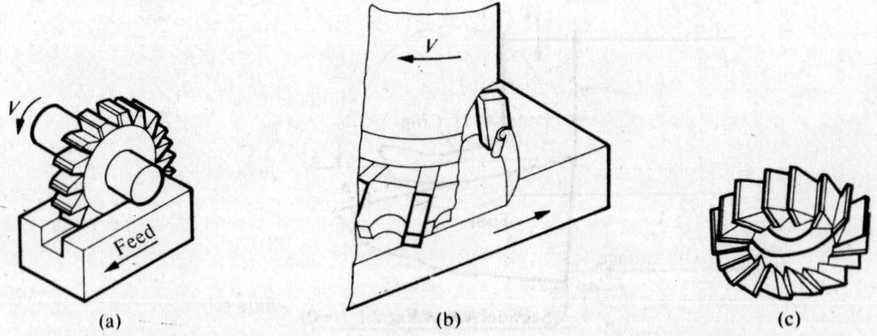

FIG. 2.5. Principal types of milling cutters. (a) Plane. (b) Face. (c) End.

The plane milling cutter is used to produce flat surfaces in the manner shown in Fig. 2.6. The cutting edge may be parallel to or inclined to the axis of the cutter. If the cutting edge is inclined to the cutter axis, the inclination angle is referred to as a helix angle. A plane milling cutter produces a cut of variable undeformed chip thickness (t) and the simple orthogonal cutting situation (Fig. 2.3) is only a crude approximation because of the variable undeformed thickness (t) involved (Fig. 2.6). In the plane milling operation the undeformed chip width is measured parallel to the cutting edge.

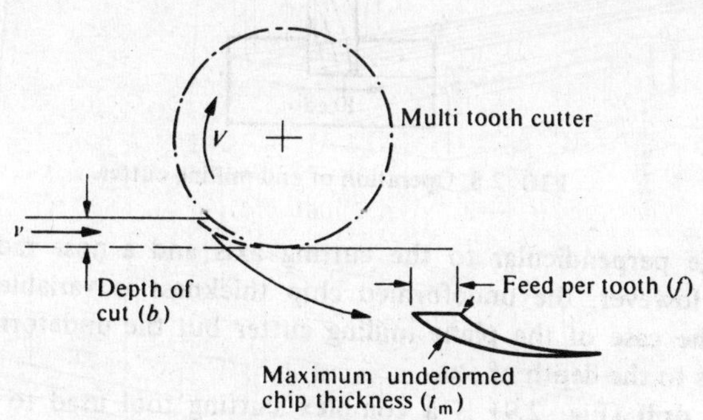

FIG. 2.6. Individual chips produced in plane milling.

A face-milling cutter (Fig. 2.5b) is also used to machine flat surfaces as shown in Fig. 2.7. However, in this case, the axis of the cutter is perpendicular to the finished surface instead of being parallel to it as in the case of a plane milling cutter. Also, the undeformed chip thickness (t) is constant throughout the cut and corresponds to the feed per tooth, while the depth of cut corresponds

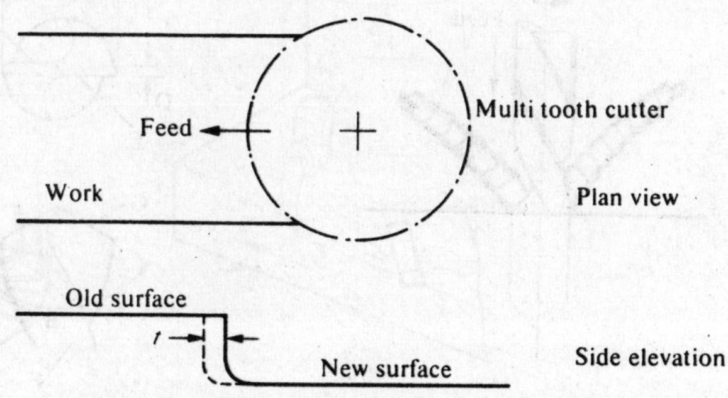

FIG. 2.7. Individual chips produced by face-milling cutter.

to the undeformed chip width (b). The inclination angle may be 0° or some value up to about 45°.

An end mill is used to produce slots and surface profiles and operates as shown in Fig. 2.8. It closely resembles a face mill in the manner it is presented to the work but is a very much smaller cutter. As in the case of the face-milling cutter, there is a major cutting edge parallel to the cutter axis, a secondary

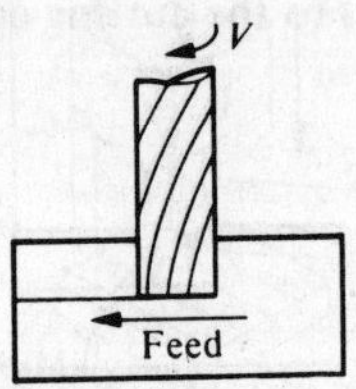

FIG. 2.8. Operation of end-milling cutter.

cutting edge perpendicular to the cutting axis and a nose radius connecting the two. However, the undeformed chip thickness is variable throughout a cut as in the case of the plane milling cutter but the undeformed chip width corresponds to the depth of cut.

A twist drill (Fig. 2.9) is a complex cutting tool used to produce rough holes. The machine tool involved in this case is a drill press (Fig. 2.10). There are two cutting edges that produce chips similar to those produced in turning on a lathe and a central web region that extrudes the metal at the center of the hole not removed by the cutting edges. Both lips (cutting edges) of a drill operate with variable rake angle, inclination angle, and clearance angle along the cutting edge. The flutes of a drill play the important role of conveying the chips out of the hole and the helix angle of the drill is important in this

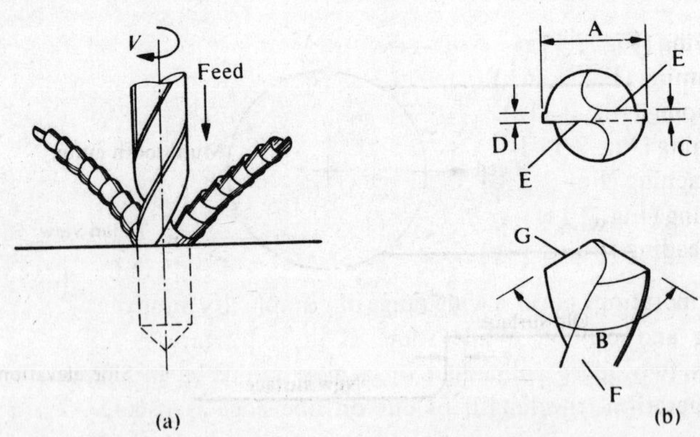

FIG. 2.9. Twist drill nomenclature.

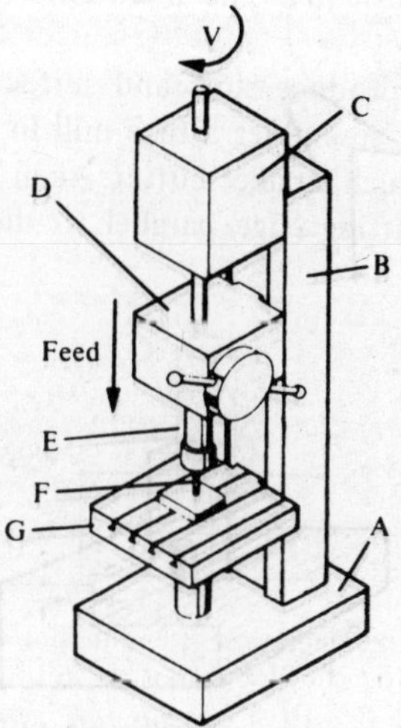

FIG. 2.10. Drill press.

connection. The fundamental quantities of importance (rake angle (α), undeformed chip thickness (t), and inclination angle (i)) are related in a complex way to be discussed in a later chapter to the quantities specified on a drill drawing (helix angle, point angle and web thickness) as well as the principal operating variables (speed and feed).

Other Cutting Operations

In addition to the three most widely used operations discussed above, there are many others including:

1. Sawing (Fig. 2.11a)
2. Reaming (Fig. 2.11b)
3. Tapping (Fig. 2.11c)
4. Planing (Fig. 2.11d)
5. Broaching (Fig. 2.11e)
6. Boring (Fig. 2.11f)
7. Threading (Fig. 2.11g)

These operations cover a wide range of complexity similar to that encountered in turning and drilling. While some of these operations involve simultaneous cutting on two edges connected by a nose radius as in the case of a straight turning operation, the length of cut on one edge is usually minor relative to the other.

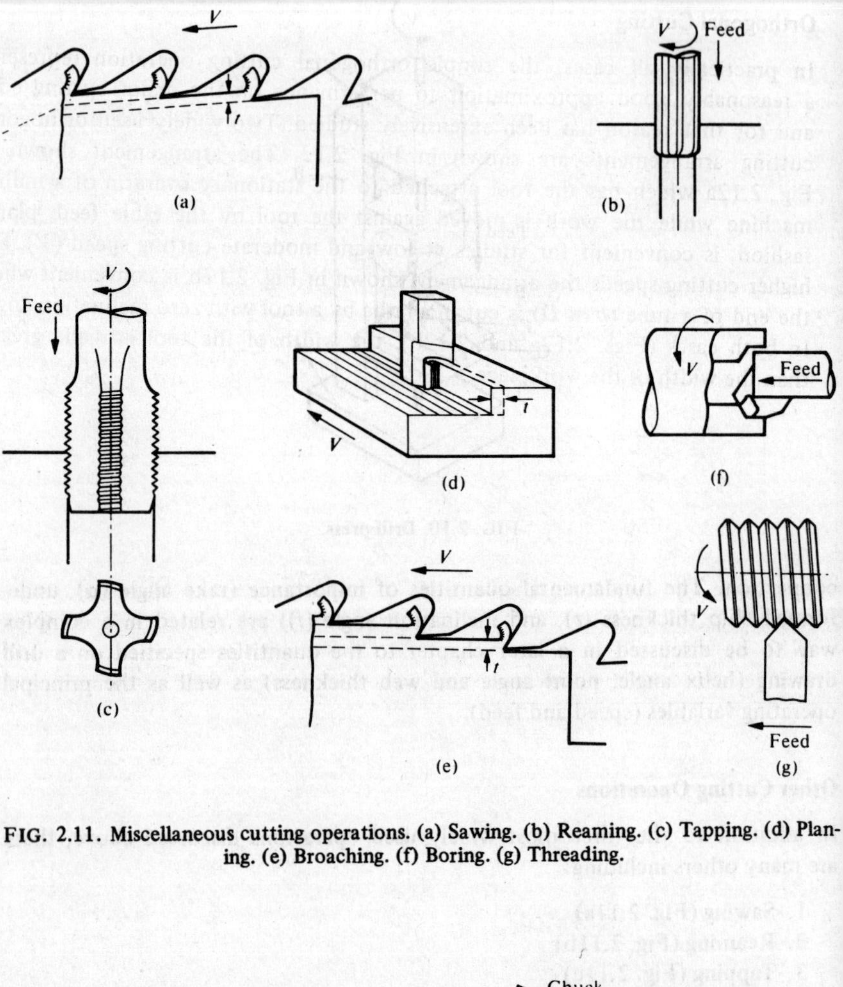

FIG. 2.11. Miscellaneous cutting operations. (a) Sawing. (b) Reaming. (c) Tapping. (d) Planing. (e) Broaching. (f) Boring. (g) Threading.

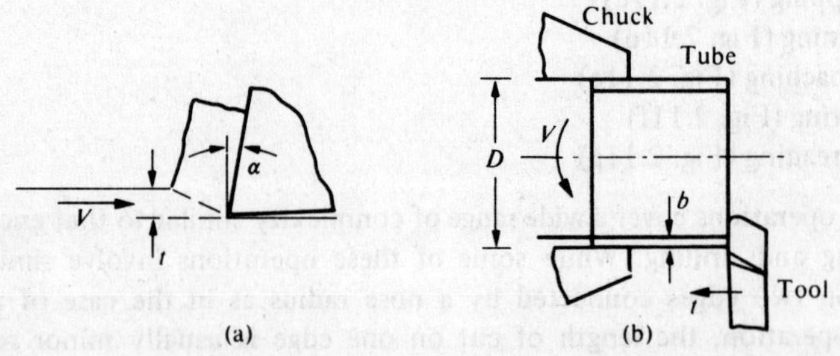

FIG. 2.12. Orthogonal cutting operations. (a) Planing. (b) Turning end of tube.

Orthogonal Cutting

In practically all cases, the simple orthogonal cutting operation represents a reasonably good approximation to performance on the major cutting edge and for that reason has been extensively studied. Two widely used orthogonal cutting arrangements are shown in Fig. 2.12. The arrangement shown in Fig. 2.12a which has the tool attached to the stationary overarm of a milling machine while the work is moved against the tool by the table feed, planer fashion, is convenient for studies at low and moderate cutting speed (V). For higher cutting speeds the arrangement shown in Fig. 2.12b is convenient where the end of a tube ($b \ll D$) is cut in a lathe by a tool with zero inclination angle. In both cases (Figs. 2.12a and 2.12b), the width of the tool is made greater than the width of the workpiece.

3
MECHANICS OF ORTHOGONAL CUTTING

As is evident in Figure 1.2, metal cutting involves concentrated shear along a rather distinct shear plane. As metal approaches the shear plane it does not deform until the shear plane is reached. It then undergoes a substantial amount of simple shear as it crosses a thin shear zone (Fig. 3.1). There is essentially no further plastic flow as the chip proceeds up the face of the tool. The small amount of secondary shear along the tool face is generally ignored in a first treatment of the cutting process and the motion of the chip along the tool face is considered to be similar to that of a friction slider of constant coefficient from A to C.

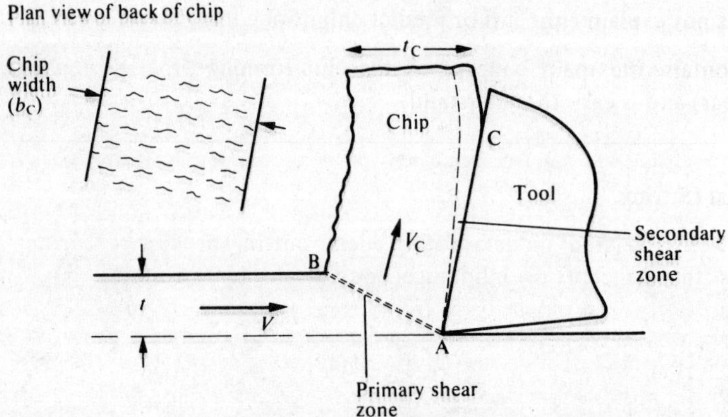

FIG. 3.1. Schematic of orthogonal cutting operation.

The back of a chip is rough due to the strain being inhomogeneous. This is due to the presence of points of weakness or stress concentration present in the metal being cut. A shear plane passing through a point of stress concentration will deform at a lower value of stress than one that does not include a point of stress concentration. Once deformation begins on a given shear plane, it tends to continue there as though the material exhibited negative strain hardening. Reasons for this apparent negative strain hardening will be discussed in Chapter 9. Thus, some metal in the chip strains more than other metal resulting in a wavy surface on the back of the chip. These waves however are not continuous clear across the back of the chip but are of limited extent in the width direction as illustrated in the insert in Fig. 3.1 which shows the back of a chip as viewed in a normal direction.

Card Model

Despite its limitations, the so-called card model of the cutting process (Fig. 3.2) due to Piispanen (1937) is very useful. This model depicts the material cut as a deck of cards inclined to the free surface at an angle corresponding to the shear angle (ϕ). As the tool moves relative to the work, it engages one card at a time and causes it to slide over its neighbor. While this model

1. exaggerates the inhomogeneity of strain,
2. depicts tool face friction as elastic rather than plastic,
3. assumes shear to occur on a perfectly plane surface,
4. ignores any BUE that may be present,
5. involves an arbitrarily assumed shear angle (ϕ),
6. does not explain chip curl or predict chip-tool contact length,

it does contain the main concept in the chip-forming process (concentrated simple shear) and is easy to understand.

Orthogonal Cutting

Figure 3.3 is a related model of the 'ideal' cutting process that completely suppresses the concept of inhomogeneous strain by assuming the material

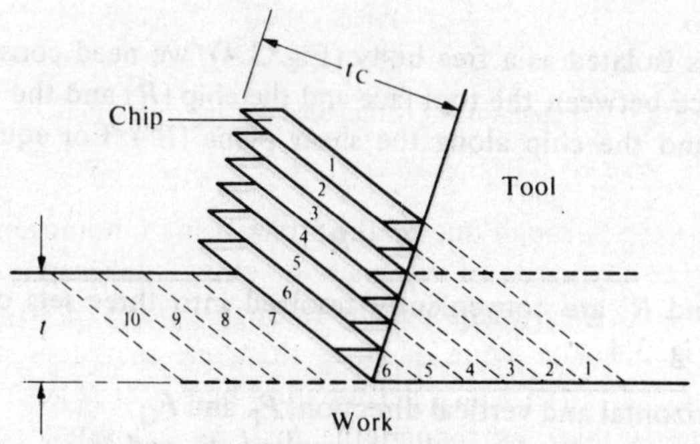

FIG. 3.2. Piispanen's idealized model of cutting process (Piispanen, 1937).

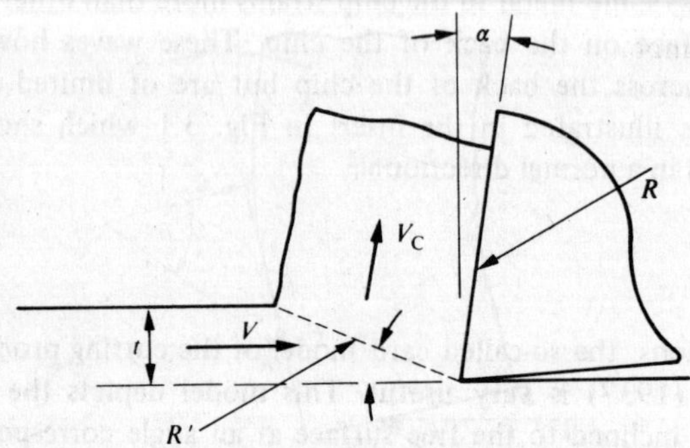

FIG. 3.3. Resultant forces R and R' in orthogonal cutting.

to behave in a completely homogeneous fashion. The assumptions on which this two dimensional model is based include

1. The tool is perfectly sharp and there is no contact along the clearance face.
2. The shear surface is a *plane* extending upward from the cutting edge.
3. The cutting edge is a straight line extending perpendicular to the direction of motion and generates a plane surface as the work moves past it.
4. The chip does not flow to either side (plane strain).
5. The depth of cut is constant.
6. The width of the tool is greater than that of the workpiece.
7. The work moves relative to the tool with uniform velocity.
8. A continuous chip is produced with no built-up edge.
9. The shear and normal stresses along shear plane and tool are uniform. (Strength of materials approach). Such an ideal two dimensional cutting operation is referred to as orthogonal cutting.

Forces

When the chip is isolated as a free body (Fig. 3.4), we need consider only two forces—the force between the tool face and the chip (R) and the force between the workpiece and the chip along the shear plane (R'). For equilibrium these must be equal

$$R = R'. \tag{3.1}$$

The forces R and R' are conveniently resolved into three sets of components as indicated in Fig. 3.4.

1. In the horizontal and vertical direction, F_P and F_Q.
2. Along and perpendicular to the shear plane, F_S and N_S.
3. Along and perpendicular to the tool face, F_C and N_C.

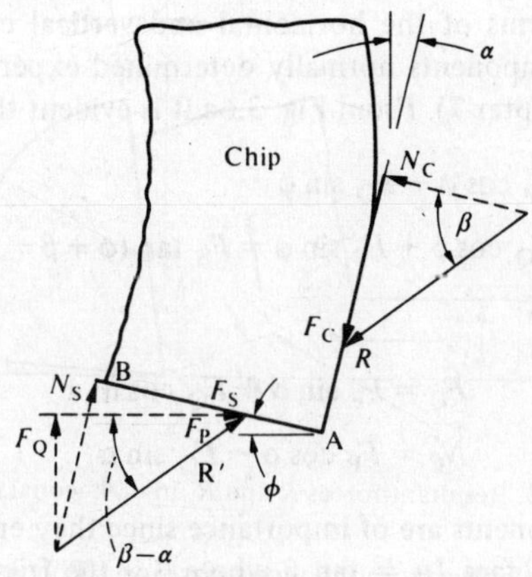

FIG. 3.4. Free body diagram of chip.

If the forces R' and R are plotted at the tool point instead of at their actual points of application along the shear plane and tool face, we obtain a convenient and compact diagram (Fig. 3.5). Here R and R' (which are equal and parallel) are coincident, and are made the diameter of the dotted reference circle shown. Then, by virtue of the geometrical fact that lines that terminate at the ends of diameter R and intersect at a point on the circle will intersect at right angles, we have a convenient means for graphically resolving R into orthogonal components in any direction. This type of plot was suggested by Merchant (1945a).

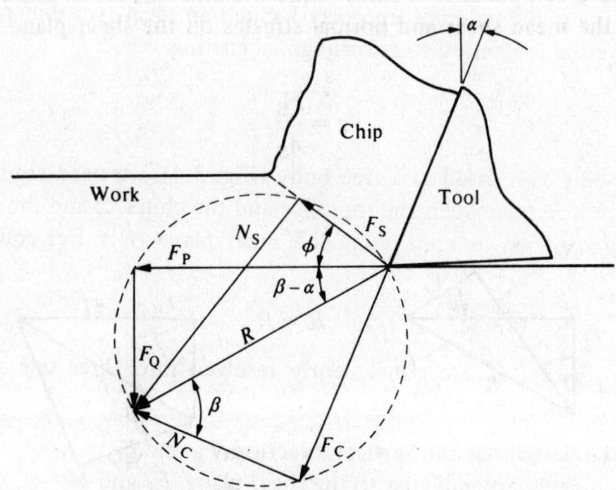

FIG. 3.5. Composite cutting force circle.

Analytical relationships may be obtained for the shear and friction components of force in terms of the horizontal and vertical components (F_P and F_Q) which are the components normally determined experimentally by means of a dynamometer (Chapter 7). From Fig. 3.6a it is evident that

$$F_S = F_P \cos\phi - F_Q \sin\phi \tag{3.2}$$

$$N_S = F_Q \cos\phi + F_P \sin\phi = F_S \tan(\phi + \beta - \alpha). \tag{3.3}$$

Similarly from Fig. 3.6b

$$F_C = F_P \sin\alpha + F_Q \cos\alpha \tag{3.4}$$

$$N_C = F_P \cos\alpha - F_Q \sin\alpha. \tag{3.5}$$

The tool face components are of importance since they enable the coefficient of friction on the tool face ($\mu = \tan\beta$ where β is the friction angle shown in Fig. 3.4) to be determined

$$\mu = \frac{F_C}{N_C} = \frac{F_P \sin\alpha + F_Q \cos\alpha}{F_P \cos\alpha - F_Q \sin\alpha} = \frac{F_Q + F_P \tan\alpha}{F_P - F_Q \tan\alpha}. \tag{3.6}$$

The nomographs given in Figs. 3.7 and 3.8 due to Merchant and Zlatin (1945) are useful for computing the friction force and the coefficient of friction. Alternatively, programs for a programmable handheld electronic calculator may be written.

Stresses

The shear plane components of force (eqns. 3.2 and 3.3) are of importance since they enable the mean shear and normal stresses on the shear plane to be determined. Thus,

$$\tau = \frac{F_S}{A_S} \tag{3.7}$$

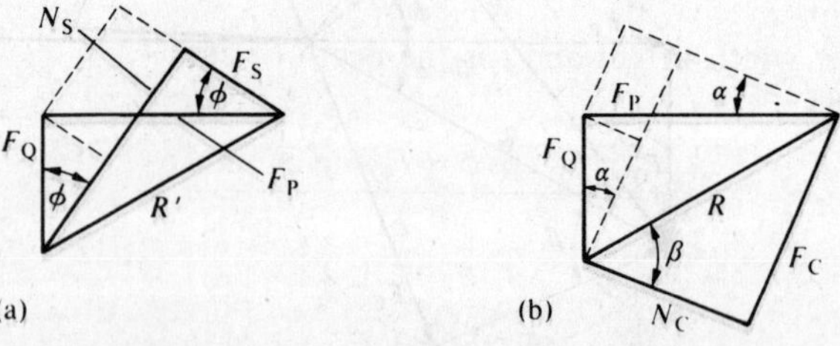

FIG. 3.6. Resolution of resultant cutting forces. (a) Shear plane. (b) Tool face.

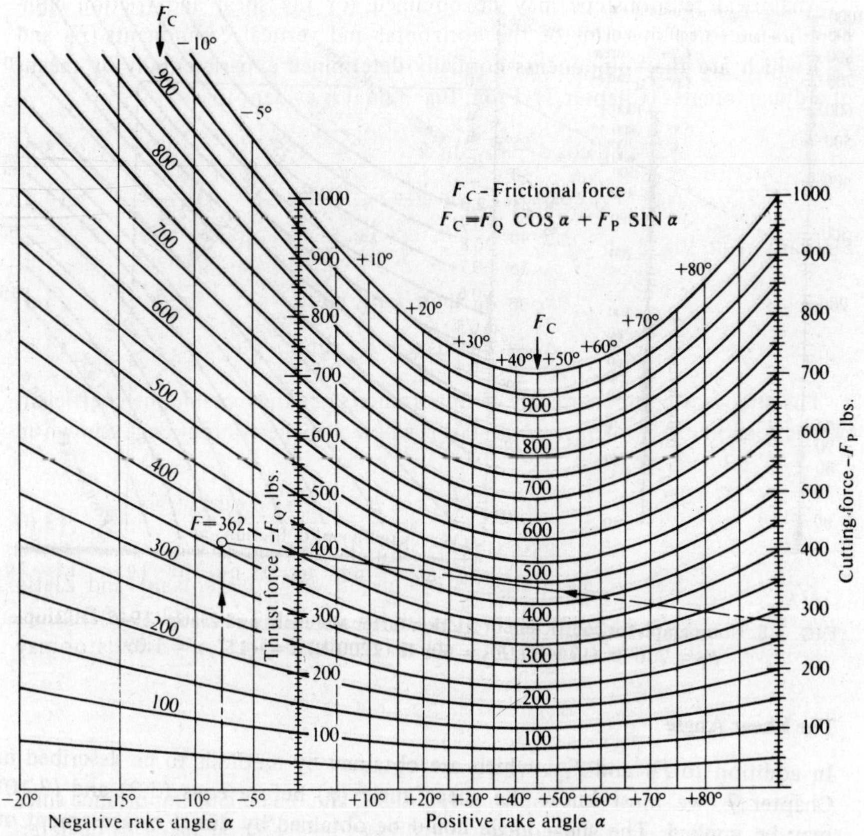

FIG. 3.7. Nomograph for friction force (after Merchant and Zlatin, 1945). Example: $F_P = 280$ lb. (1232 N) $F_Q = 400$ lb. (1760 N) $\alpha = -7.5°$, $F_C = 362$ lb. (1593 N).

where A_S is the area of the shear plane but,

$$A_S = \frac{bt}{\sin \phi} \tag{3.8}$$

where b is the width of cut and t is the depth of cut. From eqns (3.2), (3.3), (3.7), and (3.8)

$$\tau = \frac{(F_P \cos \phi - F_Q \sin \phi) \sin \phi}{bt}. \tag{3.9}$$

Similarly

$$\sigma = \frac{N_S}{A_S} = \frac{(F_P \sin \phi + F_Q \cos \phi) \sin \phi}{bt}. \tag{3.10}$$

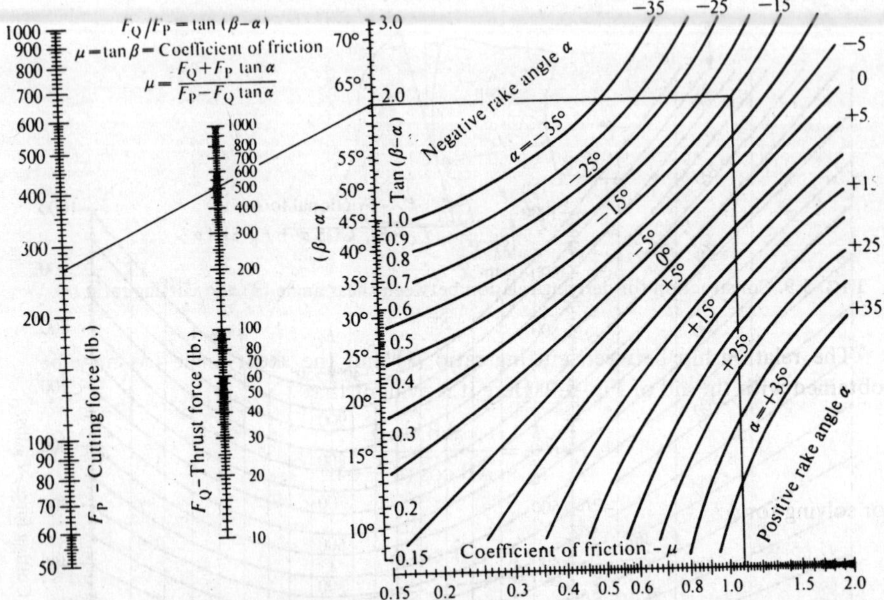

FIG. 3.8. Nomograph for coefficient of friction (after Merchant and Zlatin, 1945). Example: $F_P = 260$ lb. (1144N), $F_Q = 500$ lb (2200 N) $\alpha = -15°$, $\mu = 1.09$.

The Shear Angle

In addition to F_P and F_Q, which are obtained by methods to be described in Chapter 7, we must know the shear angle (ϕ) before eqns (3.9) and (3.10) may be applied. The shear angle could be obtained by direct measurement of a photomicrograph, but this is not very convenient. Another method of obtaining the shear angle is by means of the cutting ratio (r) which is the ratio of the depth of cut (t) to the chip thickness (t_C). This ratio may be determined by directly measuring t and t_C.

When the metal is cut it is experimentally found that there is no change in density and hence,

$$tbl = t_C b_C l_C \tag{3.11}$$

where t, b, and l are the depth of cut, width of cut, and length of cut, and the subscript C refers to the corresponding measurements on the chip. If $b/t > 5$ it is found experimentally that the width of the chip is the same as that of the workpiece and hence, from eqn (3.11)

$$\frac{t}{t_C} = \frac{l_C}{l} = r. \tag{3.12}$$

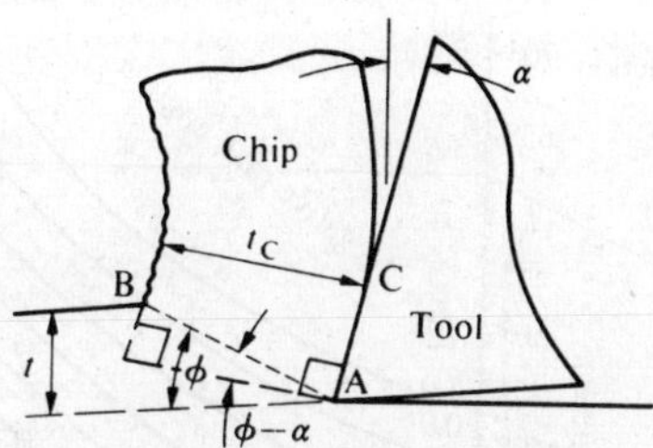

FIG. 3.9. Construction for deriving relation between shear angle (ϕ) and cutting ratio (r).

The relationship between cutting ratio (r) and the shear angle (ϕ) may be obtained with the aid of Fig. 3.9. Here it is evident that

$$r = \frac{t}{t_C} = \frac{AB \sin \phi}{AB \cos (\phi - \alpha)} \tag{3.13}$$

or solving for ϕ:

$$\tan \phi = \frac{r \cos \alpha}{1 - r \sin \alpha}. \tag{3.14}$$

This equation is most conveniently solved with the aid of the nomograph given in Fig. 3.10.

While the cutting ratio can be obtained by measuring chip thickness (t_C) and depth of cut (t), this is not the most precise procedure. Due to the roughness on the back surface of a chip it is difficult to obtain a representative chip thickness. A better procedure is to measure the chip length (l_C) and corresponding work length (l). From eqn (3.12) it can be seen that r is also equal to the chip length ratio (l_C/l). The work length l is easily determined in the case of turning. A useful technique in this type of operation is to provide the workpiece with a deep scratch which will cause a notch to be formed in the chip for each revolution. Then, the work length may be computed from the work diameter.

The work length may also be determined by weighing a chip. If the weight of a chip is w_C, then

$$1 = \frac{w_C}{tb\gamma'} \tag{3.15}$$

where γ' is the specific weight of the metal cut in pounds per cubic inch (N m^{-3}).

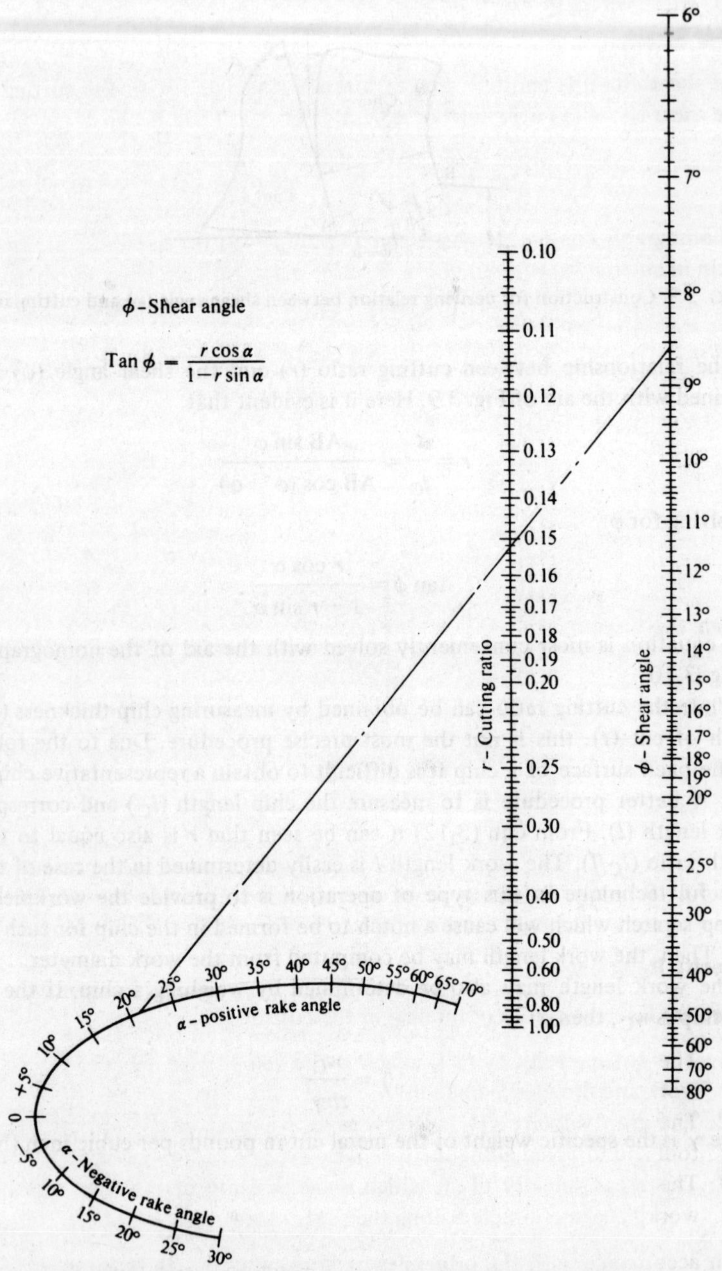

FIG. 3.10. Nomograph for shear angle (after Merchant and Zlatin, 1945). Example: Rake angle = 21°, cutting ratio = 0.152; shear angle = 8.55°.

Shear Strain

The shear strain is another item of interest associated with the cutting process. The shear strain (γ) is defined as $\Delta S/\Delta Y$ (Fig. 3.11) and hence in cutting,

$$\gamma = \frac{\Delta S}{\Delta Y} = \frac{AB'}{CD} = \frac{AD}{CD} + \frac{DB'}{CD} = \tan(\phi - \alpha) + \cot\phi. \tag{3.16}$$

A nomograph due to Merchant and Zlatin (1945) for use in determining shear strain in cutting (γ) is given in Fig. 3.12. It may also be readily shown that

$$\gamma = \frac{\cos\alpha}{\sin\phi\cos(\phi - \alpha)} \tag{3.17}$$

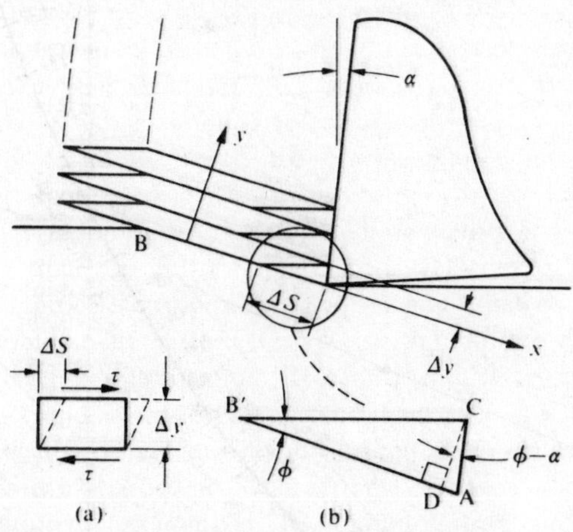

FIG. 3.11. Determination of shear strain. (a) Shear strain in general. (b) shear strain in cutting.

Velocity Relations

There are three velocities of interest in the cutting process (Fig. 3.13):

1. The cutting velocity (V), which is the velocity of the tool relative to the work and directed parallel to F_P.
2. The chip velocity (V_C), which is the velocity of the chip relative to the tool and directed along the tool face.
3. The shear velocity (V_S), which is the velocity of the chip relative to the workpiece and directed along the shear plane.

In accordance with the principles of kinematics, these three velocity vectors must form a closed velocity diagram as shown by the hodograph of Fig. 3.13. The vector sum of the cutting velocity and the chip velocity is equal to the

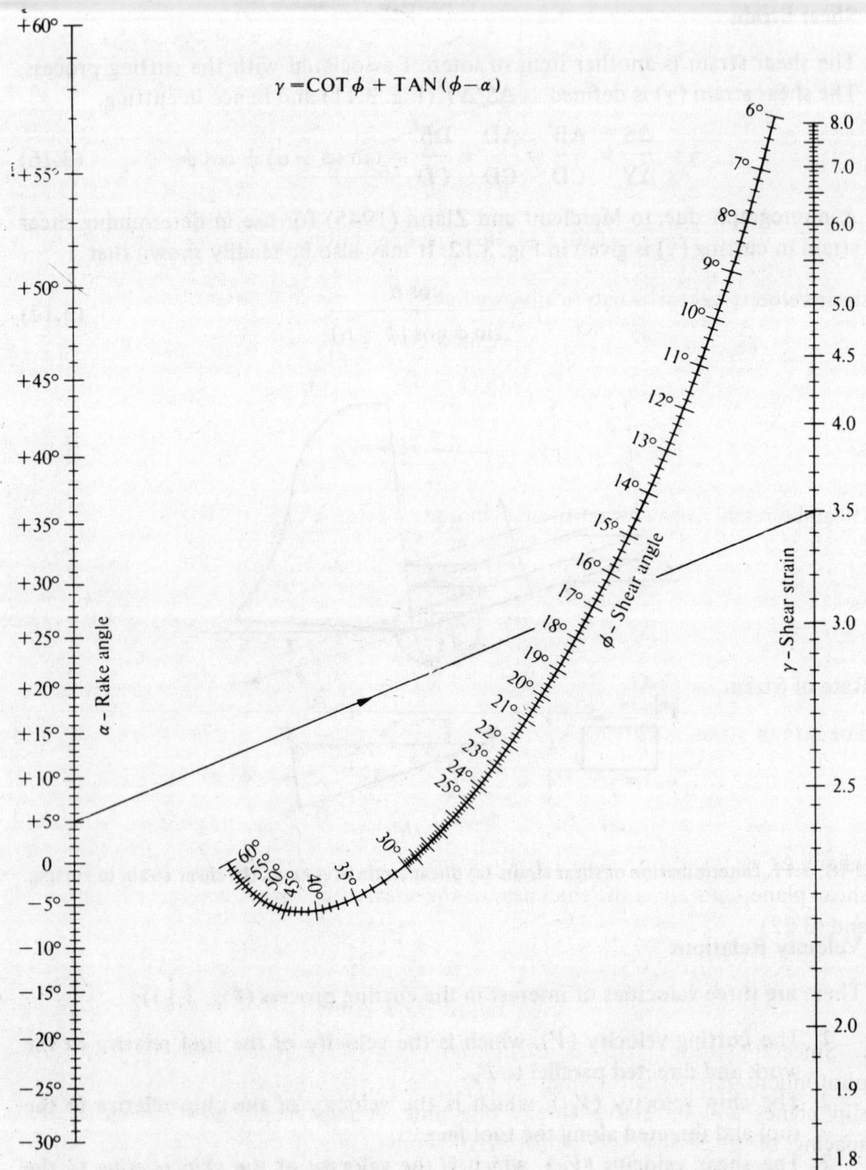

FIG. 3.12. Nomograph for shear strain (after Merchant and Zlatin, 1945). Example: Rake angle = 5°, shear angle = 17°, shear strain = 3.5.

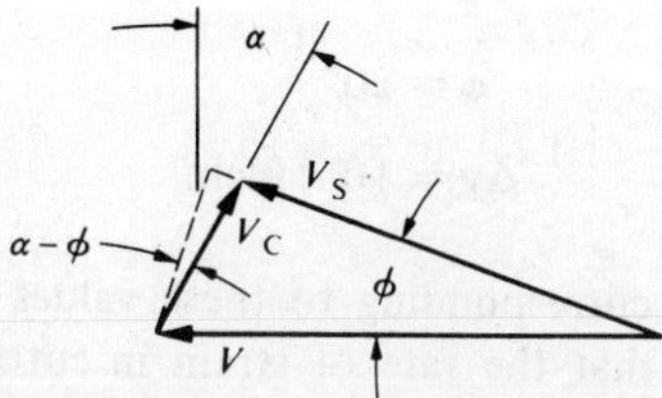

FIG. 3.13. Hodograph for cutting.

shear velocity vector. Analytically, we have

$$V_C = \frac{\sin \phi}{\cos (\phi - \alpha)} V = rV \tag{3.18}$$

$$V_S = \frac{\cos \alpha}{\cos (\phi - \alpha)} V \tag{3.19}$$

It can be readily shown that these equations may also be written

$$V_C = rV \tag{3.20}$$

$$V_S = \gamma \sin \phi \ V \tag{3.21}$$

Rate of Strain

The rate of strain in cutting is

$$\dot{\gamma} = \frac{\Delta S}{\Delta y \Delta t} = \frac{V_S}{\Delta y} \tag{3.22}$$

where Δt is the time elapsed for the metal to travel a distance ΔS along the shear plane, and Δy is the thickness of the shear zone. From eqns (3.17), (3.21), and (3.22)

$$\dot{\gamma} = \frac{\cos \alpha}{\cos (\phi - \alpha)} \frac{V}{\Delta y} \tag{3.23}$$

Since the shear 'plane' is usually very thin (Kececioglu 1958) as shown by photomicrographs such as that of Fig. 1.2, it is to be expected that very few slip planes will be active at any one time. A reasonable mean value for the spacing of successive slip planes (Δy) would appear to be about 10^{-3} inch (~ 25 μm).

An idea of the rate of strain to be expected in cutting may be had by substituting representative values into eqn 3.23 (Drucker 1949):

$$V = 100 \text{ f.p.m.}$$

$$\alpha = 0^\circ$$

$$\phi = 20^\circ$$

$$\Delta y = 10^{-3} \text{ inch.}$$

The mean rate of strain corresponding to these values is found to be 21 000 sec^{-1}. We may thus say that the rate of strain in cutting will be of the order of 10^4 s^{-1} or higher. This is recognized as a very high value when it is realized that the rate of strain in an ordinary tensile test is only about 10^{-3} s^{-1} and 10^3 s^{-1} in the most rapid impact test.

Energy Considerations

When metal is cut in a two-dimensional cutting operation, the total energy consumed per unit time is

$$U = F_P V \tag{3.24}$$

The *total energy per unit volume* of metal removed will therefore be

$$u = \frac{U}{Vbt} = \frac{F_P}{bt} \tag{3.25}$$

where b and t are the width and depth of cut respectively.

This total energy per unit volume will be consumed in several ways (Shaw, 1954):

1. As shear energy per unit volume (u_S) on the shear plane.
2. As friction energy per unit volume (u_F) on the tool face.
3. As surface energy per unit volume (u_A) due to the formation of new surface area in cutting.
4. As momentum energy per unit volume (u_M) due to the momentum change associated with the metal as it crosses the shear plane.

The shear energy per unit volume may be obtained as follows:

$$u_S = \frac{F_S V_S}{Vbt} = \tau\left(\frac{V_S}{V \sin\phi}\right). \tag{3.26}$$

However, from eqn (3.21)

$$u_S = \tau\gamma. \tag{3.27}$$

The friction energy per unit volume may be similarly found

$$u_F = \frac{F_C V_C}{Vbt} \tag{3.28}$$

or from eqn (3.20)

$$u_F = \frac{F_C r}{bt}. \tag{3.29}$$

When a new surface is generated in a solid substance, sufficient energy must be supplied to separate the ions at the interface. A certain energy is thus associated with the formation of new surface and this is usually referred to as the surface energy of the substance (T). The surface energy of a solid is analogous to the surface tension of a liquid. The surface energy per unit volume in cutting will be

$$u_A = \frac{(T)(2Vb)}{Vbt} = \frac{2T}{t}. \tag{3.30}$$

The quantity 2 in eqn (3.30) comes from the fact that two surfaces are generated simultaneously when a cut is made. The value of T for most metals is about 0.006 in lb in^{-2} (1000 dyne cm^{-1}).

The resultant momentum force associated with the change in momentum of the metal as it crosses the shear plane will obviously be in the direction of slip or along the shear plane. This resultant momentum force (F_M) may, therefore, be obtained by applying the linear momentum law of applied mechanics along the shear plane:

$$F_M = \frac{\gamma'}{g}(Vbt)\,[V_C \sin(\phi - \alpha) + V \cos\phi] \tag{3.31}$$

where γ'/g is the mass density of the metal. By use of eqns (3.13), (3.17), and (3.18) this may be written

$$F_M = \gamma'/gV^2\, bt\, \gamma \sin\phi. \tag{3.32}$$

Hence, the momentum energy per unit volume will be

$$u_M = \frac{F_M V_S}{Vbt} = \gamma'/gV\, \gamma \sin\phi\, V_S. \tag{3.33}$$

From eqn (3.21) this becomes

$$u_M = \gamma'/gV^2\, \gamma^2 \sin^2\phi. \tag{3.34}$$

As will be shown in the next section, the surface energy per unit volume (u_A) and momentum energy per unit volume (u_M) are negligible relative to the other two components and hence to a very good approximation we have

$$u = u_S + u_F. \tag{3.35}$$

Practically all of the energy associated with a cutting operation is consumed in either plastic deformation or friction, and essentially all of this ends up as thermal energy. Further details concerning the mechanics of orthogonal cutting may be found in the literature (Merchant 1945, Piispanen 1948, Drucker 1949).

TABLE 3.1

Representative metal cutting data (a) English units

The following data were obtained by Merchant (1945a), using two-dimensional tools. The work material was NE 9445 steel having a Brinell hardness of 187. The tool material was sintered carbide. The width of cut was 0.25 in (6.35 mm). All chips were of the continuous type and no cutting fluids were used.

t in (× 10^3)	V ft $\min^{-1}$	α deg	Cutting ratio r	ϕ deg	γ	F_P lb	F_Q lb	μ	τ p.s.i. (× 10^{-3})	σ p.s.i. (× 10^{-3})	u in lb in^{-3} (× 10^{-3})	u_S in lb in^{-3} (× 10^{-3})	u_S/u
3.70	197	+10	0.29	17.0	3.4	370	273	1.05	85	115	400	292	0.73
3.70	400	+10	0.33	19.0	3.1	360	283	1.11	88	137	390	266	0.68
3.70	642	+10	0.37	21.5	2.7	329	217	0.95	90	129	356	249	0.70
3.70	1186	+10	0.44	25.0	2.4	303	168	0.81	93	130	328	225	0.69
3.70	400	−10	0.32	16.5	3.9	416	385	0.64	89	153	450	342	0.76
3.70	637	−10	0.37	19.0	3.5	384	326	0.58	90	152	415	312	0.75
3.70	1160	−10	0.44	22.0	3.1	356	263	0.51	94	154	385	289	0.75
1.09	542	+10	0.33	19.0	3.1	127	101	1.12	103	164	465	308	0.66
2.34	542	+10	0.32	18.5	3.1	242	186	1.08	93	140	414	287	0.69
3.70	542	+10	0.37	21.5	2.7	336	226	0.96	91	131	363	249	0.69
7.88	542	+10	0.44	25.0	2.4	605	315	0.76	89	116	307	214	0.70
1.09	542	−10	0.23	12.5	5.0	181	198	0.78	103	179	664	523	0.79
2.34	542	−10	0.30	16.0	4.0	295	291	0.70	96	172	505	385	0.76
3.70	542	−10	0.37	19.0	3.5	401	350	0.60	94	164	434	306	0.71
7.88	542	−10	0.46	22.5	3.1	698	472	0.46	90	138	354	270	0.76

(b) *Metric units*

t mm	V m min^{-1}	α deg	Cutting ratio r	ϕ deg	γ	F_P N	F_Q N	μ	τ MPa	σ MPa	u Nm m^{-3}	u_S Nm m^{-3}	u_S/u
0.094	60	+10	0.29	17.0	3.4	1646	1214	1.05	586	793	2758	2013	0.73
0.094	122	+10	0.33	19.0	3.1	1601	1259	1.11	607	945	2689	1834	0.68
0.094	196.	+10	0.37	21.5	2.7	1463	965	0.95	621	889	2455	1717	0.70
0.094	362	+10	0.44	25.0	2.4	1348	747	0.81	641	896	2262	1551	0.69
0.094	122	−10	0.32	16.5	3.9	1850	1712	0.64	614	1055	3103	2358	0.76
0.094	194	−10	0.37	19.0	3.5	1708	1450	0.58	621	1048	2861	2151	0.75
0.094	354	−10	0.44	22.0	3.1	1583	1170	0.51	648	1062	2655	1993	0.75
0.001	165	+10	0.33	19.0	3.1	565	449	1.12	710	1131	3206	2124	0.66
0.059	165	+10	0.32	18.5	3.1	1076	827	1.08	638	965	2855	1979	0.69
0.094	165	+10	0.37	21.5	2.7	1495	1005	0.96	627	903	2503	1717	0.69
0.200	165	+10	0.44	25.0	2.4	2691	1401	0.76	614	800	2117	1476	0.70
0.001	165	+10	0.73	12.5	5.0	805	881	0.78	710	1234	4578	3606	0.79
0.059	165	+10	0.30	16.0	4.0	1312	1294	0.70	662	1186	3482	2655	0.76
0.094	165	+10	0.37	19.0	3.5	1784	1557	0.60	648	1131	2992	2110	0.71
0.200	165	+10	0.46	22.5	3.1	3105	2099	0.46	621	952	2441	1862	0.76

TABLE 3.2

Representative metal cutting data. (a) English units

The following data were obtained by Lapsley, Grassi and Thomsen (1950), using a two-dimensional tool. The work material was pierced seamless tubing (probably SAE 4130). The tool material was high-speed steel. The width of cut was 0.475 in (12.1 mm) and the cutting speed in all tests was 90 ft min^{-1} (27 m min^{-1}). All chips were continuous and no cutting fluids were used.

t in (× 10^3)	α deg	Cutting ratio r	ϕ deg	γ	F_P lb	F_Q lb	μ	τ p.s.i. (× 10^{-3})	σ p.s.i. (× 10^{-3})	u in lb in^{-3} (× 10^{-3})	u_S in lb in^{-3} (× 10^{-3})	u_S/u
2.5	25	0.358	20.9	2.55	380	224	1.46	82.8	104.0	320	209.0	0.65
3.5	25	0.366	21.5	2.48	475	281	1.46	74.6	96.0	286	180.0	0.63
5	25	0.407	24.0	2.23	643	357	1.38	75.5	100.0	270	169.0	0.63
6	25	0.345	20.1	2.65	728	398	1.36	65.8	86.5	255	175.0	0.69
8.5	25	0.383	22.4	2.38	992	551	1.38	66.5	83.6	246	159.0	0.65
2.5	35	0.527	31.6	1.56	254	102	1.53	71.4	96.0	214	112.0	0.52
3.5	35	0.528	31.9	1.55	306	122	1.52	61.2	84.0	184	96.4	0.52
5	35	0.529	32.0	1.55	433	166	1.48	62.4	82.9	182	96.5	0.53
6	35	0.533	32.2	1.54	507	184	1.43	61.8	88.6	178	95.5	0.54
8.5	35	0.532	32.0	1.55	675	234	1.38	58.5	80.9	167	91.3	0.55
2.5	40	0.585	35.7	1.32	232	71	1.54	72.0	94.5	195	94.0	0.48
3.5	40	0.580	35.4	1.33	296	87	1.50	66.5	84.3	178	87.8	0.49
5	40	0.611	37.5	1.26	412	112	1.44	66.1	87.0	173	83.1	0.48
6	40	0.606	37.2	1.27	475	127	1.42	63.5	90.0	167	81.1	0.49
8.5	40	0.606	37.2	1.27	634	153	1.35	61.5	83.1	157	78.1	0.50
2.5	45	0.670	41.9	1.06	232	68	1.83	71.1	14.6	195	74.8	0.38
3.5	45	0.670	41.9	1.06	285	77	1.74	64.2	98.8	171	68.5	0.40
5	45	0.649	40.2	1.10	386	94	1.64	63.8	87.2	162	69.6	0.43
6	45	0.642	39.6	1.01	443	102	1.60	61.7	80.8	155	62.7	0.40
8.5	45	0.646	39.9	1.11	581	117	1.51	58.7	73.6	144	65.0	0.45

(*b*) *Metric units*

t mm	α deg	Cutting ratio r	ϕ deg	γ	F_P N	F_Q N	μ	τ MPa	σ MPa	u Nm m^{-3}	u_S Nm m^{-3}	$U_{S/u}$
0.064	25	0.358	20.9	2.55	1690	996	1.46	571	717	2206	1441	0.65
0.089	25	0.366	21.5	2.48	2113	1250	1.46	514	662	1972	1241	0.63
0.127	25	0.407	24.0	2.23	2860	1588	1.38	521	690	1862	1165	0.63
0.152	25	0.345	20.1	2.65	3238	1770	1.36	454	596	1758	1207	0.69
0.216	25	0.383	22.4	2.38	4412	2451	1.38	459	576	1696	1096	0.65
0.064	35	0.527	31.6	1.56	1130	454	1.53	492	662	1476	772	0.52
0.089	35	0.528	31.9	1.55	1361	543	1.52	422	579	1269	665	0.52
0.127	35	0.529	32.0	1.55	1926	738	1.48	430	572	1255	665	0.53
0.152	35	0.533	32.2	1.54	2255	818	1.43	426	611	1227	658	0.54
0.216	35	0.532	32.0	1.55	3002	1041	1.38	403	558	1151	630	0.55
0.064	40	0.585	35.7	1.32	1032	316	1.54	496	652	1345	648	0.48
0.089	40	0.580	35.4	1.33	1317	387	1.50	459	581	1227	605	0.49
0.127	40	0.611	37.5	1.26	1833	498	1.44	456	600	1193	573	0.48
0.152	40	0.606	37.2	1.27	2113	565	1.42	438	621	1151	559	0.49
0.216	40	0.606	37.2	1.27	2820	681	1.35	424	573	1083	539	0.50
0.064	45	0.670	41.9	1.06	1032	302	1.83	490	790	1345	516	0.38
0.089	45	0.670	41.9	1.06	1268	343	1.74	443	681	1179	472	0.40
0.127	45	0.649	40.2	1.10	1717	418	1.64	440	601	1117	480	0.43
0.152	45	0.642	39.6	1.01	1970	454	1.60	425	557	1069	432	0.40
0.216	45	0.646	39.9	1.11	2584	520	1.51	405	507	993	448	0.45

Representative Data

Representative orthogonal cutting data are given in Tables 3.1 and 3.2 to illustrate the nature of the results obtained when cutting dry. The values of strain (γ) are found to be much larger in cutting than corresponding values in ordinary materials tests. The reason for this lies in the fact that gross fracture is postponed by the presence of a large normal stress on the shear plane. This and other materials aspects will be discussed in detail in Chapter 6.

The magnitudes of the several energy components discussed in the foregoing section may be compared by use of the data in the first line of Table 3.1. From eqns (3.25), (3.27), (3.29), (3.30), and (3.34) respectively,

$$u = \frac{F_P}{bt} = 400\,000 \text{ p.s.i.}$$

$$u_S = 292\,000 \text{ p.s.i.}$$

$$u_F = \frac{F_C r}{bt} = 108\,000 \text{ p.s.i.}$$

$$u_A = \frac{2T}{t} = \frac{2(0.006)}{0.0037} = 3.2 \text{ p.s.i.}$$

$$u_M = \gamma'/gV^2\gamma^2\sin^2\phi = \frac{0.28}{386}\frac{197(12)^2}{60}(3.4)^2(\sin 17°)^2 = 1.1 \text{ p.s.i.}$$

It is evident from this example that u_A and u_M are negligible compared with u and that the sum of u_F and u_S is therefore equal to the total energy per unit volume (u). In the foregoing example shear energy is seen to account for 73% of the energy, while 27% is due to friction. In general the friction energy is found to be about $\frac{1}{3}$ the shear energy in a turning operation.

While the treatment of cutting mechanics presented in this chapter is highly idealized and based on many simplifying assumptions, it does represent a useful first approximation to what is taking place when metal is cut so as to produce a continuous chip. The dominant action occurs on the shear 'plane' and this may be characterized as concentrated simple plastic shear. Since the plastic behavior of the material in front of the tool plays such an important role in metal cutting, this topic will be considered in some detail in Chapter 5. In subsequent chapters, frictional behavior under conditions pertaining at the chip-tool interface will be considered as well as the character of the finished surface.

Principal Chip Types

Up to this point, it has been assumed that the chip produced is a continuous ribbon of uniform thickness (t_C) with a well-defined thin shear plane (Figs. 1.2, 3.1). However, in practice, this is not always the case (Shaw 1967).

At low cutting speeds or when cutting a material containing points of stress concentration (such as the graphite flakes in cast iron or the manganese sulfide inclusions in a free-machining steel) discontinuous chips may form (Fig. 3.14).

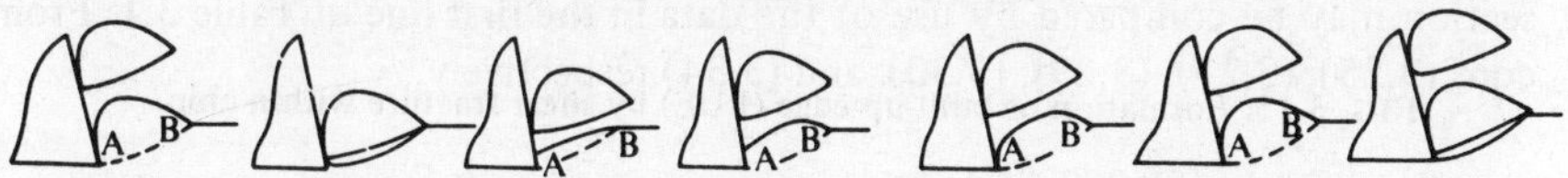

FIG. 3.14. Discontinuous chip formation. Sketches of beta-brass chips photographed by motion picture camera. Undeformed chip thickness, 0.007 in (0.18 mm); cutting speed, 0.5 ipm (12.7 mm min^{-1}); rake angle, 15°; width of cut, 0.161 in (4.09 mm); fluid, none.

In this case, cutting begins at a relatively high shear angle which decreases as the cutting proceeds (Field and Merchant 1949). When the strain in the chip reaches a critical value, the chip fractures and the process begins over again. The result is a series of discrete chip segments that have been produced by a process that resembles an extrusion process more than that of the concentrated shear obtained in orthogonal cutting (Cook, Finnie, and Shaw 1954). The orthogonal model and subsequent calculations represent a very poor approximation for cutting with discontinuous chips.

Another deviation from simple orthogonal cutting is continuous chip formation with a built-up edge. At speeds where the temperature at the chip-tool interface is relatively low, fracture may occur within the chip along a plane approximately at right angles to the shear plane (Nakayama 1957), leaving behind a portion of the chip attached to the tool face (Figs. 1.4 and 3.15). This detached material then acts as the cutting edge and is called a built-up edge (BUE). A BUE tends to grow until it reaches a critical size and then passes off with the chip. This gives rise to a cyclic variation in size of BUE. Since the BUE grows outward and downward, this gives rise to a variation in depth of the cut surface which represents a major component of surface roughness when cutting with a BUE (Fig. 3.16). As cutting speed (and hence chip-tool interface temperature) increases, the size of the BUE decreases. The BUE disappears when thermal softening at the interface causes a lower flow stress at the interface than in the main body of the chip. Under ordinary turning conditions, the BUE will usually have its largest value at a cutting speed of about 50 f.p.m. (0.3 m s^{-1}) and will disappear at about 200 f.m.p. (1.2 m s^{-1}).

The simple orthogonal model represents only a fair approximation when cutting with a continuous chip and BUE. The main influences of the BUE are

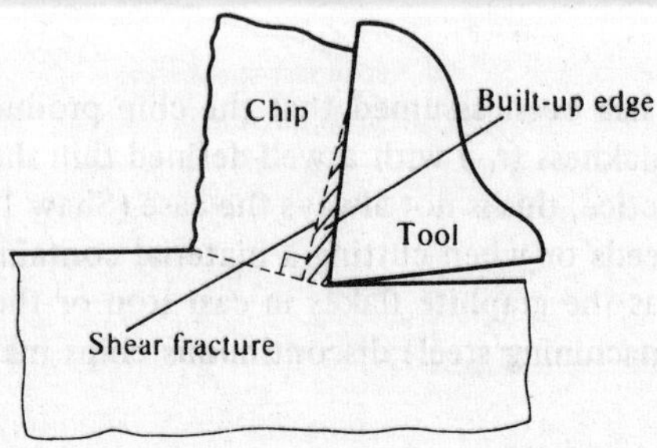

FIG. 3.15. Formation of built-up edge (BUE) by shear fracture within chip.

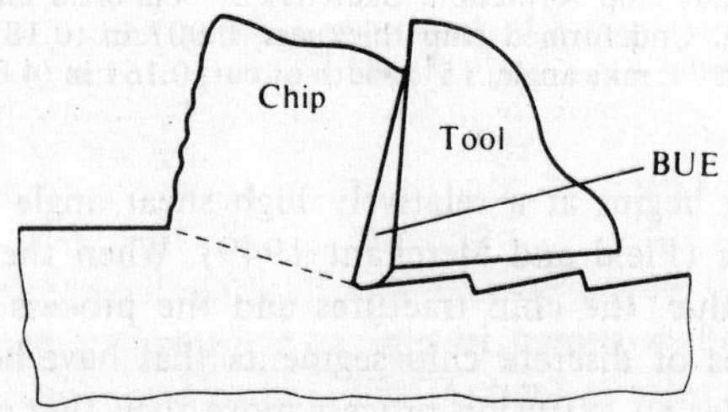

FIG. 3.16. Surface roughness due to built-up edge (BUE).

to give an unusually high friction stress (fracture vs. sliding stress) at the tool point but an increase in the effective rake angle. These two effects influence cutting forces in opposite directions and hence tend to cancel each other. Similarly, the BUE has opposing actions relative to tool wear. The rough particles of hard highly worked BUE passing off with the chip tend to increase abrasive wear while the BUE actually protects the cutting edge from wear. On the other hand, cutting with a large variable BUE always give rise to a substantial increase in surface roughness.

Still another deviation from the simple orthogonal model involves secondary shear on the tool face. When the friction stress on the tool face reaches a value equal to the shear flow stress of the chip material, flow occurs internally within the chip adjacent to the tool face. This is called secondary shear flow to distinguish it from shear flow occurring on the shear plane (primary shear). Fig. 1.2 is a partially formed chip where secondary shear flow is clearly evident. Fig. 3.17 is a diagrammatic sketch corresponding to Fig. 1.2. The secondary shear zone is seen to increase in thickness as the chip travels from A to C. Since secondary shear is due to high shear stress on the tool face, it is accompanied by an increase in cutting forces. One way of considering the influence of secondary shear is in

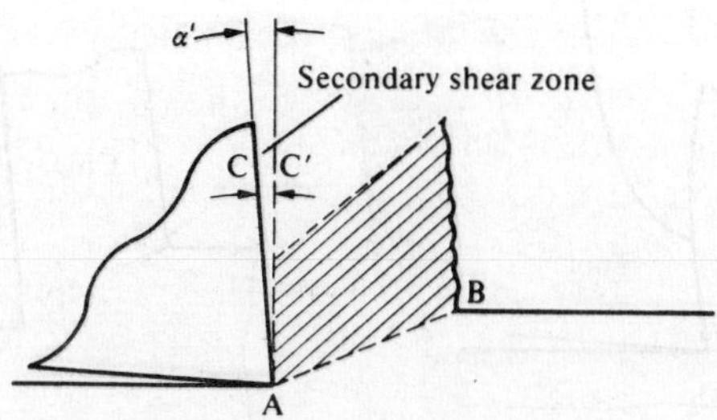

FIG. 3.17. Secondary shear.

terms of a decrease in rake angle by an amount α' (Fig. 3.17). AC' in Fig. 3.17 corresponds to an equivalent rake face when cutting with sticking friction (i.e. with shear stress on the tool face = flow stress of the chip material in shear). Any action that tends to increase shear angle ϕ will tend to decrease the magnitude of secondary shear (α'). Since angle (α') is usually small, the orthogonal model represents a good approximation for cutting with secondary shear flow provided the chip is continuous.

When the material cut is soft and unstrainhardened, the shear zone will tend to be pie-shaped as shown in Fig. 3.18 instead of being uniformly thin as in Fig. 3.1. An increase in feed (t) will tend to favor a pie-shaped shear zone. When cutting with a pie-shaped shear zone, there is no definite shear angle. Use of the dotted line in Fig. 3.18 for the equivalent shear angle in the orthogonal analysis represents a good first approximation.

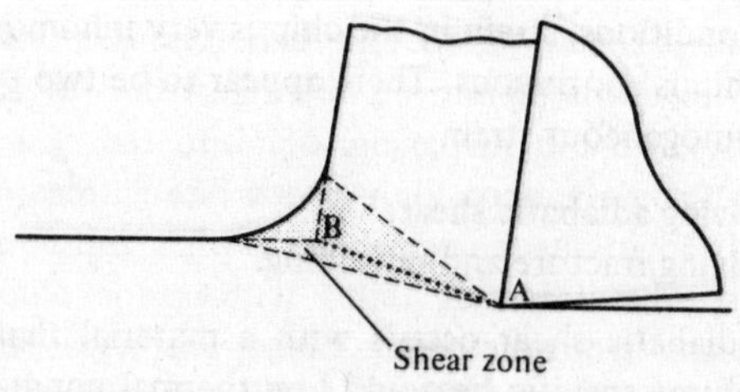

FIG. 3.18. Pie-shaped shear zone.

Controlled contact cutting (Fig. 3.19a) and cutting with a small negative rake honed on the cutting edge (Fig. 3.19b) represent further departures from the ideal orthogonal model. The dead zone shown in Fig. 3.19b is a trapped volume of stationary metal. In these cases there is a composite rake angle and the only reasonable approximation appears to be the use of an equivalent rake angle that represents a compromise between the extremes.

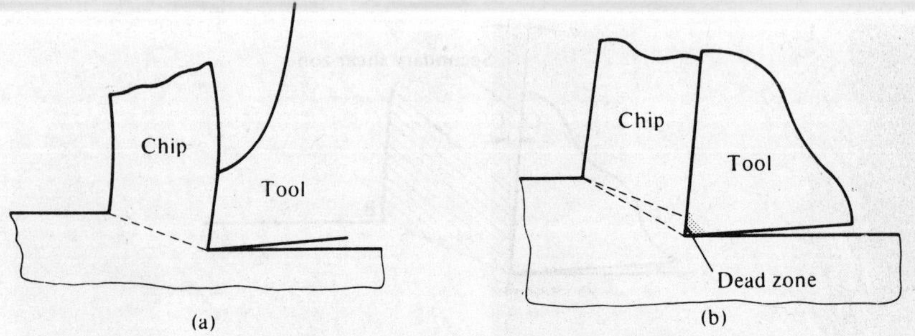

FIG. 3.19. (a) Reduced contact tool, (b) Tool with chamfered edge.

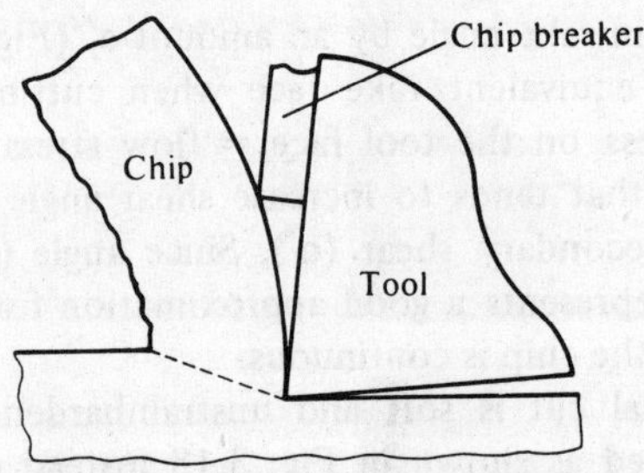

FIG. 3.20. Clamped chip breaker.

Cutting with chip curlers (Fig. 3.20) of different types designed to break the chip periodically represents still another departure from the idealized model (Fig. 3.1). Cutting under such conditions is very complex and it is difficult to suggest a simple equivalent that approximates the idealized model.

Under certain conditions, strain in the chip is very inhomogeneous (Fig. 3.21) even though the chip is continuous. There appear to be two types of chip formation involving inhomogeneous strain:

1. Cutting involving adiabatic shear.
2. Cutting involving fracture and rewelding.

Cutting with adiabatic shear occurs with a material that has poor thermal properties (low volume specific heat and low thermal conductivity), reasonably high specific energy, and a tendency to soften at elevated temperatures. A titanium alloy cut at a relatively high speed represents such a case (Fig. 3.21a, b). Here shear begins at a point of stress concentration. The shear work does not have time to escape and raises the temperature on the shear plane which in turn causes thermal softening. This causes more strain to occur on the initial plane than would ordinarily occur. The shear plane then jumps to the next point of stress concentration leaving the material between relatively unstrained.

Fracture and rewelding (Fig. 3.21c, d) gives a similar result except the region

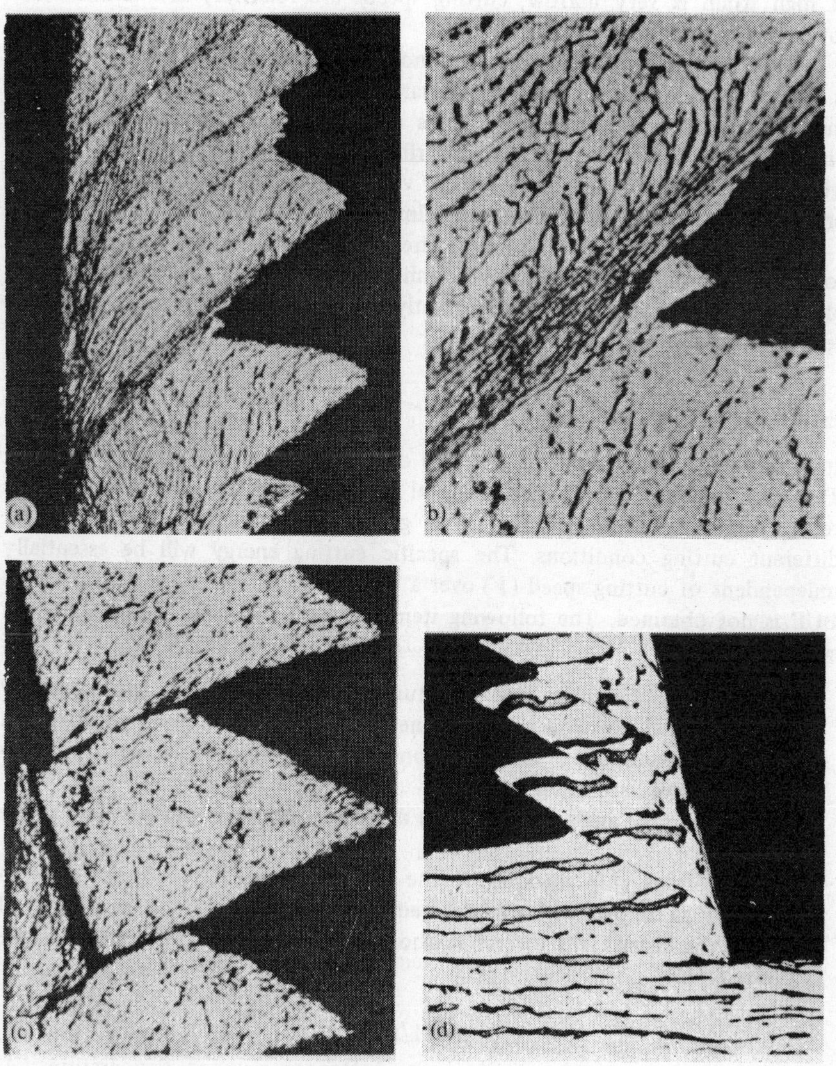

FIG. 3.21. Continuous chips with inhomogeneous shear. (a) Titanium cut at high speed (175 f.p.m., 53 m min^{-1}), adiabatic shear. (b) Enlargement of (a). (c) Titanium cut at low speed (1 i.p.m., 25 mm min^{-1}), periodic fracture, gross sliding and rewelding. (d) 60–40 cold-rolled brass (60% reduction in area) cut with HSS tool having minus 15° rake angle. Cutting speed, 0.075 m min^{-1} (2.95 i.p.m.); undeformed chip thickness, 0.16 mm (0.0063 in). (Brass photomicrograph courtesy K. Nakayama, Yokohama National University.)

of high strain is very narrow, cutting speeds are relatively low and thermal softening does not appear to be involved.

In both types of cutting with inhomogeneous strain the back of the chip is very wavy, the cutting ratio as ordinarily obtained is without meaning and the simple orthogonal cutting model is about as useless as for cutting with discontinuous chips. The shift from a ribbon-like to a segmental chip occurs when the shear strain reaches a critical value dependent on the normal stress on the shear plane (Sharma, Rice, and Salmon 1971).

Cooke and Rice (1973) have used a directionally solidified gold-silver eutectic to provide a built-in grid to study the inhomogeneous shear in cutting. Internal plastic flow was found to be significantly different from that more easily observed on the free surface of a chip.

Estimation of Cutting Forces

It is frequently important to estimate cutting forces F_P and F_Q in practice. This may best be done in terms of total specific energy (u) since this tends to remain approximately constant for a given work material operating under different cutting conditions. The specific cutting energy will be essentially independent of cutting speed (V) over a wide range of values, provided a large BUE is not obtained. The following items do have the influence indicated on specific energy:

1. Workpiece chemistry and structure (specific energy is approximately proportional to workpiece hardness and Table 3.3 gives approximate average values for a few common classes of work materials cut at representative hardness levels).
2. Effective rake angle of tool α_e (u decreases about 1% per degree increase in α_e).
3. Undeformed chip thickness t (the specific cutting energy varies approximately as follows with undeformed chip thickness, t in the usual range of chip thickness. This inverse relationship is sometimes referred to as the size effect).

$$u \sim 1/t^{0.2}. \tag{3.36}$$

The principal cutting force responsible for the power consumed (F_P) may be estimated as follows

$$F_P = \frac{u\, Vbt}{V} = ubt. \tag{3.37}$$

The cutting force component in the feed (t) direction is not easily computed but to a first approximation may be assumed to be one half of F_P.

TABLE 3.3
Approximate values of specific energy for different materials cut with $\alpha_e = 0°$ and $t = 0.010$ in (0.25 mm) for cutting with continuous chip and no BUE

Material	u_0 in lb in^{-3}†	u_0 J m^{-3} ‡
Aluminum alloy	100 000	7.02×10^8
Gray cast iron	150 000	10.53×10^8
Free-machining brass	150 000	10.53×10^8
Free-machining steel (AISI 1213)	250 000	17.55×10^8
Mild steel (AISI 1018)	300 000	21.06×10^8
Titanium alloys	500 000	35.10×10^8
Stainless steel (18–8)	700 000	49.14×10^8
High-temperature alloys (Ni or Co base materials)	700 000	49.14×10^8

† In the workshop specific cutting energy is sometimes expressed in terms of the curious units hp/(in³/min) as a matter of convenience. (1 h.p./in³/min) = 99 000 in lb in^{-3}.

‡ In converting from English to SI units it is convenient to know that 1 in lb in^{-3} = 7020 N m m^{-3} = 7020 J m^{-3}.

$$F_Q = \frac{F_P}{2}. \tag{3.38}$$

The horsepower consumed at the tip of the tool will be

$$\text{h.p.} = \frac{F_P V}{33\,000} \quad (F_P \text{ in lb, } V \text{ in f.p.m.}) \tag{3.39}$$

or

$$\text{kW} = \frac{F_P V}{1000} \quad (F_P \text{ in N, } V \text{ in m s}^{-1}). \tag{3.40}$$

The foregoing method of estimating cutting forces is only very approximate and is for cutting with a continuous chip and no BUE. With more complex chip formation, cutting forces and power may be quite different as follows:

1. Forces will generally be less with discontinuous chip formation.
2. Forces will in general be less with a BUE, the amount of decrease depending on the size of the BUE. Thus, cutting forces will in general decrease with decrease in cutting speed in the range of cutting speeds where the size of BUE increases with decrease in speed.

3. Cutting forces will tend to increase with the extent of secondary shear (conveniently measured by α').
4. Cutting forces will increase with increased strain hardening during cutting. When cutting an annealed material that strain hardens substantially during cutting, a fan-shaped shear zone will tend to be produced as already discussed in connection with Fig. 3.18.
5. Cutting forces will usually decrease with a controlled contact length tool design (but not always—this depends on whether optimum contact length is larger or smaller than that obtained with a conventional tool).
6. Cutting forces will be greater when cutting with a tool tip protecting chamfered cutting edge.
7. Cutting with adiabatic shear will be strain rate sensitive and cutting forces will be unusually high at high cutting speeds with adiabatic shear chip formation (titanium alloys).

The influence of prior cold working of the work material on cutting forces is quite complex. On the one hand forces will tend to be higher due to an increase in initial hardness of the work. On the other hand there will be less tendency for a pie-shaped shear zone to develop. While these two tendencies tend to cancel each other, either may predominate.

Example—Single Point Turning

Figure 3.22 shows a cutting tool having zero side-cutting edge angle in a straight turning situation. The depth of cut is b and the feed rate t (i.p.r. or mm rev^{-1}). There are two cutting edges (of extent b and t respectively). In most cases $b/t \sim 10$ and therefore cutting on the end of the tool will be negligible compared with cutting along the main cutting edge of extent b. As a first approximation therefore this three-dimensional chip formation situation may be approximated by an equivalent two-dimensional orthogonal case where

α = side rake angle of tool
t = feed per revolution (undeformed chip thickness)
b = depth of cut (conventional workshop nomenclature)
V = surface speed of work at o.d.

The cutting ratio may be obtained by taking a chip of any convenient length (l_C) and calculating the corresponding undeformed length (l) from the weight of the chip, the specific weight of the work material and the known values of b and t.

Consider the following specific turning case:

back rake angle	$(\alpha_b) = 15°$
side rake angle	$(\alpha_S) = 10°$

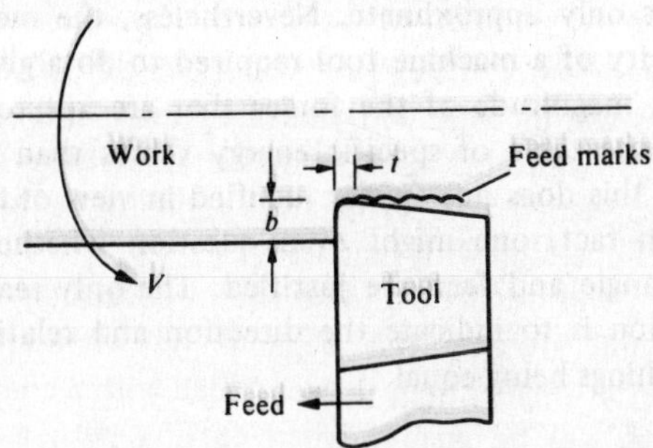

FIG. 3.22. Single-point turning tool showing depth of cut (b) and feed (t).

side cutting edge angle	$(C_s) = 10°$
work material	AISI 1020 steel
workpiece r.p.m.	$(N) = 200$
cutting speed	$(V) = 200$ f.p.m. (61 m min^{-1})
feed rate	$(t) = 0.015$ i.p.r. (0.38 mm rev^{-1})
depth of cut	$(b) = 0.150$ in (3.8 mm)

From Table 3.3, eqn (3.36), and assuming a continuous chip without BUE and small secondary shear

$$u = (300\,000)\left(1 - \frac{15 - 0}{100}\right)\left(\frac{0.010}{0.015}\right)^{0.2} = 235\,100 \text{ p.s.i. (34.1 MPa).}$$

From eqn (3.37)

$$F_P = ubt = (235\,100)\,(0.150)\,(0.015) = 529 \text{ lb (2328N).}$$

and eqn (3.38)

$$F_Q = \frac{F_P}{2} = 265 \text{ lb (1166N)}$$

and eqn (3.39)

$$\text{h.p.} = \frac{F_P V}{33\,000} = \frac{(529)(200)}{33\,000} = 3.2 \text{ h.p. (2.39 kW)}$$

The h.p. required to feed the tool would be

$$(\text{h.p.})_t = \frac{(F_Q)\,(t/12\ N)}{33\,000} = \frac{(529)(0.015/12)(200)}{33\,000} = 0.004 \text{ h.p. (0.003 kW)}$$

This is seen to be negligible compared with the power required to turn the work (3.2 h.p. or 2.39 kW).

It should be kept in mind that the foregoing method of estimating cutting forces and power is only approximate. Nevertheless, the method is useful in estimating the capacity of a machine tool required to do a given job or in estimating the order of magnitude of the forces that are apt to be encountered. While a more elaborate table of specific energy values than that of Table 3.3 could be developed, this does not appear justified in view of the other approximations involved. In fact, one might even question whether the corrections for change in rake angle and feed are justified. The only reason for including these in the discussion is to indicate the direction and relative magnitudes of these effects, other things being equal.

References

Cook, N. H., Finnie, I., and Shaw, M. C. (1954). *Trans. Am. Soc. mech. Engrs* **76**, 153.
Cooke, W. B. H. and Rice, W. B. (1973). *J. Engng Ind.* **95**, 844.
Drucker, D. C. (1949). *J. appl. Phys.* **20**, 1013.
Field, M. and Merchant, M. E. (1949). *Trans. Am. Soc. mech. Engrs* **71**, 421.
Kececioglu, D. (1958). *Trans. Am. Soc. mech. Engrs* **80**, 158.
Lapsley, J. T., Grassi, R. C., and Thomsen, E. G. (1950). *Trans. Am. Soc. mech. Engrs* **72**, 979.
Merchant, M. E. (1944). *Trans. Am. Soc. mech. Engrs* **66**, A-168.
Merchant, M. E. (1945). *J. appl. Phys.* **16**, 267(a) and 318(b).
Merchant, M.E. and Zlatin, N. (1945). *Mech. Engng* **67**, 737.
Nakayama, K. (1957). *Bull. Fac. Engng Yokohama natn. Univ.* **6**, 1.
Piispanen, V. (1937). *Eripaines Teknillises1ä Aikakauslehdeslä* **27**, 315.
Piispanen, V. (1948). *J. appl. Phys.* **19**, 876.
Sharma, C. S., Rice, W. B., and Salmon, R. (1971). *Annals of CIRP* **29**, 545.
Shaw, M. C. (1954). *Metal cutting principles* (3rd edn). MIT Press, Cambridge, MA.
Shaw, M. C. (1967). Report 94, Iron and Steel Institute, London.

4
ELASTIC BEHAVIOR

The mechanical behavior of a material being cut plays a very important role and it is therefore useful at the outset to review briefly the behavior of materials relative to elastic and plastic behavior and fracture. This will be done in the next three chapters.

When a metal is loaded in uniaxial tension, uniaxial compression, or simple shear (Fig. 4.1), it will behave elastically until a critical value of normal stress (S) or shear stress (τ) is reached and then it will deform plastically. In the

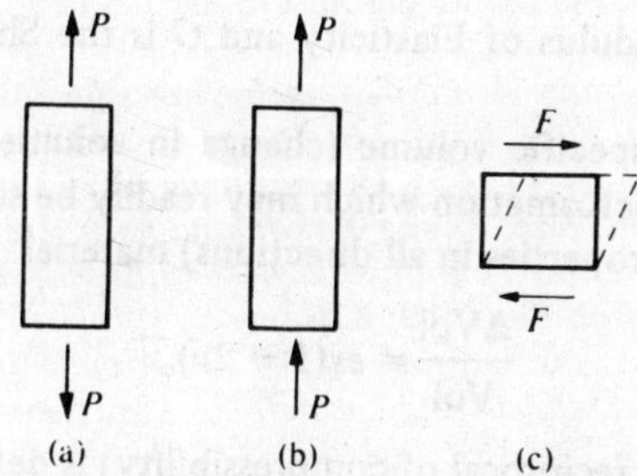

FIG. 4.1. (a) Uniaxial tension. (b) Uniaxial compression. (c) Simple shear.

elastic region, the atoms are temporarily displaced but return to their equilibrium positions when the load is removed. Stress (S or τ) and strain (e or γ) in the elastic region are defined as indicated in Fig. 4.2: the usual convention of considering tensile stress and strain positive and compressive stress and strain negative is followed here. Poisson's ratio (ν) is the ratio of transverse to direct strain in tension or compression:

$$\nu = -\frac{e_2}{e_1} \tag{4.1}$$

where e_2 is the transverse strain that accompanies strain in the direction of the applied load (e_1 = direct strain). In the elastic region ν is between $\frac{1}{4}$ and $\frac{1}{3}$ for metals. The relation between stress and strain in the elastic region is given by Hooke's Law:

$$S = Ee \text{ (tension or compression)} \tag{4.2}$$

$$\tau = G\gamma \text{ (simple shear)} \tag{4.3}$$

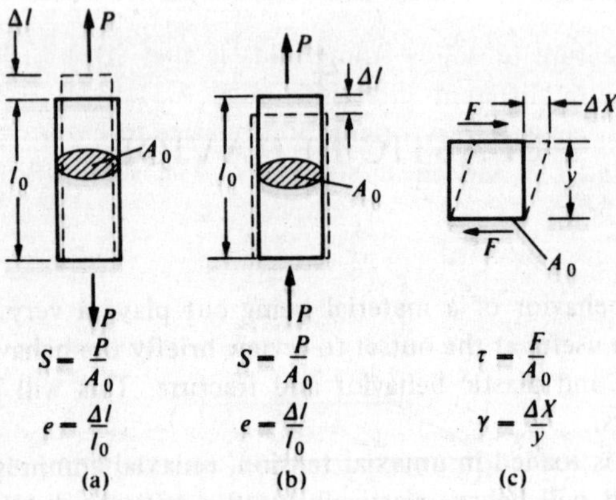

FIG. 4.2. Elastic stress and strain for: (a) Uniaxial tension; (b) Uniaxial compression; (c) Simple shear.

where E is Young's modulus of Elasticity and G is the Shear Modulus of Elasticity.

A small change in specific volume (change in volume per unit volume) is associated with elastic deformation which may readily be shown to be as follows for an isotropic (same properties in all directions) material

$$\frac{\Delta \text{Vol}}{\text{Vol}} = e_1(1-2\nu). \tag{4.4}$$

The bulk modulus (K = reciprocal of compressibility) is defined as follows

$$K = \frac{\Delta p}{\dfrac{\Delta \text{Vol}}{\text{Vol}}} \tag{4.5}$$

where Δp is the pressure acting at a particular point. For an elastic solid loaded in uniaxial compression (S)

$$K = \frac{S}{\dfrac{\Delta \text{Vol}}{\text{Vol}}} = \frac{S}{e_1(1-2\nu)} = \frac{E}{1-2\nu}. \tag{4.6}$$

Thus, an elastic solid is compressible as long as ν is less than $\frac{1}{2}$ which is normally the case for metals.

Stress vs. Direction

A distinguishing feature of solids from fluids is that stresses and strains at a point are different in different directions. Fig. 4.3 shows stresses acting on a small cube having an arbitrary orientation relative to external loads. Normal stresses are designated S and shear stresses τ. The subscript xy designates a

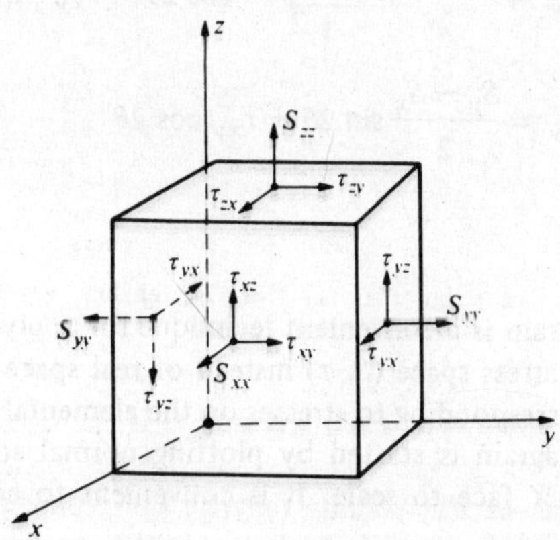

FIG. 4.3. Stress on infinitesimal cube at a point.

stress in a plane normal to the x-axis but in the y-direction (a shear stress—hence τ_{xy}). Fig. 4.4a is the two dimensional counterpart of Fig. 4.3 and Fig. 4.4b is a free-body diagram of the part of the square of Fig. 4.4a below oblique plane AB. The normal to plane AB makes an angle θ with the x-axis. The normal (S') and shear (τ') stresses on plane AB may be found relative to stresses on the faces normal to the x- and y-axes and angle θ by writing the three equations

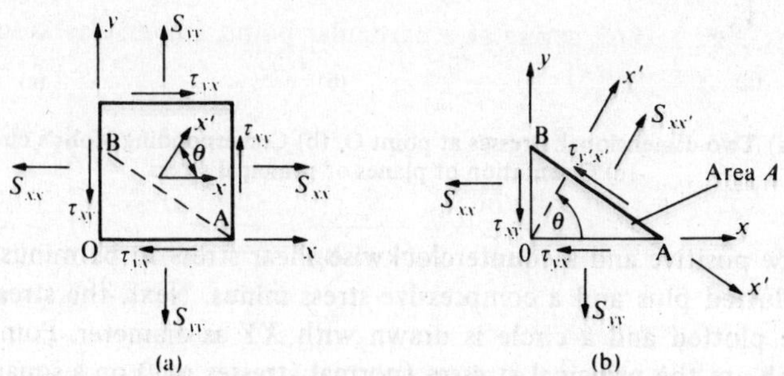

FIG. 4.4. Two-dimensional stresses at a point. (a) Oblique plane AB oriented with its normal angle θ to x-axis. (b) Free-body diagram corresponding to (a).

for static equilibrium for the free body of Fig. 4.4b (for static equilibrium relative to rotation ($|\tau_{xy}| = |\tau_{yx}|$). After simplification, the transformation stress equations become:

$$S_{x'x'} = \frac{S_x + S_y}{2} + \frac{S_x - S_y}{2} \cos 2\theta + \tau_{xy} \sin 2\theta$$

$$S_{y'y'} = \frac{S_x + S_y}{2} - \frac{S_x - S_y}{2} \cos 2\theta - \tau_{xy} \sin 2\theta \qquad (4.7)$$

$$\tau_{x'y'} = \frac{S_y - S_x}{2} \sin 2\theta + \tau_{xy} \cos 2\theta$$

Mohr Stress Diagram

The Mohr stress diagram is a convenient technique for applying these equations. This is a diagram in stress space (S, τ) instead of real space (x, y). Figure 4.5b is the Mohr's circle corresponding to stresses on the elemental square of Fig. 4.5a. The Mohr's stress diagram is started by plotting normal stress (S_x) and shear stress (τ_{xy}) for the X face to scale. It is convenient to consider a clockwise

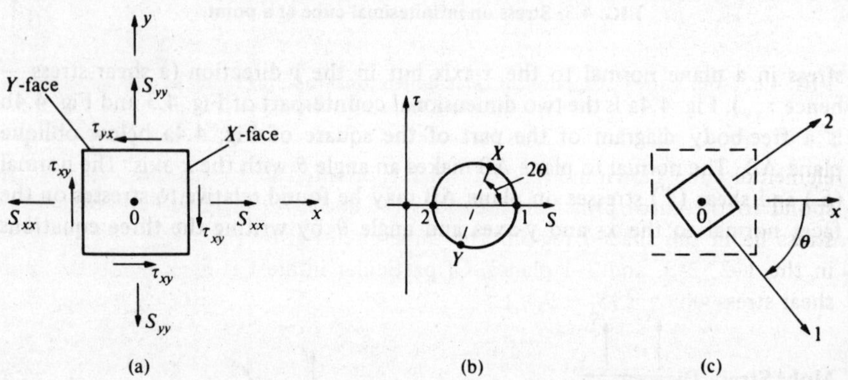

FIG. 4.5. (a) Two-dimensional stresses at point O. (b) Corresponding Mohr's circle diagram. (c) Orientation of planes of principal stress.

stress to be positive and a counterclockwise shear stress to be minus. A tensile stress is plotted plus and a compressive stress minus. Next, the stresses on the Y-face are plotted and a circle is drawn with XY as diameter. Points 1 and 2 in Fig. 4.5b are the principal stresses (normal stresses at 0 on a square oriented such that shear stresses τ are zero). The orientation of the planes of principal stress may be found by noting that angles turned through in the stress space diagram are twice as large as those turned through in the real space diagram and in the same direction provided the above sign convention is adopted for shear stresses. Figure 4.5c shows the orientation of principal planes (1, 2) corresponding to the Mohr's circle of Fig. 4.5b. The double angle relationship in stress space arises from the fact that only (2θ) appears in eqns (4.7).

The Mohr's circle technique is widely used to convert stresses at a point with one orientation to another. However, it is not usually used by plotting to scale but simply as a nemonic (memory aid) device for conveniently applying eqns (4.7).

The Mohr's circle concept may be readily extended to three dimensions. Figure 4.6b shows the three-dimensional Mohr's circle representation for the

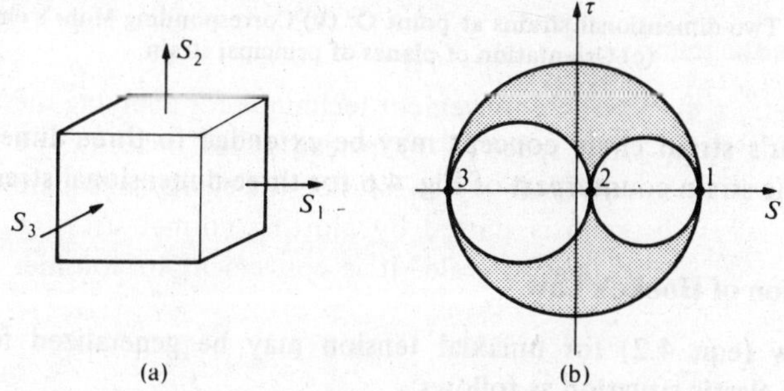

FIG. 4.6. Mohr's stress diagrams for three-dimensional case. (a) Principal stresses at point 0. (b) Corresponding Mohr's circle diagrams.

elemental cube shown in Fig. 4.6a which is oriented so that the cube faces correspond to the three principal directions (1, 2, 3). All possible combinations of stress lie in the shaded region. The three circles give values of S and τ that lie in the 1–2, 2–3, and 3–1 planes. Of particular interest is the overall maximum shear stress which is $(S_1 - S_3)/2$.

Mohr Strain Diagram

A parallel treatment of strain at a point leads to a Mohr's circle of strain. In this case the strain coordinates are e and $\gamma/2$ in place of S and τ, otherwise all details of Mohr's circle for stress have a direct counterpart in Mohr's circle for strain. Figure 4.7a shows the strains at a point A for the X and Y planes while Fig. 4.7b is the corresponding Mohr's strain diagram and Fig. 4.7c shows the corresponding directions of principal strain ($\gamma = 0$). Angles measured about 0′ in the Mohr diagram again correspond to twice the angle turned through in the same direction in real space, provided a clockwise shear strain is considered positive.

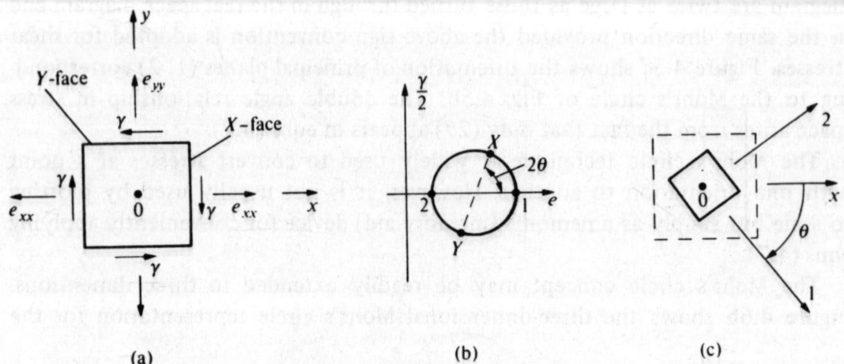

FIG. 4.7. (a) Two-dimensional strains at point O. (b) Corresponding Mohr's circle diagram. (c) Orientation of planes of principal strain.

The Mohr's strain circle concept may be extended to three dimensions and Fig. 4.8 is the strain counterpart of Fig. 4.6 for three-dimensional stress.

Generalization of Hooke's Law

Hooke's law (eqn 4.2) for uniaxial tension may be generalized for a three-dimensional elastic situation as follows:

$$\begin{aligned} e_1 &= 1/E\,[S_1 - \nu(S_2 + S_3)] \\ e_2 &= 1/E\,[S_2 - \nu(S_3 + S_1)] \\ e_3 &= 1/E\,[S_3 - \nu(S_1 + S_2)] \end{aligned} \tag{4.8}$$

where e_1, e_2, e_3 and S_1, S_2, S_3 are the principal strains and stresses for an isotropic homogeneous material that is elastically loaded.

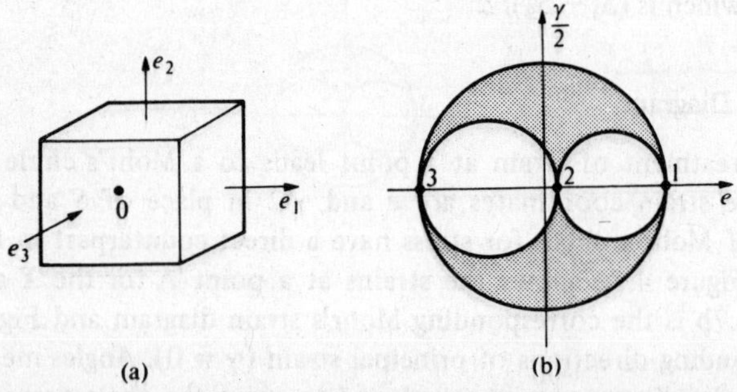

FIG. 4.8. Mohr's strain diagrams for three-dimensional case. (a) Principal strains at point O. (b) Corresponding Mohr's circle diagrams.

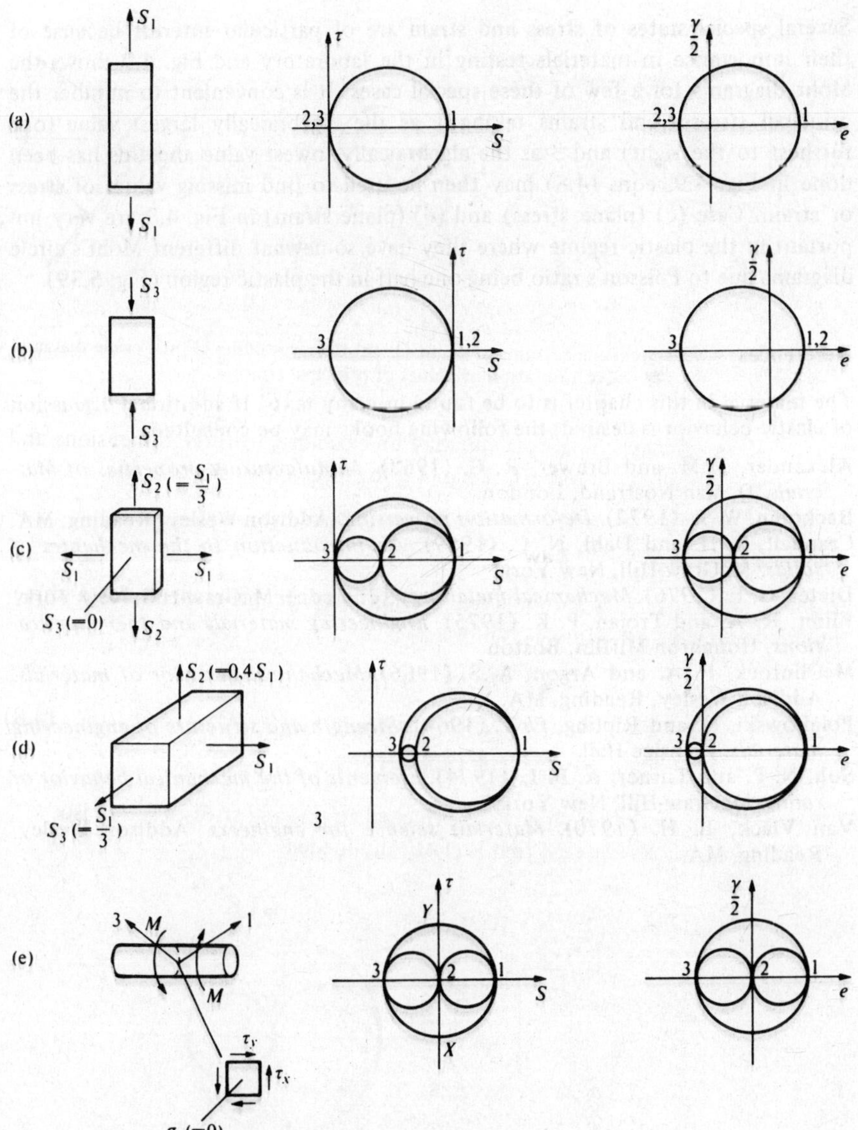

FIG. 4.9. Mohr's circle diagrams for special states of stress and strain. (a) Uniaxial tension. (b) Uniaxial compression. (c) Biaxial stress ($S_2 = S_1/3$) for thin specimen (plane stress, $S_3 = 0$). (d) Biaxial stress ($S_3 = S_1/3$) for wide specimen (plane strain, $e_2 = 0$). (e) Torsion (stress and strain in outer surface of round bar). Poisson's ratio has been assumed to be 0.30 in all cases.

Examples

Several special states of stress and strain are of particular interest because of their importance in materials testing in the laboratory and Fig. 4.9 shows the Mohr diagrams for a few of these special cases. It is convenient to number the principal stresses and strains taking 1 as the algebraically largest value (one furthest to the right) and 3 as the algebraically lowest value and this has been done in Fig. 4.9, eqns (4.8) may then be used to find missing values of stress or strain. Case (c) (plane stress) and (d) (plane strain) in Fig. 4.9 are very important in the plastic regime where they have somewhat different Mohr's circle diagrams due to Poisson's ratio being one half in the plastic region (Fig. 5.39).

References

The material in this chapter is to be found in many texts. If additional discussion of elastic behavior is desired, the following books may be consulted:

Alexander, J. M. and Brewer, R. C. (1963) *Manufacturing properties of Materials.* D. Van Nostrand, London.

Backofen, W. A. (1972). *Deformation processing.* Addison Wesley, Reading, MA.

Crandall, S. H. and Dahl, N. C. (1959). *An introduction to the mechanics of solids.* McGraw-Hill, New York.

Dieter, G. E. (1976). *Mechanical metallurgy* (2nd edn). McGraw-Hill, New York.

Flinn, R. A. and Trojan, P. K. (1975). *Engineering materials and their applications*, Houghton Mifflin, Boston.

McClintock, F. A. and Argon, A. S. (1966). *Mechanical behavior of materials.* Addison Wesley, Reading, MA.

Polakowski, N. and Ripling, E. T. (1964). *Strength and structure of engineering materials,* Prentice Hall.

Suh, N. P. and Turner, A. P. L. (1974). *Elements of the mechanical behavior of solids.* McGraw-Hill, New York.

Van Vlack, L. H. (1970). *Materials science for engineers.* Addison Wesley, Reading, MA.

5

PLASTIC BEHAVIOR

At a certain value of strain ($e \cong 10^{-3}$) called the yield point, a material ceases to behave elastically—does not return to its original condition when the load is removed but undergoes permanent deformation. This is called plastic behavior. It is found experimentally that there is negligible change in specific volume during plastic flow and this suggests that more than the stretching of atoms is involved. A suitable model to explain plastic flow and its negligible change in density is slip in which one layer of atoms slides over another in shear. However, when the shear stress required to cause one layer of atoms in a perfect lattice to slide over an adjacent layer is estimated, the required value far exceeds the flow stress of ordinary materials.

Theoretical Shear Strength

Figure 5.1a shows two layers of atoms in a perfect crystal under no-load equilibrium conditions. The atoms represented by spheres are all vertically aligned. Figure 5.1b shows the atoms displaced relative to each other due to the application of shear stress (τ). The resisting stress (τ) may be assumed to vary linearly with displacement (x). When $x = a_1$ (the horizontal atom spacing) the atoms will be realigned and $\tau = 0$. Also, when $x = a_1/2$, a displaced atom will be equally attracted to its nearest neighbors B in Fig. 5.1b and τ should again be zero. As a first approximation, the maximum resisting shear stress ($\tau_0 =$ shear strength) might be expected to occur when $x = a_1/4$. For the assumed linear relation between stress and displacement, it follows that

$$\frac{\tau}{\tau_0} = \frac{x}{a_1/4}. \tag{5.1}$$

From the definition of simple shear strain (γ)

$$\gamma = \frac{x}{a_2} \tag{5.2}$$

from eqns (5.1) and (5.2)

$$\tau_0 = \frac{\tau a_1}{4 a_2 \gamma} \tag{5.3}$$

but from eqn (4.3)

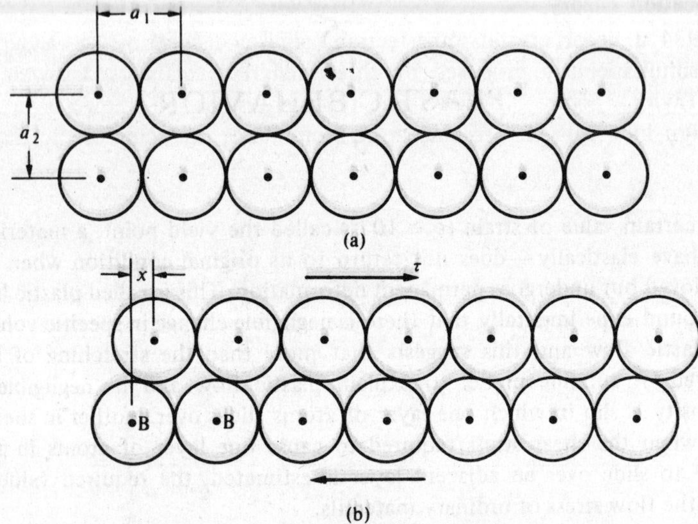

FIG. 5.1. Slip of one layer of atoms over another layer of atoms in a perfect crystal. (a) Before deformation. (b) After shear displacement x.

$$\tau_0 = \frac{G}{4}\frac{a_1}{a_2} \cong \frac{G}{4} \tag{5.4}$$

since $a_1 \cong a_2$.

When this value of theoretical shear strength (τ_0) is compared with values for ordinary materials, a very large discrepancy is found. For example, mild steel has a value of G of about 11.6×10^6 p.s.i. (80 093 MN m^{-2}). Thus from eqn (5.4)

$$\tau_0 = \frac{G}{4} = \frac{11.6 \times 10^6}{4} = 2.9 \times 10^6 \text{ p.s.i. (19 651 MPa)}.$$

However, the actual shear strength of such a material is only about 30 000 p.s.i. (207 MPa). This represents a discrepancy of $2.9 \times 10^6/0.03 \times 10^6 \cong 97$ or two orders of magnitude.

While the assumed linear variation of stress with displacement to such a large strain ($\gamma = \frac{1}{4}$) might be questioned, this is not the problem. This may be shown by substituting a sinusoidal variation of stress with displacement for a linear one. That is, when $\tau = \tau_0 \sin x/(a_1/4)\,(2\pi)$ instead of $\tau = \tau_0\, x/(a_1/4)$ (eqn 5.1), the value of τ_0 is found to be $G/2\pi$ instead of $G/4$. This changes the discrepancy to $1.85 \times 10^6/0.03 \times 10^6 = 62$ which is still a very large value. Thus, the assumed relation between stress and displacement does not appear to represent the basic difficulty.

Dislocation Theory

In 1934 a linear crystal imperfection called a dislocation was independently and simultaneously proposed by three materials scientists (Orowan, Polanyi, and Taylor). Figure 5.2 shows one type of dislocation known as an edge dislocation in which there is one more atom above the plane of slip than below.

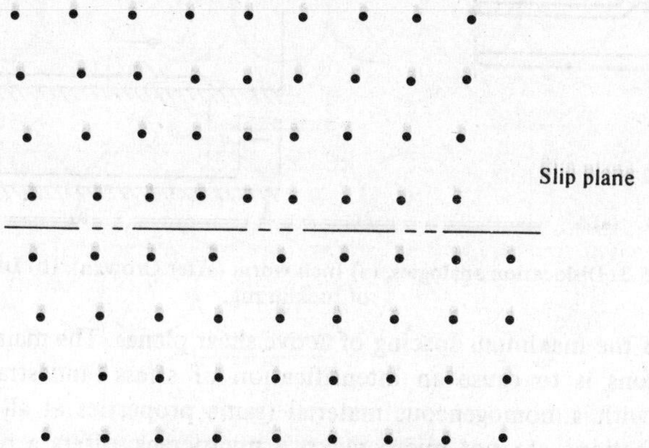

FIG. 5.2. Positive edge dislocation.

This results in some atoms above the slip plane being attracted to their nearest neighbors below the plane while others are resisting separation from their nearest neighbors. The net effect of such a configuration is to require far less energy to cause displacement of the upper row relative to the lower row than if all atoms acted in the same way as in Fig. 5.1.

A couple of useful analogies explain the difference in flow stress with and without the presence of dislocations. Figure 5.3 shows the mode of transport of an inchworm over the ground. The progressive movement of a small section of its body relative to the ground, instead of sliding all at the same time, requires much less energy. Similarly, if a large rug is to be slid across the floor, the easiest way of doing this is to form a ruck in the rug and then kick this over the length of the rug instead of sliding the entire rug over the floor at the same time (Fig. 5.3b).

When a dislocation runs across a crystal, this results in a displacement of one atom spacing (a_1). If dislocations were found in every plane of atoms in the a_2 direction of Fig. 5.1, the passage of one dislocation across a crystal would correspond to a local shear strain of unity $= a_1/a_2$. However, it is found that dislocations are located on planes that are very widely separated relative to atomic spacing. This means that the passage of a large number of dislocations is required on the active slip planes to account for observed plastic strains.

While several complex mechanisms of dislocation formation have been proposed, only one will be considered here. It is well established that real materials contain imperfections at a spacing of about one micrometer which corresponds

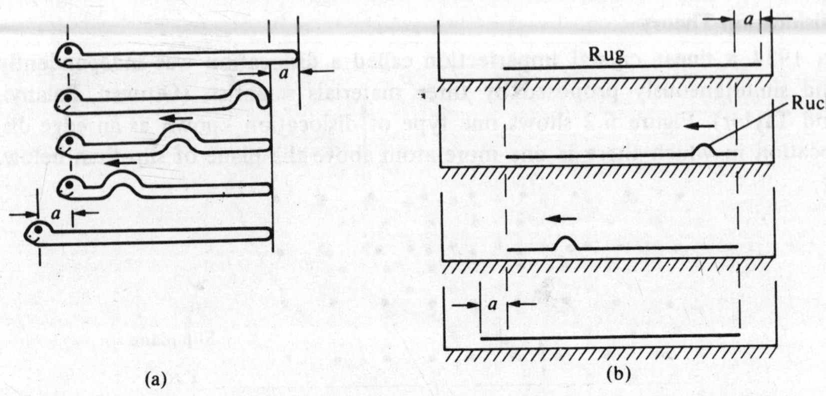

FIG. 5.3. Dislocation analogies. (a) Inch worm (after Orowan). (b) Displacement of ruck in rug.

roughly to the maximum spacing of active shear planes. The main effect of such imperfections is to cause an intensification of stress (and strain) above that obtained with a homogeneous material (same properties at all points). While all imperfections are not microcracks, a microcrack offers a useful model to explain the significance of such a stress intensifier. Figure 5.4a shows the manner in which the stress (σ_t) at the tip of a microcrack rises above the homogeneous value (σ_0) as a crack tip is approached. Figure 5.4b shows atoms along a line 45° to the direction of homogeneous stress (σ_0). The closed circles represent the unloaded positions of the atoms while the open dots represent the deformed positions. The open dots constitute a dislocation that once formed will run across the crystal at a fraction of the energy required to move all atoms simultaneously. Each time a dislocation is formed and moves off under the action of a tensile stress, the crack will grow and when the crack becomes sufficiently large, it will grow spontaneously as discussed later.

An important characteristic of plastic flow is strain hardening which is the increase in plastic flow stress with strain. Dislocations will move across a crystal until a crystal boundary or second phase or impurity atom is encountered. Stuck dislocations project a back stress to the point of dislocation formation or interfere with the motion of dislocations and these actions constitute the major sources of strain hardening.

The increase in flow stress with plastic strain can be removed by heating a material to a sufficiently high temperature so that the increased amplitude of vibration of atoms is sufficient to shake dislocations from the structure. Such a softening operation is accompanied by a rearrangement of crystal boundaries and a decrease in grain size. When a metal is plastically deformed at a temperature above the strain recrystallization temperature (≅ absolute melting point/2), this is called hot working and strain hardening is avoided because dislocations are removed by grain boundary rearrangement as fast as they are formed.

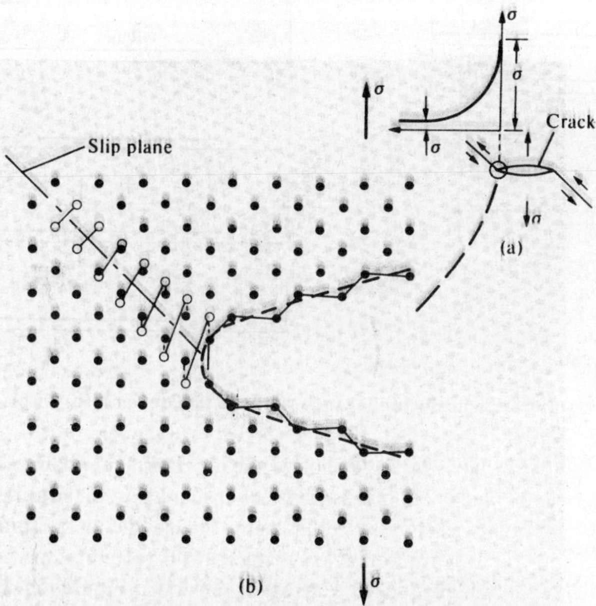

FIG. 5.4. Displacement of ions in metal to form a dislocation at the tip of a microcrack. (a) Stress intensification at crack tip. (b) Displacement of ions at crack tip. Solid points represent unloaded ions, open points represent ion centers after stressing.

While dislocations are too small to be photographed directly, their presence may be revealed by indirect evidence such as the formation of etch pits or precipitation (decoration) at dislocation pile ups. X-ray diffraction and transmission electron microscopy (TEM) of very thin sections of plastically deformed material have also proved useful in verifying the existence of dislocations.

Probably the most convincing evidence for dislocations is provided by the soap bubble analogy devised by Bragg and Nye (1947). Small air bubbles of uniform size are attracted to form a close-packed array (Fig. 5.5) due to surface tension but resist approaching beyond an equilibrium spacing due to internal pressure. There are two opposing forces as in the case of atoms in a metal that similarly give rise to an equilibrium bubble spacing. In addition to crystal boundaries, dislocations also appear in a bubble raft as accidents of growth. When such a two-dimensional analogy of a metal structure is sheared, dislocations are observed to run across single 'grains' and to be generated at points of stress concentration represented by a microcrack tip. In addition to dislocations formation and movement, the Bragg–Nye analogy also clearly demonstrates strain recrystallization. An excellent motion picture that illustrates many of the details of dislocation mechanics in terms of the bubble model has been produced by Bragg, Lomer, and Nye (1954).

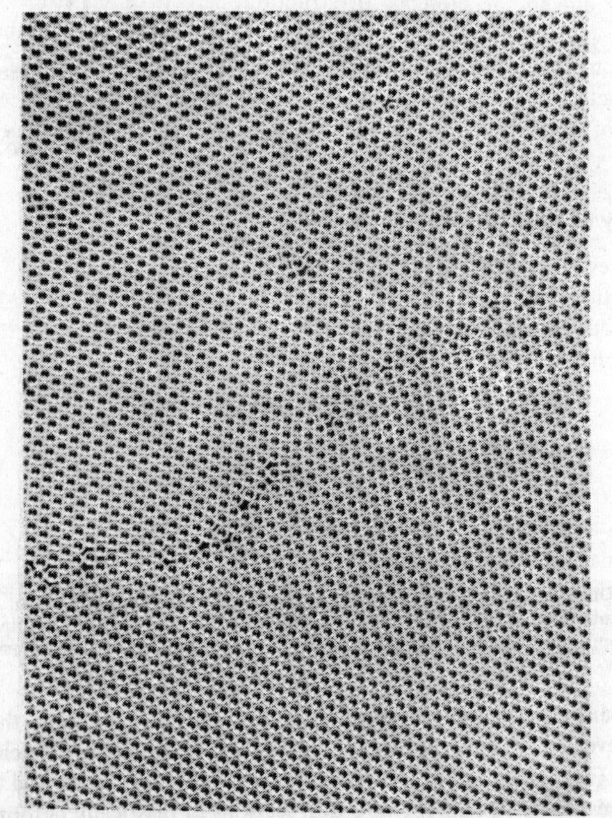

FIG. 5.5. Bubble model showing grain boundary running diagonally across center of field and dislocation just above center of field.

Metal Whiskers

It was found quite by accident (Herring and Galt 1952) that very small samples of metals free of dislocations may be produced by electrodeposition at very low values of potential. These 'whiskers' are produced so slowly there is plenty of time for each atom to find its proper place and there are no accidents of growth (dislocations). When such whiskers are tested in bending they

1. Are elastic to the point of fracture (no plastic flow),
2. Are nonlinearly elastic, the elastic stress–strain curve resembling the first quarter cycle of a sine wave,
3. Have a very large strain at fracture approaching $\gamma = \frac{1}{4}$ suggested by the derivation for theoretical strenth (τ_0).

From the foregoing discussion, it is evident that metals are ductile and flow plastically due to the presence of structural defects called dislocations. Plastic behaviour is ordinarily possible only due to dislocations. In the absence of dislocations, metals are perfectly brittle (yield strain ≧ fracture strain). The ductility characteristic of metals is due to the tendency for metals to generate large numbers of dislocations.

Constancy of Volume and Poisson's Ratio

It was previously stated that there is essentially no change in specific volume with plastic flow. From eqn (4.4), it follows that in order that $\Delta Vol/Vol = 0$, the value of Poisson's ratio (ν) must be $\frac{1}{2}$. Thus, Poisson's ratio is always $\frac{1}{2}$ for all metals in the plastic region.

Uniaxial Tensile Test

Figure 5.6 shows a stress–strain curve for a ductile metal as frequently determined in acceptance testing. Here stress (S) is the applied load (P) divided by the original area (A_0) and strain (e) is the change in length (Δl) divided by the original gage length (l_0). Also, S_Y is the yield stress (limit of linear elasticity) S_U is called the ultimate stress (P = a maximum) while S_F is the fracture stress. What appears to be negative strain hardening beyond point U is really due to a degree of confusion introduced by defining the stress in terms of the original area instead of the actual instantaneous area. Figure 5.7 shows the curve of Figure 5.6 replotted in terms of true stress (σ) and true strain (ϵ) where stress and strain are now defined in terms of the instantaneous area and gage length instead of initial values (A_0 and l_0). A metal that is highly strain-hardened before testing will yield a much flatter curve that approaches the 'ideal plastic' which

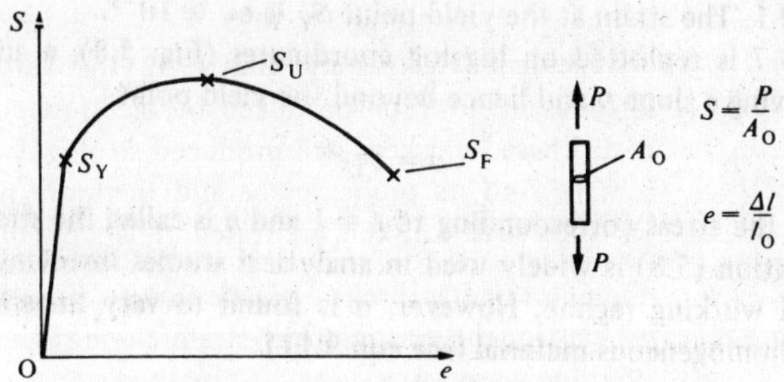

FIG. 5.6. Engineering stress–strain uniaxial tensile curve. S_Y = yield stress, S_U = ultimate stress, S_F = fracture stress.

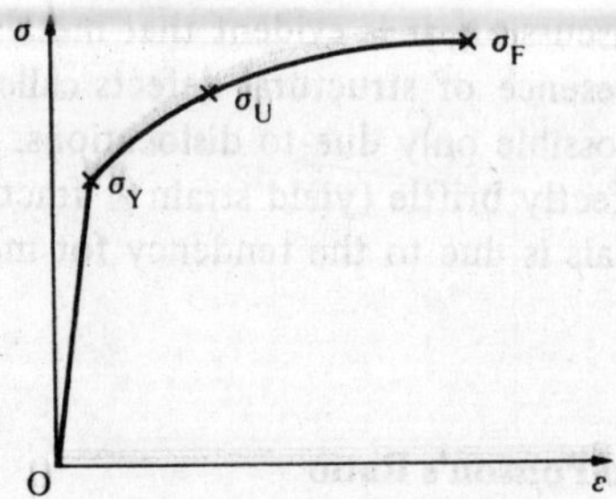

FIG. 5.7. True-stress–true-strain uniaxial tensile curve. σ_Y = yield stress, σ_U = ultimate stress, σ_F = fracture stress.

has a perfectly flat σ–ϵ curve in the plastic region. Since there is no change of volume in the plastic regime

$$Al = A_0 l_0 \tag{5.5}$$

and by definition

$$\sigma = \frac{P}{A} = \frac{Pl}{A_0 l_0} = \frac{S(l_0 + \Delta l)}{l_0} = S(1 + e). \tag{5.6}$$

Also,

$$d\epsilon = \frac{dl}{l}.$$

Integrating and evaluating the constant of integration:

$$\epsilon = \ln \frac{l}{l_0} = \ln \left[\frac{l_0(1 + e)}{l_0}\right] = \ln(1 + e). \tag{5.7}$$

Equations 5.6 and 5.7 are useful in converting S and e into σ and ϵ. From eqn (5.7) it is evident that the difference in ϵ and e is negligible for values of e less than 0.1. The strain at the yield point S_Y is $\epsilon_Y \cong 10^{-3}$.

If Fig. 5.7 is replotted on log–log coordinates (Fig. 5.8), a straight line is obtained having a slope n and hence beyond the yield point

$$\sigma = \sigma_1 \epsilon^n \tag{5.8}$$

where σ_1 is the stress corresponding to $\epsilon = 1$ and n is called the strain hardening index. Equation (5.8) is widely used in analytical studies involving plastic flow in the cold working regime. However, σ is found to vary linearly with ϵ for $\epsilon > 1$ for a homogeneous material (see eqn 9.11).

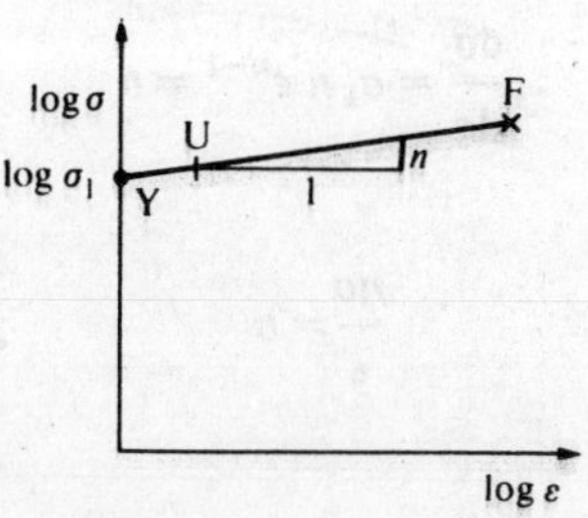

FIG. 5.8. Log–Log true-stress–true-strain uniaxial tensile curve. Y = yield point, U = ultimate point, F = fracture point.

Necking

In the initial plastic region (between Y and U in Figs. 5.6 and 5.7) the area decreases uniformly along the gage length and the increase in load (P) required due to strain hardening exceeds the decrease in P required due to reduction in area. However, at point U, a localized neck begins to form and then the change in P required due to area reduction exceeds the change in P due to strain hardening and what appears to be negative strain hardening occurs if stress is measured in terms of original area A_o instead of actual area A. The necking strain may be obtained as follows:

(by definition)
$$P = \sigma A. \tag{5.9}$$

Differentiating both sides

$$dP = \sigma dA + A d\sigma. \tag{5.10}$$

At point U, $dP = 0$ and hence

$$\frac{d\sigma}{\sigma} = -\frac{dA}{A}. \tag{5.11}$$

Since

$$Al = \text{constant} \tag{5.12}$$

$$A\,dl + l dA = 0 \tag{5.13}$$

or

$$-\frac{dA}{A} = \frac{dl}{l} = d\epsilon. \tag{5.14}$$

Hence from eqns (5.11) and (5.14)

$$\frac{d\sigma}{d\epsilon} = \sigma \tag{5.15}$$

and from eqn (5.8)

$$\frac{d\sigma}{d\epsilon} = \sigma_1 n \, \epsilon^{n-1} = \sigma \tag{5.16}$$

or

$$\frac{n\sigma}{\epsilon} = \sigma \tag{5.17}$$

or at point U

$$\epsilon_U = n. \tag{5.18}$$

Since the value of n is about 0.2 for a ductile metal (annealed) the natural strain at necking (ϵ_U) is about 0.2.

When a concentrated neck forms in a tensile specimen added complications are involved. First of all, the strain cannot be measured in terms of length since the strain is nonuniform along the gage length. However, due to the constance of volume

$$l d^2 = l_o d_o^2 \tag{5.19}$$

$$\epsilon = \ln \frac{l}{l_o} = \ln \left(\frac{d_o}{d}\right)^2 = 2 \ln \left(\frac{d_o}{d}\right). \tag{5.20}$$

Thus, after necking the strain must be measured in terms of the minimum diameter at the neck (d).

Stress Distribution in Neck

The stress is no longer uniformly distributed across the minimum area once a neck forms. The reason for this becomes evident when we consider two fibers A and B in the necked tensile specimen shown in Fig. 5.9. While the stress is axially directed in the region of uniform area, a considerable radial component of stress will be present in the neck which will increase as the center of the specimen is approached. An increase in radial stress must be accompanied by an equal increase in axial stress as a result of a flow criterion (Tresca criterion) to be discussed presently. Thus, the state of stress across the minimum section of a necked specimen will vary from a uniaxial stress (σ_c) at the periphery of the neck to ($\sigma_c + \sigma_H$) at the axis where σ_H is a hydrostatic component of stress due to the shape of the neck.

Bridgman has found the variation of axial stress (σ_a) with radial distance (r) from the specimen centerline to be

$$\sigma_a = \sigma_c \left[1 + \ln\left(\frac{a^2 + 2aR - r^2}{2aR}\right)\right] \tag{5.21}$$

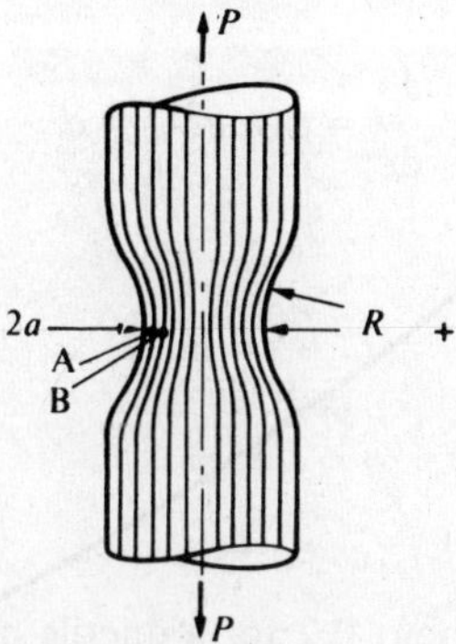

FIG. 5.9. Neck in tensile specimen.

where a is the section radius and R is the profile radius at the neck (Fig. 5.9). Thus, the maximum tensile stress occurs at the center of the specimen and equals

$$\sigma_{a,max} = \sigma_c \left[1 + \ln\left(1 + \frac{a}{2R}\right)\right] \tag{5.22}$$

The corrected uniaxial tensile stress (σ_c) is equal to the stress that would obtain in the absence of a neck and in turn is equal to the stress at the outer periphery of the neck

$$\sigma_c = \frac{\bar{\sigma}}{\left(1 + 2\frac{R}{a}\right)\ln\left(1 + \frac{a}{2R}\right)} \tag{5.23}$$

where $\bar{\sigma}$ is the mean stress

$$\bar{\sigma} = \frac{P}{\pi a^2}.$$

The quantity in the denominator of eqn (5.23) is known as the Bridgman correction factor and is shown plotted against a/R in Fig. 5.10. Figure 5.11 illustrates the manner in which the stress varies across the neck and the relation between $\bar{\sigma}$, σ_c, and $\sigma_{c,max}$. Figure 5.12 is an empirical curve due to Bridgman (1944) for estimating $\sigma_c/\bar{\sigma}$ for different values of strain (ϵ) as defined in eqn 5.20. Also shown on this curve are representative experimental values for mild steel and copper. The Bridgman approximation is found to be much better for steel than for nonferrous metals such as copper. The validity of the Bridgman correction has been verified by Marshall and Shaw (1952) by use of tensile specimens into which necks of known curvature had been machined.

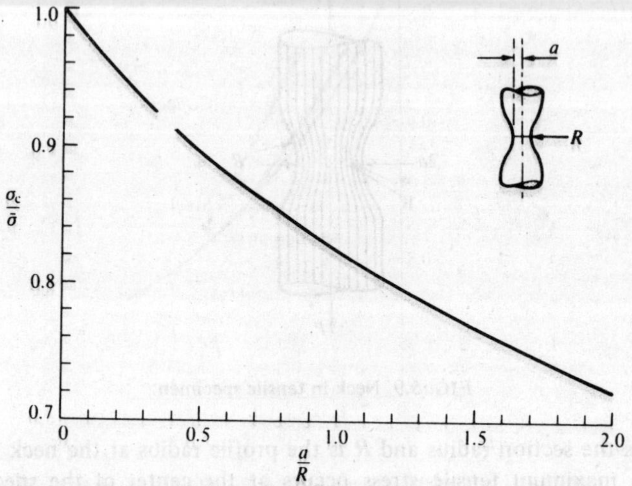

FIG. 5.10. Variation of Bridgman correction factor ($\sigma_c/\bar{\sigma}$) with a/R ratio.

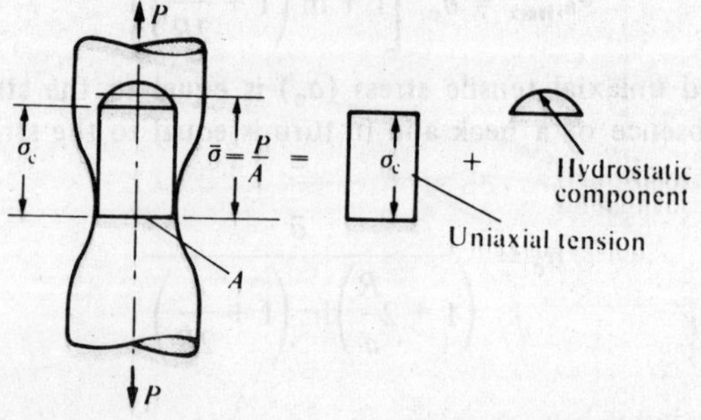

FIG. 5.11. Stresses in the neck of a tensile specimen.

Figure 5.13 shows the relationship between the so-called engineering stress-strain curve based on the original area, the true stress–strain curve and the corrected true stress–strain curve where the stress plotted (σ_c) is the uniaxial tensile stress in the absence of the hydrostatic component introduced by curvature of the neck.

From the foregoing discussion, it is evident that interpretation of tensile test results is really quite involved despite the apparent simplicity of the test.

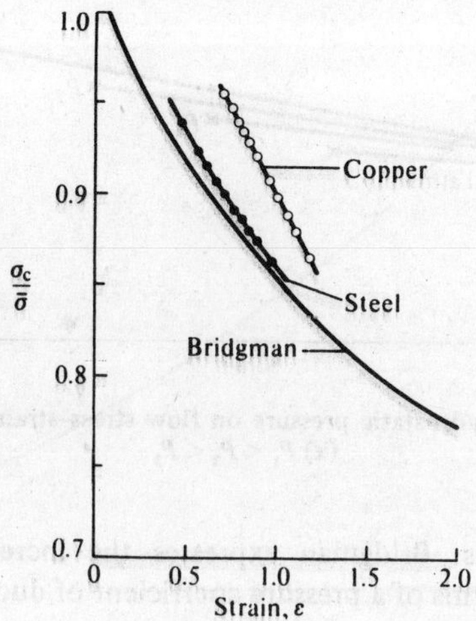

FIG. 5.12. Bridgman's empirical correction factor vs. true strain (ϵ).

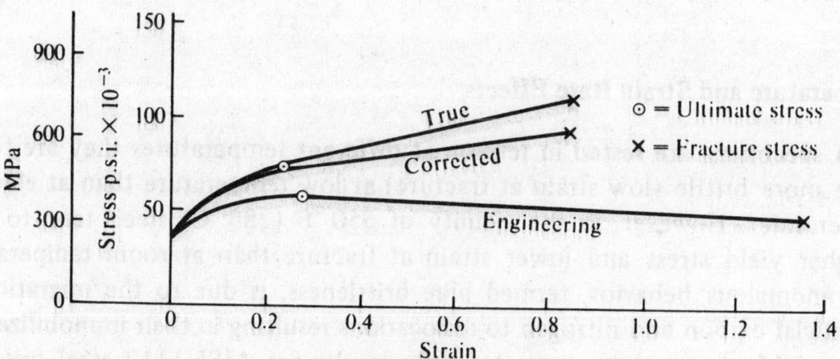

FIG. 5.13. Relationship between engineering, true, and corrected tensile stress–strain curves for AISI 1112 steel.

Hydrostatic Stress

When tensile specimens are tested under different values of hydrostatic pressure, a negligible effect is found upon the yield stress and the post yield shape of the true stress strain curve. For example, Bridgman (1952) found an increase of less than 2 percent in the ultimate stress ($\epsilon = 0.2$) of AISI 1045 steel when subjected to a hydrostatic pressure of 100 000 p.s.i. (690 MPa). He similarly found an increase in Brinell hardness of less than 4 percent with a hydrostatic pressure of 100 000 p.s.i. However, hydrostatic stress has an appreciable influence on fracture stress, a compressive hydrostatic stress increasing fracture stress and a tensile hydrostatic stress decreasing fracture stress. Figure 5.14

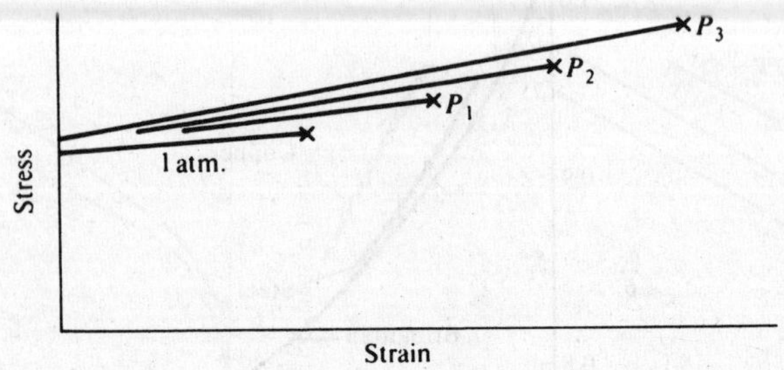

FIG. 5.14. Influence of hydrostatic pressure on flow stress–strain curve and fracture stress (x) $P_1 < P_2 < P_3$.

illustrates these effects. Bridgman expresses the increased fracture strength (σ_f) with pressure in terms of a pressure coefficient of ductility (K):

$$\sigma_f = \sigma_{f0} + K\sigma_H \tag{5.24}$$

where σ_H is the hydrostatic pressure (+ for compression).

Temperature and Strain Rate Effects

When specimens are tested in tension at different temperatures they are found to be more brittle (low strain at fracture) at low temperature than at elevated temperatures. However, in the vicinity of 550 °F (288 °C) steels tend to have a higher yield stress and lower strain at fracture than at room temperature. This anomalous behavior, termed blue brittleness, is due to the migration of intersticial carbon and nitrogen to dislocations resulting in their immobilization. Figure 5.15 shows some typical tensile results for AISI 1112 steel tested at different temperatures but ordinary strain rate (10^{-3}sec^{-1}). As a first approximation, the blue brittle anomaly is ignored and the yield stress is considered to decrease with increase in temperature.

When metals are tested in tension at different strain rates, the flow stress corresponding to a given strain is found to increase with strain rate. Figure 5.16 illustrates this behavior. The following equation resembling eqn (5.8) for strain hardening is frequently used to relate flow stress and strain rate at a given strain and temperature

$$\sigma = \sigma_1 \dot{\epsilon}^m \tag{5.25}$$

where $\dot{\epsilon} = d\epsilon/dt$ and σ_1 and m are material constants. The exponent m (strain rate sensitivity) is found to increase with temperature, especially above the strain recrystallization temperature. In the hot working region, metals tend to approach the behavior of a Newtonian liquid for which $m = 1$, particularly

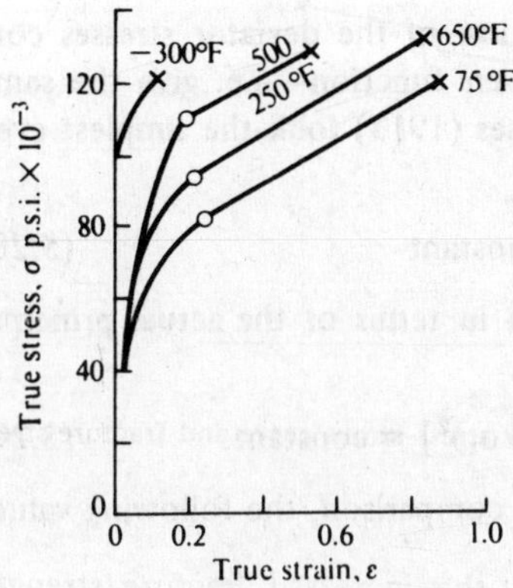

FIG. 5.15. True-stress–true-strain tensile curves for AISI 1112 steel at different temperatures, Strain rate = 10^{-3} sec^{-1} (after MacGregor, 1944).

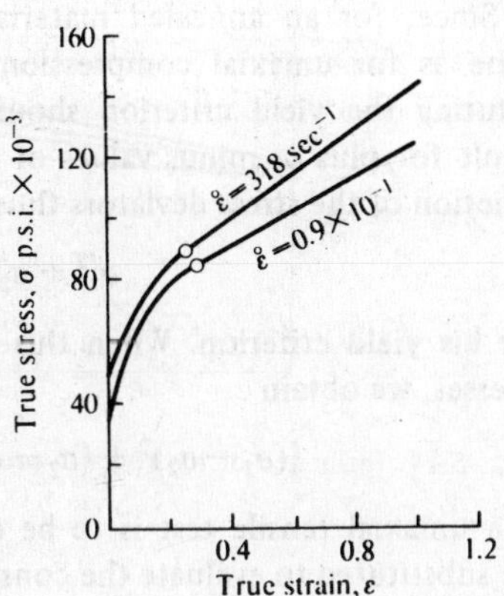

FIG. 5.16. True stress–strain curves for AISI 1112 steel at different strain rates. Temperature = 70 °F (21 °C) (after MacGregor, 1944).

metals of very small crystal size (~ 1 μm) which are superplastic (capable of deformation to very large strains at high temperature and low strain rate).

In general, increased temperature and strain rate have opposite effects on the flow stress of a material.

Von Mises Flow Criterion

A basic problem in design is to predict the yield stress (or flow stress corresponding to a given plastic strain) under a complex state of stress given the yield stress (or flow stress curve) in uniaxial tension or other standardized materials test. There is considerable experimental evidence that plastic flow is not influenced by a purely hydrostatic component of stress. Since the mean principal stress given by

$$\sigma_H = \frac{\sigma_1 + \sigma_2 + \sigma_3}{3} \tag{5.26}$$

approximates hydrostatic stress, it is convenient to use this in formulating a flow criterion. The first step in formulating a flow criterion is to subtract (σ_H) from the principal stresses pertaining and then to formulate a criterion based on these altered principal stresses which are termed deviator stresses. The principal deviator stresses are

$$\begin{aligned} \sigma_1' &= \sigma_1 - \sigma_H \\ \sigma_2' &= \sigma_2 - \sigma_H \\ \sigma_3' &= \sigma_3 - \sigma_H. \end{aligned} \tag{5.27}$$

Since, for an annealed material, the yield stress in uniaxial tension is the same as for uniaxial compression, the function of the deviator stresses constituting the yield criterion should be an even function—i.e. give the same result for plus or minus values of σ'. Von Mises (1913) took the simplest even function of the stress deviators thus

$$\sigma_1'^2 + \sigma_2'^2 + \sigma_3'^2 = \text{constant} \tag{5.28}$$

for his yield criterion. When this is rewritten in terms of the actual principal stresses, we obtain

$$[(\sigma_1 - \sigma_2)^2 + (\sigma_2 - \sigma_3)^2 + (\sigma_3 - \sigma_1)^2] = \text{constant.} \tag{5.29}$$

If a uniaxial tensile test is to be the basis of comparison, the following values are substituted to evaluate the constant

$$\sigma_1 = \sigma_y$$
$$\sigma_2 = 0$$
$$\sigma_3 = 0.$$

When these values are substituted into (5.29), we obtain

$$\sigma_y = \frac{1}{\sqrt{2}} [(\sigma_1 - \sigma_2)^2 + (\sigma_2 - \sigma_3)^2 + (\sigma_3 - \sigma_1)^2]^{\frac{1}{2}} = \text{constant.} \tag{5.30}$$

This is known as the von Mises yield criterion. The stress σ_y is the equivalent of any complex state of stress for which the principal stresses are σ_1, σ_2, σ_3 as far as yield or flow at the same strain is concerned.

This equation which is in good agreement with experiment has also been obtained by Nadai (1950) by a more physical approach. He reasoned that flow would begin when the shear stress on a plane oriented such that its normal stress was equal to σ_H reached a particular value. The so-called octahedral plane is one that makes equal angles with the three principal stress coordinates and Nadai reasoned that flow would occur when the shear stress on this plane reached a critical value. The result is the same as eqn (5.30). Maxwell (1856) and later Hencky (1924) arrived at eqn (5.30) from still another point of view. They assumed that flow will occur when a material absorbs a certain shear strain energy per unit volume. The strain energy per unit volume associated with an elastic body subjected to principal stresses σ_1, σ_2, σ_3 is

$$u_S = \frac{1+\nu}{3E} [(\sigma_1 - \sigma_2)^2 + (\sigma_2 + \sigma_3)^2 + (\sigma_3 - \sigma_1)^2]. \tag{5.31}$$

For a tensile specimen

$$u_S = \frac{1+\nu}{3E} (2\epsilon_y^2). \tag{5.32}$$

Equating eqns (5.31) and (5.32) again leads to eqn (5.30). The Maxwell-Hencky criterion may also be derived in terms of principal strains and when this is done, the flow criterion becomes

$$\epsilon_y = \frac{1}{\sqrt{2}(1+\nu)} [(\epsilon_1 - \epsilon_2)^2 + (\epsilon_2 - \epsilon_3)^2 + (\epsilon_3 - \epsilon_1)^2]^{\frac{1}{2}} \tag{5.33}$$

where ϵ_y is the principal yield strain in a tensile test.

Tresca Flow Criterion

Tresca (1864) and Mohr (1914) took a different approach to the flow criterion problem. They assumed that flow in a polycrystalline material will occur when the greatest of the principal shear stresses reaches a critical value. For plane stress

$$\frac{\sigma_1 - \sigma_2}{2} = \frac{\sigma_y}{2} \tag{5.34}$$

where $\sigma_y/2$ is the maximum shear stress in a tensile test at the onset of plastic flow.

Equation (5.34) may be written

$$(\sigma_1 - \sigma_2)^2 - \sigma_y^2 = 0 \tag{5.35}$$

or when generalized for the three-dimensional case

$$[(\sigma_1 - \sigma_2)^2 - \sigma_y^2]\,[(\sigma_2 - \sigma_3)^2 - \sigma_y^2]\,[(\sigma_3 - \sigma_1)^2 - \sigma_y^2] = 0 \tag{5.36}$$

or when written in terms of deviator stresses:

$$[(\sigma_1' - \sigma_2')^2 - \sigma_y^2]\,[(\sigma_2' - \sigma_3')^2 - \sigma_y^2]\,[(\sigma_3' - \sigma_1')^2 - \sigma_y^2] = 0 \tag{5.37}$$

It is thus seen that the Tresca criterion leads to an even function of the stress deviator components just as the von Mises relationship did. While eqn (5.37) is more complex than that arbitrarily assumed by von Mises, there is no theoretical reason why it is any more or less correct.

The results predicted by eqns (5.30) and (5.37) are usually in close agreement (Fig. 5.17). While the Tresca criterion is more appealing on physical grounds, those analytically inclined seem to favor the von Mises approach. However, the two criteria appear to be equally useful and the one to use is that which is more convenient for the problem involved.

Effective Stress and Strain

The yield criterion is sometimes extended to cover the entire flow curve. When this is done, equivalent stress (σ_e) and equivalent strain (ϵ_e) are defined as follows based on the von Mises approach:

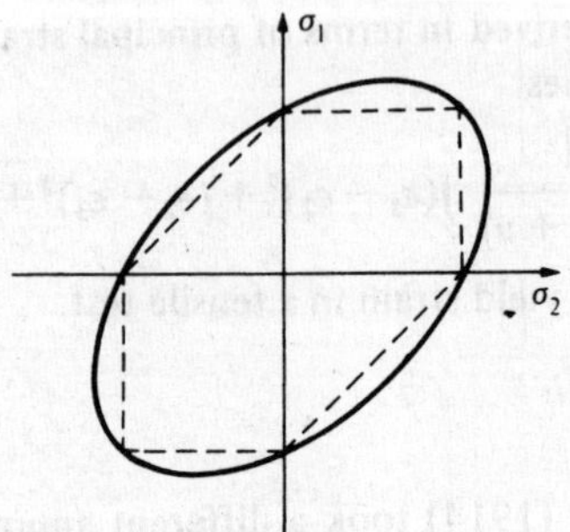

FIG. 5.17. Yield loci for plane stress. Solid line = von Mises criterion, dotted line = Tresca criterion. Maximum difference = 15.5%.

$$\sigma_e = \frac{1}{\sqrt{2}} \, [(\sigma_1 - \sigma_2)^2 + (\sigma_2 - \sigma_3)^2 + (\sigma_3 - \sigma_1)^2]^{\frac{1}{2}} \tag{5.38}$$

$$\epsilon_e = \frac{\sqrt{2}}{3} \, [(\epsilon_1 - \epsilon_2)^2 + (\epsilon_2 - \epsilon_3)^2 + (\epsilon_3 - \epsilon_1)^2]^{\frac{1}{2}}. \tag{5.39}$$

The numerical multipliers in each case have been chosen to yield the uniaxial tensile values for σ_e and ϵ_e when values for principal stress and strain corresponding to uniaxial tension are substituted

$$(\sigma_1 = \sigma_y, \sigma_2 = \sigma_3 = 0; \epsilon_1 = \epsilon_y, \epsilon_2 = \epsilon_3 = -\epsilon_y/2).$$

Similar equations based on the Tresca approach are

$$\sigma_e = \sigma_1 - \sigma_3 \tag{5.40}$$

$$\epsilon_e = \frac{2}{3}(\epsilon_1 - \epsilon_3) \tag{5.41}$$

where σ_1, ϵ_1 and σ_3, ϵ_3 are the algebraically largest and smallest principal stresses and strains respectively.

Uniaxial Compression Test

The simple uniaxial compression test (Fig. 5.18) offers another way of obtaining the plastic flow curve of a material. In performing such a test, the height of the specimen (H_o) should be about 1.5 times the diameter (D_o) and friction on the loading surface should approach zero. The role of friction in this test is to cause barrelling (Fig. 5.19) by making it more difficult for the material near the loading surfaces to flow outward than that in the central region. Barrelling plays a similar role in the compression test to necking in the tensile test—that is, it causes the radial stress distribution to be nonuniform (Fig. 5.20)

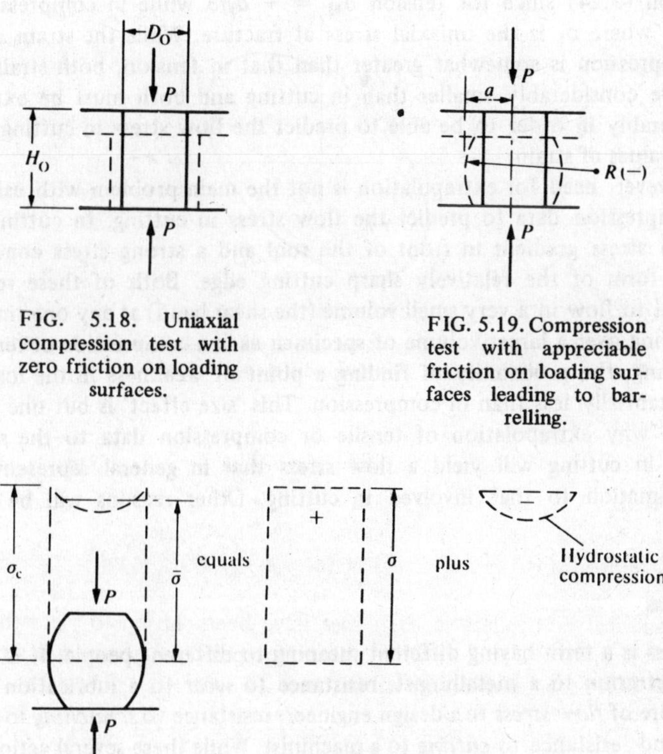

FIG. 5.18. Uniaxial compression test with zero friction on loading surfaces.

FIG. 5.19. Compression test with appreciable friction on loading surfaces leading to barrelling.

FIG. 5.20. Stress distribution across a compression specimen after barrelling.

and the Bridgman correction may be applied by use of a negative value for R (where R = the curvature of the barrel). However, the effect of barrelling is usually avoided and the corrected flow curve is directly obtained. This is done by either arranging for negligible friction on the die faces or by periodically returning the specimen to cylindrical shape by machining the outer surface of the specimen in a lathe.

Friction on the die surface may be reduced to a very low value by providing a helical scratch in the die face filled with a grease containing a solid lubricant (MoS_2 or graphite) or by use of a thin (~ .004 in or 0.1 mm) film of teflon. Where the specimen must be deformed to a very large strain before fracture, it is advisable to eliminate friction as much as possible and then remove the residual bulge by machining the specimen periodically.

The corrected flow curves for uniaxial tension and compression will be the same for an annealed material, except that the strain at fracture will be greater in compression than in tension. This is in accordance with Bridgman's equation (5.24) since for tension $\sigma_H = + \sigma_f/3$ while in compression $\sigma_H = - \sigma_f/3$ where σ_f is the uniaxial stress at fracture. While the strain at fracture in compression is somewhat greater than that in tension, both strains at fracture are considerably smaller than in cutting and both must be extrapolated considerably in order to be able to predict the flow stress in cutting at appropriate values of strain.

However, need for extrapolation is not the main problem with using tensile or compression data to predict the flow stress in cutting. In cutting there is a steep stress gradient in front of the tool and a strong stress concentration in the form of the relatively sharp cutting edge. Both of these require the material to flow in a very small volume (the shear band) at any one time instead of flowing over a larger volume of specimen as in a compression or tension test. In cutting, the probability of finding a point of weakness in the loaded zone is substantially less than in compression. This 'size effect' is but one of several reasons why extrapolation of tensile or compression data to the strain pertaining in cutting will yield a flow stress that in general represents a poor approximation to that involved in cutting. Other reasons will be discussed later.

Hardness

Hardness is a term having different meaning to different people. It is resistance to *penetration* to a metallurgist, resistance to *wear* to a lubrication engineer, a measure of *flow stress* to a design engineer, resistance to *scratching* to a mineralogist, and resistance to *cutting* to a machinist. While these several actions appear to differ greatly in character they are all related to the plastic flow stress of the material (Y).

The wide variety of hardness test procedures that have been used may be classified as follows:

1. *Static indentation* tests in which a ball, cone, or pyramid is forced into a surface and the load per unit area of impression is taken as the measure of hardness. The Brinell, Vickers, Rockwell, Monotron, and Knoop tests are of this type.
2. *Scratch* tests in which we merely observe whether one material is capable of scratching another. The Mohs and file hardness tests are of this type.
3. *Plowing* tests in which a blunt element (usually diamond) is moved across a surface under controlled conditions of load and geometry and the width of the groove is the measure of hardness. The Bierbaum test is of this type.
4. *Rebound* tests in which an object of standard mass and dimensions is bounced from the test surface and the height of rebound is taken as the measure of hardness. The Shore Scleroscope is an instrument of this type.
5. *Damping* tests in which the change in amplitude of a pendulum having a

a hard pivot resting on the test surface is the measure of hardness. The Herbert pendulum test is of this type.

6. *Cutting* tests in which a sharp tool of given geometry is caused to remove a chip of standard dimensions.

7. *Abrasion* tests in which a specimen is loaded against a rotating disc and the rate of wear is taken as a measure of hardness.

8. *Erosion* tests in which sand or abrasive grain is caused to impinge upon the test surface under standard conditions and the loss of material in a given time is taken as the measure of hardness. The hardness of grinding wheels is measured in this way.

The equipment and detailed test conditions for most of the hardness tests in use today may be found in texts by O'Neill (1934), Williams (1942), Tabor (1951), and von Weingraber (1952).

Scratch Hardness

One of the oldest scales of hardness, and one that is still in use by geologists, is that due to Mohs (1822) in which the scratch resistance of a material is compared with the scratch resistances of a standard series of ten minerals. Each of these minerals is assigned a number from 1 to 10. A Mohs hardness of 1 corresponds to a very soft material while 10 refers to the most scratch resistant of all known materials—the diamond. Mohs' scale of hardness is given in Table 5.1.

In applying the Mohs scale practically, the standard minerals are not employed, but the simple tests given in Table 5.1 under the heading *working scale* are used instead. The Mohs hardness of engineering metals falls within the narrow range from 4 to 7.

Tabor (1956) has shown that whereas a stylus must be 20% harder than a flat surface in order to scratch the flat surface, the minerals constituting the Moh's scale increase in hardness in geometric ratio by a factor of approximately 1.4. Diamond is an exception since it is approximately three times as hard as corundum.

As can be seen Mohs' method of describing hardness is little more than qualitative. While attempts have been made to put the scratch technique on a firmer quantitative basis by measuring the width of a scratch made with a diamond under standard conditions (Bierbaum 1930), such tests are difficult to carry out with the necessary precision. The indentation method is by far the most practical and useful method of measuring hardness.

TABLE 5.1
Mohs' scale of hardness

Mineral	Mohs hardness number	Working scale
Talc	1	Very easily scratched by fingernail; greasy feeling
Gypsum	2	Easily scratched by fingernail
Calcite	3	Scratched by copper coin
Fluorite	4	Easily scratched by knife
Apatite	5	Scratched with difficulty by knife
Orthoclase	6	Easily cut by file
Quartz	7	Hardly cut by file—will scratch glass
Topaz	8	Scratches glass
Corundum	9	Scratches glass
Diamond	10	Scratches glass

Indentation Hardness Tests

Two important types of indentation tests are in use. In one type (Brinell, Vickers, Knoop), the diameter of the impression left after a hard indentor is pressed into the surface is measured. In the other type (Rockwell), the depth to which the indentor penetrates the specimen is the measure of hardness.

The indentation hardness test measures the resistance material offers to plastic indentation. The quantity measured is the flow stress at relatively small strain ($\epsilon < 0.1$), low strain rate ($\sim 10^{-3}$ $\sec^{-1}$), and negligible friction. The main distinguishing feature is the presence of a large constraint relative to plastic deformation which causes the flow stress to be several times that involved in uniaxial compression.

Indentation hardness is usually expressed in units of pressure (kg mm^{-2} or p.s.i.), obtained by dividing the maximum applied load (P) by the area of the indentation measured either over the surface of the indentor (Brinell hardness) or in the plane of the surface indented (Meyer hardness). In both the Brinell and Meyer tests, the indentor is a sphere (Fig. 5.21a) and:

$$H_B = \text{Brinell hardness} = \frac{2P}{\pi D^2\{1 - \sqrt{[1-(2a/D)^2]}\}} \tag{5.42}$$

$$H_M = \text{Meyer hardness} = \frac{P}{\pi a^2} \tag{5.43}$$

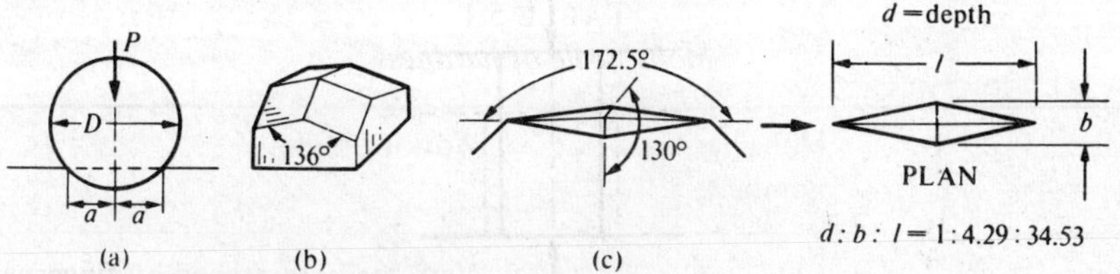

FIG. 5.21. Hardness indentors. (a) Brinell. (b) Vickers. (c) Knoop.

Brinell (1901) and Meyer (1908) hardness values to not differ greatly, but the Meyer value is often preferred because of its simplicity and correspondence to the true mean stress over the area of contact.

Vickers (Smith and Sandland 1922) (Fig. 5.21b) and Knoop (1939) (Fig. 5.21c) indentors are blunt pyramids. Faces of the Vickers indentors are inclined at an angle of 136°, while the ridges of the Knoop indentor have angles of 130° and $172\frac{1}{2}$°, respectively. Because the Knoop indentor penetrates only about half as deeply as the Vickers for the same load, it is frequently preferred for studies of superficial hardness. The Vickers hardness is the load divided by the contacting surface area, while the Knoop hardness is the load divided by the projected area, and hence corresponds to the Meyer value.

The hardness test is very easily conducted, but not so easily interpreted. Action beneath the indentor is complex and must be understood if full use is to be made of hardness values.

The simple compression test, Fig. 5.22, provides another measure of resistance to plastic flow that is more widely used in design analysis. If friction is kept to a low value on the die faces, a compression specimen will deform as shown by the dotted lines in Fig. 5.22 without barrelling, and the uniaxial flow stress will be:

$$\sigma_1 = \frac{P}{A} \tag{5.44}$$

where A is the cross-sectional area of the specimen.

It is important to relate Meyer hardness to uniaxial flow stress, a term with which most engineers are accustomed. The plastic zone beneath a hardness indentation is surrounded by elastic material which acts to hinder plastic flow in a manner similar to the die surfaces in a closed die forging. In the simple compression test the entire specimen goes plastic, and there is no resistance to side flow because the specimen is surrounded by air. Therefore, a greater mean stress is required to cause plastic flow in the hardness test than in the simple compression test.

The relation between the Meyer hardness and the uniaxial flow stress may be expressed as follows:

$$H_M = C\sigma_1 \tag{5.45}$$

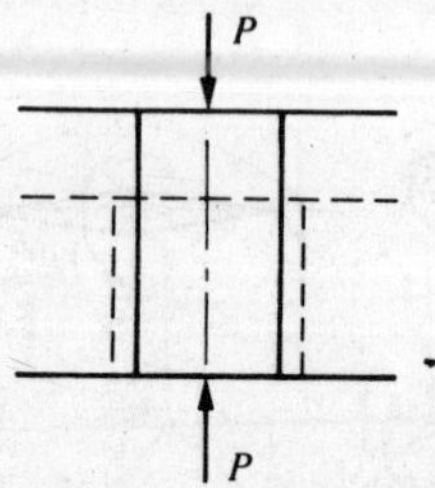

FIG. 5.22. Plane strain uniaxial compression test.

where C is called the constraint factor for the hardness test. Experimentally, C approximates three for the Brinell, Vickers, and Knoop hardness tests. A central problem in the theory of hardness is to explain the origin of constraint factor, C.

Slip Line Field Theory

The generally adopted explanation of indentation hardness is given in terms of the slip line field theory (SLF). According to this theory, the material beneath a punch flows plastically in plain strain over a region consistent with the material dsiplaced by the punch. At all other points, the specimen is considered rigid. The SLF is a network of curves along which the shear stress or shear strain rate is maximum. A suitable flow pattern need only be consistent from the point of view of velocities; when this is so, it is said to be a kinematically admissible solution.

The first kinematically admissible solution for a flat two-dimensional punch is that shown in Fig. 5.23a, due to Prandtl (1920). The solid lines are directions of constant maximum shear stress. They constitute a set of orthogonal shear stress coordinates designated α and β. The normal stress on shear planes α and β will be constant as long as the slip lines are straight but will change when traversing a curved slip line. The normal stress on surface AB (σ_1) will be zero since this communicates with air at zero gage pressure.

The Mohr's stress circle for all points in region ABC is shown in Fig. 5.23b. The convention adopted in labeling the α and β directions is such that the directions of the algebraically greatest stress (σ_1) lies in the first quadrant of the α–β coordinate system. Thus the positive α and β directions at point D are directed as shown in Fig. 5.23a according to the usual convention. The Mohr's circle of Fig. 5.23b will hold for all points within region ABC. The normal stress on the shear plane for any point within region ABC will be $p_D = k$.

In going from C to E along the curved α line there will be no change in shear stress but a change in normal stress according to the following Hencky (1924) equation

$$p + 2\phi k = \text{constant} \tag{5.46}$$

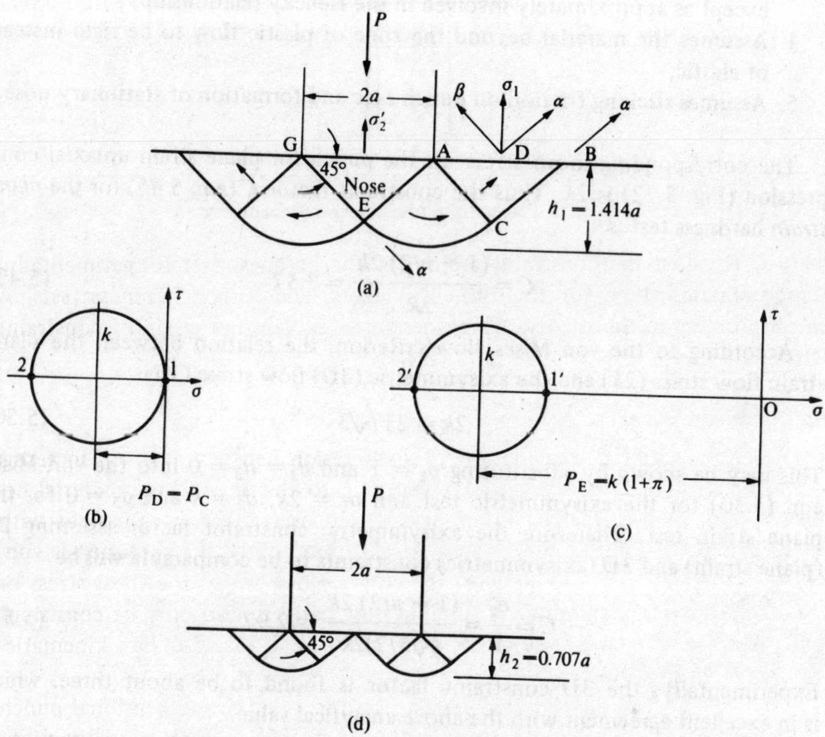

FIG. 5.23 Slip line field solutions for flat two dimensional punch: (a) Prandtl (1920) SLF for punch face with high friction. (b) Mohr's circle for field ABC. (C) Mohr's circle for field AEG. (d) Hill (1950) for frictionless punch face.

where ϕ is the angle between the α and x axes. At point C (Fig. 5.23a) $\phi = \pi/4$ and at point E, $\phi = -\pi/4$. From eqn (5.46)

$$p_E = p_c + 2k\frac{\pi}{2} = k(1 + \pi) \tag{5.47}$$

The Mohr's circle diagram for point E and all points in region AEG is shown in Fig. 5.23c. The normal stress on the face of the punch (σ'_2) in Fig. 5.23c will be

$$\sigma'_2 = k + p_E = 2k\left(1 + \frac{\pi}{2}\right) = 2k\,(2.57) \tag{5.48}$$

It should be noted that the foregoing slip line field analysis

1. Is for plane strain,
2. Ignores strain hardening (k constant) and strain history (Bauschinger effect),

3. Involves a kinematically admissible flow pattern but ignores equilibrium except as approximately involved in the Hencky relationship,
4. Assumes the material beyond the zone of plastic flow to be rigid instead of elastic,
5. Assumes sticking friction on punch face and formation of stationary nose.

The corresponding mean stress on the punch for plane strain uniaxial compression (Fig. 5.22) is $2k$. Thus the constraint factor C (eqn 5.45) for the *plane strain* hardness test is

$$C = \frac{(1+\pi/2)\,2k}{2k} = 2.57 \tag{5.49}$$

According to the von Mises flow criterion, the relation between the plane strain flow stress ($2k$) and the axisymmetric (3D) flow stress (Y) is

$$2k = 2Y/\sqrt{3} \tag{5.50}$$

This may be shown by substituting $\sigma_1 = Y$ and $\sigma_2 = \sigma_3 = 0$ into the von Mises eqn (5.30) for the axisymmetric test and $\sigma_1 = 2k$, $\sigma_2 = k$ and $\sigma_3 = 0$ for the plane strain test. Therefore the axisymmetric constraint factor assuming 2D (plane strain) and 3D (axisymmetric) constraints to be comparable will be

$$C = \frac{\sigma_2'}{Y} = \frac{(1+\pi/2)\,2k}{(\sqrt{3}/2)2k} = 2.97 \tag{5.51}$$

Experimentally, the 3D constraint factor is found to be about three, which is in excellent agreement with the above analytical value.

A solution that satisfies flow kinematics but not equilibrium will lead to results on the high side of truth and is termed an upper bound solution. One which ignores flow considerations but satisfies equilibrium leads to a lower bound solution. Thus, a SLF solution should yield an upper bound since it does not take equilibrium *fully* into account. According to the SLF approach, upward flow accounts for the material displaced by the punch and therefore constraint factor C may be termed a flow constraint from this point of view.

The SLF of Fig. 5.23a assumes a dead zone of material on the punch face and hence corresponds to maximum (sticking) friction. Hill (1950) has suggested a second SLF solution (Fig. 5.23d) that corresponds to zero punch face friction. When a slip line field analysis is performed on Fig. 5.23d, the same result as that for Fig. 5.23a is obtained. It is found experimentally that friction has a negligible influence on the constraint factor C associated with indentation by a punch. When applying the SLF technique, it is important to check that the flow pattern assumed is reasonably close to that actually obtained in practice. More details concerning the SLF approach to plasticity problems may be found in Johnson and Mellor (1962), and Rowe (1977).

Upper Bound Approach

The SLF technique is a quasi upper bound approach since equilibrium is taken into account in part by the Hencky equation but not completely on a point-to-point basis. Solutions that ignore equilibrium completely are simpler but more approximate than SLF solutions and these are generally referred to as upper bound solutions. An example of an upper bound solution for the indentation hardness problem is given in Fig. 5.24. In this solution, deformation is

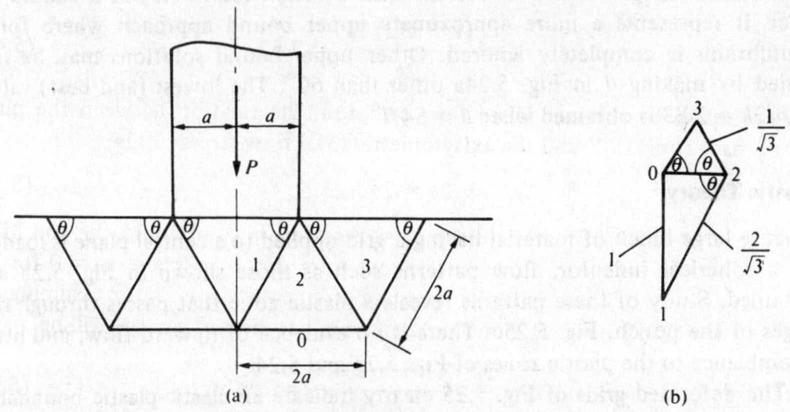

FIG. 5.24. Upper bound solution for plane strain punch. (a) Slip lines for $\theta = 60°$. (b) Corresponding hodograph for flow to right for unit punch velocity ($v = 1$).

assumed to occur by slip only along the solid lines shown in Fig. 5.24a. At all other points, the material is considered to be rigid. The shear flow stress at all points is assumed to be k (uniaxial plane strain flow stress in shear) as in the SLF approach. The external work per unit time is then equated to the internal work per unit time. The external work per unit time is the product of the punch force P and the punch velocity V, while the internal work per unit time is the sum of the products of shear force on each flow surface and the appropriate velocity. The appropriate velocity is obtained from a hodograph (velocity diagram) that satisfies velocity continuity. Figure 5.24b shows the synthesis of the partial hodograph for the right half of Fig. 5.24a. Here, distance 0–1 corresponds to the absolute velocity of the punch (taken to be unity), 1–2 to the relative velocity of region 2 of Fig. 5.24a relative to region 1, and 2–3 the velocity of region 3 relative to region 2. Velocity vectors 0–2 and 0–3 correspond to the absolute velocities of blocks 2 and 3. Fig. 5.24b satisfies continuity of flow to the right since the rate of flow into the deformation zone equals the rate of flow from the deformation zone ($a \times 1 = 2a \times \frac{1}{2}$).

Equating the external and internal work per unit time per unit distance perpendicular to the paper for flow to the right:

$$\frac{P}{2}(1) = k(2a \times 1)\frac{2}{\sqrt{3}} + 3k(2a \times 1)\frac{1}{\sqrt{3}} = \frac{5k(2a)}{\sqrt{3}}$$

or

$$\frac{P}{(2a)(2k)} = \frac{p}{2k} = \frac{5}{\sqrt{3}} = 2.89.$$

This value is larger than that for the SLF solution (eqn 5.49) as it should be since it represents a more approximate upper bound approach where force equilibrium is completely ignored. Other upper bound solutions may be obtained by making θ in Fig. 5.24a other than 60°. The lowest (and best) value of $p/2k = 2.83$ is obtained when $\theta = 54.7°$.

Elastic Theory

When a large block of material having a grid applied to a central plane is loaded by a spherical indentor, flow patterns such as those shown in Fig. 5.25 are obtained. Study of these patterns reveals a plastic zone that passes through the edges of the punch, Fig. 5.25c. There is no evidence of upward flow, and little resemblance to the plastic zones of Figs 5.23 and 5.24.

The deformed grids of Fig. 5.25 clearly indicate an elastic-plastic boundary, which has a shape resembling that of a line of constant maximum shear stress beneath a sphere pressed against a flat surface. This elastic problem was first studied by Hertz (1895). He found that the contact stress was distributed in a hemispherical pattern over the surface, and that lines of constant maximum shear stress were as shown in Fig. 5.26, where:

$$M' = \frac{\tau_{max}}{\bar{p}} \tag{5.52}$$

and

$$\tau_{max} = \text{maximum shear stress}$$

$$\bar{p} = \text{mean pressure on punch face (Meyer hardness).}$$

In Fig. 5.26 the punch face is shown flat for simplicity, whereas in reality it is the surface of a large-radius sphere. The elastic-plastic boundary of Fig. 5.25c closely resembles a line between $M' = 0.15$ and 0.20.

An alternative approach to hardness has been presented by Shaw and DeSalvo (1970) in which the material is assumed to be plastic-elastic instead of plastic-rigid.

When applied to the indentation hardness problem, the alternative theory suggests that, if there is sufficient material beneath an indentor, the displaced

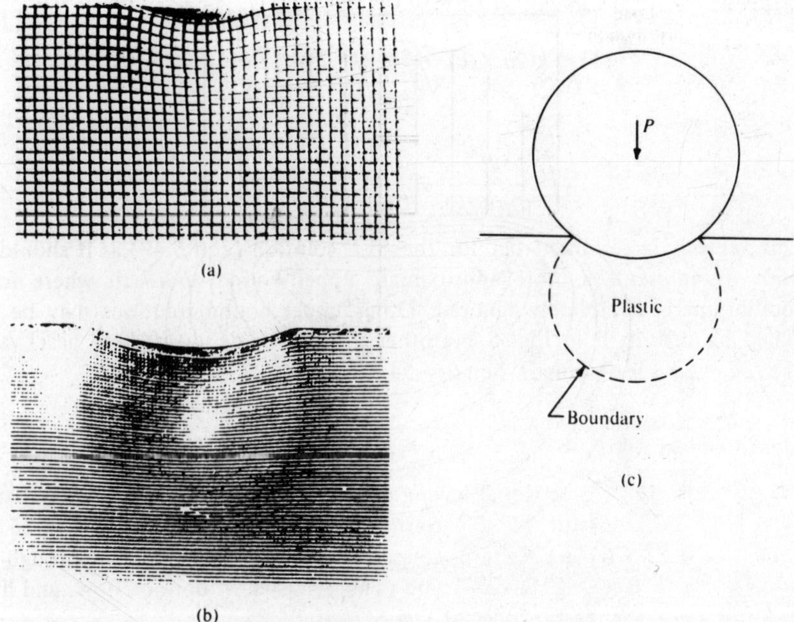

FIG. 5.25. Deformation of grid on meridional plane in a Brinell test. (a) Plasticene. (b) Mild steel. (c) Interpretation of (a) and (b).

material may be completely accounted for by the decrease in volume of the material elastically loaded in compression. There is then no upward flow, as called for in all slip line field solutions. The constraint factor that arises in this way is termed an elastic constraint since the displaced volume is accommodated by an elastic decrease in volume (instead of by upward flow, as in the slip-line field approach).

Since there is no evidence of upward flow in the Figs of 5.25, the constraint involved is of the elastic variety in both instances.

By maximum shear theory of plasticity, one of the lines of Fig. 5.26 should correspond to the elastic-plastic boundary. The shear stress on this particular line should be $Y/2$ (where Y is the uniaxial flow stress of the material). Analysis reveals that the curve for $M' = 0.177$ corresponds to the elastic-plastic boundary and

$$M' = \frac{\tau_{\max}}{\bar{p}} = \frac{Y}{2\bar{p}} = 0.177.$$

Or the corresponding value of constraint factor C is:

$$C = \frac{\bar{p}}{Y} = \frac{1}{2(0.177)} = 2.82.$$

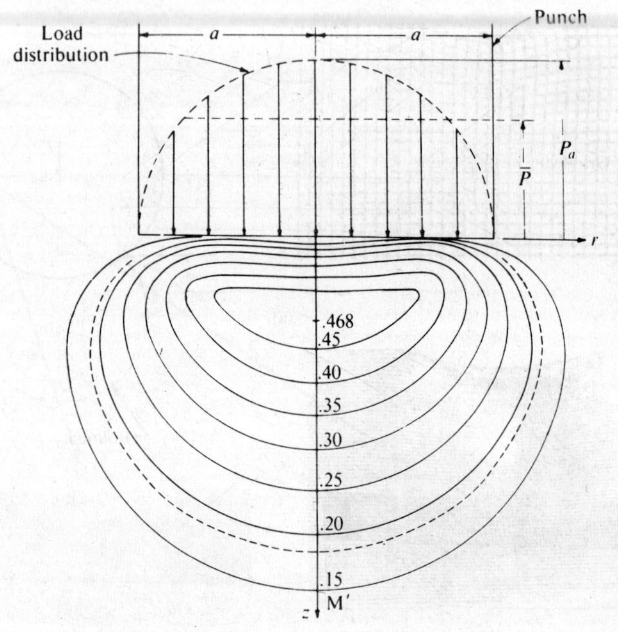

FIG. 5.26. Hertz lines of constant maximum shear stress for a frictionless spherical indentor. The dotted line of constant maximum shear stress is the elastic–plastic boundary ($M' = 0.17$).

However, this constraint factor is referred to the area of the punch in actual contact during indentation (radius a in Fig. 5.26) instead of the area of the plastic impression that remains after the test. These two areas will differ because the edge of the indentor is elastically loaded. When an adjustment is made for the elastically loaded area, the constraint factor based on the plastic impression is found to be $2.82/0.94 = 3.0$, which is in excellent agreement with experiment.

The amount of material required to enable an elastic constraint to be fully developed in a Brinell test of mild steel corresponds to a hemisphere of radius $10(2a) = 20a$. The coefficient 10 is directly proportional to the ratio Young's modulus/Meyer hardness $= E/\bar{p}$. For the standard Brinell test of mild steel (3000 kg load on a 10 mm ball with $2a/D = 0.4$), the impression should be surrounded by a sphere of material of radius 40 mm (1.58 in). If less material surrounds the impression there must be some upward flow. In fact, complete upward flow may be demonstrated by use of a thin layer of modelling clay (low E) on steel (very high E). The material beyond the plastic zone (steel) then has such a high Young's modulus relative to that of the clay that the plastic-rigid assumption holds, producing flow patterns almost identical in appearance to those in Figs. 5.23a and 5.23d as shown in the plane strain flow patterns of Fig. 5.27.

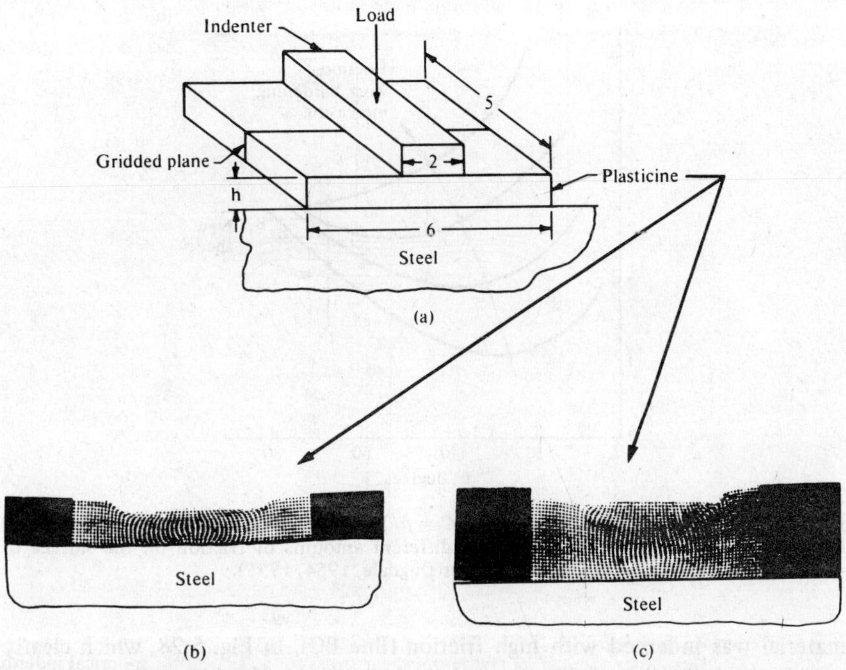

FIG. 5.27. Photographs of grid patterns on transverse planes of plasticene indented in plane strain by flat punch. (a) Test arrangement. Pattern (b) when $h = 1.414a$ as in Fig. 5.23a and (c) when $h = 0.707a$ as in Fig. 5.23d. In both cases the specimens were supported on a steel plate (rigid relative to plasticene).

In order that there be no upward flow, an indentor must be blunt (large cone angle). Increased friction will postpone upward flow as will an increased tendency for strain hardening. The effective cone angle for most indentors, however, is such that some upward flow results even when there is sufficient material surrounding the indentor to provide a full elastic constraint. Thus, most hardness tests correspond to a constraint that is predominantly elastic, but with a small flow component.

Cone Angle Effect

The effective cone angle is a very important variable in hardness testing. Dugdale (1954, 1955) has extensively studied the constraint factor of indentors of different cone angle; his experimental results are summarized in Fig. 5.28. The constraint value was found to decrease markedly with decrease in cone semiangle (θ) when a fully-worked specimen was indented with very low friction (line AB). However, the constraint increased with decrease in θ when an annealed

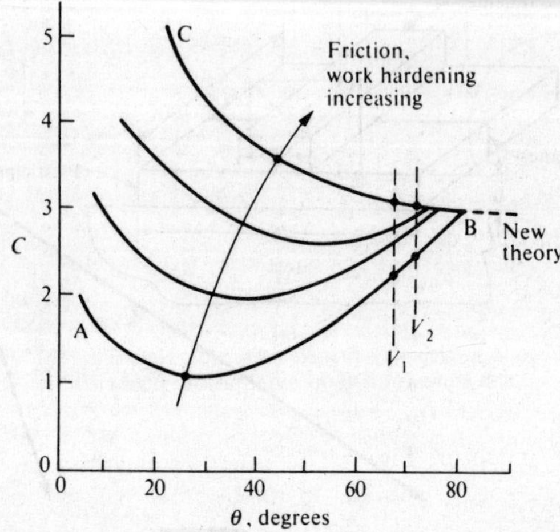

FIG. 5.28. Variation in constraint factor $C = \bar{p}/Y$ for cone semi angles (θ) for metals work hardened to different degrees and with different amounts of friction on the surface of the indentor (after Dugdale, 1954, 1955).

material was indented with high friction (line BC). In Fig. 5.28, which clearly reveals the influence of friction and strain-hardening tendency, the horizontal dotted line shows the fully elastic constraint value from the elastic constraint point of view.

While most of the practical indentors are not conical, they may be assigned an effective cone half angle (θ_e). The effective cone half angle for the sphere is the angle that the tangent to the sphere makes with the vertical at the edge of the indentation, Fig. 5.29a. For the sphere:

$$\theta = \cos^{-1} \frac{2a}{D}. \tag{5.53}$$

When $2a/D = 0.4$, $\theta_e = 66.4°$.

For the Vickers indentor, the cone half angle θ is $136/2 = 68°$ for flow in direction 1 in Fig. 5.29b. For flow in a direction inclined at an angle α to direction 1

$$\theta = \tan^{-1} \left(\frac{\tan 68°}{\cos \alpha} \right). \tag{5.54}$$

For direction 2, this gives $\theta = 72.18°$. The mean value of θ may be obtained as follows:

$$\theta_e = \frac{4}{\pi} \int_0^{\pi/4} \tan^{-1} \left(\frac{\tan 68°}{\cos \alpha} \right) d\alpha = 70.3°. \tag{5.55}$$

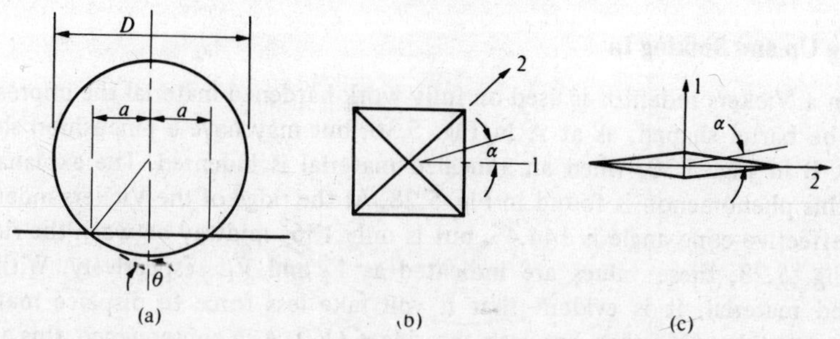

FIG. 5.29. Effective cone angles. (a) Brinell test, side elevation. (b) Vickers indentation, plan view. (c) Knoop indentation, plan view.

For the Knoop indentor, Fig. 5.27c the cone angle (θ) is $130/2 = 65°$ in direction 1, and $172.5/2 = 86.25°$ in direction 2. The mean value $\bar{\theta}$, found as above is 72.12°.

These values are summarized as follows:

Indentor	$2\bar{\theta}$	$2\theta_{min}$	$2\theta_{max}$
Sphere ($2a/D = 0.4$)	132.8	–	–
Vickers	140.6	136	144.4
Knoop	144.2	130	172.5

A sphere loaded such that $2a/D = 0.4$ is seen to be less blunt than either the Vickers or Knoop indentors.

Strain in Hardness Test

Tabor (1951) has empirically estimated the strain at the outer edge of a Brinell impression to be:

$$\epsilon \cong 0.2 \frac{2a}{D} \tag{5.56}$$

From eqn 5.53 this will be

$$\epsilon \cong 0.2 \cos \theta \tag{5.57}$$

For a value of θ of 70°, ϵ corresponds to 0.068, (a relatively low value for plasticity).

Piling Up and Sinking In

When a Vickers indentor is used on fully work-hardened material the impression will be barrel shaped, as at A in Fig. 5.30, but may have a pincushion shape, as at B in Fig. 5.30, when an annealed material is indented. The explanation for this phenomenon is found in Fig. 5.28. At the ridge of the Vickers indentor, the effective cone angle is 144.4°, but is only 136° midway between the ridges. In Fig. 5.28, these values are indicated as V_2 and V_1, respectively. With annealed material, it is evident that it will take less force to displace material along the ridge (V_2) than between the ridges (V_1). As a consequence, this effect will lead to a flow pattern that is more extensive at the corners than between the corners (the pin cushion pattern). Similarly, if the material is fully cold-worked, line AB in Fig. 5.28 will pertain instead of BC. Then it will take more force to displace material along the ridge (V_2) than between ridges (V_1). This effect, of course, leads to the barrel-shaped pattern.

Due to the cone angle being about 140° for most practical indentors, there will be a small upward flow. This upward flow is accentuated if the elastic constraint is kept from developing due to insufficient material being present beneath the indentor. The upward flow will lie close to the surface of the indentor, and will 'pile up' if the material is fully work-hardened and friction is low. For an annealed material, the upward flow will extend farther from the indentor, and will give the appearance of 'sinking-in' at the indentor if friction is high.

If there were no elastic action, residual stresses would not arise. The large residual compressive stresses produced by shot peening to extend fatique life are unexplained by the SLF theory of indentation hardness, which assumes the material to act in a plastic-rigid manner. The elastic constraint enables the magnitude and extent of residual stresses to be estimated.

A spherical indentor develops the full constraint of three gradually with load, and should not be used until $2a/D$ has reached a value of about 0.4 for a spherical indentor made of steel. Since Vickers and Knoop indentors show no such gradual approach to full plasticity, they can be used over a wider range of loads.

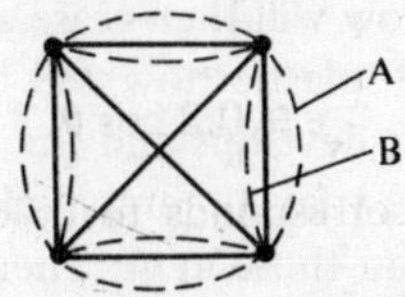

FIG. 5.30. Plan view of Vickers indentation showing barrel-shaped pattern (A) and pin cushion shaped pattern (B).

Relative 'Bluntness' of Indentor

The more blunt the indentor, the smaller will be the amount of upward flow. In the absence of upward flow, the hardness will be the same for an annealed and fully strain-hardened specimen, and friction will play no role. Under such conditions, the hardness measured corresponds to the initial yield point of the material.

An indentor that is blunt enough to prevent upward flow produces an impression that is difficult to see and measure accurately. The effective cone angle must be decreased to a value that approximates that for the Vickers and Knoop indentors for accurate measurement. Presence of the ridges in Vickers and Knoop indentations improves visibility, and hence accuracy of measurement.

With indentors having effective cone angles as small as those in the Vickers and Knoop indentors, there will be a small amount of upward flow. The resulting constraint will still be essentially of elastic origin with a small flow component included. This flow component causes the hardness value to correspond to the flow stress at a small plastic strain (approximately 0.05). It also alters the constraint factor upward or downward, depending upon whether the metal is fully strain-hardened, or annealed, and whether the indentor friction is high or low.

Flow on Unloading Indentor

Initial plastic flow occurs when load is applied to a hardness indentor and a second flow occurs on unloading. Removal of the load may be approximated by superimposing the Hertz solution to an upward force on the ball equal to the downward force in the hardness test. This will correspond to a mean tensile stress on the contact area $= 3Y$ (point A in Fig. 5.31) where Y equals the uniaxial flow stress of the material. The approximate vertical tensile stresses at points B and C will be $2Y$ and Y respectively. Thus after the load has been removed, the state of stress beneath the indentor will be approximately as shown in Fig. 5.32.

However, this is not consistent with the Tresca flow criterion and hence plastic flow will occur until the difference between principal stresses at A, B, and C are each equal to Y. The material at A will flow upward until the lateral stresses drop from $2Y$ to Y. Flow will then cease and the residual biaxial stresses at A will be equal to Y. At B, plastic flow will take place until the difference between the vertical stress that develops and the lateral biaxial stress equals Y. The residual stress at B will be triaxial compression and the transverse residual stresses at B will be greater than those at A. There will be no secondary plastic flow on unloading at point C. Hence, the second plastic flow volume associated with unloading will be smaller than the initial flow volume associated with loading.

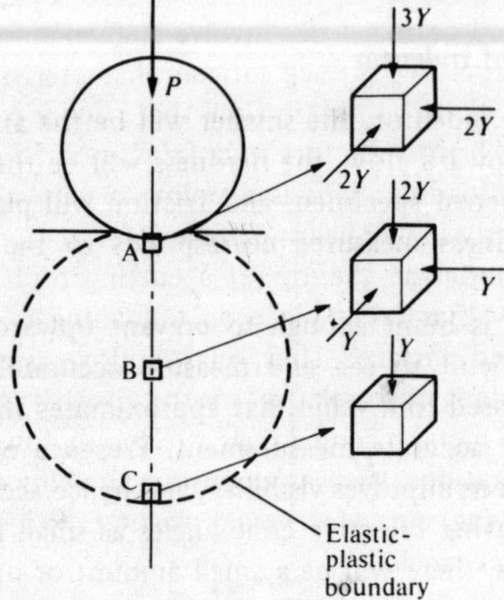

FIG. 5.31. States of stress at different points beneath a hardness indentor when load is still applied.

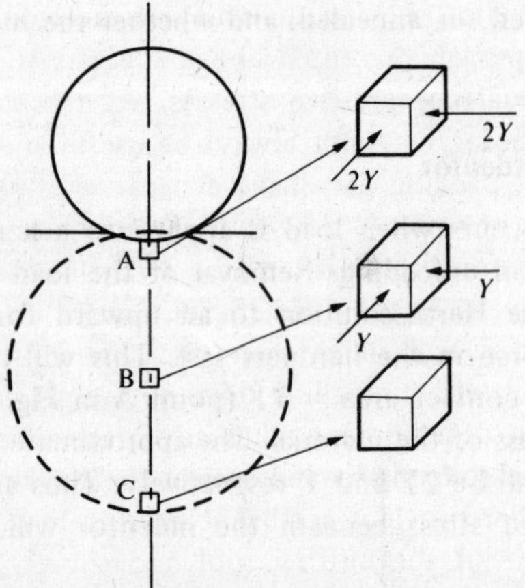

FIG. 5.32. States of stress at different points beneath a hardness indentor just after the load is removed but before the second plastic flow takes place.

The concept of a second plastic flow associated with unloading has several important practical implications.

1. In fundamental studies of indentation hardness, the zone of plastic flow after unloading is often observed and related to the applied load. In interpreting such results, it is important to realize that a second plastic flow will have taken place on unloading; hence care must be taken in inferring the shape of the free surface when the load is still applied from that observed after the load has been relaxed.

2. Ball bearings and gears usually involve a small amount of plastic flow near the surface. In interpreting surface fatigue characteristics of such items, it is important to realize that the plastic flow that occurs on unloading is far more likely to cause local brittle fracture and fatigue crack growth than the plastic flow that occurs on loading. This is a point of view that apparently has not been pursued previously in ball-bearing life analysis.

3. The work material at the tip of a cutting tool is plastically deformed and the adverse flow that takes place on unloading is apt to result in surface cracks in the finished surface. Thus, plastic flow associated with unloading may play an important role relative to the integrity of machined surfaces, particularly when the metal machined has a relatively low fracture strain.

4. In forming operations, such as extrusion, drawing, rolling, forging, thread-rolling, peening, swaging, etc., the second plastic flow associated with unloading may give rise to surface cracks.

5. In shot peening, to increase fatigue life, the surface is bombarded by hard spheres which make a large number of small Brinell impressions in the surface and leave behind residual biaxial and triaxial compressive stresses. The second plastic flow that accompanies unloading will obviously be involved in any rational approach to shot peening that may be developed in the future. The simple qualitative approach presented above suggests the origin and character of these useful residual compressive stresses. At the same time, it becomes evident why shot peening may not always be useful and why, with somewhat brittle materials, shot peening may have a negative effect on fatigue life. This will be the case when the plastic flow that accompanies unloading gives rise to surface imperfections (cracks) that are more detrimental to fatigue life than the residual compressive stresses are beneficial.

6. In burnishing and ball sizing operations, care must be taken that the plastic flow associated with unloading does not result in surface cracks. However, a recent Russian development takes advantage of cracks thus developed. In this case, a carbide roller is loaded against the workpiece just ahead of the cutting tool. This roller provides sufficient pressure to produce cracks in the material just in front of the cutting tool and substantially reduces cutting forces and increases tool life by as much as 50 percent when chilled cast-iron rolls are machined.

The best way of experimentally proving the existence of a second plastic flow on unloading is to trace the shape of the free surface with a sensitive stylus instrument first with the load applied and then after the load has been removed. However, this cannot be done with existing tracing instruments since it is not possible to get the tracing point close enough to the ball when the load is still applied.

It was therefore decided to make a replica of the surface under load and to compare the shape of the free surface of the replica with that of the metal surface after the load had been removed.

A $\frac{1}{4}$ inch diameter steel sphere was loaded into a flat steel surface as shown in Fig. 5.33 at loads varying from 1000 to 4000 kg. In each case, a casting was made with the load on the ball held constant throughout the period of solidification. The load was then removed and the surfaces of the steel and plastic were traced in four directions oriented 90° from each other. In all cases, there was a substantial difference between the loaded and unloaded surfaces.

Figure 5.34. is a representative pair of tracings showing the difference between the loaded and unloaded surface shapes.

It appears that a substantial reverse plastic flow occurs when a Brinell ball is unloaded.

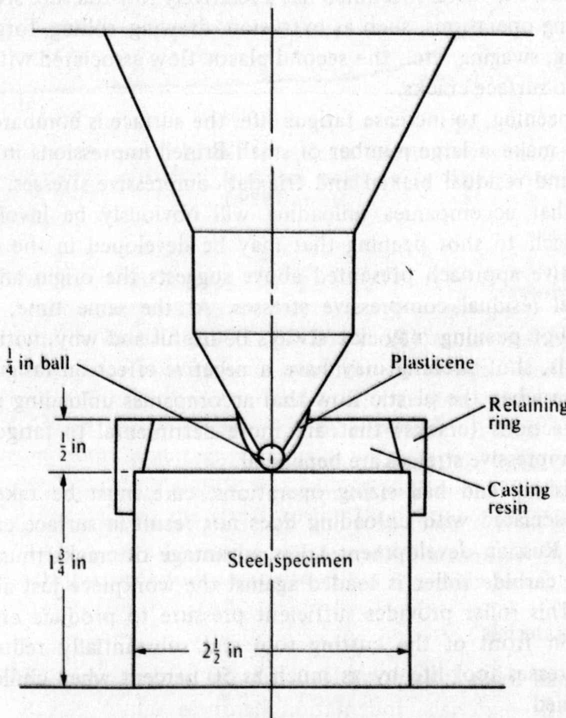

FIG. 5.33. Test arrangement used to obtain a replica of the shape of a loaded surface. (After Shaw, Hoshi and Henry 1979.)

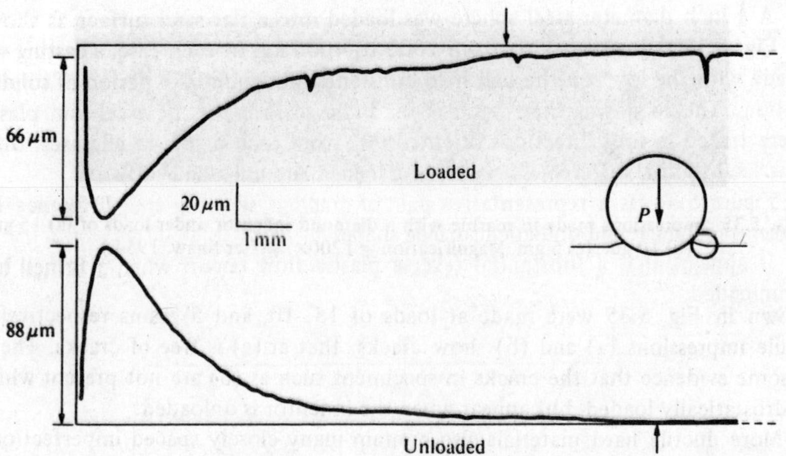

FIG. 5.34. Representative Tallysurf tracings of loaded (upper—plastic replica) and unloaded (lower—steel surfaces in region indicated in insert). (After Shaw, Hoshi, and Henry 1979.)

Rockwell Hardness

Rockwell hardness values (Rockwell 1922) are also in common use. In this case the hardness value is an arbitrary number that varies inversely with the depth of penetration under standard loading conditions. Indentors of different geometries are used for materials of different ranges of hardness. The B-scale (B for ball) is used for relatively soft materials while the C-(for cone) and A-scales are used for materials of increasing ranges of hardness. Rockwell hardness determinations leave a small dent and hence represent relatively nondestructive tests (along with Vickers and Knoop) compared with the standard Brinell test.

Hard Brittle Materials

By use of small loads it is possible to obtain hardness readings of extremely hard and brittle materials. Indentation hardness values may be obtained for materials such as glass, which is normally considered to be completely brittle, if the size of the impression (load) is sufficiently small.

Brittle materials such as glass contain many surface cracks and if the indentor engages one or more of these the specimen will be borken out before a fully developed impression can be made. By using an impression that is smaller than the mean crack spacing it is possible to produce a perfect indentation in a brittle material. Brittleness is a function of specimen size. Marble, which is so brittle that it will break into many pieces if a Brinell test is attempted, may be indented successfully if the load is sufficiently small. The three impressions

FIG. 5.35. Impressions made in marble with a diamond indentor under loads of (a) 15 gm; (b) 10 gm; (c) 5 gm. Magnification = 1200×. (After Shaw, 1954.)

shown in Fig. 5.35 were made at loads of 15, 10, and 5 grams respectively. While impressions (a) and (b) show cracks, that at (c) is free of cracks. There is some evidence that the cracks in specimens such as (b) are not present while hydrostatically loaded, but appear when the indentor is unloaded.

More ductile hard materials also contain many closely spaced imperfections and likewise exhibit a *size effect*. As long as the size of the impression is large compared with the mean imperfection spacing, the normal hardness value will be obtained. However, as the load is decreased to the point where the impression size approaches imperfection spacing, hardness values begin to increase. Values of Knoop hardness versus load and depth are shown in Fig. 5.36 for several hard

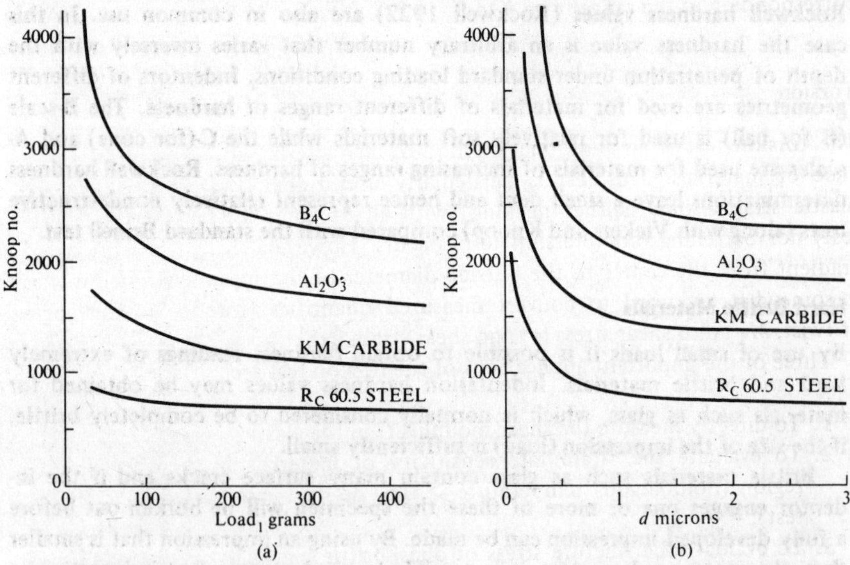

FIG. 5.36. Variation of Knoop microhardness (kg mm^{-2}) with (a) load in grams and (b) depth of impression (d) in microns produced by loads on indentor of varying magnitude. (After Thibault and Nyquist 1947.)

materials. The hardest material (B_4C) is seen to give values which begin to increase with decreased load at a larger value of load than the softer material (tool steel). This is because it takes a higher load to produce the same size impression in B_4C than it does in tool steel. When hardness is plotted against depth of impression in microns, Fig. 5.36b, all curves are found to be of similar shape and to turn upward for depths of impression less than about 2 microns.

Representative microhardness values for a variety of hard materials and metallographic constituents are given in Table 5.2.

Conversion Curves

The curves of Fig. 5.37 enable relative magnitudes of hardness, expressed in several systems, to be compared. Brinell hardness has been used as the standard of comparison. The Knoop hardness is seen to be very nearly equivalent to Brinell hardness over the entire range while the other hardness scales are significantly nonlinear relative to the Brinell scale. The Mohs scale of hardness is seen to be very nonlinear, particularly at the upper end of the scale. The equivalent Brinell hardness values for Mohs hardness values of 8, 9, and 10 are as follows: 1150, 1900, and 7000. From this it is evident that the range from 9 to 10 on the Mohs scale covers more than four times the range of hardness corresponding to the range from 1 to 9.

Torsion

The torsion test is an additional method of characterizing the plastic performance of materials. This test is particularly useful for studies involving large plastic strains since neither necking (with all its complications in the tensile test) nor barrelling (compression test) occurs. However, there is a large strain gradient from the center to the outside diameter of the specimen which makes it somewhat awkward to convert measured quantities (torque, θ, and angle of twist, M_T) into shear stress (τ) and shear strain (γ).

Three of the important applications of the torsion test are

1. To directly measure the shear modulus G in the elastic region.
2. To evaluate the strain rate sensitivity of materials in the hot working region (homologous temperature > 0.5) at high rates of strain (Gleeble test).
3. To evaluate the fracture stress of quasi-brittle materials such as quenched and tempered tool steels. Since the mean principal stress (σ_H) in the torsion test is zero (Fig. 4.9e) instead of being tensile ($\sigma_H = \sigma_1/3$) as in the uniaxial tensile test there is a greater spread of fracture values in torsion for specimens of different quality than in tension.

TABLE 5.2.
Vickers hardness values for a variety of materials

Material	kg mm^{-2}
Structural constituents of steel:	
Austenite	400
Cementite	1100
Pure iron	70
Ferrite	80
Graphite	10
Martensite	800
Tempered martensite	250–800
Pearlite (eutectoid)	250
Spheroidite	175
Sorbite	275
Bainite	485
Troostite	550
Fully hardened tool steel	350–700
Carbides and oxides:	
B_4C	2800
Cr_3C	1200
Chrom-vanadium carbide	2700
Chromium-tungsten carbide	2000
Molybdenum-tungsten carbide	2100
SiC	2400
MO_2C	2000
TaC	1800
TiC	2400
WC	1600
WC + 6% Co	1400
WC + 13% Co	1300
VC	2800
Al_2O_3	2100
Fe_2O_3	1100
Fe_3O_4	650
Diamond	8000
Quartz	1100
Glass	400–600
Hard Chromium plate	1000
Nickel plate	340

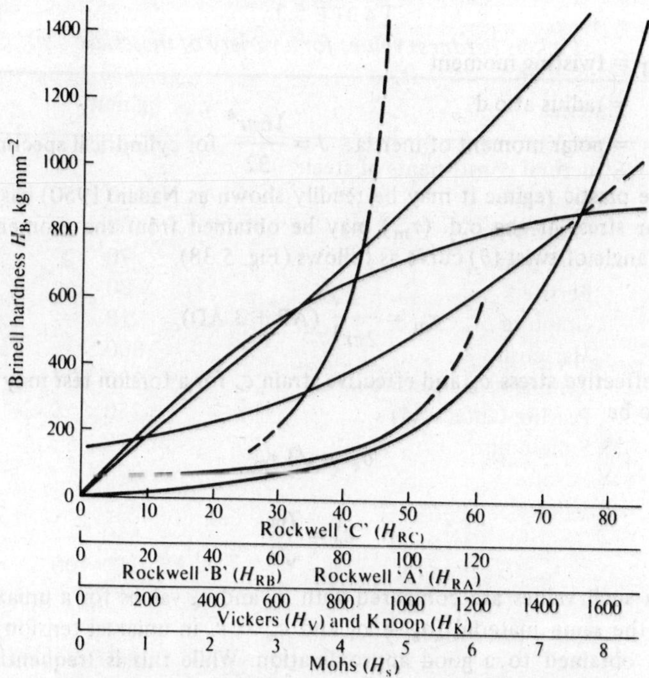

FIG. 5.37. Conversion curves for several hardness scales.

The stress–strain curve in torsion has the same appearance as a true-stress-true-strain tensile curve (with yield point, strain hardening, and finally fracture).

In both the elastic and plastic regimes, the shear strain increases linearly with radius, being zero at the center of the specimen and a maximum (γ_m) at the o.d. where

$$\gamma_m = \frac{r\theta}{L} \tag{5.58}$$

where

r = radius at o.d.

θ = angle of twist of one reference plane relative to the other

L = axial distance between reference planes.

In the elastic regime the shear stress at the o.d. is

$$\tau = \frac{M_T r}{J} \tag{5.59}$$

where

M_T = twisting moment
r = radius at o.d.
J = polar moment of inertia $J = \dfrac{16\pi r^4}{32}$ for cylindrical specimen

In the plastic regime it may be readily shown as Nadai (1950) has done that the shear stress at the o.d. (τ_m) may be obtained from the moment of twist (M_T) vs angle of twist (θ) curve as follows (Fig. 5.38).

$$\tau_m = \frac{1}{2\pi r^3}\,(AB + 3\,AD) \tag{5.60}$$

The effective stress σ_e and effective strain ϵ_e for a torsion test may be readily shown to be

$$\sigma_e = \sqrt{3}\,\tau_m \tag{5.61}$$

$$\epsilon_e = \frac{\gamma_m}{\sqrt{3}} \tag{5.62}$$

When such values are compared with σ_e and ϵ_e values for a uniaxial tensile test on the same material ($\sigma_e = \sigma_1$ and $\epsilon_e = \epsilon_1$ in uniaxial tension) the same curve is obtained to a good approximation. While this is frequently cited as evidence for the general applicability of the concept of equivalent stress and strain, it should be kept in mind that there is no support for the equivalent strain concept in dislocation theory.

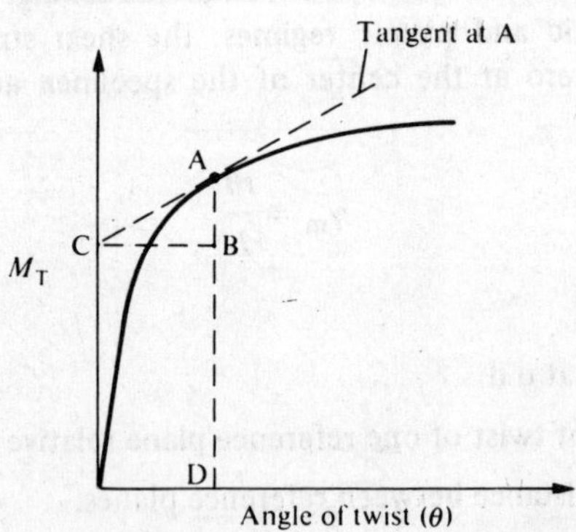

FIG. 5.38. Construction for converting torsional moment vs angle of twist curve to shear-stress–shear strain curve. (After Nadai 1950.)

Mohr's Circle in Plastic Region

Strictly speaking, only Mohr's circle for stress may be applied in the plastic region. Mohr's strain circle only holds exactly for infinitesimal strain increments. However, as Backofen (1972) has shown, Mohr's strain circle also holds providing proportional straining obtains which is often the case. Mohr's circles for stress and strain are frequently employed in the plastic region (first approximation for strain). Equations 5.63 which are the plastic equivalent of elastic equations (4.8) may also be employed as a first approximation where plastic normal stress is designated σ instead of S, plastic normal strain is designated ϵ instead of e:

$$\begin{aligned} \epsilon_1 &\sim \sigma_1 - \nu(\sigma_2 + \sigma_3) \\ \epsilon_2 &\sim \sigma_2 - \nu(\sigma_3 + \sigma_1) \\ \epsilon_3 &\sim \sigma_3 - \nu(\sigma_1 + \sigma_2) \end{aligned} \tag{5.63}$$

As previously mentioned, it is found experimentally that the change in volume in the plastic region is essentially zero and therefore

$$\epsilon_1 + \epsilon_2 + \epsilon_3 = 0 \tag{5.64}$$

When eqns (5.63) are substituted into eqn (5.64), it is found that

$$\nu = 1/2 \tag{5.65}$$

Plane stress and plane strain are important states of stress in plasticity and Fig. 5.39 gives the Mohr's circle diagrams for these two cases. In the example

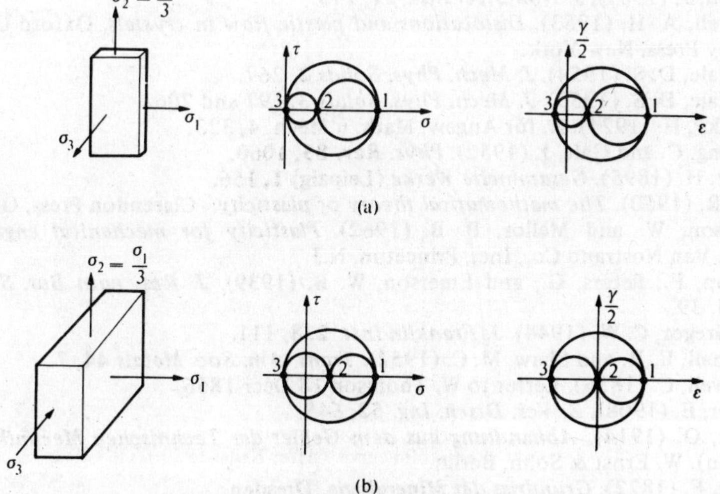

FIG. 5.39. Mohr's circle diagrams for (a) plane stress; (b) plane strain ($\sigma_2 = \sigma_1/3$).

of Fig. 5.39, σ_2 has been taken to be $\sigma_1/3$ and eqns 5.63 and 5.65 have been used to determine the unknown strain and stress respectively.

As mentioned in the previous section, another relatively crude approximation in plasticity is to use an effective flow strain (eqn 5.39) as well as an effective flow stress (eqn 5.38). When this is done, it is equivalent to assuming that strain hardening is a scalar type effect. In reality, strain hardening is strongly dependent on the direction of the strain. The Bauschinger (1881) effect in which the flow stress in compression following plastic flow in tension is found to be substantially reduced (and vice versa) is direct evidence that strain hardening not only depends upon the magnitude of the strain but also the direction of the strain. Considering all strain to be equivalent relative to strain hardening is equivalent to assuming the Bauschinger effect does not exist.

References

Backofen, W. A. (1972). *Deformation processing.* Addison Wesley, Reading, MA.
Bauschinger, J. (1881). *Zivilingur* **27**, 289.
Bierbaum, C. H. (1930). *Trans. Am. Soc. Steel Treat.* **18**, 1009.
Bragg, Sir Lawrence and Nye, J. F. (1947). *Proc. R. Soc.* **A190**, 474.
Bragg, Sir Lawrence and Lomer, Wm., and Nye, J. F. (1954). *Experiments with the bubble model of a metal structure* (16 mm silent film). Brent Laboratories Ltd, North Circular Road, Cricklewood London, NW2.
Bridgman, P. (1944). *Trans. Am. Soc. Metals* **32**, 553.
Bridgman, P. (1952). *Studies in large plastic flow and fracture.* McGraw-Hill Book Co., New York.
Brinell, J. (1901), *J. Iron Steel Inst.* **51**, 243.
Cottrell, A. H. (1953). *Dislocations and plastic flow in crystals.* Oxford University Press, New York.
Dugdale, D. S. (1954). *J. Mech. Phys. Solids* **2**, 267.
Dugdale, D. S. (1955). *J. Mech. Phys. Solids* **3**, 197 and 206.
Hencky, H. (1924). Z. für Angew. Math. u Mech. **4**, 323.
Herring, C. and Galt, J. (1952). *Phys. Rev.* **85**, 1060.
Hertz, H. (1895). *Gesammelte Werke* (Leipzig) **1**, 156.
Hill, R. (1950). *The mathematical theory of plasticity.* Clarendon Press, Oxford.
Johnson, W. and Mellor, P. B. (1962). *Plasticity for mechanical engineers.* D. Van Nostrand Co., Inc., Princeton, N.J.
Knoop, F., Peters, G., and Emerson, W. B. (1939). *J. Res. natn Bur. Stands.* **23**, 39.
MacGregor, C. W. (1944). *J. Franklin Inst.* **238**, 111.
Marshall, E. R. and Shaw, M. C. (1952). *Trans. Am. Soc. Metals* **44**, 7.
Maxwell, C. (1856). Letter to W. Thomson 18 Decr 1856.
Meyer, E. (1908). *Z. Ver. Dtsch. Ing.* **52**, 645.
Mohr, O. (1914). *Abhandlung aus dem Gebiet der Technischen Mechnik* (2nd edn). W. Ernst & Sohn, Berlin.
Mohs, F. (1822). *Grundriss der Mineralogie*, Dresden.
Nabarro, F. R. N. (1967). *Theory of crystal dislocations.* Oxford University Press, New York.

Nadai, A. (1950). *Theory of flow and fracture of solids* Vol. I. McGraw-Hill Book Co., New York.
Nadai, A. (1963). *Theory of flow and fracture of solids* Vol. II, McGraw-Hill Book Co., New York.
O'Neill, H. (1934). *The hardness of metals and its measurement.* Chapman and Hall, London.
Orowan, E. (1934). *Z. Phys.* **89**, 605.
Polanyi, M. (1934). *Z. Phys.* **89**, 660.
Prandtl, L. (1920), *Nachr. Ges. Wiss. Gottingen, Math. Phys. K1.*, 74.
Rockwell, S. R. (1922). *Trans. Am. Soc. Steel Treat.* **2**, 1013.
Rowe, G. W. (1977). *Principles of industrial metal working* (2nd edn). E. Arnold, London.
Shaw, M. C. (1954). *Proc. natn. Acad. Sci. U.S.A.* **46**, 394.
Shaw, M. C. and DeSalvo, G. J. (1970). *J. Engng Ind.* **92**, 469 (Part I) and 480 (Part II).
Shaw, M. C., Hoshi, T., and Henry, D. (1979). *J. Engng Ind.* **101**, 104.
Smith, R. and Sandland, G. (1922). *Proc. Inst. mech. Engrs* (London) **623**.
Smith, R. and Sandland, G. (1925). *J. Iron Steel Inst.* **111**, 285.
Tabor, D. (1951). *The hardness of metals.* Clarendon Press, Oxford.
Tabor, D. (1956). *Br. J. appl. Phys.* **7**, 159.
Taylor, G. I. (1934). *Proc. R. Soc.* **A145**, 362.
Thibault, N. W. and Nyquist, H. S. (1947). *Trans. Am. Soc. Metals* **38**, 271.
Tresca, H. (1864). *C. r. hebd. Séanc. Acad. Sci. Paris* **59**, 754.
Tresca, H. (1867). *C. r. hebd. Séanc. Acad. Sci. Paris* **64**, 809.
von Mises, R. (1913). *Nachr. Ges. Wiss. Göttingen, Math. Phys. K1.* 582.
von Weingraber, N. (1952). *Technische Hartemessung.* Carl Hauser Verlag, Munich.
Williams, S. R. (1942). *Hardness and hardness measurement.* Am. Soc. Met., Cleveland.

6

FRACTURE

There are in general two types of fracture—ductile fracture and brittle fracture. Ductile fracture occurs after appreciable plastic strain and the criterion usually employed is the same as for yield (Tresca or von Mises). Brittle fracture occurs at a strain that is below the yield stress (perfectly brittle material) or relatively close to the yield point (quasi-brittle material) and the criterion usually employed in such a case is the maximum tensile stress criterion. This states that fracture occurs on a plane perpendicular to the direction of maximum tensile stress (σ_1) when the maximum tensile stress reaches a critical value (σ_1 = constant). The fracture surface is usually in the direction of maximum shear stress in the case of ductile fracture. Brittle or quasi-brittle materials exhibit a greater 'pressure coefficient of ductility' (K in eqn 5.24) than do ductile materials.

In some cases both types of fracture are involved. For example, in the tensile test of a ductile metal, initial fracture occurs at the center of the neck where the hydrostatic tensile stress is a maximum. As load is further increased, this initial tensile crack grows radially outward until the remaining area is insufficient to support the shear stress pertaining at which time final fracture occurs along a conical surface inclined approximately 45° to the axis of the specimen. This is called a cup-cone fracture due to the appearance of the fracture surfaces.

Figure 6.1 shows the fracture obtained in tension for (a) a perfectly brittle material such as hard high-carbon steel, (b) a relatively ductile material such as low-carbon steel that necks before fracture, and (c) a perfectly ductile material such as polycrystalline lead.

Griffith's Theory

The first rational approach to fracture was due to Griffith (1924) for perfectly brittle materials. Griffith reasoned that all real materials contain microcracks or equivalent regions of stress concentration that will grow with tensile strain until the elastic strain energy stored at the tip of the crack becomes sufficient to satisfy the surface energy associated with the generation of the new surface area accompanying crack growth, when the crack will spread spontaneously. The microcrack visualized by Griffith was elliptical in shape as shown in Fig. 6.2a for a surface crack or in Fig. 6.2b for an internal crack in a thin plate (plane stress). The stress concentration factor at the tip of such a crack is $(1 + 2c/w) \cong 2c/w$. The elastic energy per unit width stored at the tip of the

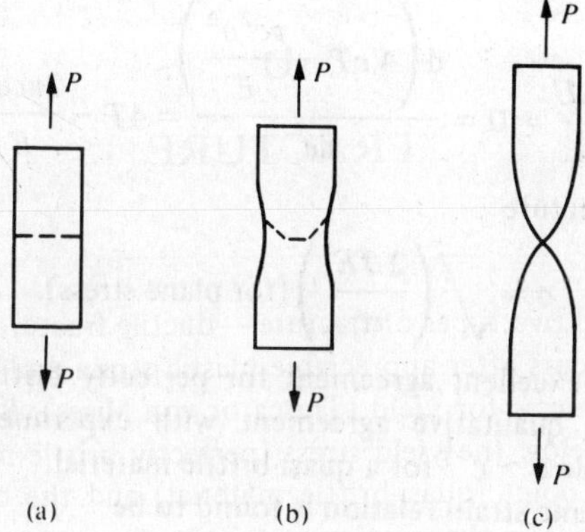

FIG. 6.1. Types of fracture. (a) Perfectly brittle material. (b) Partially ductile material showing cup-cone fracture pattern. (c) Perfectly ductile material.

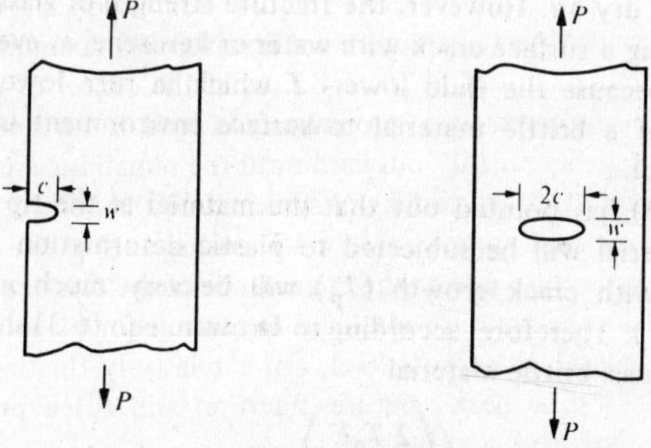

FIG. 6.2. Elliptical crack in plane stress tensile specimen. (a) Crack at surface. (b) Internal crack.

crack of Fig. 6.2b will be $\pi c^2\sigma^2/E$ while the surface energy associated with the two areas of the crack will be $2(2c \times 1)(T)$ where T is the surface energy of the metal. The change in energy ΔU due to growth of the crack will be

$$\Delta U = \left(4\,cT - \frac{\pi c^2\sigma^2}{E}\right) \tag{6.1}$$

where energy absorbed is considered to be (+) and energy released during crack growth is (−). For the crack to grow according to Griffith's thesis

$$\frac{\mathrm{d}\Delta U}{\mathrm{d}c} = 0 = \frac{\mathrm{d}\left(4\,cT - \dfrac{\pi c^2 \sigma^2}{E}\right)}{\mathrm{d}c} = 4T - \frac{2\pi c \sigma^2}{E} \tag{6.2}$$

or solving for σ at fracture

$$\sigma = \sqrt{\left(\frac{2\,TE}{\pi c}\right)} \text{(for plane stress).} \tag{6.3}$$

This equation is in excellent agreement for perfectly brittle materials such as glass, but only in qualitative agreement with experiment for quasi-brittle materials, for example $\sigma \sim c^{-\frac{1}{2}}$ for a quasi-brittle material.

The equivalent plane strain relation is found to be

$$\sigma = \sqrt{\left[\frac{2\,TE}{(1-\nu)^2 \pi c}\right]} \tag{6.4}$$

For glass, the value of $2c$ is estimated to be about 1 μm and T is about 1000 erg cm^{-2} in dry air. However, the fracture strength of glass is significantly reduced by coating a surface crack with water or kerosene, as every glass worker knows. This is because the fluid lowers T which in turn lowers σ (eqn 6.3). The sensitivity of a brittle material to surface environment is known as the Joffe (1928) effect.

Orowan (1950) has pointed out that the material at the tip of a crack in a quasi-brittle material will be subjected to plastic deformation and the energy thus associated with crack growth (T_{p}) will be very much greater than the surface energy (T). Therefore, according to Orowan, eqn (6.3) should be written as follows for a quasi-brittle material

$$\sigma = \sqrt{\left(\frac{2\,T_{\mathrm{p}}E}{\pi c}\right)} \text{ (plane stress)} \tag{6.5}$$

The units of T_{p} are the same as those for T (erg cm^{-2}).

Fracture Mechanics

In the middle 1950s, Irwin and co-workers laid the foundations for what has since become known as *fracture mechanics* (Irwin 1957). The most important contribution of this development has been to introduce an experimentally determined material constant called the stress intensity factor (K) that chara terizes the significance of the defects present in a material from th unt of view of brittle crack growth. The stress intensity factor is defined as follows:

$$K = \sigma\sqrt{(\pi c)}, \text{(p.s.i.)(in)}^{\frac{1}{2}} \text{ or } \mathrm{Nm}^{\frac{1}{2}} \tag{6.6}$$

where σ is the fracture stress and c is the half length of an internal flaw. From eqns (6.5) and (6.6), it is evident that

$$K = \sqrt{(2\, T_p E)}. \tag{6.7}$$

Thus, in effect, K is simply a measure of T_p in the Griffith–Orowan approach to brittle fracture. However, fracture mechanics has contributed a standardized procedure for measuring K for a given material whereas previously T_p was only a qualitative concept. Determination of K involves finding the slopes (m) of the load (P)–deflection (δ) curves for several values of crack half length (c) in the experiment shown in Fig. 6.3. The crack is started by making a saw cut in the thin specimen which is then sharpened by growth in fatigue. The strain clip gage is used to measure the deflection (δ) that results for different values of P.

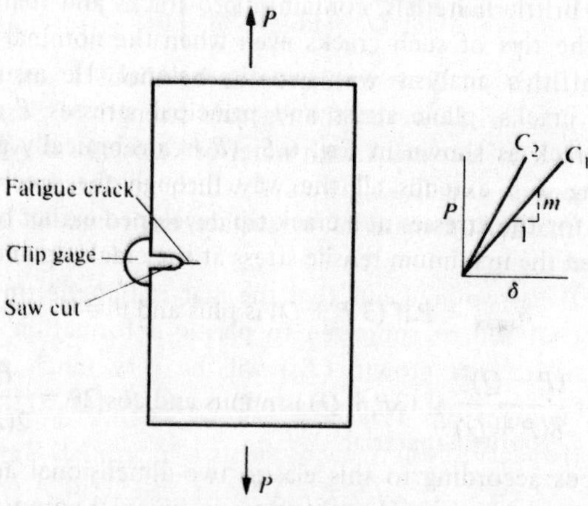

FIG. 6.3. Fracture mechanics test for determining mode I stress intensity factor (K_{IC}).

As may be seen from eqn (6.6), the variation in fracture stress with crack length may be determined as soon as the stress intensity factor (K) has been determined for a given material.

The stress intensity factor described above involves a crack opening mode of loading (mode I) and is therefore frequently designated K_{IC} (C for critical). This is to distinguish this most widely used value of K from those obtained by other modes of loading (II = shear perpendicular to crack front, and III = shear parallel to crack front as shown in Fig. 6.4).

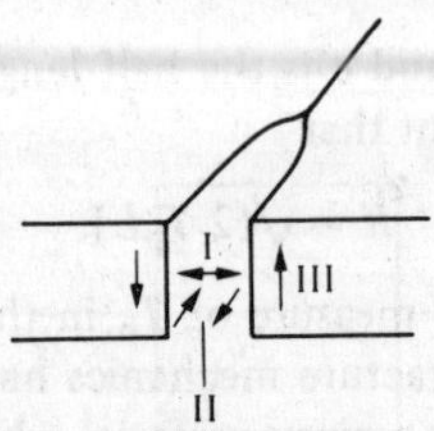

FIG. 6.4. Modes of loading for different critical stress intensity factors K_{IC}, K_{IIC}, K_{IIIC}.

Crack Initiation in Compression

If brittle materials fail according to a maximum tensile stress criterion, it is a fair question to ask how it is that material such as rock, concrete, plaster, and glass can fail brittly when subjected to uniaxial compression. Griffith (1924) provided the answer to this many years ago. Griffith reasoned that all materials, particularly very brittle materials, contain micro-cracks and that tensile stresses can develop at the tips of such cracks even when the nominal stress is purely compressive. Griffith's analysis was two-dimensional. He assumed very thin two-dimensional cracks, plane stress and principal stresses P and Q oriented relative to the crack as shown in Fig. 6.5. (P is algebraically greater than Q.) The crack in Fig. 6.5 extends all the way through the specimen. Using the elastic solutions for the stresses at a crack tip developed earlier by Inglis (1913), Griffith computed the maximum tensile stress at the crack tip and found that

$$\sigma_{max} = P \text{ if } (3P + Q) \text{ is plus and } \theta = 0 \tag{6.8}$$

$$\sigma_{max} = \frac{(P-Q)^2}{8(P+Q)} \text{ if } (3P+Q) \text{ is minus and } \cos 2\theta = \frac{P-Q}{2(P+Q)}$$

The fracture locus according to this elastic two-dimensional analysis is shown in Fig. 6.6. It is seen that the fracture stress in uniaxial compression (Point C) is 8 times the fracture stress in uniaxial tension (Point T).

Paul and Mirandy (1975) have extended the Griffith approach to three dimensions assuming very thin three-dimensional microcracks having an elliptical plan form. Like Griffith, they assumed a linearly elastic material and a maximum tensile stress fracture criterion. The surfaces of the thin flaws were assumed not to touch and the microcracks were assumed to be far enough apart so that there was no interaction. Fig. 6.7 shows fracture loci for very thin defects of different plan shape (b/a, where a and b are the semi major and minor diameters of the assumed elliptical defect and ν is Poisson's ratio. In these plots S_t is the uniaxial elastic tensile stress at fracture. It is evident that the ratio of uniaxial compressive strength to uniaxial tensile strength is between about 4.5 and 8 depending upon the values of b/a and ν. Paul and Mirandy stress the fact that the fracture loci presented are for initial fracture and that initial cracks

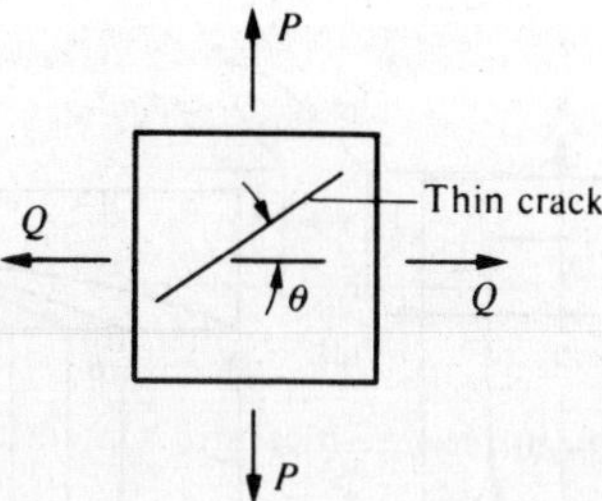

FIG. 6.5. Orientation of thin crack in biaxial stress field in Griffith analysis.

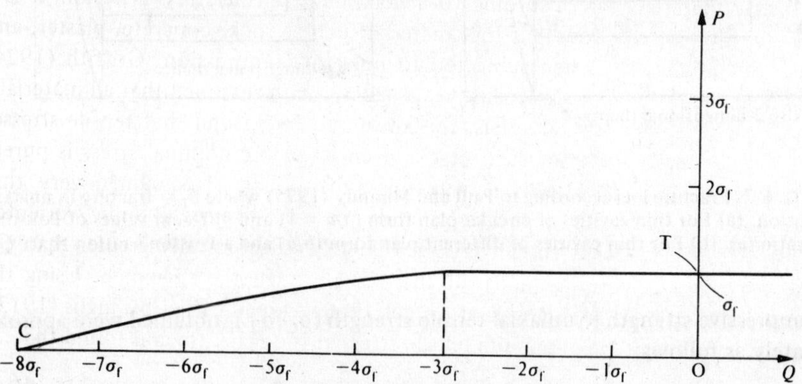

FIG. 6.6. Fracture locus according to Griffith (1924) in tension–compression regime. Fracture stress in uniaxial tension (T) = σ_f. Fracture stress in uniaxial compression (C) = 8 σ_f.

which form may not be self propagating.† This analysis, like Griffith's is based on a maximum tensile fracture stress criterion and hence the results are independent of the intermediate stress and should hold for either plane stress or plane strain.

Usui, Ihara, and Shirakashi (1979) have applied the analysis of Paul and Mirandy to the fracture of cutting tool materials, incorporating a Weibull probability distribution into the analysis. In order to evaluate the Weibull constants, four-point bending and uniaxial compression tests were performed on K10 (5.5 $^w/_o$ Co + 94.5 $^w/_o$ WC) and P20 (8 $^w/_o$ Co and 92 $^w/_o$ mixed carbides) sintered tungsten carbide tool materials. The ratio of median values of uniaxial

† For brittle materials there is a relatively small difference between crack initiation and propagation and the size of the crack plays a minor role. For ductile materials, however, crack initiation does not ensure crack propagation since sufficient energy must be stored elastically in the vicinity of the crack tip to satisfy the substantial plastic energy involved in crack growth. The size of the crack plays a major role in crack propagation in ductile materials since crack size is directly related to the amount of stored energy available for crack propagation.

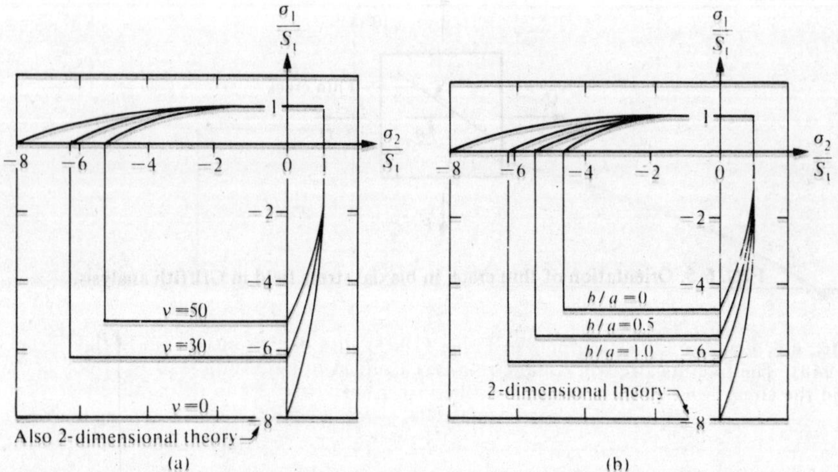

FIG. 6.7. Fracture loci according to Paul and Mirandy (1975) where S_t = fracture in uniaxial tension. (a) For thin cavities of circular plan form ($b/a = 1$) and different values of Poisson's ratio (ν). (b) For thin cavities of different plan form (b/a) and a Poisson's ratio $\nu = 0.30$.

compressive strength to uniaxial tensile strength $(\sigma_C/\sigma_T)_f$ obtained were approximately as follows:

Material	$(\sigma_C/\sigma_T)_f$
K10	1.95
P20	2.64

Fisher (1953) has suggested a somewhat different approach to brittle fracture. He assumed that brittle materials yield just before fracture and therefore the initial flow stress is a good approximation for the fracture stress of a brittle material such as cast iron. Like Griffith, Fisher assumed a random distribution of defects. However, his defect was not a void but a graphite platelet which he assumed would intensify tensile stresses but that compressive stress would be transmitted across the defect without intensification. The flow criterion used was that due to von Mises (distortion energy criterion). In applying his theory, the intensification factor for tensile stress (K) due to the defect and the flow stress of the material were quantities to be adjusted to give the best fit between theory and experiment. An example from Fisher's paper (Fig. 6.8) shows experimental fracture data due to Grassi and Cornet (1948) who tested gray cast-iron tubes to fracture that were subjected to internal pressure and end loading. The curve shown is based on Fisher's theory with $K = 3.2$ and the yield stress = 90 000 p.s.i. (622 MPa), which were the values found to give the best fit. The

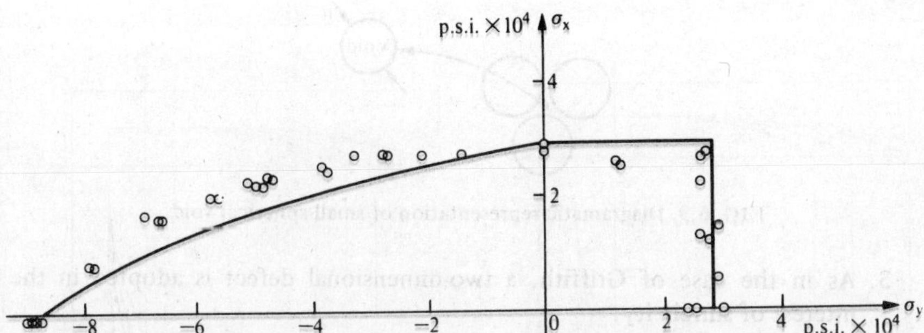

FIG. 6.8. Fracture locus according to Fisher (1953) with data points of Grassi and Cornet (1948). The fracture stress in compression was assumed to be 90 000 p.s.i. (622 MN m^{-2}) and the stress concentration factor K in the Fisher analysis = 3.2 in fitting the theoretical curve to the experimental data. (1 p.s.i. = 6905 N m^{-2}.)

agreement between theory and experiment is excellent but in saying this, it must be kept in mind that the theory contains two floating constants.

When the experimental values of σ_C/σ_T of Usui *et al.* (1979) for tungsten carbide and Grassi *et al.* (1948) for cast iron are compared with those from the Griffith/Paul and Mirandy analysis (4.5 to 8), a very large discrepancy is observed. This suggests that some of the assumptions in the analysis do not hold for tungsten carbide or cast iron. In examining these assumptions, it is found that the value of $(\sigma_C/\sigma_T)_f$ depends on the shape of the defect assumed. By Assuming a thin, hair-line defect, Griffith obtained a predicted value of $(\sigma_C/\sigma_T)_f$ of 8 for the two-dimensional case and Paul and Mirandy a range of values from 4.5 to 8 for three-dimensional defects. As will be shown below, a circular defect leads to a predicted value of $(\sigma_C/\sigma_T)_f$ of 3 which is in very much better agreement with experimental values for quasi-brittle materials such as cemented tungsten carbide and cast iron than the values of Griffith or Paul and Mirandy.

Theory Assuming Spherical Voids

The theory of brittle fracture for materials subjected to biaxial stresses presented here is based on the following assumptions:

1. Following Griffith and Paul and Mirandy (but not Fisher), fracture is assumed to occur when the maximum tensile stress reaches a critical value (maximum tensile stress criterion for fracture).
2. Like Griffith and Paul and Mirandy, defects are assumed to play an important role in intensifying local stresses. However, the defects are assumed to be small, circular voids instead of the thin, hair-line cracks assumed by Griffith and Paul and Mirandy (Fig. 6.9.)

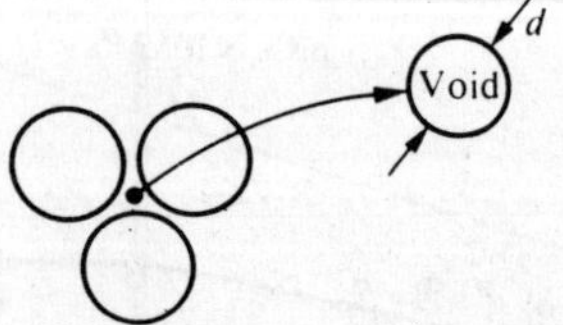

FIG. 6.9. Diagramatic representation of small spherical void.

3. As in the case of Griffith, a two-dimensional defect is adopted in the interest of simplicity.
4. Voids are considered to be far enough apart so they may be considered in isolation.

Figure 6.10a shows the tangential stress at point A on a circular void when the nominal applied stress is σ_x in tension. The intensified stress at A will be $3\sigma_x$. Figure 6.10b shows the tangential stress at A when the nominal applied stress is σ_y (compressive). The stress at A will then be σ_y (tension). Thus, in uniaxial tension (Fig. 6.10a), the fracture stress at A should be reached when the nominal stress is one third of that for uniaxial compression (Fig. 6.10b).

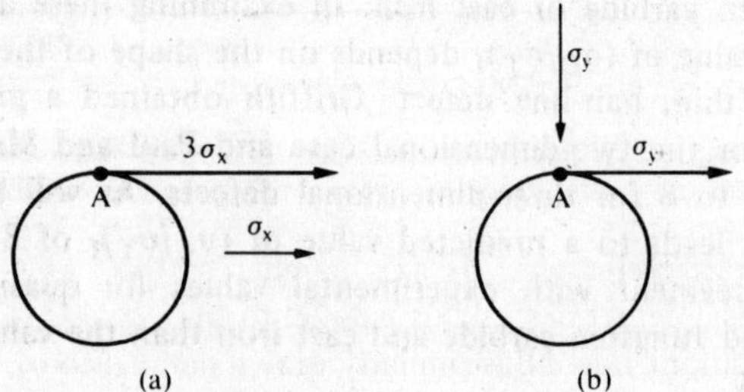

FIG. 6.10. Tangential tensile stress at surface of circular void. (a) Due to nominal tensile stress σ_x. (b) Due to nominal compressive stress σ_y.

Figure 6.11 shows the fracture locus to be expected from the maximum tensile stress criterion when nominal stresses are intensified by a two-dimensional defect (void). Point T is for uniaxial tension and point C for uniaxial compression. The fracture stress for uniaxial compression is three times that for uniaxial tension.

The complete fracture locus is shown by the solid line in Fig. 6.12. In the third quadrant, CB′ is drawn parallel to σ_x because the third stress σ_z is zero (plane stress) and the stress combination σ_y, σ_z will lead to fracture at a lower value of applied load than the combination σ_x, σ_y. In other words, the second

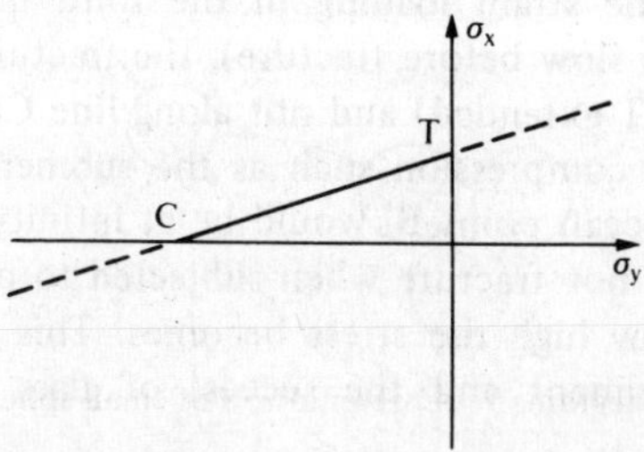

FIG. 6.11. Fracture locus in second quadrant for specimen subjected to biaxial plane stress.

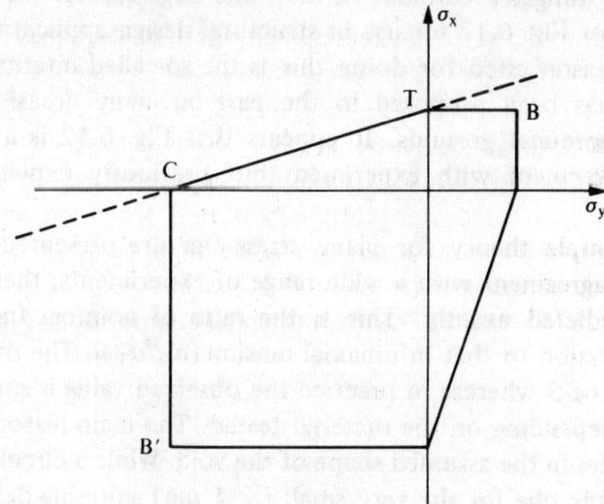

FIG. 6.12. Complete fracture locus for quasi-brittle material such as tungsten carbide. T = uniaxial tensile strength, C = uniaxial compressive strength.

quadrant, the critical defect has its axis parallel to the *z*-axis but in the third quadrant, the critical defect has its axis parallel to the *y*-axis.

In the first quadrant line TB is similarly drawn parallel to σ_y instead of being directed along CT extended. This is because the third stress σ_z is again zero (plane stress) and the stress combination σ_x, σ_z will lead to fracture at a lower value of applied load than the combination σ_x, σ_y.

It is well established experimentally that for a biaxial tensile loading in plane stress, brittle fracture will occur when the major nominal tensile stress reaches a critical value regardless of the magnitude of the minor tensile stress. The explanation for this lies in the fact that for plane stress, the third stress $\sigma_3 = 0$ and hence values in the first quadrant are equivalent to uniaxial tension (point T in Fig. 6.12). This ignores the secondary effect sometimes termed triaxiality which reflects the fact that the critical maximum tensile stress (a function of the mean principal stress $\sigma_H = (\sigma_1 + \sigma_2 + \sigma_3)/3$) is somewhat less when σ_H is positive (tensile) than when σ_H is negative (compressive).

In the case of plane strain loading in the third quadrant with Poisson's ratio = $\frac{1}{2}$ (slight plastic flow before fracture), the fracture locus would be along the dotted line (line CT extended) and not along line CB′. In the case of a true hydrostatic loading in compression such as the submergence of a material to a great depth in the ocean point B′ would be at infinity. Glass which contains numerous defects will not fracture when subjected to purely hydrostatic compression no matter how high the stress becomes. This is a result in excellent agreement with experiment and the success of glass for deep submergence vessels.

The fracture criterion of Fig. 6.12 is in excellent agreement with experimental data for tungsten carbide. In fact, one large carbide manufacturer suggests the locus of Fig. 6.12 for use in structural design applications of carbide. However, the reason cited for doing this is the so-called internal friction flow theory which has been proposed in the past on many occasions but always rejected on theoretical grounds. It appears that Fig. 6.12 is a fracture locus in excellent agreement with experiment but previously explained by an incorrect theory.

While the simple theory for plane stress fracture presented here is in excellent general agreement with a wide range of experiments, there is one detail that is not predicted exactly. This is the ratio of nominal fracture stress in uniaxial compression to that in uniaxial tension $(\sigma_C/\sigma_T)_f$. The theory presented predicts a ratio of 3 whereas in practice the observed value is somewhat greater or less than 3 depending on the material tested. The main reason for this deviation probably lies in the assumed shape of the void. While a circular void appears to be a reasonable one for the very small ($\stackrel{\sim}{<}$ 1 μm) sintering defects for a tungsten carbide tool material, other mateials may have defects of different characteristic shape.

Other reasons for the small variation from 3 are:

1. The circular void assumed has been considered to be two dimensional whereas in reality, it will be three dimensional with implications similar to those observed when comparing the Griffith (2D) and Paul and Mirandy (3D) solutions. This should tend to decrease the predicted value of $(\sigma_C/\sigma_T)_f$ to a value slightly below 3.
2. Strain hardening and deviation from the elastic stress distribution assumed in the analysis due to the small amount of plastic flow that usually precedes fracture in quadrants 2 and 3.
3. The possibility of a high pressure or an equivalent residual stress existing in or around sintering voids. Due to surface tension, a sintering void at equilibrium at high temperature must have a high internal pressure. As the body cools, this pressure will remain or be converted to an equivalent residual stress pattern in the vicinity of the void. The influence of a high void pressure or its residual stress equivalent has been ignored in the

analysis presented. While this is believed to be a secondary-order effect, it could account in part for the observed deviation of $(\sigma_C/\sigma_T)_f$ from 3.

4. Plane stress has been assumed but in reality there may be a small third stress in situations that approximate plane stress.
5. Triaxiality which acknowledges the fact that tensile stress at fracture is less when the mean principal stress $(\sigma_H = (\sigma_1 + \sigma_2 + \sigma_3)/3)$ is positive than when it is negative has been ignored.

The last two of these are relatively unlikely. While considerable emphasis has been attributed to triaxiality in the past, if this were a strong influence, TB in Fig. 6.12 would be inclined downward which is not in agreement with experiment. This suggests that a large degree of what has been attributed to triaxiality is in reality due to a difference in stress intensification with different stress states in the presence of defects.

Weibull Statistics

In any operation that involves fracture there is considerable unavoidable test-to-test variation and a number of samples must be tested in order to gain a picture of the performance of a given material.

While one could average such results and compare averages, this would not be making full use of the data collected. A better method is to plot the percent of a batch failed at a stress σ against this stress (σ) (Fig. 6.13). The range of breaking stresses and the distribution of values is then evident at a glance as well as the mean and median (equal number of test values above and below) values.

Probability paper has coordinate scales designed to make data such as that of Fig. 6.13 plot as a straight line.

Weibull paper is a special kind of probability paper that is very convenient

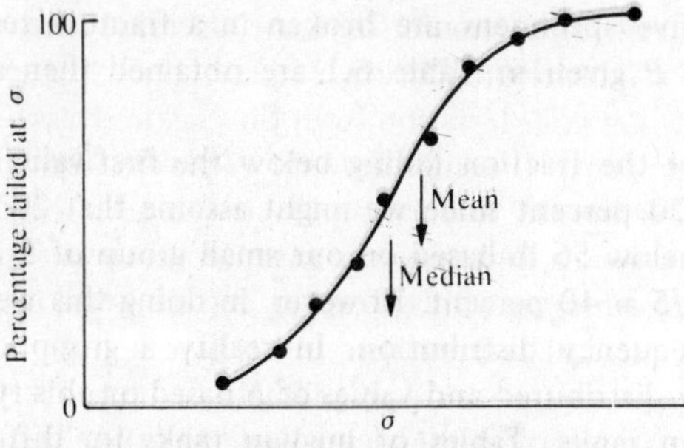

FIG. 6.13. Variation of percentage of specimens failed vs. stress level σ.

for analyzing a wide variety of industrial data, particularly that involving fracture and fatigue. This is based on some special statistics developed by W. Weibull (1939, 1951) to analyze fatigue data and the life of rolling contact bearings.

Weibull found empirically that the percentage (S) of a group of bearings that would survive N cycles at a given load and set of operating conditions was given by

$$S = e^{-(N/N_0)^m} \tag{6.9}$$

where

N_0 is a constant for the test series called the characteristic life.

m is a constant for the test series called the Weibull slope.

It has subsequently been found more convenient to work with the fraction failed (F) instead of the fraction surviving (S). Thus,

$$F = 1 - S = 1 - e^{-(N/N_0)^m} \tag{6.10}$$

Rearranging eqn (6.10)

$$e^{(N/N_0)^m} = \frac{1}{1-F} \tag{6.11}$$

and therefore,

$$\log \ln \frac{1}{1-F} = m \log N - m \log N_0 \tag{6.12}$$

Weibull paper is ruled with log ln $(1/1 - F)$ as ordinate and log N as abscissa (Fig. 6.14). When experimental data (F, N) is plotted on this paper a straight line is usually found having a slope m and a characteristic value N_0. From eqn (6.12) it is evident that when $N = N_0$, $F = 63.2\%$.

In using Weibull paper the data are first ordered (arranged in increasing order of main dependent variable = fracture load or number of cycles to failure, depending on the problem) and the values of F determined for each rank order. For example, if five specimens are broken in a fracture test and the results for breaking load P given in Table 6.1 are obtained then after ordering the data we have:

For F_1 we want the fraction failing below the first value (56). For this we could use $1/5 = 20$ percent since we might assume that 20 percent of a large batch would fail below 56 lb based on our small group of 5 tests. Similarly for F_2 we could use $2/5 = 40$ percent. However, in doing this we would be assuming a constant frequency distribution. In reality a group of test results are generally normally distributed and values of F based on this type of distribution are termed median ranks. Tables of median ranks for different sample sizes (number of tests) are available in the literature.

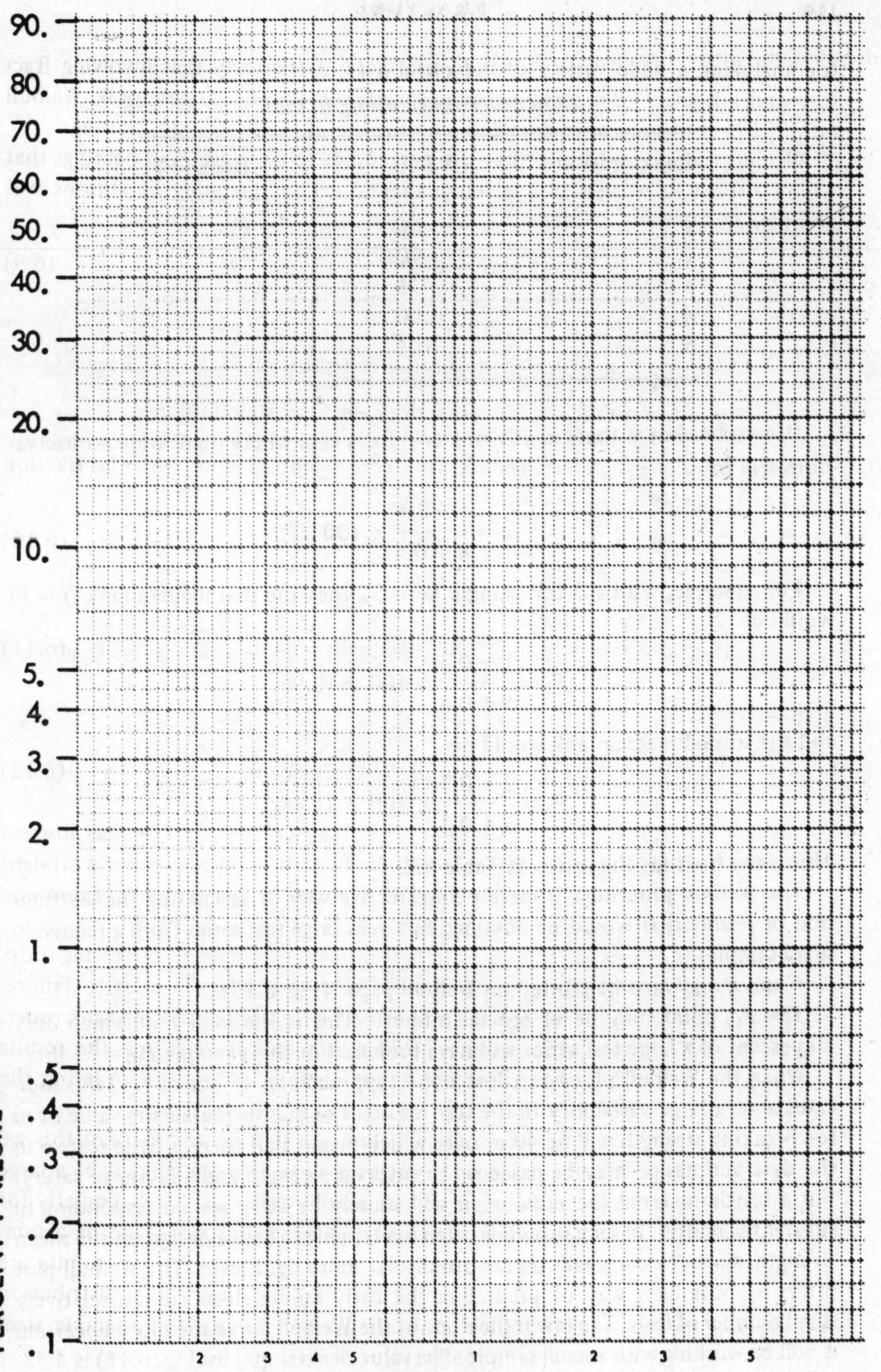

FIG. 6.14. Weibull statistical paper.

TABLE 6.1

Data ordered according to rank

Rank order	P lb	Rank (F)
1	56	F_1
2	70	F_2
3	75	.
4	90	.
5	105	.

It may be shown that the jth rank (F_j) for a case where there were n observations will be

$$F_j = \frac{j - 0.3}{n + 0.4} \times 100$$

For example, with a 5-test sample ($n = 5$), then the first median rank ($j = 1$) would be

$$F_1 = \frac{1 - 0.3}{5 + 0.4} \times 100 = 12.96$$

and the second median rank would be

$$F_2 = \frac{2 - 0.3}{5 + 0.4} \times 100 = 31.48$$

This is the basis for the values in Table 6.2.

The Weibull procedure is merely an effective way of estimating the distribution of results that would be obtained in a very large test series from a relatively small sample.

F and P may now be plotted on Weibull paper (Fig. 6.15).

The P_{10} point may be of special interest. This is the load P at which only 10 percent of a large test series will have broken. For this example $P_{10} = 53$ lb.

When the Weibull plot has a low slope (approaching 45° or $m = 1$) this indicates a very large variability in the test results. For ductile metals the value of m for breaking strength will be 40 or over which means that there is little scatter in the data and design may be based on the average strength and a factor of safety. For a brittle material the value of m will usually be 20 or less corresponding to substantial scatter. In such cases it is often inadvisable to base a design on the mean strength, since the early failures are usually far more significant. The Weibull plot offers a convenient way of predicting the early failure stress from a relatively small number of tests. The lower the slope of the Weibull curve the more uncertain it will be working with a small sample. The value of m (slope) for Fig (6.15) is 4.53.

TABLE 6.2.
Rank values

Rank No.	*P* lb	Rank (*F*)
1	56	12.96
2	70	31.48
3	75	50.00
4	90	68.51
5	105	87.03

The reason that brittle materials exhibit such a large range of fracture values is that they contain flaws covering a large range of intensity (holes, grain boundaries, impurities, etc.). Brittle materials are incapable of flowing plastically to relieve the stress at a point of stress concentration and hence fracture over a wide range of values as a result of the variable probability of finding a large or small stress concentration in a given specimen. Brittle materials also break at higher values of stress when small (size effect) or when loaded to have a stress

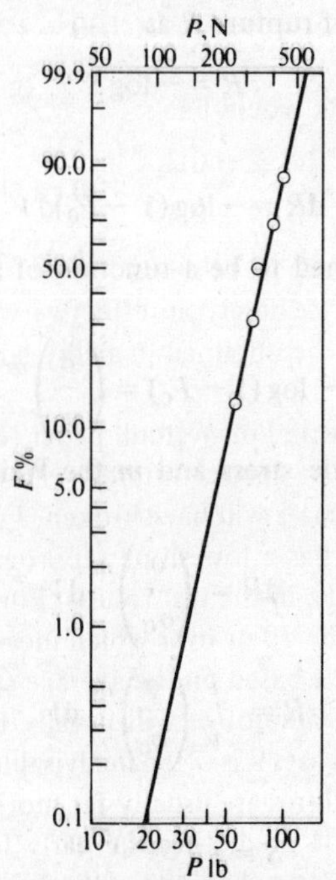

FIG. 6.15. Weibull plot for example. (After Shaw, Braiden, and DeSalvo 1975.)

gradient across the specimen. This is because the probability of finding a large imperfection decreases as the volume of specimen under high stress decreases.

Ductile materials (metals) flow plastically to reduce the stress at points of stress concentration and hence do not exhibit nearly as much scatter in breaking stress as brittle materials. They also do not tend to give different results for different specimen sizes or for loading conditions involving a stress gradient.

Weibull statistics may also be used to estimate the effects of specimen size and stress gradient on fracture stress. In this treatment it is assumed that a specimen fails when a critical tensile stress is reached. It is also assumed that the flaws causing variation in the fracture stress are uniformly distributed.

If F_0 is the probability of a failure occurring at a given stress level per unit volume then the probability of survival per unit volume will be

$$S_0 = 1 - F_0. \tag{6.13}$$

For V units of volume the probability of survival will be

$$S = (1 - F_0)^V \tag{6.14}$$

or

$$\log S = V \log (1 - F_0). \tag{6.15}$$

Weibull defines the risk of rupture R as

$$R = -\log S \tag{6.16}$$

and therefore,

$$\mathrm{d}R = -\log (1 - F_0)\mathrm{d}V \tag{6.17}$$

since $-\log(1 - F_0)$ is assumed to be a function of stress alone, Weibull experimentally found that

$$-\log (1 - F_0) = \left(\frac{\sigma}{\sigma_0}\right)^m \tag{6.18}$$

where σ_0 is the characteristic stress and m the Weibull slope. Thus, from eqns (6.17) and (6.18)

$$\mathrm{d}R = \left(\frac{\sigma}{\sigma_0}\right)^m \mathrm{d}V \tag{6.19}$$

$$R = \int_V \left(\frac{\sigma}{\sigma_0}\right)^m \mathrm{d}V. \tag{6.20}$$

From eqn (6.16)

$$S = \mathrm{e}^{-\int(\sigma/\sigma_0)^m \mathrm{d}V}. \tag{6.21}$$

It may be shown that in general R (eqn 6.20) will integrate as follows:

$$R = KV\left(\frac{\sigma}{\sigma_0}\right)^m \tag{6.22}$$

where K is a factor depending on the stress gradient involved and V is the volume of the specimen subjected to tensile stress. From eqns (6.21) and (6.22)

$$F = 1 - S = 1 - e^{-KV(\sigma/\sigma_0)^m} \tag{6.23}$$

which may be rewritten as

$$\log \ln (1/1 - F) = m \log \sigma - m \log \sigma_0 + \log KV. \tag{6.24}$$

This is seen to be equivalent to eqn (6.12).

In order to see how eqn (6.20) may be used to show the influence of specimen size consider two tensile specimens of volume V_1 and V_2 with rupture stress values of σ_1 and σ_2, respectively. Then, since σ is constant for a tensile specimen, eqn (6.21) gives

$$R_1 = \left(\frac{\sigma_1}{\sigma_0}\right)^m V_1 = R_2 = \left(\frac{\sigma_2}{\sigma_0}\right)^m V_2 \tag{6.25}$$

or

$$\frac{\sigma_1}{\sigma_2} = \left(\frac{V_2}{V_1}\right)^{1/m}. \tag{6.26}$$

For example, if $V_2/V_1 = 10$ and $m = 10$, $\sigma_1/\sigma_2 = 1.26$.

It should be noted that the size effect depends on the Weibull slope (m).

Also, the value of K in eqn (6.21) is seen to be unity for the tensile test.

Consider next a transverse rupture test in three point loading (Fig. 6.16).

In this case eqn (6.20) is evaluated only in the tensile region shown shaded in Fig. 6.16. This is because brittle materials are much stronger in compression than tension. In fact, a brittle material will be about three times as strong in compression as in tension. The tensile stress at any point x is

$$\sigma = \frac{MC}{I} = \frac{\left(\frac{Px}{2}\right)y}{bh\frac{3}{12}} = \frac{6Pxy}{bh^3} \tag{6.27}$$

Substituting into eqn (6.20):

$$R_B = 2\int_0^{1/2}\int_0^{h/2}\left(\frac{6Pxy}{bh^3\sigma_0}\right)^m b\,dx\,dy \tag{6.28}$$

$$= \frac{1}{(m+1)^2} V\left(\frac{\sigma}{\sigma_0}\right)^m \tag{6.29}$$

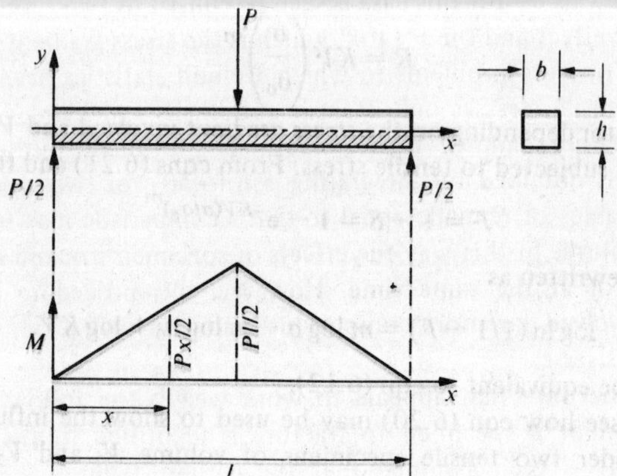

FIG. 6.16. Three-point loading transverse rupture test and resultant bending moment (M).

where

$$V = \text{volume subjected to tension} = \frac{lbh}{2} \tag{6.30}$$

$$\sigma_m = \text{maximum tensile stress} = \frac{3\,Pl}{2\,bh^2} \tag{6.31}$$

Equation (6.29) is also seen to correspond to eqn (6.22) where

$$K = \frac{1}{(m+1)^2}$$

If we write R_T for a tensile specimen then from eqn (6.20)

$$R_T = (1)\,V_T\,\frac{\sigma}{\sigma_0}^{m} \tag{6.32}$$

For equal risk of rupture $(R_T) = (R_B)$

$$\frac{\sigma_B}{\sigma_T} = \left[(m+1)^2\,\frac{V_T}{V_B}\right]^{1/m} \tag{6.33}$$

For example, if the volume of the tensile specimen (V_T) is equal to the volume of the transverse rupture specimen in tension (V_B), and $m = 10$

$$\frac{\sigma_B}{\sigma_T} = 1.614$$

This explains why transverse rupture experiments give higher values of tensile stress at rupture than tensile tests even when great pains are taken in the alignment of the tensile specimen.

The ratio of σ_B to σ_T for this case is seen to consist of two terms $(V_T/V_B)^{1/m}$ which is the size effect and $(m + 1)^{2/m}$ which is the stress gradient effect.

Most of the fracture problems to which Weibull statistics have been applied have involved tensile stresses only (tensile specimens and slender beams). Such problems are readily handled by the methods previously outlined. The risk of rupture (R) is obtained by integrating eqn (6.20) for two cases to be compared and the values of R made equal so that a comparison may be made at the same risk of rupture. In this way the effects of specimen size and stress gradient are taken care of at the same time. However, this procedure becomes very difficult and involved far more complex situations involving transverse stresses as well as direct stresses.

A simpler procedure that appears to work well is the following. The most vulnerable region of a specimen to fracture is identified following an elastic stress analysis. It is then assumed that the earliest failures occur by cracks which propagate from this region. The volume of this critical region is estimated by considering the space over which the effective stress causing the first fractures is within 10 percent of the maximum value.

The effective stress for one type of specimen need not be adjusted for a stress gradient when determining the effective stress for another type of loading, since the analysis is confined to fracture that occurs in a small critical volume over which the stress may be considered constant. It is necessary however to make an adjustment for volume difference, and this is done by use of eqn (6.26).

The Disc Test

The disc test (Fig. 6.17) introduced for testing concrete (Carniero and Barcellos, 1953) involves a thin disc that is diametrically loaded to failure. It is often used to measure the tensile strength of brittle materials, particularly concrete, rock and coal. This test has been used and improved by many research workers (for example: Peltier 1954; Mitchell 1961; Rudnick, Hunter, and Holden 1963; Spriggs, Brissette, and Vasilos 1964; and Berenbaum and Brodie 1959).

The stresses pertaining at all points will correspond to the plane stress elastic solution given in Fig. 6.18. This is a finite element solution provided by Usui of Tokyo Institute of Technology and his colleagues for a thin (plane stress) disc uniformly loaded over flat surfaces extending 20% of the diameter of the disc. This solution is in excellent agreement over the critical central region with that of Frocht (1948) for a plane stress disc subjected to concentrated diametrically opposed loads. If the load is well distributed at the top and bottom of the specimen, as by use of flats extending about 20% across the specimen and thin brass shims, no fracture will occur in the vicinity of the loading points and instead, fracture will occur at the center of the specimen (A in Fig. 6.17). At a critical value of load, a crack will run from A upward and downward

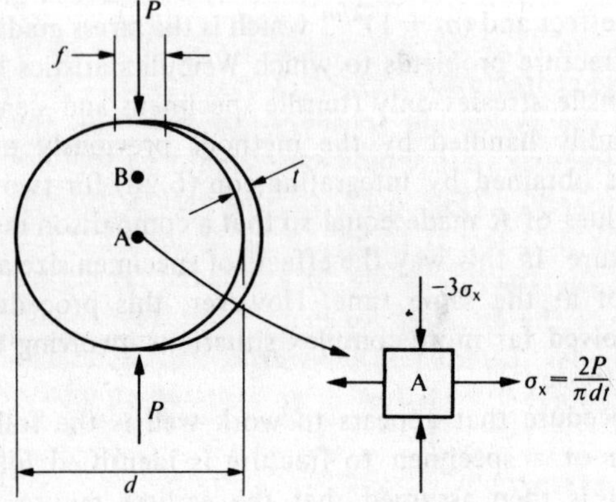

FIG. 6.17. Disc test: $d = 0.500$ in (12.7 mm), $t = 0.010$ in (0.25 mm), $f = 0.100$ in + 0.010 in thick brass shim (2.5 mm + 0.25 mm thick brass shim).

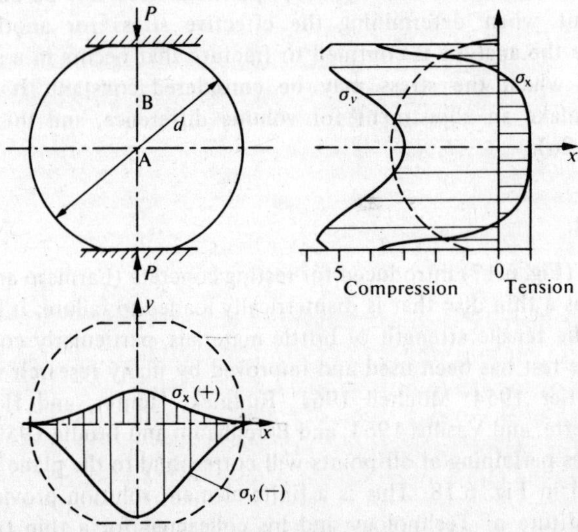

FIG. 6.18. Stress distribution for transversely loaded disc ($\nu = 0.3$). (Courtesy of E. Usui, Tokyo Inst. of Technology.)

diametrically across the specimen. The tensile stress at A perpendicular to the diametral crack at fracture will be

$$\sigma_x = \frac{2P}{\pi dt} \text{ (tension)} \tag{6.34}$$

while the radial stress at fracture will be

$$\sigma_y = -3\,\sigma_x \text{ (compression)} \tag{6.35}$$

where P is the load at fracture.

According to the maximum tensile stress criterion, the value of nominal tensile stress at fracture (σ_x) should correspond to that for a uniaxial tensile test (Four-point bending (Fig. 6.19) or ring test†). However, the value of σ_x at fracture in the disc test is found to be appreciably lower (by a factor of about two) than the fracture stress in a uniaxial tensile test. This discrepancy has baffled users of the disc test for many years. (Berenbaum and Brodie 1959; Wright 1955; Snyder, Rudnick, Duckworth, and Hyde 1962; Akazawa 1953; Lecrivain and Lambert 1961). Just as an assumed circular defect explains why

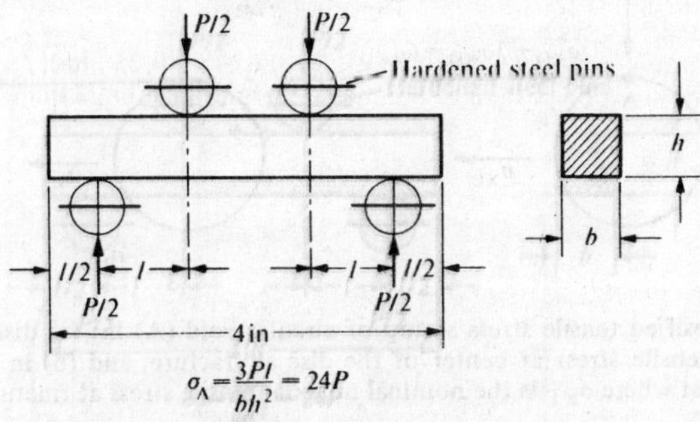

FIG. 6.19. Four-point bending test (pin dia = 0.500 in [12.7 mm], l = 1.000 in [25.4 mm], $h = b$ = 0.500 in [12.7 mm]).

a quasi-brittle material such as tungsten carbide can fail brittly in uniaxial compression it also explains why the nominal tensile stress in the disc test at fracture is about half of the fracture stress in a uniaxial tensile test. Figure 6.9 shows a circular defect at the critical point of crack initiation in the disc test (point A in Fig. 6.17). The plastic strain at fracture will be very small and an elastic analysis which leads to the following stresses at A represents a good approximation

† The ring test (Sedlacek and Holden 1962) is one of the best tests for determining the uniaxial tensile strength of a brittle material. The main disadvantage of the test lies in the high cost of test specimens. The test consists of a thin ring typically 2 in (50 mm) o.d. (d), 0.1 in (2.5 mm) wall thickness (t); and 0.5 in axial width (12.5 mm) (b). The thin ring is loaded internally by pressure acting through a rubber bag until fracture occurs. The uniaxial tensile stress is then

$$\sigma_T = \frac{pd}{2t} \tag{6.36}$$

where p is the internal pressure at fracture.

$$\sigma_x \text{ (tensile)}$$
$$\sigma_y = -3\,\sigma_x \text{ (compressive).}$$

The tensile stress at the top of the circular defect will thus be $6\sigma_{xD}$ for the disc test (Fig. 6.20a) where σ_{xD} is the nominal tensile stress at the center of the disc at fracture. Figure 6.20b shows the nominal tensile stress at fracture (σ_{xT}) for a uniaxial tensile test together with the intensified stress at the top of a circular defect ($3\sigma_{xT}$). When the intensified tensile stresses at the tops of the circular defects are equated for the disc and uniaxial tensile tests it is evident that the nominal stress in the tensile test at fracture $(\sigma_{xT})_f$ should be twice that for the disc test $(\sigma_{xD})_f$.

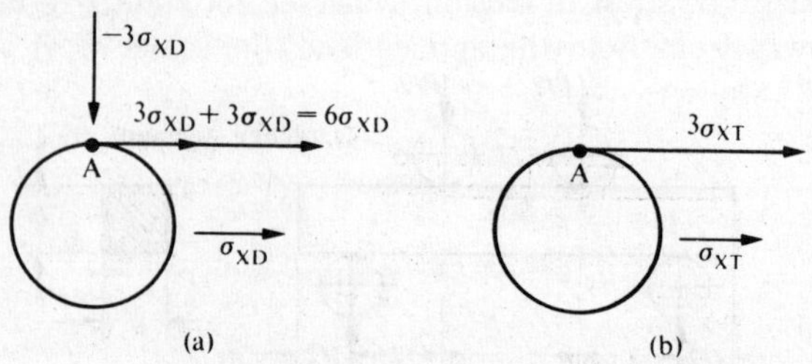

FIG. 6.20. Intensified tensile stress at top of circular void (A) in: (a) disc test where σ_{xD} is the nominal tensile stress at center of the disc at fracture; and (b) in a uniaxial tensile test where σ_{xT} is the nominal uniaxial tensile stress at fracture.

Figure 6.21 shows the plane stress fracture locus with the uniaxial tensile value (T) disc test value (D) and uniaxial compression (C) values indicated. Point D is seen to lie at the point of intersection of line CT and a line having a 3:1 slope corresponding to the ratio of $|\sigma_y|$ to $|\sigma_x|$ at the center of the disc.

The assumption of a circular defect is thus seen to represent an excellent

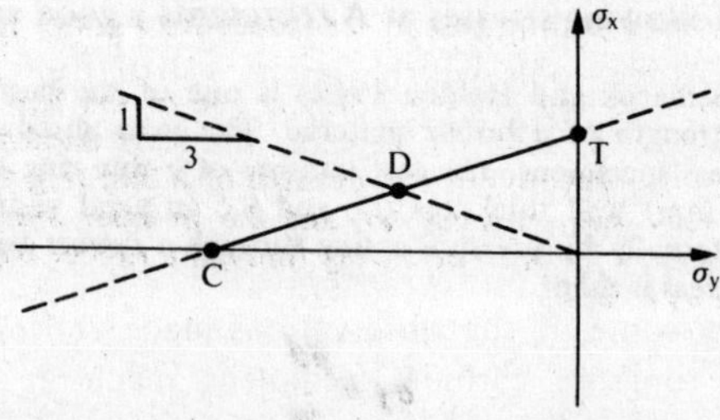

FIG. 6.21. Fracture point D for disc test on fracture locus of Fig. 6.11.

approximation in that it not only explains why $(\sigma_C/\sigma_T)_f$ is about 3 but also why $(\sigma_T/\sigma_D)_f$ is about 2. The latter ratio is not found to be exactly 2 for the same reasons $(\sigma_C/\sigma_T)_f$ is not exactly 3 as discussed above.

The reason that fracture occurs first at the center of the disc (A in Fig. 6.17) and not at points closer to the applied load (for example B in Fig. 6.17) is due to the earlier onset of plastic flow at points closer to the applied load.

It has been found by use of strain gages that whereas tungsten carbide remains elastic to the point of fracture in uniaxial tension, it undergoes plastic flow before fracture in the disc test. For example, at the center of the disc (A in Fig. 6.17), the elastic stresses are given by eqns (6.34) and (6.35) and by the Tresca criterion, the material at A should go plastic when $\sigma_x = Y/4$ where Y is the uniaxial flow stress in tension. However, for point B in Fig. 6.17 which is about half-way from A to the applied load, the values of σ_x and σ_y will be approximately as follows:

$$\begin{aligned} \sigma_x &\cong \frac{2P}{\pi dt} \text{ (tension)} \\ \sigma_y &\cong -4.5\sigma_x \text{ (compression)} \end{aligned} \tag{6.37}$$

By the Tesca criterion, plastic flow should occur when $\sigma_x = Y/5.5$. Thus, plastic flow will occur at B at a lower value of P than that required at A and thus limit the stress that develops at B. The nominal and intensified tensile stresses at fracture at A will be greater than those at B and hence fracture should originate close to point A.

Fracture of Cemented Tungsten Carbide

Brittle fracture is an important mode of tool failure particularly for tungsten carbide and other hardmetals and ceramics. Materials are generally considered to fail by brittle fracture when the maximum principal tensile stress reaches a critical value. The most widely employed test to characterize the fracture strength of tungsten carbide tool materials is the tensile stress as measured in a three- or four-point bending test at fracture (Fig. 6.19). The tensile strength of most tungsten carbide tools will be about 150 000 p.s.i. (1030 MPa) and the tensile strength is found to increase as the amount of binder employed is increased. For example, the mean tensile strength of a tungsten carbide containing 6 $^{w}/_{o}$ cobalt is about 115 000 p.s.i. (790 MPa), while one containing 12 $^{w}/_{o}$ cobalt is about 165 000 p.s.i. (1140 MPa).

In order to verify some of the observations made above, tests were run on specially produced tungsten carbide specimens made in the form of bars ($\frac{1}{2}$ in × $\frac{1}{2}$ in × 4 in (12 × 12 × 100 mm)), discs ($\frac{1}{2}$ in D × 0.10 in (12.50 × 2.5 mm)), and $\frac{1}{4}$ in (6.4 mm) square × $\frac{1}{2}$ in (12.5 mm) high compression specimens. These samples were made from the same batch of powder that was compacted and sintered under identical conditions. Two compositions were involved—6 w/o Co, 94 w/o WC and 12 w/o Co, 88 w/o WC. The specimens were diamond ground all over before testing and flats extending 20% of the diameter were provided on the discs. Figure 6.22 shows Weibull plots of tensile stresses at fracture for the bending tests, disc tests and compression tests. Elastic behavior was assumed to fracture in all cases. It is found that

1. A considerable difference in the tensile median fracture stresses at fracture exists for the disc (σ_{50} = 122 000 p.s.i., 840 MPa) and bending tests (σ_{50} = 190 000 p.s.i., 1310 MPa) for 6 w/o Co.
2. The high cobalt material is strongest in bending but weakest in the discs and compression tests.
3. The compressive stress at fracture is considerably higher than the uniaxial tensile stress at fracture.

Figure 6.23 shows stress–strain results for typical bending, disc, and compression tests. The strain readings are from very small gage length strain gages made from foil having an unusually high yield strain since they were to be used to measure strains extending slightly beyond the yield strain of the tungsten carbides employed. The gages were mounted as shown by the insets in Fig. 6.23. The curves of eqn (6.36) are slightly nonlinear all the way to the origin (zero applied load) and it is difficult to identify an exact yield point. An attempt to

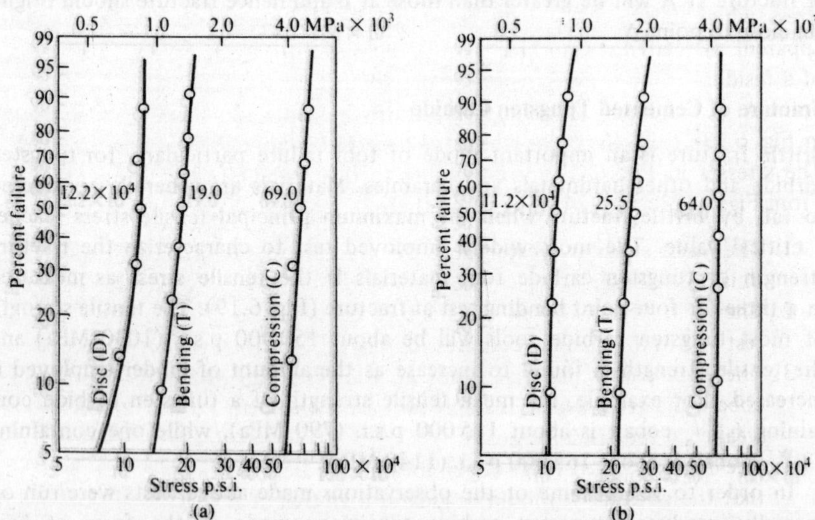

FIG. 6.22. Weibull plots of fracture stresses for disc bending and compression fracture stresses for (a) 6 w/o Co (b) 12 w/o Co (1 p.s.i. = 6905 N m^{-2}). (After Takagi and Shaw 1981, 1983).

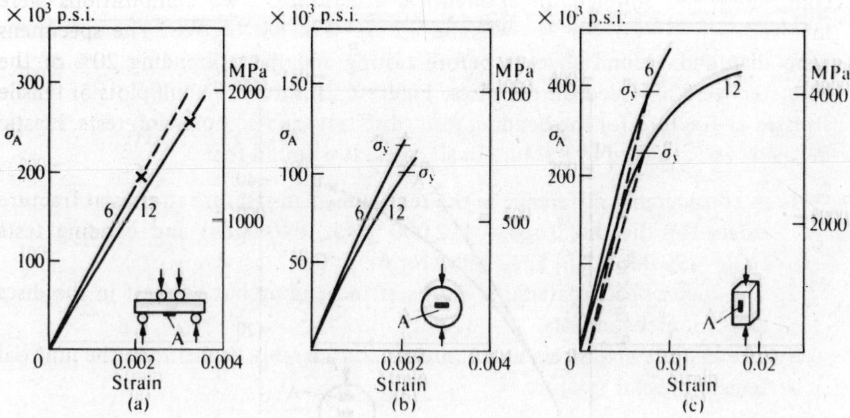

FIG. 6.23. Stress–strain curves. (a) Bending (uniaxial tensile stress). (b) Disc test with strain gage mounted transverse to load at center of disc. (c) Compression test with strain gage mounted parallel to load. X marks the point of fracture and σ_y the yield stress. (After Takagi and Shaw 1981.)

do this was made by loading and unloading and looking for the onset of a permanent set after unloading. This did not prove to be useful since before the material at A (center of disc) went plastic, that nearer the load at B went plastic (see Fig. 6.24). Upon loading, the plastic zone gave rise to a residual compressive stress which results in a slight tensile permanent set at A. Since there is no apparent way of distinguishing a small permanent set due to the development of a residual compressive stress near the applied load and actual plastic flow at point A, it was finally decided to arbitrarily consider the yield point at point A in Fig. 6.23b to be the point where the observed strain under load was 2×10^{-3}. The stresses in Fig. 6.23 corresponding to this yield point are designated σ_y. From Figs. 6.23 and 6.24 it is evident that:

1. The yield stress is greater than the fracture stress for both materials in uniaxial tension.
2. The yield stress is below the fracture stress for both materials in the disc test and in the uniaxial compression test.
3. The high cobalt material yields at a lower value of stress than the low cobalt material which is consistent with the high cobalt material being weaker in the disc test.
4. The material at B (Fig. 6.24) goes plastic before that at A which means that the stress at A should be greater than that at B at fracture. As already mentioned, this explains why the critical crack should initiate at A and not at some point a greater distance from the center if the material is reasonably homogeneous.

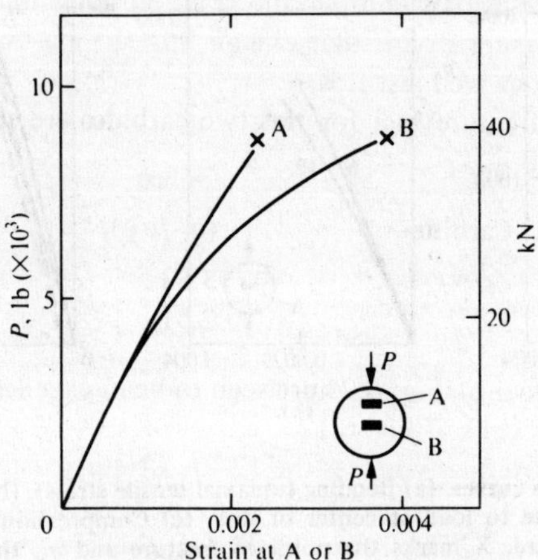

FIG. 6.24. Load (P) strain curves for 12 $^{w}/_{o}$ Co cemented tungsten carbide with strain gage mounted transversely at center of disc (A) and approximately at midpoint between point A and the load (B). The material at B goes plastic before fracture occurs at A while fracture at A occurs very near the yield point. (After Takagi and Shaw 1981.)

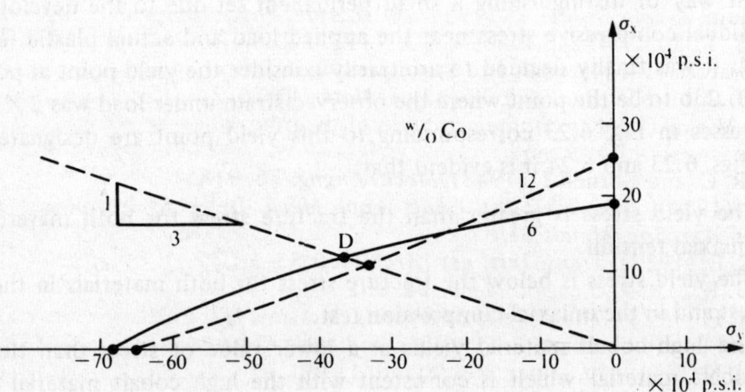

FIG. 6.25. Fracture loci for 6 $^{w}/_{o}$ and 12 $^{w}/_{o}$ tungsten carbide in the tension-compression regime. (Experimental median values from Fig. 6.22 (1 p.s.i. = 6905 MPa). (After Takagi and Shaw 1981, 1983.)

5. The material in the uniaxial compression test undergoes appreciable plastic strain before fracture.

While the exact values of strain pertaining are in question since strain gages have been used to measure strains that extend into the plastic region, the above observations appear to be well justified.

The experimental fracture loci for the two carbides are shown in Fig. 6.25 where it is evident that

Carbide	$(\sigma_C/\sigma_T)_f$
6 $^w/_o$	3.63
12$^w/_o$	2.51

The experimental loci of Fig. 6.25 are seen to be in excellent agreement with the analytical fracture locus of Fig. 6.12.

It is of interest to note that whereas the 12 $^w/_o$ Co tungsten carbide is stronger than 6 $^w/_o$ tungsten carbide in uniaxial tension, the reverse is true in the disc and uniaxial compression tests. This is due to the fact that the more brittle 6 $^w/_o$ Co material undergoes a greater increase in strength with increase in mean (hydrostatic) compressive stress than the more ductile 12 $^w/_o$ material does.

References

Akazawa, T. (1953). International Union of Testing and Research Laboratories for materials and Construction, RILEM, Paris, Bull. No. 16.

Berenbaum, R. and Brodie, L. (1959). *Br. J. appl. Phys.* **10**, 281.

Carnerio, F. L. L. B. and Barcellos, A. (1953). Union of Testing and Research Laboratories for Materials and Structures (Brazil) Bull. No. 13.

Fisher, J. C. (1953). *Bull. Am. Soc. Test. Mater.* **TP76**, 74.

Frocht, M. M. (1948), *Photoelasticity* Vol. ii, John Wiley and Sons, Inc., New York.

Grassi, R. C. and Cornet, I. (1948). *Mech. Engng.* **70**, 918.

Griffith (1924). *Proc. First Int. Cong. Appl. Mech.* Delft, 55 (reprinted (1968) *Trans. Am. Soc. Metals* **61**/87.

Inglis, C. E. (1913). *Trans. Inst. nav. Architects* **55-1**, 219.

Irwin, G. R. (1957). *J. Appl. Mech.* **24**, 361.

Joffe, A. F. (1928). *The physics of crystals,* McGraw-Hill Book Co., New York.

Lecrivain, L. and Lambert, B. (1961). *L'Industrie Ceramique* No. **534**, 367.

Mitchell, N. B. (1961). Mater. Res. Stand. **1**, 780.

Orowan, E. (1950). In *Fatigue and fracture of metals,* John Wiley and Sons Inc., New York.

Paul, B. and Mirandy, L. (1975). *J. Engng Mater. Tech.* **98**, 153.

Peltier, H. (1954). Union of Testing and Research Laboratories for Materials and Structures (Brazil) Bull. No. 19.

Rudnick, A., Hunter, A. R., and Holden, F. C. (1963). *Mater. Res. Stand.* **3**, 283.

Sedlacek, R. and Holden, F. C. (1962). *Rev. Sci. Instrum.* **33**, 298.
Shaw, M. C., Braiden, P. M., and Desalvo, G. J. (1975). *J. Engng Indus.* **96**, 77.
Snyder, M. J., Rudnick, A., Duckworth, W. H., and Hyde, C. (1962). Battelle Tech. Doc. Rep. ASD–TDR–62–7.
Spriggs, R. M., Brisette, L. A., and Vasilos, T. (1964). Mat. Res. and Stand. **4**, 218.
Takagi, J. and Shaw, M. C. (1981). *Ann. Int. Inst. Prod. Eng. Res.* CIRP, Paris 30–1, 53.
— — (1983). *J. Eng. Industry* **105**, 143.
Usui, E., Ihara, T., and Shirakashi, T. (1979). *Jap. Soc. Precis. Engrs* **13**, 189.
Weibull, W. (1939). *Ing. Vetenskaps Adakmien-Handlinger* (Stockholm) No. **151**, 1, and No. **153**, 1.
Weibull, W. (1951). *J. appl. Mech.* **73**, 293.
Wright, P. J. F. (1955). *Mag. Concr. Res.* (July), 87.
Zener, C. (1948). *The micromechanism of fracture*, Am. Soc. Met.

7

DYNAMOMETRY

In order to put the analysis of the metal cutting operation on a qualitative basis, certain observations must be made before, during, and after a cut. The number of observations that can be made during the cutting process is rather limited, one of the more important measurements of this type being the determination of cutting force components. In this chapter tool force-measuring apparatus will be considered. Design criteria will be discussed first, followed by a review of several types of force measuring techniques. Finally, a variety of useful dynamometers will be described. Although each dynamometer must be specially designed to meet specific requirements, the examples presented should serve to illustrate general considerations.

Dynamometer Requirements

Requirements of importance in dynamometer design are rigidity, sensitivity, and accuracy. The sensitivity of a good research dynamometer should be such that determinations are sensitive and accurate to within ±1%. That is, if a dynamometer is designed for a mean force of 100 pounds, one pound increments should be easily readable and accurately indicated.

Some deformation is associated with the operation of every dynamometer. However, a dynamometer should be rigid enough so that the cutting operation is not influenced by the accompanying deflections. In general, the static stiffness of a metalcutting dynamometer should be about 10^6 lb in^{-1} (10^8 N m^{-1}). Frequently, the dominating stiffness criterion is the natural frequency of the dynamometer. All machine tools operate with some vibrations, and in certain cutting operations these vibrations may have large amplitudes, (i.e. milling, grinding, shaping). In order that the recorded force (or the actual cut) be not influenced by any vibrating motion of the dynamometer, its natural frequency must be large (at least four times as large) compared to the frequency of the exciting vibration. For purpose of analysis, any dynamometer can be reduced approximately to a mass supported by a spring. The natural frequency (f_n) of such a system is equal to

$$f_n = \frac{1}{2\pi}\sqrt{\frac{K}{m}} \text{ c.p.s.} \tag{7.1}$$

where K is the spring constant in lb in^{-1}, (MN m^{-1}) and m is the mass in lb s^2 in^{-1}

(kg). In terms of the supported weight of the dynamometer (W)

$$f_n = \frac{1}{2\pi}\sqrt{\frac{386\,K}{W}}\ \text{c.p.s.} \tag{7.2}$$

where W is in lb (N). If, for example, a grinding machine is being run at $N =$ 3600 r.p.m., machine vibrations are apt to be present with an exciting frequency of

$$f_e = \frac{N}{60}\ \text{c.p.s. (Hz).} \tag{7.3}$$

In this case, the natural frequency of the dynamometer (f_n) should be at least 240 c.p.s. (Hz).

In general, a dynamometer must measure at least two force components. In a three dimensional cutting operation three force components are necessary, while in drilling or tapping, only a torque and a thrust are required. It is usually most convenient to measure force relative to a set of rectangular coordinates (x, y, z) and it is advisable that there be no *cross sensitivity* between these components. That is, an applied force in the x-direction should give no reading in the y- or z-directions. If mutual interference of the force-measuring elements exists, determination of the force components requires the solution of simultaneous equations. This prevents the immediate interpretation of the data or access to a computer. When certain of the electric transducers are suitably located and connected, unwanted strain components can often be cancelled electrically. How this may be done will be subsequently illustrated.

It is convenient to use a system having a linear calibration. In such a case, the force is determined with the precision with which a strain increment can be measured relative to an arbitrary datum. If the system is not linear, it is then necessary that the zero load point be accurately known as well as the strain increment. This introduces an additional quantity that must be carefully measured.

A dynamometer should be stable with respect to time, temperature, and humidity. Once a calibration is made, it should only have to be checked occasionally. Many existing dynamometers use devices for the separation of force components which involve friction, (i.e. rollers, balls, and sliding surfaces). As friction conditions are usually variable due to dirt or environmental changes, such instruments are of limited usefulness.

Force readings can be obtained from recording or indicating type dynamometers. Although recording dynamometers cost more than a direct-reading type, the added convenience and other obvious advantages of a recording system usually offset the added expense.

There are many other special dynamometer requirements that must be frequently met, such as size, ruggedness, and adaptability to several jobs. Such special requirements assume different degrees of importance in different particular applications. For example, dynamometer requirements are quite different for an adaptive control sensor used under production conditions than they are for a unit used in the laboratory for research purposes.

Force Measurement

In most dynamometers the force is applied to some sort of spring and the deflection thus produced is measured. Since dynamometers must be stiff, only deflection measuring devices that are capable of measuring small deflections (10^{-6} to 10^{-4} in or $\sim 1\ \mu$m) are of interest here. Some of the devices that have been used to measure such small deflections in dynamometers will now be briefly described.

The dial indicator is capable of reading deflections to about 10^{-4} in ($\sim 2\ \mu$m) when functioning properly. However, dial indicators are subject to sticking and are not to be relied upon for absolutely static readings. A lever system may be used in conjunction with a dial indicator as shown in Fig. 7.1. This is a relatively crude planing dynamometer designed and used by the author about 1940.

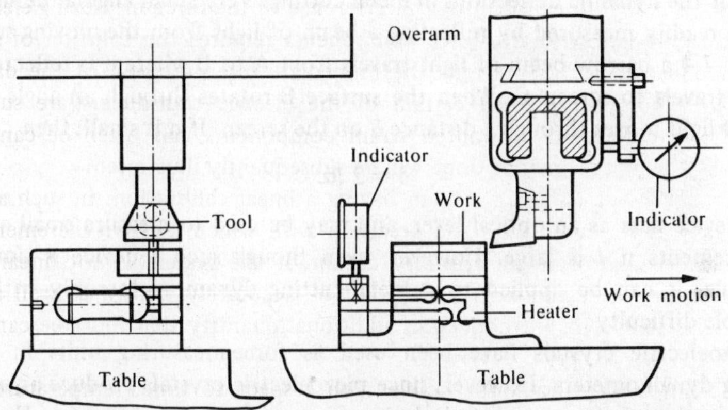

FIG. 7.1. Two component tool dynamometer which utilizes dial indicators. The vertical component of force (F_Q) acts through the lever system shown attached to the work holder. The horizontal force (F_P) is obtained by measuring the deflection of the overarm of the machine to which the tool is attached.

Hydraulic pressure cells have also been used in conjunction with pressure gages to measure or record the force on a tool. The manner in which this may be done is illustrated diagramatically in Fig. 7.2. When this method is used, the force may be read at a distance from the pressure cell.

Pneumatic devices such as the Solex micrometer in which the change in back pressure that occurs when a flat surface is brought into closer contact

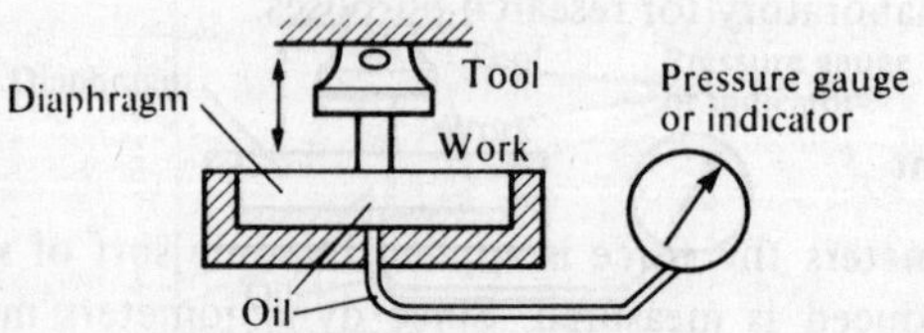

FIG. 7.2. Basic principle of hydraulic dynamometer.

with a sharp-edged orifice are sometimes used to measure deflections in tool dynamometers (Fig. 7.3). Such systems are simple and reliable if carefully supplied with clean, constant pressure air. There is, however, a limited region over which they are linear, and dynamometers of this type tend to be bulky.

Optical devices of several types have also been employed. Interferometric methods can be used to give very precise measurements using the wavelength of light as a unit. However, this principle is not easily adapted to the measurement of the dynamic deflections in metal cutting. Very small angular deflections can be readily measured by reflecting a beam of light from the moving surface. In Fig. 7.4 a narrow beam of light travels from A to B where it is reflected and hence travels to screen C. When the surface B rotates through an angle α, the spot of light moves through a distance δ on the screen. If α is small, then

$$\delta = 2\, l\alpha. \tag{7.4}$$

This device acts as an optical lever, and may be used to measure small angular displacements if l is large. However, even though such a device is simple in principle, it can be applied to a metal-cutting dynamometer only with considerable difficulty.

Piezoelectric crystals have been used as force-measuring units in metal-cutting dynamometers. However, since piezoelectric crystals produce an electric charge rather than a current, leakage effects can be troublesome. However,

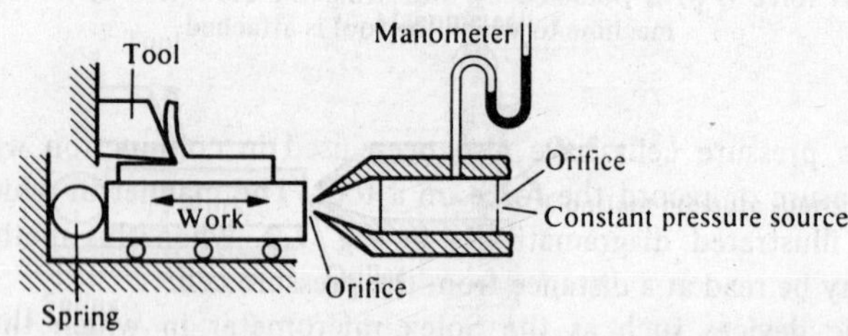

FIG. 7.3. Basis principle of pneumatic type dynamometer.

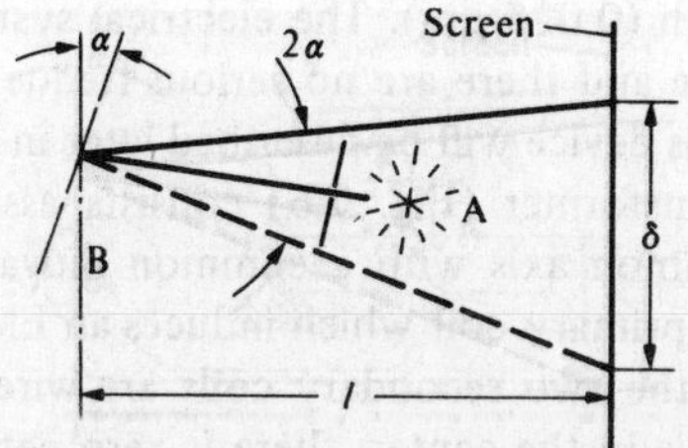

FIG. 7.4. Principle of optical lever shown diagrammatically.

piezoelectric dynamometers are well suited for making dynamic measurements since in general they can be designed to have a higher natural frequency of vibration than other types of dynamometers.

Electric Transducers

Several transducers that transform a physical displacement into an electrical signal are commercially available:

1. The electronic transducer tube.
2. The differential transformer.
3. The magnetic strain gage.
4. The unbonded wire resistance strain gage.
5. The bonded wire resistance strain gage.

An electronic transducer tube is shown schematically in Fig. 7.5. It is essentially a very small ($\frac{1}{4}$ in × $\frac{1}{2}$ in or 6 × 12 mm) triode vacuum tube with a movable plate, the tube characteristics being changed as the plate moves. The pin at A can be rotated approximately $\pm\frac{1}{2}$ degree about the diaphragm and the minimum motion of the end of the pin that can be accurately measured is of the order of

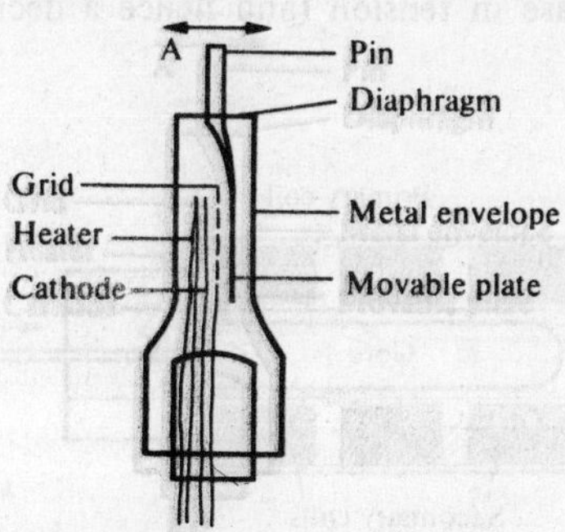

FIG. 7.5. Schematic diagram of electronic transducer tube.

one-millionth of an inch (0.025 μm). The electrical system associated with this tube is relatively simple and there are no serious frequency limitations. A dynamometer employing this device will be described later in this chapter.

The differential transformer (Fig. 7.6) consists essentially of three transformer coils on a common axis with a common movable core. AC current is supplied to the center primary coil which induces an EMF in the two secondary coils. The outputs of the two secondary coils are wired to oppose each other so that when the core is in the center, there is zero net output. When the core is displaced, an output is obtained which is proportional to the displacement, with a phase depending upon the direction of motion. The electrical system associated with this instrument is rather complex because of the necessary high amplification and phase sensitivity. However, AC carrier wave type strain gage amplifiers and recorders are commercially available and are quite satisfactory, giving a sensitivity of about 1 μin (0.025 μm).

The magnetic type strain gage shown in Fig. 7.7 allows point A to move with respect to B thus changing the inductance of coil B. If initially the balancing unit C is adjusted to give zero current at G, any motion of A will cause a current flow through G. This flow can be measured directly, or the balancing unit (at C) can be adjusted to give zero flow at G. The sensitivity of such a gage is about 10 μin (0.25 μm). This lack of sensitivity is somewhat offset by the simplicity of the associated electrical system. Since the input can safely be made quite high, the output can be measured directly without amplification.

The fact that the electrical resistance of a thin conductor is changed when it is stretched led to the development of extremely useful resistance strain gages about 1940. These may be of two types. In the unbonded type of gage (Fig. 7.8), a movable plate P is hung from two spring members. Four coils, R_1, R_2, R_3, and R_4, preloaded in tension are supported by small pins projecting from base B and plate P.

When the plate is moved relative to the base, two of the coils will undergo an increase in tension (and hence an increase in resistance) while the other two coils undergo a decrease in tension (and hence a decrease in resistance). The

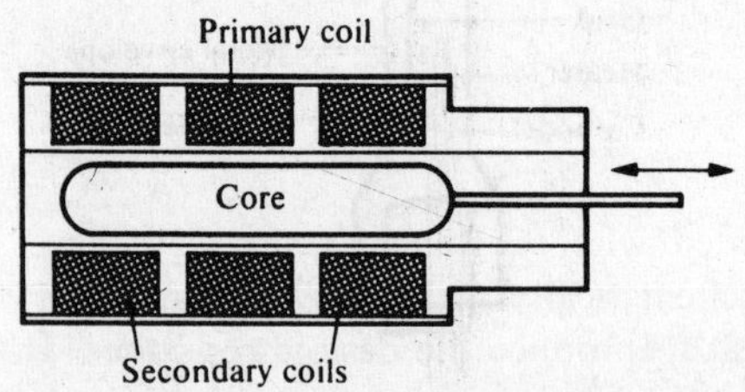

FIG. 7.6. Schematic diagram of differential transformer.

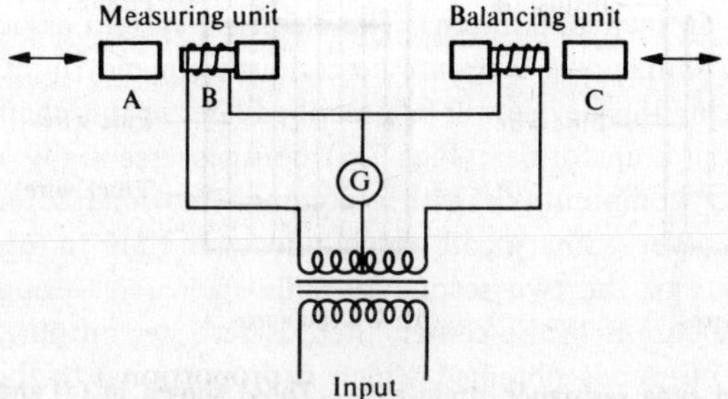

FIG. 7.7. Schematic diagram of magnetic type strain gage.

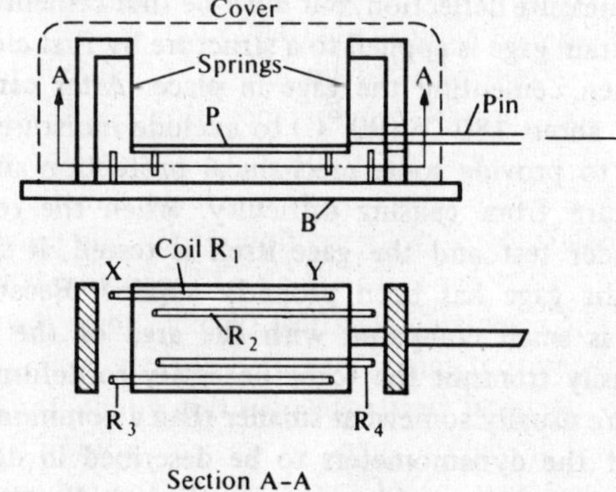

FIG. 7.8. Schematic diagram of unbonded wire resistance strain gage.

maximum deflection of an unbonded gage is usually limited by stops to about 0.0015 in ($\frac{1}{3}$ mm). When this type of gage is connected in the form of a Wheatstone bridge (to be presently described), it can measure deflections of the order of 1 μin (0.04 μm). One advantage of a unit of this type over that to be presently described (bonded type) is that it can be detached from the part under test and used over again. There are disadvantages however, not the least of which is cost.

Bonded Strain Gages

Three types of bonded wire resistance strain gages have been commercially available as shown schematically in Fig. 7.9. In each case the necessary length of gage wire (i.e. that to produce the desired resistance which is usually about 120 ohms) is present in the form of a flat coil which is cemented (bonded) between two thin insulating sheets of paper or plastic. Such a gage cannot

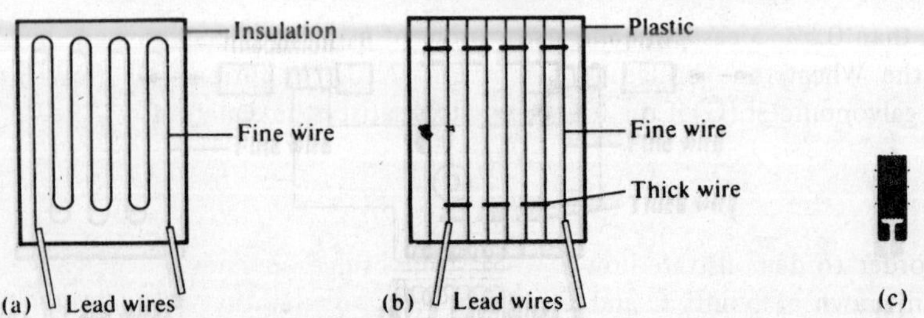

FIG. 7.9. Bonded type resistance strain gages. Those shown in (a) and (b) are wire type gages while type (c) is produced by etching a thin constantan foil (actual size, $\frac{1}{4}$ in [6.35 mm] gage length).

be used directly to measure deflection, but must be first cemented to the member to be strained. A strain gage is applied to a structure by first cleaning the surface thoroughly and then cementing the gage in place. After cementing, the unit is usually baked at about 180 °F (90 °C) to exclude moisture and then coated with wax or resin to provide some mechanical protection and also to prevent atmospheric moisture from causing difficulty. When the resistance between the metal part under test and the gage itself is tested, it should be at least 50 megaohms if the gage has been properly applied. Because the total area of all conductors is small compared with the area of the backing material, the cement can easily transmit the force necessary to deform the wire. Strain gages of this type are usually somewhat smaller than a common postage stamp.

Since several of the dynamometers to be described in detail later employ bonded wire or foil gages, it would appear justified to discuss their use in some detail. The manufacturer of a strain gage supplies two important pieces of information with every gage: the resistance (R) and the gage factor (F). The gage factor is a measure of the sensitivity of the gage as defined below

$$F = \frac{\Delta R/R}{\Delta l/l} = \frac{\Delta R}{eR} \tag{7.5}$$

where e is the normal elastic strain $\Delta l/l$. The gage factor (F) varies from about 1.75 to 3.5 for most gages. Bonded gages can measure strains as low as 10^{-7}, however, commercial strain gage amplifiers limit the practical range to about 10^{-6}. Due to fatigue, the maximum useful strain is of the order of 0.002. Thus, the practical range of operation is about 2000 to 1.

The Wheatstone Bridge

From eqn (7.5) it is evident that in order to measure strains of the order of 1 μ in (0.025 μm) it is necessary to measure small changes of resistance per unit resistance. The change in resistance of a bonded strain gage is usually less than 0.5%. Measurements of this order of magnitude may be made by means of the Wheatstone bridge shown in Fig. 7.10. No current will flow through the galvonometer (G) if the four resistances satisfy the equation

$$\frac{R_1}{R_4} = \frac{R_2}{R_3}. \tag{7.6}$$

In order to demonstrate how a Wheatstone bridge operates, a voltage scale has been drawn at points C and D of Fig. 7.10. Assume that R_1 is a bonded gage and that initially eqn (7.6) is satisfied. If R_1 is now stretched so that its resistance increases by one unit ($+\Delta R$), the voltage at point D will be increased from zero to plus one unit of voltage ($+\Delta V$), and there will be a voltage difference of one unit between C and D which will give rise to a current through G. If R_4 is also a bonded gage, and at the same time that R_1 changes by $+\Delta R$, R_4 changes by $-\Delta R$, the voltage at D will move to $+2\,\Delta V$. Also, if at the same time R_2 changes by $-\Delta R$, and R_3 changes by $+\Delta R$, then, the voltage of point C will move to $-2\,\Delta V$, and the voltage difference between C and D will now be $4\,\Delta V$. It is then apparent that although a single gage can be used, the sensitivity can be increased fourfold if two gages are used in tension while two others are used in compression.

It should be noted that the four active arm bridge is fully compensating for any change in resistance due to temperature providing all gages experience the same temperature change. This is important, since a single gage is frequently so sensitive to temperature changes as to make it useless for strain measurements in a region of steep temperature gradients. The output voltage (E) from a strain gage bridge is given by the following expression

$$E = \frac{a}{4}(F)(V)(e) \tag{7.7}$$

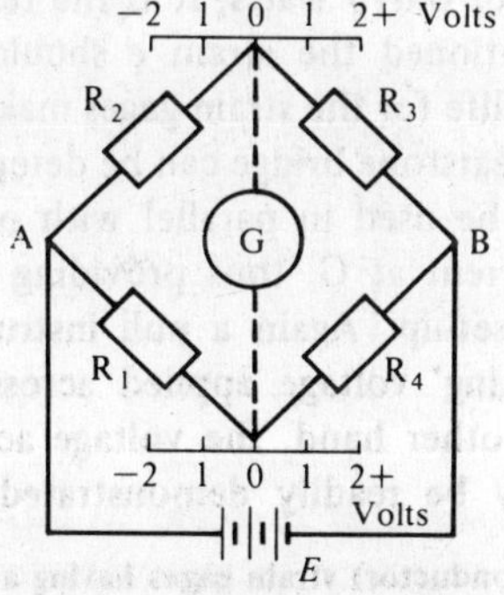

FIG. 7.10. The Wheatstone bridge.

where a is the number of active arms in the bridge (from 1 to 4)

F is the gage factor of the gages
V is the input voltage to the bridge
e is the longitudinal strain on the gages (neglecting transverse effects which are very small).

This equation is based on the assumption that the instrument used to measure E has a high impedance and hence does not draw an appreciable current from the bridge. If this is not so, the output (E) will no longer be a linear function of the strain e. It is again evident from eqn (7.7) that when maximum sensitivity (E/e) is required, (the usual case with metal-cutting dynamometers) it is desirable to make all four arms of the bridge active. It is further advantageous to use gages having a high gage factor.†

The impressed bridge voltage V is another quantity which may be varied to alter the sensitivity of a strain gage bridge circuit. With some commercial instruments this voltage is fixed at from 3 to 6 volts for reasons of stability. However, higher voltages may often be used, especially with the larger gages whose greater area allows better heat dissipation. In this respect aluminum makes a better material to which to cement gages than steel because of its higher rate of heat conductivity. The use of two strain gages in each arm of the bridge in place of one enables the impressed voltage to be doubled without overheating the gages. The maximum safe voltage in each case is determined by the desired stability. Too high a voltage will cause zero drift. When a very stable zero setting is required over a long time, the voltage must be reduced. The following expression may be used as a rough guide in determining the maximum permissible impressed voltage per gage when the gage is cemented to steel.

$$V = 2\,(WR)^{\frac{1}{2}} \tag{7.8}$$

where W is the limiting wattage which a gage may dissipate as heat, an average value being on the order of 0.075 watts; R is the resistance of a single gage in ohms. As previously mentioned the strain e should be kept below 0.002 to insure an adequate fatigue life for the strain gages making up the bridge.

The unbalance of a Wheatstone bridge can be determined by several methods. A variable resistance can be used in parallel with one arm of the bridge and adjusted to give zero current at G, thus providing a null instrument. This is the usual potentiometer set-up. Again a null instrument can be provided by use of a calibrated 'bucking' voltage applied across CD to give zero current flow through G. On the other hand, the voltage across CD can be measured directly. However, it may be readily demonstrated that if the bridge circuit

† While solid state (semi-conductor) strain gages having a gage factor an order of magnitude greater than metallic gages are available, this increase in gage factor is more than offset by the much greater adverse temperature sensitivity of the solid state gages.

is used in an unbalanced condition a microammeter without amplification cannot give the sensitivity that is directly available from a good potentiometer.

Design Considerations

In order to use any of the foregoing deflection measuring devices for force measurement, it is necessary to construct a dynamometer which will deflect the proper amount under load; and then mount the deflection detectors in such a manner that they will be unaffected by forces or moments other than the ones which they are supposed to detect. In general, dynamometers can be divided into two types shown schematically in Fig. 7.11. In the first type (Fig. 7.11a) the springs which support the external load (*P*) are independent of the deflection meter (M); while in the second type (Fig. 7.11b) the spring and the meters are integral. Any of the aforementioned deflection meters are suitable for use in the first way while only the bonded gages are effective in the second.

FIG. 7.11. Schematic views showing two types of dynamometers.

In many cases the force measuring element can be attached to either a stationary or moving member. Wherever possible the dynamometer should be placed on the stationary element inasmuch as this makes it unnecessary to use moving electrical contacts which may exhibit variable contact resistance. If slip rings have to be used it has been found advisable to employ mercury cups in which amalgamated copper discs rotate. The contact resistance is then practically zero and absolutely constant, as it must be in any successful strain gage circuit.

The design of a dynamometer may be influenced by a variety of special operating conditions that must be met. For example, if a cutting fluid is to be used, provisions must be made to keep the unit watertight. Many other special problems frequently arise in dynamometry. The dynamometers that are described in the following sections should be useful, however, in illustrating designs that have been successful.

Strain Gage Lathe Dynamometer

For simplicity of analysis, it is frequently convenient to reduce the lathe operation to a two-dimensional process (orthogonal cutting). In this case, the resultant force will act in a known plane and only two force components are required (F_P and F_Q) as shown in Fig. 7.12.

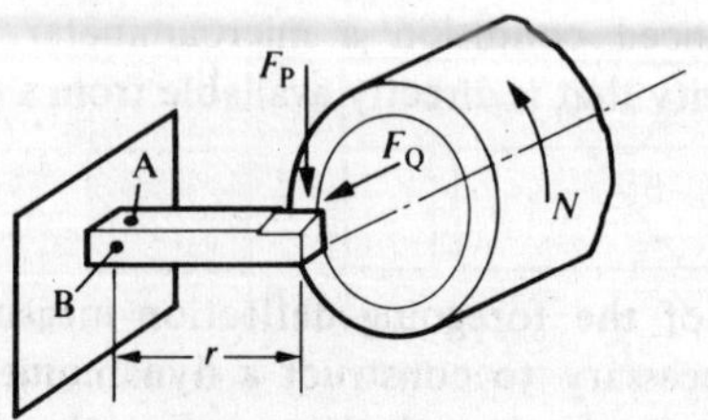

FIG. 7.12. Schematic view of dynamometer for measuring cutting forces in two-dimensional turning operation.

It is apparent that the force F_P will cause a bending moment M_P at a distance (r) from the cutting edge and that F_Q will cause a corresponding moment M_Q where

$$M_P = F_P r \tag{7.9}$$

$$M_Q = F_Q r. \tag{7.10}$$

The distribution of strain at a section through A and B, caused by bending moment M_P is as shown in Fig. 7.13. At the surface the strain is a maximum and equal to

$$e_{P1,2} = \frac{M_P\left(\dfrac{b}{2}\right)}{EI_P} \tag{7.11}$$

where e_{P1} is a tensile strain (+), e_{P2} is a compressive strain (−), E is Young's modulus of elasticity and I_P is the area moment of inertia of the section about axis C–C. For a rectangular section such as that shown in Fig. 7.13.

$$I_P = \frac{ab^3}{12} \tag{7.12}$$

and eqn (7.11) becomes

$$e_{P1,2} = \frac{6M_P}{ab^2E} \tag{7.13}$$

$$e_{Q1,2} = \frac{6M_Q}{ba^2E} \tag{7.14}$$

In order to measure the moment M_P, two strain gages are applied to the top of the measuring section (T_1 and T_2 in Fig. 7.14a) and two gages are applied directly below (C_1 and C_2) thus, when M_P is applied, two gages are put in tension, while two others receive an equal amount of compression, thereby satisfying the requirements for a complete Wheatstone bridge (Fig. 7.14b). In a similar manner M_Q is measured by the four gages T_3, T_4, C_3, C_4, which are connected

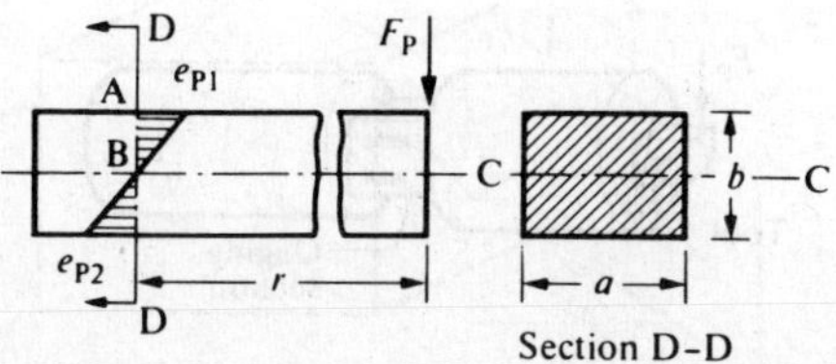

FIG. 7.13. Distribution of strain due to bending moment $M_P = rF_P$.

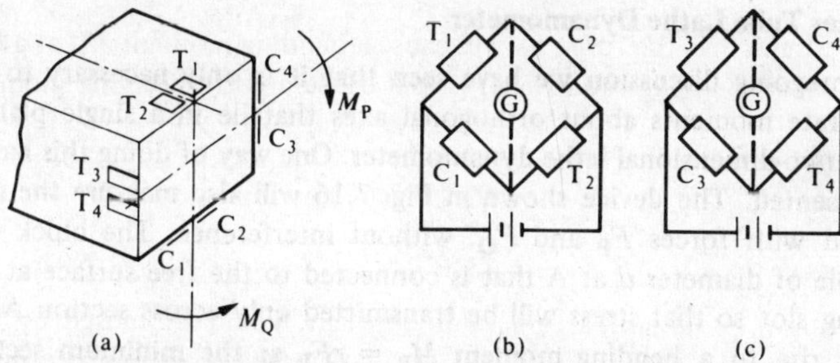

FIG. 7.14. Strain gage locations and connections. (a) Gage locations. (b) Gage circuit for measuring F_P. (c) Gage circuit for measuring F_Q.

as shown in Fig. 7.14c. It will be recalled that one of the important design criteria was that there be no cross sensitivity. When a moment M_P is applied, gages T_3 and C_4 will be subjected to tension while T_4 and C_3 are subjected to compression (Fig. 7.14a). However, if the gages are symmetrically placed with respect to the axis of the section, the strains will be of equal magnitude, and there will be no net output from the Wheatstone bridge (see Fig. 7.14c).

The foregoing discussion indicates how turning forces may be measured; we shall now consider a practical design employing such principles. Since it is inconvenient to place the gages directly on the tool, a tool-holder with a built-in measuring section is employed. A dynamometer of this type is shown in Fig. 7.15. It is important that the cutting edge be kept on the axis of the measuring section at a known distance from the gages. When the particular unit is connected to commercial strain recording equipment, a force of 3 pounds (13.4 N) gives a 1 mm pen deflection, and this represents adequate sensitivity for turning studies. The stiffness of the unit shown is actually greater than the customary tool holder for the $\frac{5}{8}$ inch square tool bit that is used.

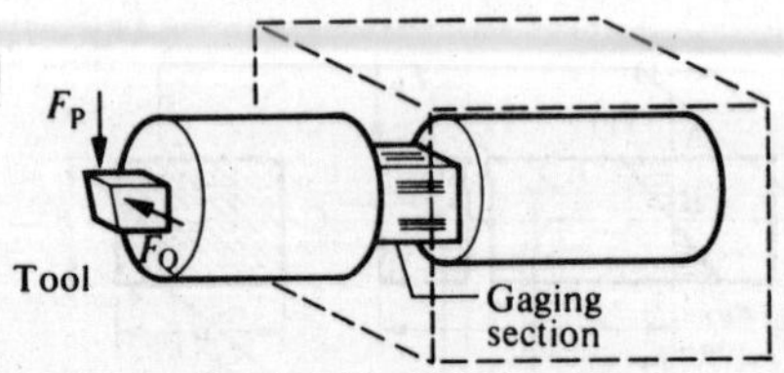

FIG. 7.15. Two-dimensional turning dynamometer. Force range: 2 to 2000 lb (10 to 10^4 N); stiffness: 500 000 lb in^{-1} (87 600 N mm^{-1}); tool size $\frac{5}{8} \times \frac{5}{8}$ in (15.9 × 15.9 mm). (After Cook, Loewen, and Shaw, 1954.)

Transducer Tube Lathe Dynamometer

In the foregoing discussion we have seen that it is only necessary to measure and separate moments about orthogonal axes that lie in a single plane in the case of a two-dimensional lathe dynamometer. One way of doing this has already been presented. The device shown in Fig. 7.16 will also measure the moments associated with forces F_P and F_Q, without interference. The block of metal has a hole of diameter d at A that is connected to the free surface at B and C by a long slot so that stress will be transmitted only across section AD. Force F_P gives rise to a bending moment $M_P = rF_P$ at the minimum section that will cause points B and C to approach each other. However, the distance BC will be unchanged by force F_Q due to the very large difference in the section moduli at section AD about horizontal and vertical axes. By use of the second slot and hole, distance EF may be used to measure force F_Q and this distance will be uninfluenced by force F_P. By placing a sensitive transducer across BC and another across EF, we have a convenient two-component dynamometer. The previously described electronic transducer tube offers a suitable means for measuring changes in BC and EF.

The choice of the slot and hole sizes in the design of the foregoing dynamometer is not readily obtained from elementary elasticity theory, but an analysis of the slotted member may be made by a combination of dimensional analysis, simple logic, and the results of a few easily performed experiments.

This is a convenient point to stop and review the principles of dimensional analysis which will be used to complete the design of the dynamometer of Fig. 7.16 and 7.17 as well as for other purposes in subsequent chapters.

Dimensional Analysis

Before discussing the details of dimensional analysis, we shall examine the nature of the end result. This will be done by considering a familiar elementary problem—the vertical displacement of a freely falling body. If we ignore air resistance, the only force acting upon a freely falling body is that due to gravity and the body will experience a downward acceleration ($g = 32.2$ ft sec^{-2} or

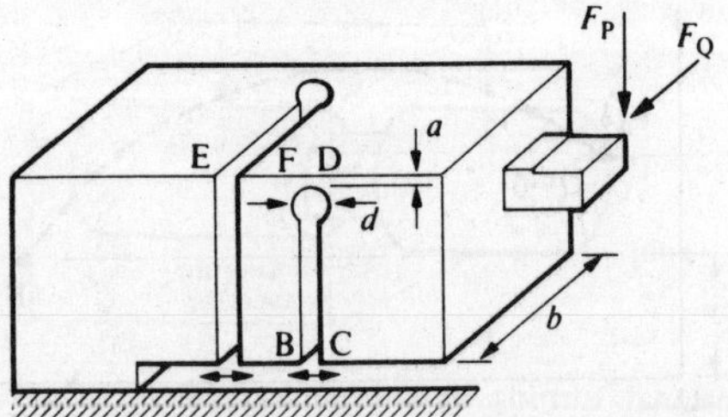

FIG. 7.16. Diagrammatic sketch of transducer tube lathe dynamometer.

9.81 m s^{-2}). Thus, if a body is projected from point A with an initial velocity V_0; (Fig. 7.17), its acceleration immediately after release will be

$$\frac{d^2y}{dt^2} = -g \tag{7.15}$$

(where t designates time in seconds) and will remain at this constant value until it strikes another object or until some additional force acts on the body.

Integrating eqn (7.15)

$$\frac{dy}{dt} = -gt + C_1 \tag{7.16}$$

and again

$$y = \frac{-gt^2}{2} + C_1t + C_2 \tag{7.17}$$

where C_1 and C_2 are constants of integration which may be evaluated by the following *boundary conditions*:

$$\frac{dy}{dt} = V_0 \quad \text{when } t = 0$$

$$y = y_0 \quad \text{when } t = 0$$

After the constants are evaluated, eqn (7.17) becomes

$$y = y_0 + V_0t - \frac{gt^2}{2}. \tag{7.18}$$

This is the complete solution in closed form.

In considering this problem by the method of dimensions, we would first observe that the main dependent variable (y) will be some function (ψ) of the variables y_0, V_0, t, and g

$$\text{i.e. } y = \psi(y_0, V_0, t, g). \tag{7.19}$$

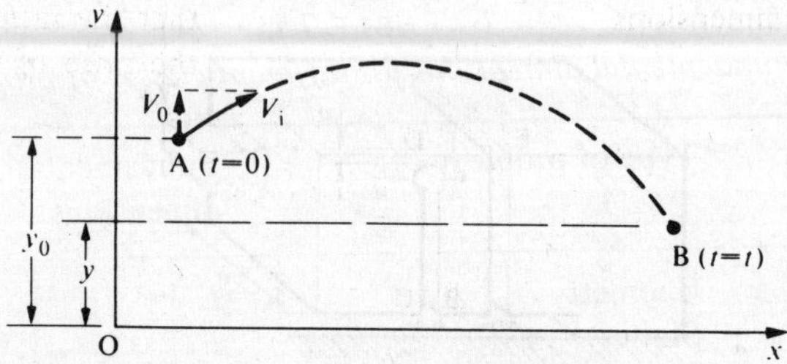

FIG. 7.17. Trajectory of projectile acted on by gravity when released from point A with initial velocity V_i.

After performing the dimensional analysis, we would write

$$\frac{y}{gt^2} = \psi_1\left(\frac{y_0}{gt^2}, \frac{V_0}{gt}\right) \tag{7.20}$$

where ψ_1 is some function of the two quantities in parentheses. The number of variables has been reduced from 5 to 3 but ψ_1 must still be evaluated either experimentally or by further analysis.

Dimensional analysis thus represents a partial solution to a problem, but one which is very easily obtained. In the solution of complex problems it can play a very valuable role.

Before considering the method of proceeding from eqn (7.19) to (7.20), a few quantities will be defined.

Fundamental (or primary) dimensions are properties of a system under study that may be considered independent of the other properties of interest. For example, there is but one fundamental dimension in any geometry problem and this is length $[L]$. The fundamental dimensions involved in different classes of mechanics problems are outlined in Table 7.1, where L stands for length, F for force, and T for time. Dimensions (other than F, L, and T) which are considered fundamental in areas other than mechanics include: temperature (θ), heat (H), electrical charge (Q), magnetic pole strength (P), chemical yield (Y), and unit cost (¢).

Dimensional equations relate the dimensions of the fundamental quantities entering a problem to nonfundamental or secondary quantities. For example, acceleration (a) is a quantity of importance in many kinematic problems, and the dimensions of acceleration are related to fundamental dimensions L and T by the following dimensional equation

$$a = [LT^{-2}] \tag{7.21}$$

It is customary to enclose the combination of fundamental dimensions employed in writing a dimensional equation in brackets.

Dimensional units are the basic magnitudes used to specify the size of a fundamental quantity. In engineering problems, lengths are frequently measured

TABLE 7.1.
Fundamental dimensions for different classes of problems

Type of problem	Fundamental dimensions
Geometry	L
Statics	F
†Temporal	T
Kinematics	L, T
Work or Energy	F, L
Momentum or Impulse	F, T
Dynamics	F, L, T

† Temporal problems are those involved in time study, frequency analysis, time tables.

in feet (m) and times in seconds. In this system of dimensional units, an acceleration would be expressed in units of feet per (second)^2 (m s^{-2}) which may be verified by reference to eqn (7.21)

A nondimensional quantity is one whose dimensional equation has unity or $F^0L^0T^0$ on the right-hand side. The quantities (y/gt^2) and (V_0/gt) in eqn (7.20) are clearly nondimensional. A group of variables that cannot be combined to form a nondimensional group are said to be dimensionally independent.

The choice of fundamental quantities is somewhat arbitrary as the following discussion will reveal. In Table 7.1 the fundamental quantities for the general dynamics problem were stated to be F, L, and T. The quantities M (mass), L, and T could also have been used and then F would be a secondary variable related to fundamental variables M, L, and T by a dimensional equation based on Newton's law:

$$F = ma \tag{7.22}$$

$$\text{i.e. } F = [MLT^{-2}]. \tag{7.23}$$

While any group of dimensionally independent quantities entering a dynamics problem may in fact be chosen as the fundamental quantities (such as mass, velocity, and acceleration), it is customary to use either F, L, and T or M, L, and T and we will use the former since most engineering dimensions are expressed in units of force and not in units of mass.

It is not permissible to consider both mass (M) and force (F) as fundamental quantities in dynamics problems as long as Newton's law holds.

At this point we will introduce the important properties of dimensional and physical equations on which dimensional analysis is based.

The ratio characteristic that most systems of measurement satisfy requires that the ratio of two measurements be independent of the units adopted. Thus, the ratio of two lengths will be the same whether the individual measurements be made in terms of inches, centimeters, cubits, or Swedish miles. Some common systems of measurement that do not satisfy this ratio requirement are the pH scale in chemistry, the Mohs' scale of hardness in minerology, and the decibel scale of acoustical engineering.

Bridgman (1931) has observed that a restriction of form for a dimensional equation exists when the fundamental quantities satisfy the ratio requirement.

This is best illustrated by an example. Let A, B, and C be three dimensional quantities that are related by some function (ψ_1) as follows:

$$A = \psi_1 (B, C) \tag{7.24}$$

For example, eqn (7.24) could be the following specific function which is in agreement with eqn (7.24)

$$A^2 = 3B + BC \tag{7.25}$$

If eqn (7.24) is a physically correct formulation then according to Bridgman the three variables A, B, and C may be replaced by three other variables which consist of products of A, B, and C raised to powers. One of the infinite number of such possibilities is

$$A^2B^2 = \psi_2 (BC, C^2). \tag{7.26}$$

That eqn (7.26) is valid for the specific example of eqn (7.25) may be seen by multiplying both sides of this equation by B^2:

$$A^2B^2 = 3B^3 + B^3C \tag{7.27}$$

which may obviously be written as follows, according to Bridgman's observation.

$$A^2B^2 = \psi_3 \left[\frac{B^3C}{(3B^3)^{\frac{2}{3}}}, \frac{(B^3C)^2}{(3B^3)^2} \right] \tag{7.28}$$

or

$$A^2B^2 = \psi_4 [BC, C^2] \tag{7.29}$$

which is identical to eqn (7.26).

The principle of dimensional homogeneity first expressed by Fourier states that the dimensions of each term of a physically correct equation must reduce to the same value. For example, eqn (7.18) is readily found to meet this test

$$[L] = [L] + [L\,T^{-1}\,T] - [LT^{-2}\,T^2]. \tag{7.30}$$

A homogeneous equation is one that contains no dimensional constants and holds true for all systems of units.

We are now ready to return to the displacement problem and to consider the steps involved between eqns (7.19) and (7.20).

Since this is a kinematic problem there are but two fundamental dimensions (L, T). Table 7.2 gives the dimensional equations for all quantities.

TABLE 7.2.
Dimensional equations for illustrative dimensional analysis problem

Quantity	Dimensional equation
Vertical displacement, y	$[L]$
Initial vertical displacement, y_0	$[L]$
Acceleration due to gravity, g	$[L\,T^{-2}]$
Time, t	$[T]$
Initial vertical velocity, V_0	$[L\,T^{-1}]$

The quantities g and t are dimensionally independent (may not be combined by raising to powers and multiplying together to form a nondimensional group). If a function exists as stated in eqn (7.19), then according to Bridgman we may also write

$$y\,g^{-1}\,t^{-2} = \psi_2(g,\, t,\, y_0 g^{-1} t^{-2},\, V_0 g^{-1} t^{-1}). \tag{7.31}$$

The left-hand side of this equation is nondimensional and by the principle of dimensional homogeneity, all terms of ψ_2 must also be nondimensional. The quantities g and t therefore cannot appear in ψ_2 except where combined with other quantities to form a nondimensional group. Thus,

$$\frac{y}{gt^2} = \psi_2\left(\frac{y_0}{gt^2}, \frac{V_0}{gt}\right) \text{ which is eqn (7.20)}$$

The foregoing procedure may be followed in any case where all units satisfy the ratio requirement.

Exponents a and b, required to make yg^at^b nondimensional, may be found by writing simultaneous equations as follows, although usually these exponents may be written directly by inspection:

$$y\,g^a t^b = [L\,(LT^{-2})^a\,T^b] = [L^0 T^0].$$
$$\text{I} \qquad\qquad \text{II}$$

Equating exponents of L in I and II yields

$$1 + a = 0$$

while equating the exponents for T in I and II gives

$$-2a + b = 0.$$

When these equations are solved simultaneously, a and b are found to be -1 and -2 respectively, in agreement with the left-hand side of eqn (7.20).

The result of a dimensional analysis is sometimes written in symbolic form in terms of Pi quantities as follows

$$\pi_1 = \psi(\pi_2, \pi_3, \text{etc.}) \tag{7.32}$$

where π_1 is a nondimensional group involving the main dependent variable (y in the displacement problem) and the other Pi quantities represent the remaining nondimensional quantities entering the problem.

The dimensionally independent quantities can usually be chosen in more than one way and hence there will often be several correct answers to a dimensional analysis. One of these answers may prove to be more convenient for a given purpose than the others. In the displacement problem all of the pairs of dimensionally independent quantities listed in Table 7.3 could have been used to obtain the dimensionless equations listed opposite each pair.

While there are no hard and fast rules concerning choice of dimensionally independent quantities, the following considerations serve as a useful guide:

1. The main dependent variable should not be chosen.
2. Variables having the greatest significance should be chosen.
3. The variables chosen should represent as many physically different aspects of the problem as possible.

In performing a dimensional analysis, it is important to include all quantities of importance to the problem. Otherwise, an incorrect and misleading result will be obtained. It is less important to include a variable about which doubt exists than to omit one which proves to be significant. Frequently, combinations of variables that are known to appear in a given class of problems in a unique association can be treated as a single quantity. This will result in fewer Pi quantities in the final result. Another method of reducing the resulting Pi quantities to a minimum is to limit the scope of the problem. Results of auxiliary or approximate analysis can sometimes be combined with conventional dimensional reasoning to greatly increase the power of dimensional analysis. Such auxiliary analysis includes

1. Arguments involving symmetry.
2. Assumptions of linearity or known behavior.
3. Special solutions for large or small values of an important variable.

The meaning of some of these more or less abstract statements will become clearer when specific applications are discussed in later chapters. Additional details concerning dimensional analysis may be found in Van Driest (1946) and Taylor (1974).

TABLE 7.3.
Different dimensional analysis solutions for different choices of dimensionally independent quantities

Dimensionally independent quantities	Resulting equation
y_0, g	$\frac{y}{y_0} = \psi\left(\frac{gt^2}{y_0}, \frac{V_0^2}{y_0 g}\right)$
y_0, t	$\frac{y}{y_0} = \psi\left(\frac{gt^2}{y_0}, \frac{V_0 t}{y_0}\right)$
y_0, V_0	$\frac{y}{y_0} = \psi\left(\frac{gy_0}{V_0^2}, \frac{V_0 t}{y_0}\right)$
g, t	$\frac{y}{gt^2} = \psi\left(\frac{y_0}{gt^2}, \frac{V_0}{gt}\right)$
g, V_0	$\frac{yg}{V_0^2} = \psi\left(\frac{gt}{V_0}, \frac{y_0 g}{V_0^2}\right)$
t, V_0	$\frac{y}{V_0 t} = \psi\left(\frac{gt}{V_0}, \frac{y_0}{V_0 t}\right)$

Dimensional Analysis of Transducer Tube Lathe Dynamometer

Returning now to the transducer tube dynamometer design, the angular deflection α between points B and C of Fig. 7.16 will be a function of the variables indicated in eqn (7.33):

$$\alpha = \psi_1(a, b, d, M_P, E) \tag{7.33}$$

where M_P is the bending moment at the minimum section. Table 7.4 gives the dimensions of each of the quantities in eqn (7.33) in terms of F, L, T.

The quantities E and a are dimensionally independent and hence we may write

$$\alpha = \psi_2\left(\frac{b}{a}, \frac{d}{a}, \frac{M_P}{Ea^3}\right). \tag{7.34}$$

TABLE 7.4.
Dimensional equations for transducer tube lathe dynamometer

Quantity	Dimensional equation
Angular displacement, α	
Linear dimension, a	$[L]$
Linear dimension, b	$[L]$
Linear dimension, d	$[L]$
Bending moment, M_P	$[FL]$
Young's modulus, E	$[FL^{-2}]$

Experience with elastic structures teaches that α should vary directly with (M_P) and inversely with b. When this constraint is applied to eqn (7.34), we obtain

$$\alpha = \left(\frac{M_P}{Ea^3}\right)\left(\frac{a}{b}\right)\psi_3\left(\frac{d}{a}\right) \tag{7.35}$$

In order to determine the function ψ_2 it is necessary to perform several experiments on simple bars (Fig. 7.18a) with various d/a ratios obtained by gradually increasing d. The results of such a series of tests is shown plotted in Fig. 7.18b. The fact that all points lie on a smooth curve is consistent with a correct dimensional analysis. The experimental curve can be well approximated by the following exponential function:

$$\alpha\frac{E\,a^2b}{M} = 17\left(\frac{d}{a}\right)^{0.395} \tag{7.36}$$

By use of this equation, the slotted member may be designed to give the proper stiffness for the force range expected and the type of transducer available. The above example illustrates how dimensional reasoning may sometimes be combined with other known facts to yield a solution to a problem that is otherwise not readily tractable.

Strain Rings

In the foregoing lathe dynamometers, the force must always be applied at the same point since such an instrument is essentially a moment-measuring device. For many purposes, it is necessary to measure the force applied to the work piece as this force moves with the progress of a cut. In order to measure such an arbitrarily placed three-dimensional force or torque, the strain ring has proven to be a useful device. We will first consider the characteristics of several strain rings and then their application to dynamometry.

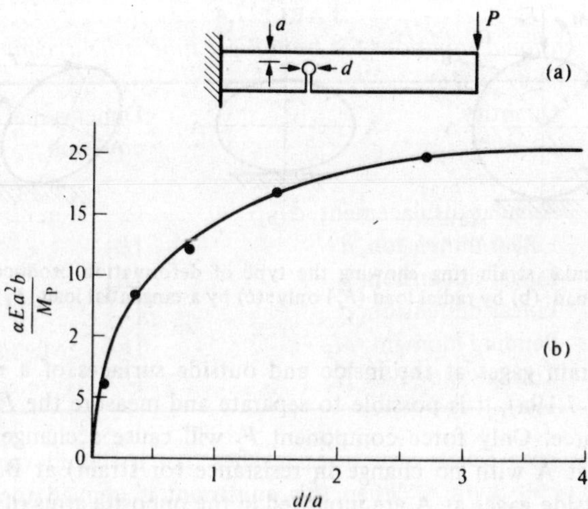

FIG. 7.18. Nondimensional plot for elastic bending characteristics of slotted bar shown at (a).

Strain rings (Fig. 7.19a) provide a high ratio of sensitivity to stiffness at the same time having adequate stability against buckling. A ring is easily produced and simple to mount. The fact that the inside is always in an opposite state of strain from the outside allows four active arms to be effectively used in a bridge circuit as already described. The symmetry of a ring provides parallel paths for heat flow and hence it is to be expected that equivalent points on opposite sides of a ring will be at the same temperature. This enables drift due to temperature gradient in the vicinity of the dynamometer to be eliminated by connecting the gages to form a complete bridge circuit. Aluminum is a good material from which rings may be made since it will not corrode as readily as steel, is light in weight, easily machined, and has excellent heat conductivity. A thin metal ring of radius r, thickness t, and axial width b is shown in Fig. 7.19a. The ring is fixed at the bottom while a radial force F_r and a tangential force F_t are applied at the top. The point of load application is maintained horizontal by a suitable moment M. When only F_r is applied, the ring will deform as in Fig. 7.19b. Thin ring elastic theory shows the strain at the inside and outside surfaces of the ring at points A to be

$$e_A = \pm \frac{1.09 F_r r}{Ebt^2} \tag{7.37}$$

while the strain at the point B (39.6° from the vertical axis) is zero. When only F_t is applied (Fig. 7.19c) the strain at A is zero while the strain at B is

$$e_B = \pm \frac{2.18 F_t r}{Ebt^2}. \tag{7.38}$$

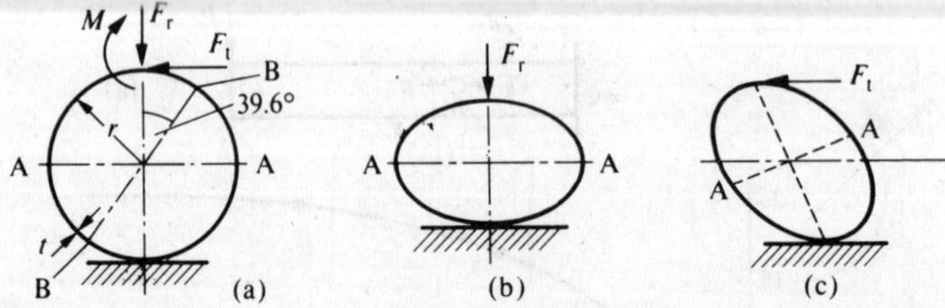

FIG. 7.19. Circular strain ring showing the type of deformation produced (a) by combined load; (b) by radial load (F_r) only; (c) by a tangential load (F_t) only.

By placing strain gages at the inside and outside surfaces of a ring at points A and B (Fig. 7.19a), it is possible to separate and measure the F_r and F_t components of force. Only force component F_r will cause a change in resistance in the gages at A with no change in resistance (or strain) at B. Thus, if the inside and outside gages at A are mounted in the opposite arms of a Wheatstone bridge circuit we have a sensitive means for detecting changes in F_r. In a similar way the gages at B are only sensitive to changes in F_t. As long as the strains at A and B are within the elastic limit of the ring material, it is evident from eqns (7.37) and (7.38) that the strains will vary linearly with the force, and hence the strain ring satisfies the desired requirement of linearity.

The values of stiffness of a thin circular ring in the radial and tangential directions are as follows:

$$K_r = \frac{F_r}{\delta_r} = \frac{Ebt^3}{1.8r^3} \tag{7.39}$$

$$K_t = \frac{F_t}{\delta_t} = \frac{Ebt^3}{3.6r^3} \tag{7.40}$$

While eqns (7.37) and (7.38) hold quite well even for fairly thick rings, the stiffness values given in eqns (7.39) and (7.40) are usually found to be much less reliable. This is due in part to the difficulty of preventing rolling at the points of attachment, which changes the end conditions from those assumed. Rings are usually mounted by bolts extending through holes in the top and bottom surfaces.

End conditions can be readily maintained by making the outside surface of the ring octagonal instead of circular (Fig. 7.20)†. Use of octagonal rings provides points of stress concentration at positions on the horizontal axis and at 45° to the vertical axis. If gages are placed at these points of stress concentration and the minimum value of t is used in eqns (7.37) and (7.38), good

† It has also been found important to make the top and bottom surfaces of the octagon (surfaces of attachment) slightly concave to avoid a change in end conditions under load. If this is not done, a troublesome hysteresis effect is apt to occur.

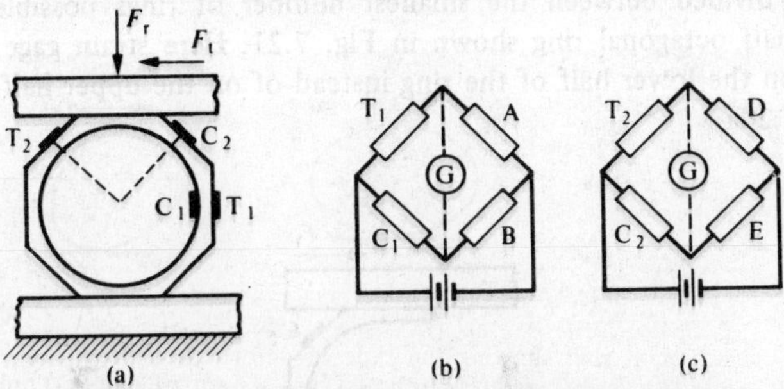

FIG. 7.20. (a) Octagonal strain ring showing gage positions. (b) Bridge connections for measuring F_r. (c) Bridge connections for measuring F_t.

results are obtained even though the inclined gages are at points 45° from the vertical instead of the 39.5° called for by the circular ring theory. The octagonal rings are stiffer than circular ones of the same minimum section.

Proper strain gage positions are shown in Fig. 7.20a while the manner of connecting the gages to independently measure F_r and F_t is shown in Figs. 7.20b and 7.20c. The resistances A and B in Fig. 7.20b might be a pair of fixed resistances or, if two or more rings are used simultaneously, they would be active strain gages at the horizontal section of another ring on the compressive and tensile sides respectively. In the latter case, the sum of the vertical loads on the two rings would be recorded. Similar remarks might be made for resistances D and E of Fig. 7.20c. In order that there be no cross sensitivity it is important that gages T_2 and C_2 be mounted exactly symmetrically with respect to the ring axes.

If both sides of eqn (7.37) are divided by force F_r we obtain

$$\frac{e_A}{F_r} = \frac{1.09\, r}{Ebt^2}. \tag{7.41}$$

If this equation and eqn (7.39) are multiplied together we have an expression for the *strain per unit deflection*

$$\frac{e_A}{\delta_r/r} = \frac{1.09}{1.8}\frac{t}{r} = .61\frac{t}{r} \tag{7.42}$$

where δ_r is the deflection in a radial direction due to load F_r. Neither the ring width (b) nor modulus of elasticity (E) is of importance with regard to the strain per unit deflection. It is evident that we should like e_A/δ_r to be as large as possible in the interest of maximum sensitivity and rigidity. This calls for r being as small as possible and t as large as possible. However, r can be reduced only to the point where it becomes too difficult to mount internal gages accurately. Thus, for a ring of given size r, we should like t to be as large as it can be consistent with the desired sensitivity. Therefore, it is clear that the total load should be divided between the smallest number of rings possible. This suggests the half octagonal ring shown in Fig. 7.21. Here strain gage T_2 must be mounted on the lower half of the ring instead of on the upper half of the ring as in Fig. 7.20a.

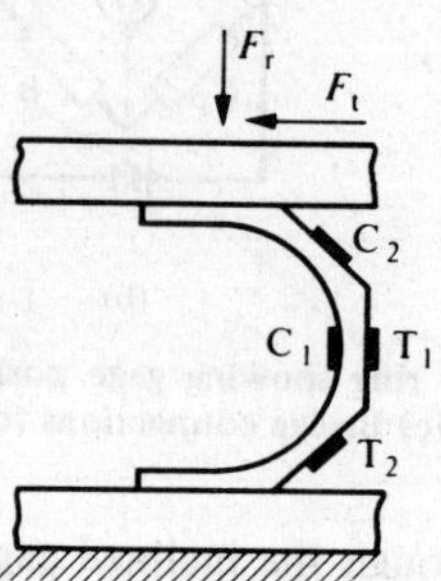

FIG. 7.21. The octogonal half-ring.

Ito, Sakai, and Ishikawa (1980) have performed a finite element analysis for the elastic behavior of octagonal rings. Figure 7.22 shows the relative radial deflection of octagonal (δ_o) and circular (δ_c) rings of different thickness (t) to mean radius (r) ratio when subjected to the same vertical load. In this analysis, E was 21 000 kg mm^{-2} ($\sim 30 \times 10^6$ p.s.i.) and Poisson's ratio was 0.28. Figure 7.22 shows that whereas the octagonal ring is substantially stiffer than the circular ring when t/r is 0.05 or less, the difference is less than 10 percent if t/r is 0.25 or greater. Figure 7.23 shows the corresponding sensitivity to stiffness ratio ($e/\delta/r$) for octagonal and circular rings of different t/r ratio. Figure 7.24 shows the angle θ measured from the direction of the radial load

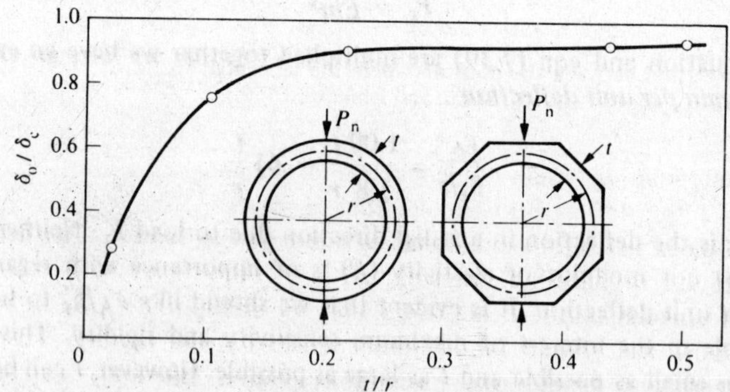

FIG. 7.22. Difference in displacement of circular ring (δ_c) and octogonal ring (δ_o). (After Ito, Sakai, and Ishikawa 1980.)

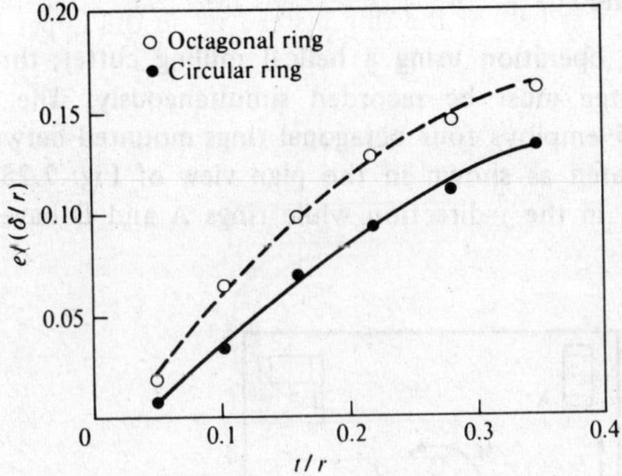

FIG. 7.23. Relation between t/r and strain output for circular and octogonal rings. (After Ito, Sakai, and Ishikawa 1980.)

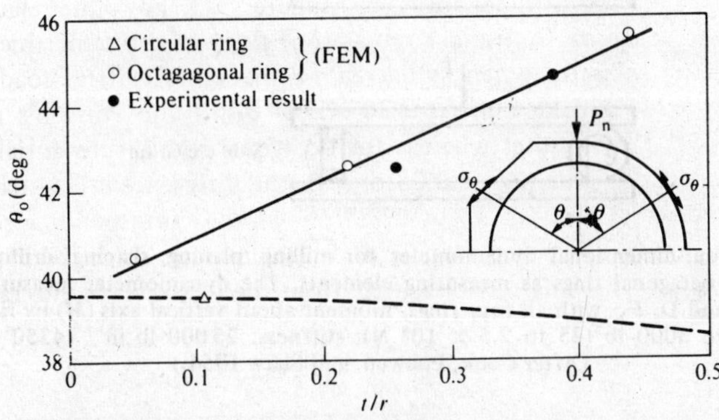

FIG. 7.24. Relation between the position where the tangential stress is zero and t/r. (After Ito, Sakai, and Ishikawa 1980.)

for the null point in strain for circular and octagonal rings for different t/r ratios. For a thin ($t/r = 0.1$) circular ring, $\theta = 39.5°$. This figure shows that the null point in strain for a radial load on an octagonal ring occurs close to 45°. These results are generally consistent with the thin ring theory considered initially and suggest that the thin ring theory is sufficiently accurate for design purposes. An exact prediction of dynamometer sensitivity is not essential since the final dynamometer must be carefully calibrated in any case.

Milling Dynamometer

In a plane milling operation using a helical milling cutter, three orthogonal components of force must be recorded simultaneously. The dynamometer shown in Fig. 7.25 employs four octagonal rings mounted between two plane rigid surfaces oriented as shown in the plan view of Fig. 7.25. The axes of rings B and C are in the *y*-direction while rings A and D have their axes in the *x*-direction.

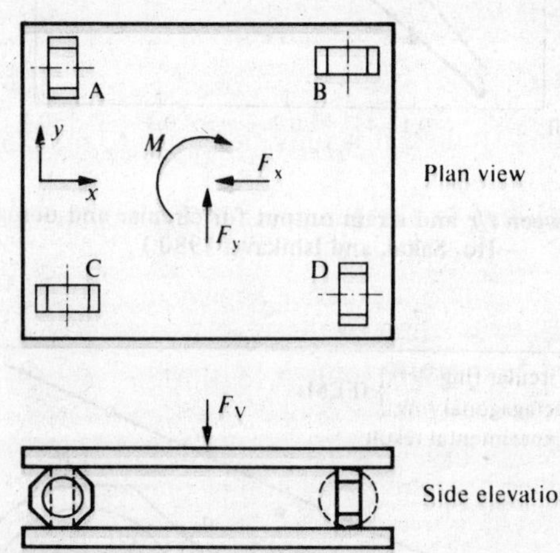

FIG. 7.25. Four-dimensional dynamometer for milling, planing, shaping, drilling studies utilizing four octagonal rings as measuring elements. The dynamometer measures F_x, F_y with rings A and D, F_V with all four rings, moment about vertical axis (*M*) by rings A and B. Range: 5 to 5000 lb (25 to 2.5×10^4 N); stiffness: 25 000 lb in^{-1} (4350 n mm^{-1}). (After Cook, Loewen, and Shaw 1954.)

When an arbitrarily placed vertical load (F_V) is applied it is carried by all four rings, and hence the two vertical measuring gages (T_1 and C_1) in Fig. 7.20 of each ring are wired into one eight-gage bridge circuit in such a way that all outputs add. A force in the *x*-direction (F_x) is measured by the gages in the inclined positions. T_2 and C_2 in Fig. 7.20 are connected in a four-gage bridge such that their deflections add. A force in the *x*-direction produces no bending in rings A and D since their axes are in the direction of the force (a very stiff direction). The gages on rings A and D are used to measure forces in the *y*-direction just as those on rings B and C are used for forces in the *x*-direction.

In designing a three-dimensional dynamometer we have need for the stiffness of a ring in the axial direction. From thin ring theory we can obtain an expression for the stiffness in the axial direction *K*, just as we did previously for the radial and tangential directions (eqns 7.39 and 7.40). However, due to the uncertainty of the end conditions this equation does not hold too well.

The following semi-empirical expression for K_a has been found to give a better estimate than that derived from thin ring theory.

$$K_a = \frac{Ebt}{40r} \tag{7.43}$$

The dynamometer shown in Fig. 7.25 can measure vertical (z-direction) loads up to 5000 pounds (22 kN) with a sensitivity of 5 pounds (22 N). The measurements are independent of the point of application of the load within the square outlined by the rings.

Drilling Dynamometer

In drilling and tapping studies, it is only necessary to measure the axial thrust force and torque on the tool.

Reversing the positions in the foregoing bridge circuit of the inclined gages on ring 1 while those on ring 3 are kept the same, enables a torque to be measured about the z-axis. Thus, the device shown in Fig. 7.25 can measure not only force components in three orthagonal directions without interference, but can also be used to measure torques by merely throwing a selector switch that reconnects the gages.

Dynamometers of the basic type shown in Fig. 7.25 may be used to study milling, planing, shaping, broaching, and drilling operations.

Surface Grinding Dynamometer

Figure 7.26 shows the manner in which the foregoing principles have been applied to produce a dynamometer for use in measuring radial and tangential forces between a grinding wheel and workpiece in a plunge (no crossfeed) surface grinding operation. Here, we have in effect two octagonal half-rings which are machined from a single piece of aluminum to provide a sensitive element of low mass and high natural frequency. Four strain gages are mounted on each half ring and connected to form two complete Wheatstone bridge circuits; one for measuring the power component of force (F_P) and one for measuring the feed component (F_Q). By using two gages from each ring in each of the bridge circuits, the instrument becomes independent of the point of application of the load between the rings.

The characteristics of this dynamometer are as follows: range, 0.1 to 100 lb (0.45 N to 445 N); stiffness, 4000 lb in^{-1} (453 N m^{-1}).

A two-component milling dynamometer having the same design but stiffer rings had the following characteristics: range, 2 to 2000 lb (8.9 to 8910 N); stiffness, 10^6 lb in^{-1} (113 kN m^{-1}).

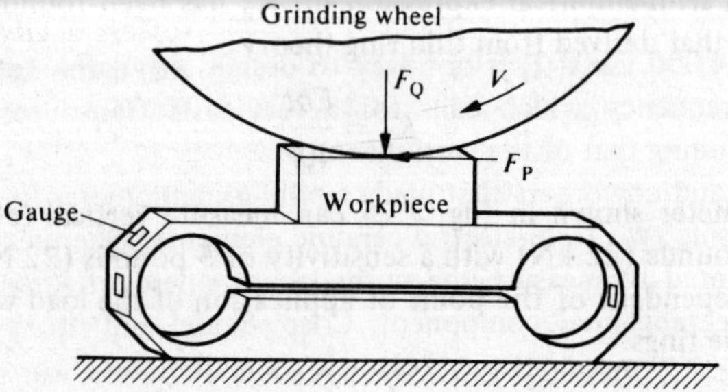

FIG. 7.26. Two-dimensional grinding dynamometer.

Three-Component Turning Dynamometer

A three-component turning dynamometer was constructed using four extended half-rings machined from a single block of steel to avoid hysteresis effects (Fig. 7.27). Force component F_R puts all four half-rings in compression, F_Q subjects all elements to a tangential force while F_P puts the lower elements in compression and the upper ones in tension. By properly connecting the gages, all three components of force may be simultaneously recorded with negligible cross sensitivity. The characteristics of this dynamometer were: range, 4 to 2000 lb (18 N to 8.9 kN); stiffness, 5×10^6 lb in^{-1} (566 kN m^{-1}).

Single Abrasive Grain Dynamometer

The characteristics of a single abrasive particle mounted in the periphery of a metal disc are frequently studied to evaluate different types of abrasive materials. At a wheel speed of 6000 f.p.m. (30 m s^{-1}), a maximum depth of cut of 0.001 in (2.5 μm), the cutting time per cut would be about 168 μs. The natural

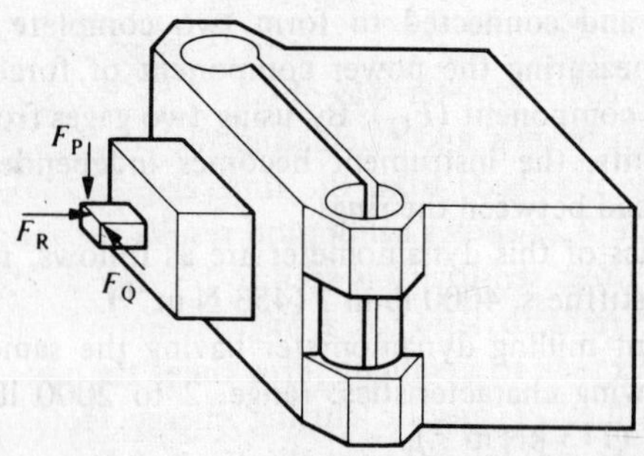

FIG. 7.27. Three-dimensional lathe dynamometer.

frequency of the dynamometer required for such a case should be about 4 × $(10^6/168)$ = 24 000 c.p.s. It is not easy to design a dynamometer with such a high natural frequency which calls for a very stiff transducer and very low total mass (including that of the workpiece).

Eiss (1967) and coworkers first made a mini-dynamometer for single abrasive grain studies using two ferro-electric ceramic elements. One piezoelectric transducer was oriented to measure the normal component of force and the other to measure the tangential component. Crisp, Seidel and Stokey (1968) built the improved two component piezo-electric dynamometer shown diagrammatically in Fig. 7.28. By substituting glass for plastic insulation between the crystals and cementing the workpiece to the dynamometer, it was possible to obtain natural frequencies of 120 000 c.p.s. (normal direction) and 60 000 c.p.s. (tangential direction) with essentially no hysteresis. Signals for the dynamometer were displayed on a dual beam oscilloscope and recorded using a polaroid camera. The oscilloscope sweep was triggered by the signal itself.

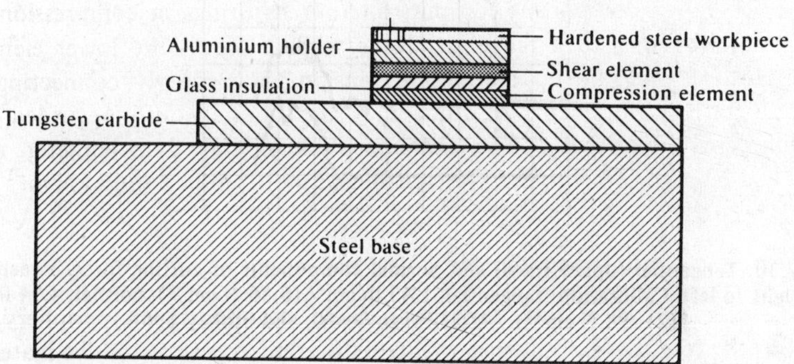

FIG. 7.28. Piezoelectric dynamometer. Steel base is 3 in (76 mm) long, other dimensions being in proportion except for thickness of piezoelectric elements and glass insulation which have been exaggerated. (After Crisp, Seidel, and Stokey 1968.)

Calibration of the dynamometer was obtained by dropping small steel balls from different known heights and calculating the height of rebound (needed to obtain the energy absorbed) from the time elapsed between the first and second impacts. Figure 7.29 shows calibration results for tangential and normal crystals respectively. Cross sensitivity is seen to be very low. Figure 7.30 shows force traces (cutting right to left) for a single diamond particle. The normal component of force is seen to be about three times the tangential component.

Table 7.5 gives values of specific cutting energy (total energy consumed divided by volume removed) for different abrasive types. Also given in Table 7.5 are values of specific energy simultaneously obtained using the inertial dynamometer described in the next section.

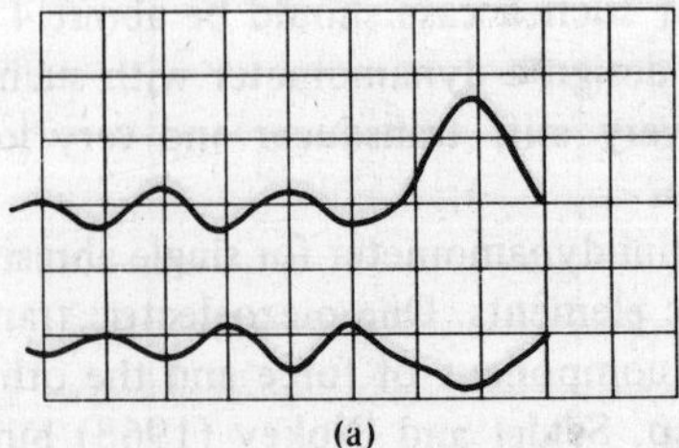
(a)

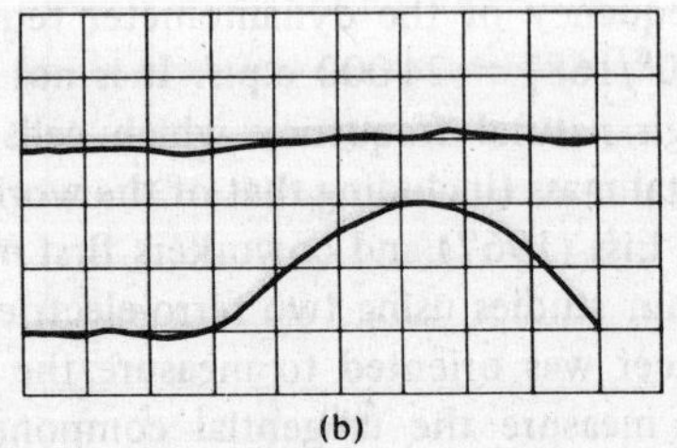
(b)

FIG. 7.29. Calibration tests. Upper traces are shear element responses and lower traces are compression. Sweep 10 $\mu s\ cm^{-1}$, right to left. (After Crisp, Seidel, and Stokey 1968.) (a) Tangential Force $\frac{1}{4}$ in ball, $\frac{1}{4}$ in drop; peak force, 11.3 lb; sensitivity: normal, 100 $mV\ cm^{-1}$; tangential, 500 $mV\ cm^{-1}$. (b) Normal Force $\frac{1}{2}$ in ball, $\frac{11}{16}$ in drop peak force, 19.6 lb; sensitivity: normal, 200 $mV\ cm^{-1}$; tangential, 500 $mV\ cm^{-1}$.

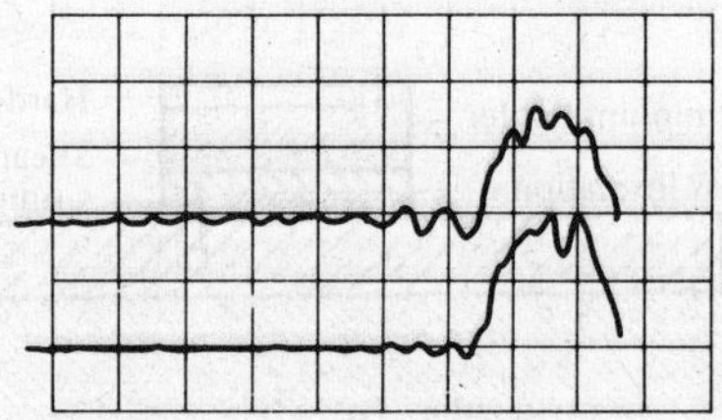

FIG. 7.30. Tangential (upper trace) and normal components of cutting force. Sweep 50 μs cm^{-1} right to left. Calibration: tangential, 1.43 lb cm^{-1} (6.36 N cm^{-1}); normal, 4.94 lb cm^{-1} (21. 98 N cm^{-1}). (After Crisp, Seidel, and Stokey 1968.)

Inertial Single Grain Dynamometer

As an alternate method of finding the specific grinding energy for single abrasive particles, the deceleration of an aluminum rotor to which a single abrasive particle was attached was measured and the mean energy consumed per cut calculated. The test arrangement is shown in Fig. 7.31. The rotor was brought up to grinding speed by use of an air-driven motor and then allowed to rotate without further energy input as it made one small cut per revolution after another. The spindle bearings were $\frac{1}{2}$ inch (12.5 mm) diameter and the 'grinding wheel' consisted of two aluminum rods—one which carried the abrasive particle and the other for counterbalance. Speed was measured by a combination light source–photocell unit and a small mirror attached to one of the arms. The photocell emits a pulse for each revolution and these pulses were recorded against time using a Visicorder. The time for each revolution may be found and from this the deceleration may be calculated.

TABLE 7.5.

Energy expended in cutting, as determined by inertial dynamometer, and piezoelectric dynamometer, for various grain types. (After Crisp, Siedell, and Stokey 1968.)

Grain type	Volume cut (average)			Specific cutting energy (in lb in^{-3} ($\times 10^{-6}$))	
	in^3 ($\times 10^6$)	mm^3 ($\times 10^3$)	$\frac{\text{width}}{\text{depth}}$	Piezoelectric dynamometer	Inertial dynamometer
Green SiC	0.065	1.065	17.0	1.73	3.9
Green SiC	0.094	1.540	12.0	4.04	1.7
Black SiC	0.044	0.721	4.8	2.95	3.2
Black SiC	0.069	1.131	10.0	3.56	4.1
Sintered bauxite	0.0093	0.152	23.0	12.02	12.5
Sintered bauxite	0.013	0.213	60.0	17.00	11.0
Al_2O_3 + 10% ZrO_2	0.13	2.130	14.0	2.13	2.2
Al_2O_3 + 10% ZrO_2	0.18	2.950	9.5	1.91	2.3

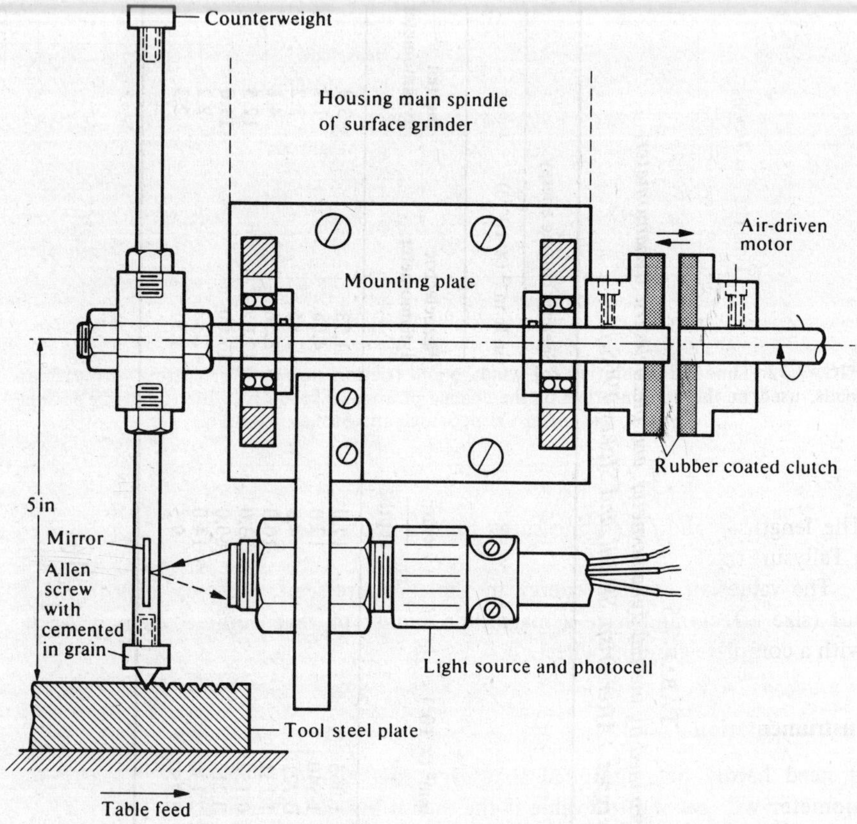

FIG. 7.31. Inertial dynamometer. (After Crisp, Seidel, and Stokey 1968.)

Figure 7.32 shows results from a typical run. The windage (and friction) curve was obtained by a preliminary run without cutting. By subtracting windage values from the total, the effect of cutting on deceleration may be calculated and hence the energy consumed per cut. Values of specific energy obtained by the inertial dynamometer simultaneously with the piezoelectric dynamometer are given in Table 7.5 where it is evident that the agreement is relatively good. The volume of material removed per cut was estimated in calculating both values of specific energy by use of the following empirical equation

$$\text{Vol} = 4/15\ blt \tag{7.44}$$

where b = maximum width of half-almond shaped groove
l = length of groove
t = maximum depth of groove.

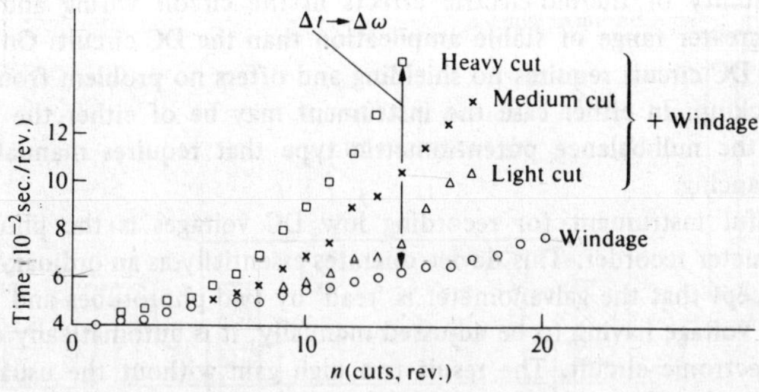

FIG. 7.32. Time per revolution for windage and friction run, and for cutting with various loads, used in the computation of the change of angular speed $\Delta\omega$ due to cutting several grooves. (After Crisp, Seidel, and Stokey 1968.)

The lengths b and l were measured using a toolmaker's microscope and t from a Tallysurf trace.

The values of specific energy in Table 7.5 are seen to decrease with size of cut (size effect) and have a magnitude similar to that obtained when cutting with a complete grinding wheel.

Instrumentation

It need hardly be mentioned that even the most efficiently designed dynamometer will be of little value if the measuring instrument used with it is not accurate or sensitive enough, or does not have a frequency response that is sufficiently high to follow significant fluctuations of force. Several instruments are available for use with strain gage bridge circuits. The one item all of these instruments have in common is an amplifier which is a necessity since strain gage output voltages are so minute. In other respects the several instruments differ widely.

A most important distinction may be made between those instruments which yield a permanent record and those which are merely indicators. Aside from the inconvenience associated with reading two or more indicators simultaneously there is the more important matter of recording an average or characteristic value of a fluctuating force. It is possible for different operators to record widely differing 'average' values for a fluctuating characteristic when an indicator is used. However, when the force curve is continuously recorded, it is then a matter of correctly interpreting the recorded data.

Another important decision which must be made concerns the choice between an AC and DC bridge circuit. The AC circuit has the advantage of eliminating the possibility of thermo-electric effects in the circuit wiring and generally offers a greater range of stable amplication than the DC circuit. On the other hand the DC circuit requires no shielding and offers no problem from 60-cycle mains pick-up. In either case the instrument may be of either the deflection type or the null-balance potentiometric type that requires manual or automatic balancing.

A useful instrument for recording low DC voltages is the photo-electric potentiometer recorder. This device operates essentially as an ordinary potentiometer except that the galvanometer is 'read' by two phototubes and instead of the back voltage having to be adjusted manually, it is automatically controlled by an electronic circuit. The result is a high gain without the usual troubles associated with DC amplifiers. With four active strain gages, this instrument is capable of measuring strains as small as 0.1 μin in^{-1}. Despite this sensitivity, it can withstand considerable overloads without damage.

If a permanent record is not required, then any DC voltmeter can be used provided it is sensitive enough and has a high input impedance. A sensitive DC vacuum tube millivoltmeter has proved quite useful. If speed is not essential any of the better thermocouple potentiometers can be used, especially the self-balancing types. These instruments are distinguished by their high sensitivity, (up to 0.003 mV per division) combined with a large scale, (up to 30 in or 0.76 m). However, the speed of response of these devices is limited (about 2 secs. for full scale). The aforementioned photoelectric potentiometer recorder can follow small variations up to about 4 cycles s^{-1} ($\frac{1}{2}$ s full scale), but like all automatic feedback potentiometers it becomes either unstable or sluggish as the bridge resistance increases. Vacuum tube voltmeters do not suffer from this restriction.

Probably the most familiar of the AC strain gage bridge instruments is the portable strain indicator, which is a manually balanced, null-type instrument. The fact that the strain indicator involves both manual balancing and reading makes it rather slow to use, which is awkward under fluctuating conditions.

Concluding Remarks

Only dynamometers with which the author has had direct experience in the laboratory are described here. Many additional dynamometer designs are to be found in the literature. However, it is felt that the units described are sufficient to demonstrate the problems involved and a few representative ways of solving them.

References

Bridgman, P. W. (1931). *Dimensional analysis* (2nd edn). Yale University Press, (New Haven).

Cook, N. H., Lowewn, E. G., and Shaw, M. C. (1954). *Am. Mach., N.Y.* **98**, 125.
Crisp, J., Seidel, J. R., and Stokey, W. F. (1968). *Int. J. Prod. Res.* **7**, 159.
Eiss, N. S. (1967). *J. Engng Ind.* **89**, 463.
Ito, S., Sakai, S., and Ishikawa, M. (1980). *Bull. Jap. Soc. Precis. Engrs* **14**, 25.
Taylor, E. S. (1974). *Dimensional analysis for engineers.* Clarendon Press, Oxford.
Van Driest, E. R. (1946). *J. appl Mech.* **68**, A34.

8

SHEAR STRAIN IN CUTTING

In Chapter 3 the mechanics of orthogonal metal cutting with a continuous chip and a sharp tool was discussed from the point of view of the interrelationships that exist between important variables. However, no attempt was made to *predict* cutting performance. In Chapter 3 it was revealed that the bulk of the energy in metal cutting is consumed on the shear plane and that the specific shear energy (u_S) accounts for about $\frac{3}{4}$ of the total. It was further shown that

$$u_S = \tau \gamma. \tag{3.27}$$

A first step in predicting cutting performance would therefore involve estimating the shear stress on the shear plane (τ) and the shear strain associated with chip formation (γ). Both of these quantities are strongly influenced by the friction on the face of the tool which has been expressed in terms of the friction angle β, (Fig. 3.4) where

$$\beta = \tan^{-1} \frac{F_C}{N_C}. \tag{8.1}$$

In this and the next two chapters, problems associated with the estimation of γ, τ, and β will be discussed in that order.

In Chapter 3, it was shown that a relatively simple relationship exists between shear strain (γ), rake angle (α), and shear angle (ϕ):

$$\gamma = \frac{\cos \alpha}{\sin \phi \cos (\phi - \alpha)} \tag{3.17}$$

Since the rake angle will be known, the problem of predicting γ reduces to the problem of predicting (ϕ).

Dimensional Analysis

To begin, we might perform a dimensional analysis in an effort to determine the nature of the relationship that is to be expected for ϕ. In addition to ϕ, two quantities are needed to define completely the geometry of a two-dimensional cutting process, namely the rake angle (α) and the depth of cut (t). Two distinct operations are involved in the cutting process: a shearing action along the shear zone, and sliding with friction along the tool face. The coefficient of friction

($\mu = \tan\beta$) is usually sufficient to characterize a friction process and the shear stress τ will usually characterize the dynamic aspects of a shear process. The cutting velocity V, together with the geometry of the process, will fix the kinematics of the system. It would then appear that the several quantities that might figure in a determination of the angle ϕ are those listed in Table 8.1, where F, L, and T refer to the fundamental dimensions of force, length, and time. When a dimensional analysis is performed in the usual manner we obtain

$$\phi = \psi(\alpha, \beta) \tag{8.2}$$

which states that the shear angle ϕ is a function of α and β only.

TABLE 8.1.
Dimensional analysis for shear angle

Quantity	Symbol	Dimensions
Shear angle	ϕ	. . .
Rake angle	α	. . .
Depth of cut	t	L
Friction angle	β	. . .
Shear stress	τ	FL^{-2}
Cutting velocity	V	LT^{-1}

Early Complete Solutions

There have been many notable attempts in the past to derive a complete physical equation corresponding to eqn (8.2). The reasoning followed in several early analyses will next be briefly outlined followed by a discussion of this work.

Piispanen (1937) presented a graphical analysis for the shear angle, which was equivalent to the second analysis of Merchant given later in this chapter. However, inasmuch as this early work did not lead to an analytical expression it will not be considered further.

Ernst and Merchant (1941) presented the first quantitative analysis. In Fig. 8.1 the forces at the point of a tool are shown. Assuming the shear stress on the shear plane τ to be uniformly distributed, it is evident that

$$\tau = \frac{F_S}{A_S} = \frac{R' \cos(\phi + \beta - \alpha)\sin\phi}{A} \tag{8.3}$$

where A_S and A are the areas of the shear plane and that corresponding to the width, b, times the depth of cut, t, respectively. Ernst and Merchant reasoned that ϕ should be an angle such that τ would be a maximum, and a relationship

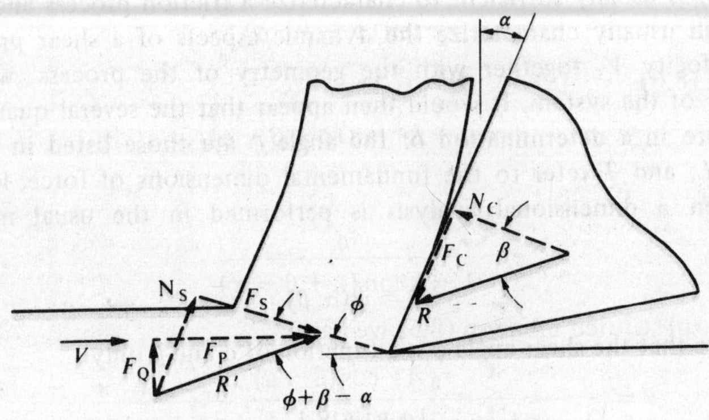

FIG. 8.1. Force relations at tool point.

for ϕ was obtained by differentiating eqn (8.3) with respect to ϕ and equating the resulting expression to zero. This led to the result

$$\phi = 45 - \frac{\beta}{2} + \frac{\alpha}{2}. \tag{8.4}$$

However, it is to be noted that in differentiating, both R' and β were considered independent of ϕ.

Merchant (1945) presented a different derivation that also led to eqn (8.4). This time, an expression for the total power consumed in the cutting process was first written

$$P = F_P V = \tau A V \frac{\cos(\beta - \alpha)}{\sin\phi \cos(\phi + \beta - \alpha)} \tag{8.5}$$

It was then reasoned that ϕ would be such that the total power would be minimum. An expression identical to eqn (8.4) was obtained when P was differentiated with respect to ϕ, this time considering τ and β to be independent of ϕ.

This is what Piispanen (1937) had done previously in a graphical way. However Piispanen immediately carried this line of reasoning one step further and assumed that the shear stress τ would be influenced directly by normal stress on the shear plane σ as follows

$$\tau = \tau_0 + K\sigma \tag{8.6}$$

where K is a material constant. Piispanen then incorporated this observation into his graphical solution for the shear angle.

Upon finding eqn (8.4) to be in poor agreement with experimental data, Merchant also assumed a relationship as given by eqn (8.6) and proceeded to work this into his second analysis as follows:

From Fig. 8.1 it may be seen that

$$\sigma = \tau \tan(\phi + \beta - \alpha) \tag{8.7}$$

or from eqn (8.6)

$$\tau = \tau_0 + K\tau \tan(\phi + \beta - \alpha) \tag{8.8}$$

hence

$$\tau = \frac{\tau_0}{1 - K\tan(\phi + \beta - \alpha)} \tag{8.9}$$

When this is substituted into eqn (8.5) we have

$$P = \frac{\tau_0 A V \cos(\beta - \alpha)}{[1 - K\tan(\phi + \beta - \alpha)] \sin\phi \cos(\phi + \beta - \alpha)} \tag{8.10}$$

Now, when P is differentiated with respect to ϕ with τ_0 and β considered independent of ϕ we obtain

$$\phi = \frac{\cot^{-1}(K)}{2} - \frac{\beta}{2} + \frac{\alpha}{2}. \tag{8.11}$$

Merchant called the quantity $\cot^{-1}K$ the 'machining constant' C. In Fig. 8.2 the quantity C is seen to be the angle the assumed line relating τ and σ makes with the τ-axis.

Stabler (1951) has presented an analysis of three-dimensional cutting operations in which an expression for shear angle is also presented. When it was assumed that the maximum shear stress on the shear plane and the resultant shear strain were collinear, then it followed from purely geometrical considerations that

$$\phi = 45 - \beta + \frac{\alpha}{2} \tag{8.12}$$

Slipline Field Solutions

Lee and Shaffer (1951) have derived still other relationships for ϕ. These authors assumed (a) that the material cut behaves as an ideal plastic which does not strain-harden, (b) that the shear plane represents a direction of maximum shear stress.

The slip-line field A–B–C in Fig. 8.3a, was assumed. Here region A–B–C will be plastically rigid and subjected to a uniform state of stress. This means that the stress at any point in A–B–C may be represented by a single Mohr's stress circle diagram. The diagram in Fig. 8.3b, in fact, satisfies all SLF requirements. Here point **a** corresponds to the stress on face **a** in Fig. 8.3a, i.e. on the shear plane. This is seen to follow the assumption that the shear plane is a direction of maximum shear stress. Line B–C in Fig. 8.3a is a line along which the stress is zero. Due to the double-angle feature of Mohr's circle, the zero

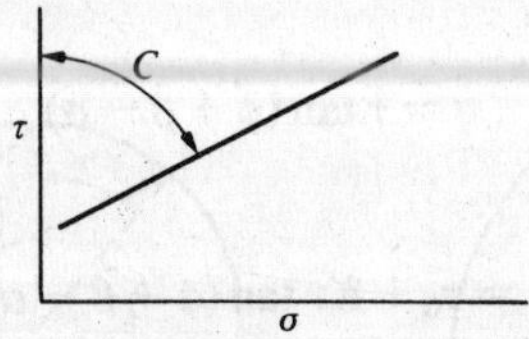

FIG. 8.2. Relation between flow stress and normal stress assumed by Piispanen (1937) and Merchant (1945).

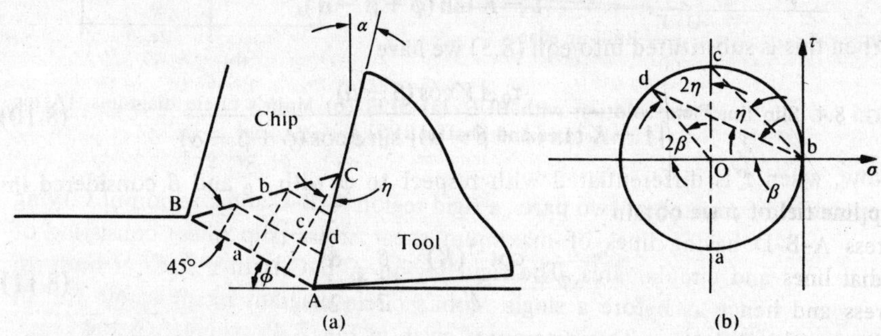

FIG. 8.3. Ideal plastic solution for stress field at tool point. (a) Slip line field. (b) Mohr's circle diagram for field ACB. (After Lee and Shaffer 1951.)

point of stress **b** in Fig. 8.3b which is 90 degress from point **a**, will be but 45 degrees from point **a** in Fig. 8.3a. Thus B–C makes an angle of 45 degrees with A–B in Fig. 8.3a. In a similar way point **c** in Fig. 8.3b represents the stress on face **c** in Fig. 8.3a, and the stress on the tool face is represented as point **d** in Fig. 8.3b.

From Fig. 8.3b it is seen that angle **dbc** $= \eta$, since angle **dOc** $= 2\eta$. Therefore

$$\eta = 45 - \beta. \tag{8.13}$$

From Fig. 8.3a

$$\eta = \phi - \alpha \tag{8.14}$$

and hence from eqns (8.13) and (8.14)

$$\phi = 45 - \beta + \alpha. \tag{8.15}$$

It is evident from Fig. 8.3 that the foregoing ideal plastic solution indicates that the shear stress on the shear plane τ, and the normal stress on the shear plane σ, should be equal. This, however, is generally not the case and hence eqn (8.15) cannot be regarded as a general solution.

Lee and Shaffer presented a second SLF solution based on the presence of a small BUE represented as a circular arc as shown in Fig. 8.4a. This time the

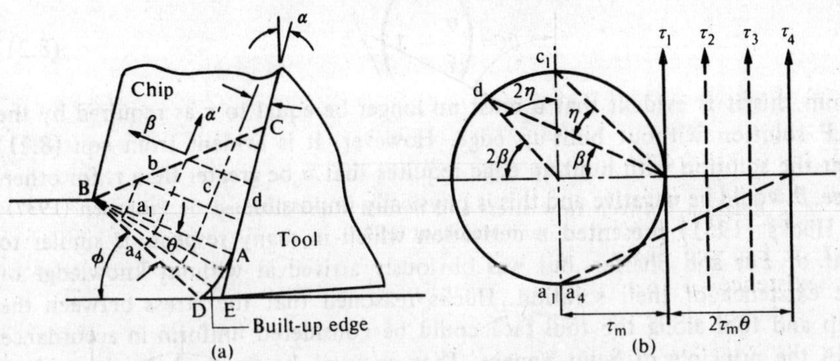

FIG. 8.4. Slip line field solution with BUE. (a) SLF. (b) Mohr's circle diagrams. (After Lee and Shaffer 1951.)

slip line field consisted of two parts, a rigid region A–B–C and a region of varying stress A–B–D having lines of maximum shear stress (slip lines) consisting of radial lines and circular arcs. The rigid region A–B–C is in a state of uniform stress and hence as before a single Mohr's circle diagram is applicable for all points within this area. This diagram is given in Fig. 8.4b where τ_1 is the shear-stress axis. The stresses for points along line D–B are given by the circle in Fig. 8.4b if the τ-axis is at τ_4, the stresses for points along line F–B are given by the circle if the τ-axis is shifted to τ_3, while the stresses along line A–B (as well as all other points in region A–B–C) are given by the circle when the τ-axis is at τ_1.

From Fig. 8.4a it is evident that

$$\phi = \alpha + \eta + \theta \tag{8.16}$$

But from Fig. 8.4b

$$\eta = 45 - \beta \tag{8.17}$$

and hence

$$\phi = 45 + \theta - \beta + \alpha. \tag{8.18}$$

This differs from eqn (8.15) only by the angle θ, which is a measure of the size of the built-up edge. When there is no built-up edge, θ is zero and eqns (8.15) and (8.18) give the same result. From Fig. 8.4b it is evident that

$$\sigma = \tau_m(1 + 2\theta) \tag{8.19}$$

$$\tau = \tau_m \tag{8.20}$$

where σ and τ are the normal and shear stresses on the shear plane, respectively. Combining eqns (8.19) and (8.20)

$$\theta = \left(\frac{\sigma}{\tau} - 1\right) \Big/ 2 \tag{8.21}$$

From this it is evident that σ must no longer be equal to τ as required by the SLF solution without built-up edge. However, it is evident from eqn (8.21) that the solution with built-up edge requires that σ be greater than τ. for otherwise, θ would be negative and this is physically impossible.

Hücks (1951) presented a derivation which in many respects is similar to that of Lee and Shaffer, but was obviously arrived at without knowledge of the existence of their solution. Hücks reasoned that the stress between the chip and tool along the tool face could be considered uniform in accordance with the principle of Saint Venant. Then as usual, he assumed the shear plane to be in the direction of maximum shear stress. Hücks further reasoned that the resistance to flow in the direction of chip motion would be relatively small and thus considered the stress on face **b** in Fig. 8.5a to be zero. The Mohr's circle diagram corresponding to this set of assumptions is shown in Fig. 8.5b.

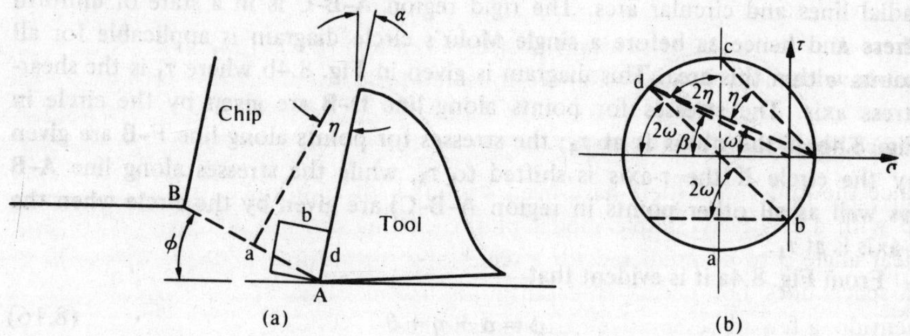

FIG. 8.5. Hücks (1951) solution for tool point stress field. (a) Stress field. (b) Mohr's circle diagram.

Here it is evident that

$$\tan 2\omega = 2\mu \tag{8.22}$$

where μ is the coefficient of friction on the tool face.

From Fig. 8.5a

$$\phi = \eta + \alpha \tag{8.23}$$

and from Fig. 8.5b

$$\eta + \omega = 45°. \tag{8.24}$$

Combining eqns (8.22) to (8.24) we obtain

$$\phi = 45 - \omega + \alpha \tag{8.25}$$

where

$$\omega = \tfrac{1}{2}\tan^{-1} 2\mu. \tag{8.26}$$

It may be seen from Fig. 8.5b that $\beta > \omega$ and thus eqn (8.25) will yield values of ϕ that are larger than those obtained from eqn (8.15).

In order that the stress on face **b** in Fig. 8.5a be zero under all conditions, it is necessary that the τ-axis intersect the stress circle at some point **b**, as in Fig. 8.5b. This requires that the shear stress on the shear plane τ always be greater than σ the normal stress on the shear plane. Inasmuch as σ is frequently observed to be greater than τ_1, Hücks' solution cannot hold, in general.

Hücks (1951) also extended the foregoing treatment to include the influence of normal stress on the shear plane, as Piispanen and Merchant had done. This led to the following equation

$$\phi = \frac{\cot^{-1} K}{2} - \frac{\tan^{-1} 2\mu}{2} + \alpha \tag{8.27}$$

where K is the material constant postulated in eqn (8.6). The several equations that have been reviewed thus far are collected for convenience in Table 8.2.

Later Shear Angle Relationships

Since the mid-1950s there have been a number of additional attempts to come up with a relatively simple model of the cutting process that will enable the shear angle ϕ to be predicted for a wide range of cutting conditions. While lack of space does not permit a detailed discussion of all of these or even brief mention, a few representative treatments follow:

1. Oxley (1961) presented a SLF solution including strain hardening.
2. Sata (1963) derived a shear angle relation in terms of the angle $\theta = \phi + \beta - \alpha$ (Fig. 8.1) between resultant force R' and the shear plane, ratio of shear stress on shear plane to that on rake face (K), and the ratio of tool chip contact length (a') to undeformed chip thickness (t) in addition to rake angle (α). The angle θ was used as a boundary condition since it was empirically found to remain constant over a wide range of cutting conditions including work material, rake angle, cutting speed, and controlled contact length. The value of K was also found to be essentially constant ($\sim$0.8 to 1).
3. Oxley (1963) presented a SLF solution including strain hardening and variation of shear stress on the tool face.
4. Usui and Hoshi (1963) presented a SLF solution for tools of controlled contact length. Figure 8.6a shows the interpretation of flow lines from quick stop photomicrographs and Fig. 8.6b gives the corresponding SLF. This solution involves a centered fan of extent θ which increases in magnitude as the tool face contact length (a') decreases. When $\theta = 0$ this solution

TABLE 8.2.
Collection of early equations for shear angle

Source	Equation	Equation
Dimensional analysis	8.2	$\phi = \psi(\alpha, \beta)$
Ernst and Merchant	8.4	$\phi = 45 - \beta/2 + \alpha/2$
Merchant	8.11	$\phi = \dfrac{\cot^{-1} K}{2} - \beta/2 + \alpha/2$
Stabler	8.12	$\phi = 45 - \beta + \alpha/2$
Lee and Shaffer	8.15	$\phi = 45 - \beta + \alpha$
Lee and Shaffer	8.18	$\phi = 45 + \theta - \beta + \alpha$†
Hücks	8.25	$\phi = 45 - \dfrac{\tan^{-1} 2\mu}{2} + \alpha$
Hücks	8.27	$\phi = \dfrac{\cot^{-1} K}{2} - \dfrac{\tan^{-1} 2\mu}{2} + \alpha$

† With built-up edge of magnitude θ.

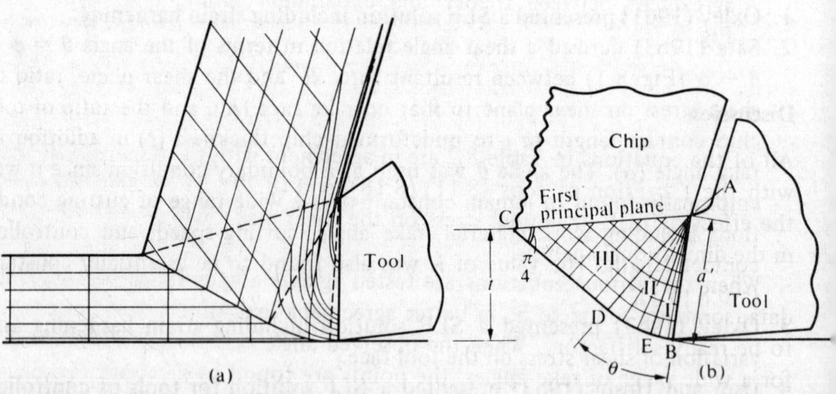

FIG. 8.6. Slip line field solution with controlled tool–chip contact length. (a) Slip line field solution for cut-away tool. (b) Flow lines corresponding to (a). (After Usui and Hoshi 1963.)

reduces to the Lee–Shaffer solution that pertains when cutting without chip curl and with a natural contact length.

5. Rowe and Spick (1967) presented a simple upper bound model of orthogonal cutting with two discontinuities (shear plane and tool face). Instead of using a coefficient of friction, friction on the tool face was expressed in terms of the relative shear stress (m) where

$$m = \frac{\text{mean shear stress on tool face}}{\text{uniaxial flow stress of material in shear}} \tag{8.28}$$

A relatively awkard trigonometric shear angle relationship was then obtained by applying the minimum energy principle.

6. De Chiffre and Wanheim (1981) have also used an upper bound approach. but instead of using a friction angle (β), they employed a reduced contact length $L = a'/t$ (where a' = tool chip contact length) and in place of ϕ, the main dependent variable is taken to be what is called the chip compression factor (λ = reciprocal of the cutting ratio = t_C/t). Applying the minimum energy principle and assuming rigid plastic work material and sticking friction on the tool face, the following relation is obtained

$$\lambda = \sqrt{(1 + L\cos\alpha)} \tag{8.29}$$

When turning tests were run on an aluminum alloy using tools of HSS with $\alpha = 12°$, $V = 50$ m min^{-1}, no fluid, and various reduced contact lengths (L), results shown in Fig. 8.7 were obtained which are in excellent agreement with eqn (8.29). In obtaining eqn (8.29), it was assumed that L does not depend on λ. While this is the case for reduced contact length tools, it is not the case for natural contact lengths and for such tools, the agreement between theory and experiment is not as good as that of Fig. 8.7.

Discussion

All of the equations in Table 8.2 are in agreement with the dimensional analysis with the exception of eqns (8.11) (8.18), and (8.27). These equations include the effect of BUE or normal stress on the shear plane, which were not included in the dimensional analysis.

When the various equations are tested against a wide range of experimental data, some are found to be in better agreement than others, but none is found to be really satisfactory. When the observed angle ϕ is plotted against $(-\beta + \alpha)$ for a wide range of rake angles, the points are found to lie along a straight line whose slope indicates that

$$\phi = \text{constant} - \beta + \alpha \tag{8.30}$$

However, the value of the 'constant' is not always the same, nor does it correspond

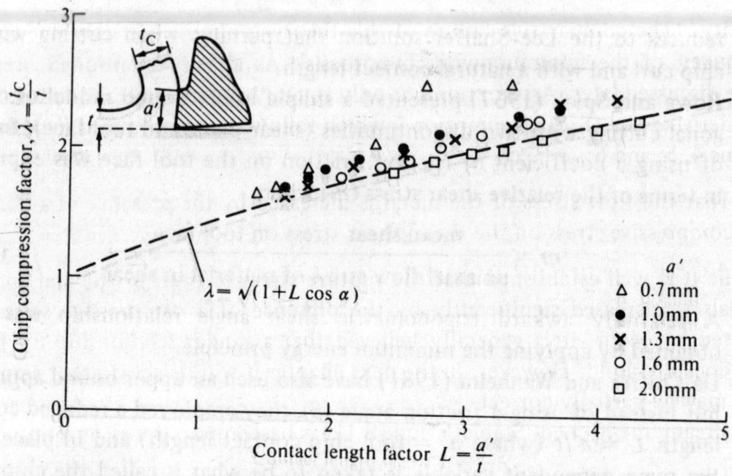

FIG. 8.7. Experimental data for restricted-contact tools. Work material, 515 WP aluminum alloy; tool material, HSS; ranke angle, 12°; cutting speed, 50 m min^{-1} (155 f.p.m.). (After De Chiffre and Wanheim 1981.)

to the values given by the equations of Table 8.2. Before proceeding it may be well to consider briefly the assumptions on which the several early derivations were based.

In Ernst and Merchant's analysis leading to eqn (8.4), the following assumptions were made:

1. That the shear stress is a maximum in the direction of the shear plane.
2. That the coefficient of friction is independent of the shear angle.
3. That the resultant force R is independent of the shear angle.

All three of these are questionable. Items 1 and 2 will be discussed in the next chapter; item 3 is obviously incorrect since an examination of cutting data reveals a strong dependence of R on ϕ.

In the second derivation of eqn (8.4), a different set of assumptions were made as follows:

4. That the angle ϕ is such that the total power is a minimum.
5. That the coefficient of friction is independent of the shear angle.
6. That the shear stress on the shear plane is independent of the shear angle.

There is no clear physical reason why item 4 should be true. Many examples may be found in nature in which nonconservative processes, such as that of metal cutting, clearly occur in such a manner that the energy consumed is not a minimum.

Speaking of the derivation of eqn (8.4), Hill stated in his book (1950) on plasticity: 'The comparative failure of this theory is almost certainly due to the inadequacy of the minimum-work hypothesis'. As already mentioned, item 5 will be discussed later. Assumption 6 is only roughly approximate.

In deriving eqn (8.11) assumption 7 which follows was introduced in addition to items 4, 5, and 6:

7. That the flow stress of the metal is increased by the presence of a normal compressive stress on the shear plane.

While it is well established that the rupture stress of both brittle and ductile materials is increased significantly by the presence of a compressive stress, as suggested by Mohr, it is generally believed that a similar relationship for flow stress does not hold. However, in the next chapter it will be shown that assumption 7 may be partially correct.

As already mentioned, Stabler's eqn (8.12) is based upon the assumption:

8. That the maximum shear stress and shear strain are colinear on the shear plane in three-dimensional cutting operations. Shear must occur on both the shear plane and tool face which are in general not at right angles to each other. This means that both cannot be planes of maximum shear stress. It is therefore to be expected that neither the shear plane nor the tool face will correspond to a direction of maximum shear stress.

In deriving eqn (8.15) it was assumed that:

9. The material cut is an ideal plastic.
10. The shear plane is along the direction of maximum shear stress.

The strain-hardening that occurs when metals are cut is usually important. However, despite this fact, assumption 9 does not lead to a poor approximation for the reason that it is applied only to a region in which the material is essentially rigid, i.e. A–B–C in Fig. 8.3. All of the strain-hardening that takes place occurs as the metal crosses line A–B, and this is before the ideal plastic assumption is applied to the material within area A–B–C.

Equation (8.18) differs from (8.15) only by the introduction of a built-up edge. In the treatment of the built-up edge the ideal plastic assumption must be considered as an approximation, since the material within zone A–B–D of Fig. 8.4 is actually highly deformed. However, if the built-up edge is small the degree of approximation associated with assumption 9 is probably slight since area A.B.D will then be small compared with A.B.C.

In deriving eqn (8.25) the following assumptions were made:

11. The stresses along the tool face and shear plane are uniformly distributed.
12. The shear plane is along the direction of maximum shear stress.
13. The normal stress on a plane perpendicular to the tool face is zero.

Item 11 will be discussed in the next chapter and item 12 is the same as 8.

Assumption 13 appears to be purely arbitrary. While we should expect the normal stress on face **b** of Fig. 8.5 to be low there is no reason why it has to be zero. If assumption 13 is adopted, there is no line corresponding to B–C in Fig. 8.3a beyond which the stress in the free chip is zero.

Equations (8.25) and (8.27) differ only by the introduction of the previously discussed concept of influence of normal stress on flow stress.

A general steady-state theory of metal cutting must incorporate or at least be consistent with the following phenomena, which are listed in approximate order of importance.

1. Chips are produced by a shear process.
2. There is a strong interaction between the shear deformation occurring in the shear zone and that occurring on the tool face.
3. Most materials strain-harden.
4. A built-up edge is usually present at low cutting speeds.
5. At high cutting speeds a secondary shear zone is present in the chip along the tool face.
6. Chips frequently curl and this in turn changes the chip–tool contact length.
7. The chip tool contact length may be controlled by tool design.
8. Dull tools or tools cutting with a large BUE will have a round cutting edge.
9. Forces exist on the relief face of a tool, especially when an appreciable wear land is present.

In addition to the steady-state problem there are several important problems associated with nonsteady state input conditions, such as variable undeformed chip thickness (t), speed (V), rake angle (α), depth of cut (b), or material properties.

All steady-state theories of chip formation that have been proposed to date have emphasized one or at most two of these items.

Most workers tend to believe that the problem is to identify the one or two key effects and that the resulting model will hold for all problems. Actually the situation is far more complex than this. All but the first of the nine items listed may be either of first-rank importance or completely insignificant depending upon cutting conditions. Practically all of the limited solutions that have been previously published represent good models over a limited range of operating conditions. The problem now appears to be to recognize the true breadth and complexity of the metal cutting problem.

Of the several attempts at predicting ϕ, the Lee–Shaffer one appears to represent the best starting point (least objectionable assumptions). When eqn (8.15) is compared with appropriate orthogonal cutting data involving a sharp tool and continuous chip by plotting ϕ vs. $(\beta-\alpha)$, the slope of the curve is in relatively good agreement but the intercept differs from 45°. This suggests an equation

$$\phi = 45° - \beta + \alpha + C_1 + C_2 + C_3 + \text{etc.}\ldots \qquad (8.31)$$

where C_1, C_2, C_3, etc. are correction terms as follows:

$C_1 =$ rake angel correction (+) for presence of BUE (increasing with decrease in cutting speed).
$C_2 =$ rake angle correction (−) for size of secondary shear zone (increasing with increase in cutting speed).
$C_3 =$ correction for controlled contact tool or chip curl (+).
$C_4 =$ correction factor for rounding of cutting edge due to wear (+).
$C_5 =$ correction factor for interaction between shear plane and tool face (−). This generally tends to make the shear plane deviate from a plane of maximum shear stress and, as Kobayashi and Thomsen (1962) have suggested, causes the total energy consumed to be greater than the minimum possible value assumed by Merchant in his analysis.
$C_6 =$ correction factor for strain hardening during chip formation.

Correction factors C_1 and C_2 result in the shear plane not being a plane of maximum shear stress. Item 9 (forces on the clearance face) should have no influence on the chip formation process since these forces arise after the chip is formed. Clearance face forces need be considered only in interpreting dynamometer data since most force-measuring systems will include these forces in what is measured.

Equation (8.31) is not useful for predicting the shear angle since it contains so many unknown quantities (seven, i.e. β, and C_1 to C_6). The only virtue of eqn (8.31) lies in its helping make clear why it is unlikely that a relatively simple general shear angle relationship is to be found.

Hill (1954) has come to a similar conclusion. He suggests that there are many possible steady state solutions due to the lack of geometrical constraint in the chip forming process. Hill then concludes that we must be content with a *range* of solutions that are permissible in that the material between the tool face and shear plane is not stressed beyond the flow stress of the material. It is suggested here that the problem is even more complex than this due to the large number of variables actually involved to a greater or lesser extent as discussed above. While it is disturbing to encounter a problem that cannot be characterized by a relatively simple model, this is not unique with the chip forming process. Another example of an intractable problem is that of predicting the buckling mode of a thin wall cylinder that is axially loaded (Fung and Sechler, 1960). This problem first considered by von Karman has been unsuccessfully studied by many of the giants of applied mechanics. Apparently, the difficulty in this case is that subtle differences in the boundary conditions at the points of load application play a major role in determining how the thin tube ends up buckling.

References

De Chiffre, L. and Wanheim, T. (1981). *Proc. 9th North American Manuf. Res. Conf.* Soc. Man. Eng. (Dearborn), p. 231.
Ernst, H. J. and Merchant, M. E. (1941). *Trans. Am. Soc. Metals* **29**, 299.
Fung, Y. C. and Sechler, E. E. (1960). *Proc. First Symp. on Struct. Mech.* Pergamon Press, New York, p. 115.
Hill, R. (1950). *The mathematical theory of plasticity*. Clarendon Press, Oxford. (Reissued 1983.)
Hill, R. (1954). *J. Mech. Phys. Solids* **3**, 47.
Hücks, H. (1951). Dissertation, Aachen T. H.
Kobayashi S. and Thomsen, E. G. (1962). *J. Engng Ind.* **84**, 63.
Lee, E. H. and Shaffer, B. W. (1951). *J. appl. Mech.* **73**, 405.
Merchant, M. E. (1945). *J. appl. Phys.* **16**, 267.
Oxley, P. L. B. (1961). *Int. J. Mech. Sci.* **3**, 68.
Oxley, P. L. B. (1963). *International research in production engineering,* p. 50. ASME, New York.
Piispanen, V. (1937). *Teknillinen Aikakauslehti* (Finland) **27**, 315.
Piispanen, V. (1948). *J. appl. Phys.* **19**, 876.
Rowe, G. W. and Spick, P. J. (1967). *J. Engng Ind.* **89**, 530.
Sata, T. (1963). In *International research in production engineering,* p. 18. ASME, New York.
Stabler, G. V. (1951). *Proc. Inst. Mech. Eng.* **165**, 14.
Usui, E. and Hoshi, K. (1963). In *International research in production engineering,* p. 61. ASME, New York.

9

SHEAR STRESS IN CUTTING

In this chapter the shear stress that occurs on the shear plane in metal cutting is considered and compared with results from other materials tests.

When shear-stress–shear-strain orthogonal cutting data obtained by use of eqns (3.9) and (3.16) are compared with conventional torsional tests data for the same material, substantial differences are observed (Fig. 9.1). The cutting data do not lie on a single curve (appreciable scatter) and stresses for a given strain are considerable higher than for torsion. Likewise, when cutting data are compared with extrapolated uniaxial tensile data on the basis of equivalent stress and strain, the correlation is relatively poor.

Plastic flow on the shear plane in orthogonal cutting with continuous chip formation may be characterized as follows:

1. Large values of strain rate and temperature.
2. Moderate normal stress on the shear plane giving rise to large values of homogeneous strain without gross fracture in the chip.
3. Progressive deformation with very small volume deformed at one time.
4. Close proximity of shear zone to rigid cutting tool which constitutes a flow constraint.

Some of these characteristics give rise to entirely different flow conditions than those met in conventional materials tests which results in the conventional tests being of limited value in predicting the flow stress to be expected in cutting. Each of the above characteristics will be discussed in order.

Strain Rate and Temperature

The influences of strain rate and temperature may be considered at the atomic level in terms of dislocations or at the macrolevel in terms of homogeneous material. Becker (1926) has suggested that the thermal energy of a material is available to assist the applied stress τ, in forming dislocations while Orowan (1934, 1935) has indicated that although the thermal energy is insufficient to close the gap between observed τ and calculated τ_0 values of flow stress, a stress concentration factor C may exist at the tip of a microcrack and thus augment thermal energy to bridge the gap between $C\tau$ and τ_0 (τ_0 = theoretical strength).

In Fig. 9.2 the stress-strain curve for a perfectly homogeneous material

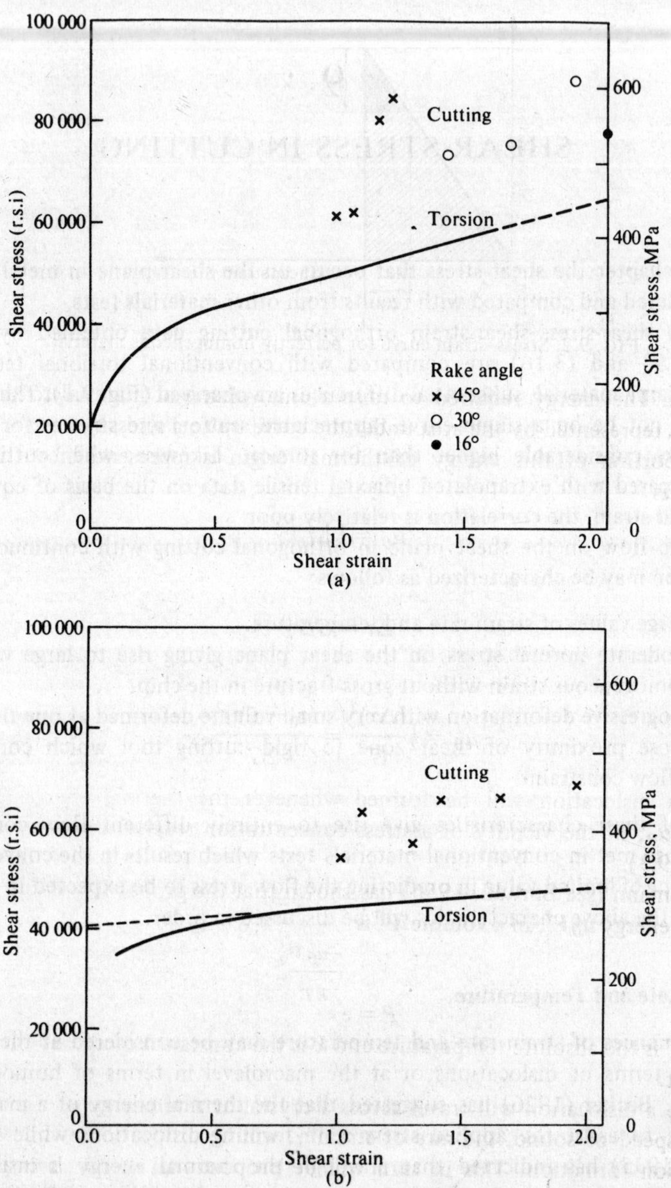

FIG. 9.1. Comparison of shear-stress–shear-strain data for cutting and torsion tests. (a) Material, AISI B1112 steel cut at low speed using several rake angles and fluids. (b) Material, leaded free machining steel cut at low speed using several rake angles and fluids. (After Shaw and Finnie 1955.)

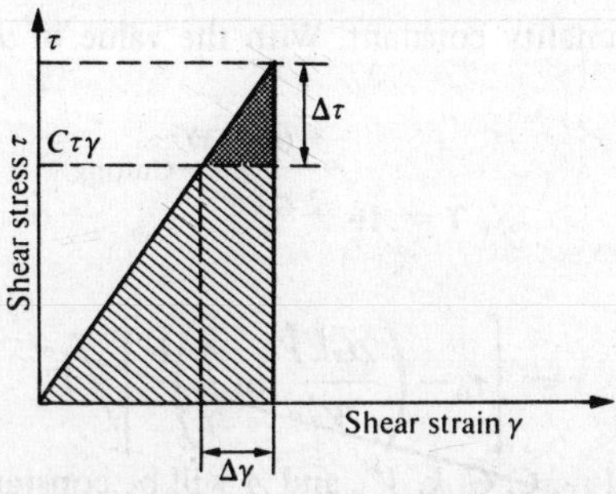

FIG. 9.2. Stress–strain curve for perfectly homogeneous material.

is shown. The energy required to raise a unit volume of material to the stress level τ_0 is represented by the area under the curve up to a stress τ_0.

The portion of this energy of thermal origin is represented by the black area and is

$$u_\theta = \tfrac{1}{2}\Delta\tau\Delta\gamma \tag{9.1}$$

but as

$$\Delta\tau = G\Delta\gamma \tag{9.2}$$

then

$$u_\theta = \tfrac{1}{2}\frac{(\Delta\tau)^2}{G} = \frac{(\tau_0 - C\tau)^2}{2G}. \tag{9.3}$$

Hence a dislocation will be formed whenever the thermal energy per unit volume u_θ, in the vicinity of a stress concentration reaches the value required by eqn (9.3).

Boltzmann (see Sproull, 1956) has shown that the probability P, of finding a thermal energy $u_\theta V'$, in a volume V' is

$$P = e^{\frac{-u_\theta V'}{kT}} \tag{9.4}$$

where T is the absolute temperature and k is the constant known as Boltzmann's constant.

Since a dislocation will travel across a crystal at a velocity approaching that of the speed of sound, the rate of strain $\dot{\gamma}$, will be proportional to the rate of dislocation formation. This in turn will be proportional to the probability of there being sufficient thermal energy to form a dislocation. Hence we may write

$$\dot{\gamma} = Ae^{\frac{-u_\theta V'}{kT}} \tag{9.5}$$

where A is a proportionality constant. With the value of u_θ given by eqn (9.3), eqn (9.5) becomes

$$\dot{\gamma} = A e^{\frac{-V'(\tau_0 - C\tau)^2}{2\,GkT}} \tag{9.6}$$

or solving for

$$\tau = \left[\tau_0 - \left(\frac{2GkT}{V'}\ \ln\frac{A}{\dot{\gamma}}\right)^{\frac{1}{2}}\right] \Big/ C. \tag{9.7}$$

For a given material τ_0, C, G, k, V', and A will be constants and we may write

$$\tau = C_1 - C_2\,[T(C_3 - \ln\dot{\gamma})]^{\frac{1}{2}} \tag{9.8}$$

where C_1, C_2, and C_3 are constants.

Table 9.1 gives some tensile data obtained by MacGregor (1944) for the ultimate stress of annealed AISI B1112 steel at different temperatures and strain rates.

With the values of Table 9.1, the constants in eqn (9.8) may be evaluated and thus

$$\tau = 85\,500 - 311\ [T(36.2 - \ln\dot{\gamma})]^{\frac{1}{2}} \tag{9.9}$$

It should be noted that the effect of temperature is greatly reduced at high strain rates as is the effect of strain rate at low temperatures.

TABLE 9.1.

Tensile data for annealed AISI B1112 steel

Absolute temperature (T), deg R	Strain rate, ($\dot{\gamma}$), s^{-1}	True ultimate stress (σ), p.s.i.	Equivalent[a] shear stress (τ), p.s.i.
530	0.0009	77 100	38 550
530	482	92 800[b]	46 400
160	0.0009	119 400	59 700

[a] Calculated according to maximum shear theory.
[b] Value corrected for difference in hardness of high- and low-speed test specimens.

At the macro level, MacGregor and Fisher (1946) combined the effects of temperature and strain rate in the tension-testing of metals by use of a velocity modified temperature

$$T_m = T\left(1 - k \ln \frac{\dot{\epsilon}}{\dot{\epsilon}_0}\right) \tag{9.10}$$

where T is the absolute test temperature, $\dot{\epsilon}$ is the constant true strain rate of the test and $\dot{\epsilon}_0$ and k are constants. While velocity modified temperature had been used previously in studies of secondary creep, MacGregor and Fisher found this concept quite satisfactory for correlation of tensile tests run at different temperatures and strain rates.

It is to be expected that since metal cutting normally involves *both* high strain rates and high temperatures on the shear plane, the two effects will tend to cancel as Drucker (1949) has suggested.

Plastic Behavior at Large Strain

There has been remarkably little work done in the region of large plastic strains. Bridgman (1952), using the hollow tubular notched specimen shown in Fig. 9.3, performed tests under combined axial compression and torsion. The specimen was loaded axially in compression as the center section was rotated relative to the ends. Strain was concentrated in the reduced sections and it was possible to crudely estimate and plot shear stress vs. shear strain with different amounts of compressive stress on the shear plane. From these tests, Bridgman concluded that the flow curve was the same for all values of compressive stress on the shear plane, a result consistent with other materials tests involving much lower plastic strains. However, the strain at gross fracture was found to be strongly influenced by compressive stress, increasing markedly with increase in compressive stress.

Lankford and Cohen (1969) were interested in the behavior of dislocations at very large plastic strains and whether there was a saturation relative to the strain hardening effect with strain or whether strain hardening continued to occur with strain to the point of fracture. Their experimental approach was an interesting and fortunate one. They performed wire drawing on iron specimens using a large number of progressively smaller dies with remarkably low

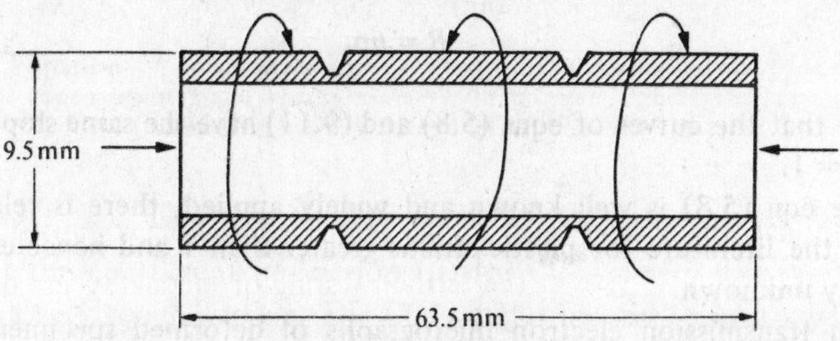

FIG. 9.3. Bridgman (1952) specimen for combined axial load and torsion.

semi die angle (1.5°) and a relatively low (~ 10%) reduction in area per die pass. After each die pass, a specimen was tested in uniaxial tension and a true stress–strain curve obtained. The drawing and tensile tests were performed at room temperature and low speeds to avoid heating and specimens were stored in liquid nitrogen between tests to avoid strain aging effects. All tensile results were then plotted in a single diagram, the strain used being that introduced in drawing (0.13 per die pass) plus the plastic strain in the tensile test. The result is shown in Fig. 9.4a. The general overlap of the tensile stress–strain curves gives an overall strain hardening envelope which indicates that the wire drawing and tensile deformations are approximately equivalent relative to strain hardening. Fig. 9.4b shows similar results on the same iron wire tested in uniaxial compression following drawing.

Fig. 9.4c shows somewhat similar results obtained earlier by Blazynski and Cole (1960) for AISI 1012 steel but carried to much lower values of total strain. Blazynski *et al.* were interested in strain hardening in tube drawing and sinking. Drawn tubes were sectioned as shown in Fig. 9.4d and tested in plane strain compression as shown in Fig. 9.4e. Like Fig. 9.4b, Fig. 9.4c shows the flow stress in compression plotted against the total strain. The curves with bent tops in Fig. 9.4c were obtained using graphite grease as a lubricant in the plane strain tests while the data points were obtained in similar tests using a more effective molybdenum disulfide lubricant. The smooth curve drawn through the molybdenum disulfide data points constitutes the flow curve essentially in the absence of friction.

Up to a strain of about 1 (Fig. 9.4a, b and c) the usual strain-hardening curve was obtained that is in good agreement with the generally accepted equation

$$\sigma = \sigma_1 \epsilon^n. \tag{5.8}$$

However, beyond a strain of 1, the strengthening curve was linear corresponding to the equation

$$\sigma = A + B\epsilon \quad (\epsilon > 1) \tag{9.11}$$

where A and B are constants. It may be shown that

$$A = (1 - n)\,\sigma_1 \tag{9.12}$$

$$B = n\sigma_1 \tag{9.13}$$

in order that the curves of eqns (5.8) and (9.11) have the same slope and ordinate at $\epsilon = 1$.

While eqn (5.8) is well known and widely applied, there is relatively little data in the literature for plastic strains greater than 1 and hence eqn (9.11) is relatively unknown.

From transmission electron micrographs of deformed specimens Lankford *et al.* found that cell walls representing concentrations of dislocations began to form at strains below 0.2 and became ribbon shaped with decreasing mean linear intercept cell size as the strain progressed. Dynamic recovery and cell wall migration resulted in only about 7 percent of the original cells remaining after a strain of 6. The flow stress of the cold-worked wires was found to vary linearly with the reciprocal of the mean transverse cell size.

Data that makes one question the wisdom of extrapolating ordinary materials test data into the larger strain regime of metal cutting is given in Fig. 9.5. These are data for the same material cut under the same conditions except for rake angle (α). The shear stress on the shear plane is obviously not constant but appears to decrease with increase in shear strain (negative strain hardening).

Toward the end of the sixties, it was decided to conduct an acoustical emission study of concentrated shear at Carnegie Mellon University. The initial acoustical studies were on specimens of the Bridgman type but fortunately, lower levels of axial compressive stress than Bridgman had used were employed in order to more closely simulate the concentrated shear process of metal cutting. The apparatus used which was capable of measuring stresses and strains as well as acoustical signals arising from plastic flow is described in the dissertation of T. J. Walker (1967). Two important results were obtained:

1. A region of rather intense acoustical activity occurred at the yield point followed by a quieter regime until a shear strain of about 1.5 was reached. At this point, there was a rather abrupt increase in acoustical activity that continued to the strain at fracture which was appreciably greater than 1.5.
2. The shear stress appeared to reach a maximum at the shear strain corresponding to the beginning of the second burst of acoustical activity ($\gamma \cong 1.5$).

The presence of the notches in the Bridgman specimen (Fig. 9.3) made interpretation of stress-strain results somewhat uncertain. Therefore, a new specimen was designed (Fig. 9.6) which substitutes simple shear for torsion with normal stress on the shear plane. By empirically adjusting distance Δx (Fig. 9.6) to a value of 0.25 mm, it was possible to confine all the plastic shear strain to the reduced area, thus making it possible to determine readily the shear strain ($\gamma = \Delta y/\Delta x$). When the width of minimum section was greater or less than 0.25 mm, the extent of the plastic strain observed in a transverse micrograph at the minimum section did not extend completely across the 0.25 mm dimension or beyond this width.

A representative set of curves is shown in Fig. 9.7 for resulfurized low carbon steel. Similar results were obtained for nonresulfurized steels and other ductile metals. There is little difference in the curves for different values of normal stress on the shear plane (σ) to a shear strain of about 1.5. This is in agreement with Bridgman. However, beyond this strain, the curves differ substantially with compressive stress on the shear plane (σ). At large strains τ was found

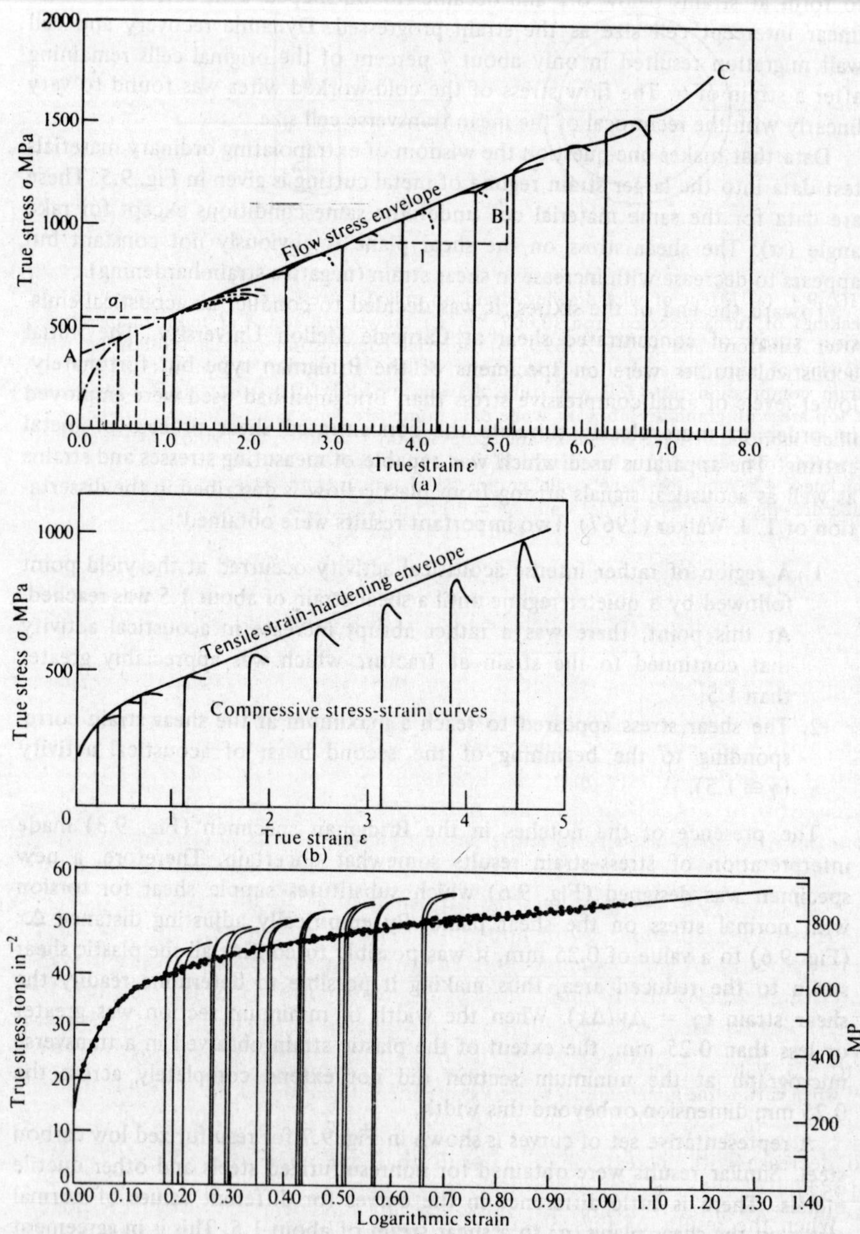
True stress σ MPa
True strain ε
Flow stress envelope
σ1
A
B
C
(a)
True stress σ MPa
True strain ε
Tensile strain-hardening envelope
Compressive stress-strain curves
(b)
True stress tons in^{-2}
MP
Logarithmic strain
(c)

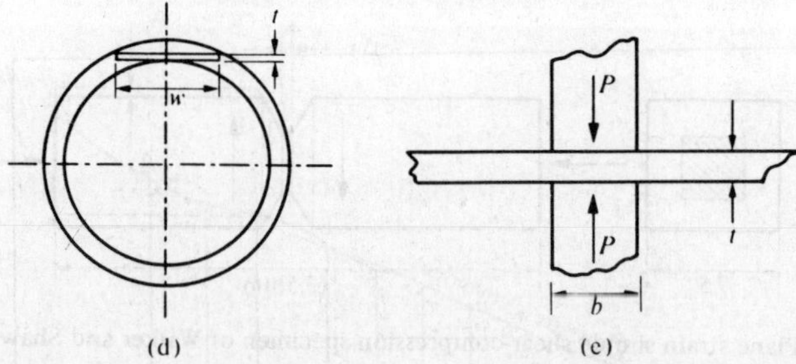

FIG. 9.4. (a) Effect of wire drawing strain on uniaxial stress–strain curves (corrected for necking) of furnace-cooled iron (0.007 w/o C) drawn and tested at room temperature. (After Lankford and Cohen 1969.) (b) Uniaxial compression test data following wire drawing of same materials (tested in tension following drawing) shown in (a). (c) Plane strain compression tests following tube drawing of AISI 1012 steel. Curves with hooks at top are with graphite lubricant while data points are for a more effective molybdenum sulfide lubricant. (After Blazynski and Cole 1960.) (d) How plane strain specimen was removed from drawn tube by Blazynski *et al.*; tube diameter approximately 55 mm, wall thickness 4.8 mm. (e) Plane strain compression test used by Blazynski *et al.* following tube drawing. $t = 0.045$ in (1.14 mm); $b = 0.115$ in (2.92 mm), w (perpendicular to paper) $= 0.690$ in (17.53 mm).

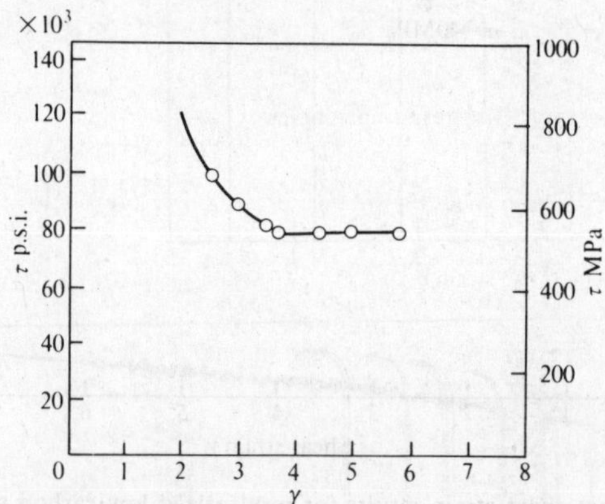

FIG. 9.5. Values of shear stress on the shear plane (τ) versus shear strain in the chip (γ) when cutting the same material with tools of different rake angle. (After Shaw, 1954.)

to decrease with increase in γ, a result that does not agree with Bridgman (1952).

When the results of Figs. 9.4a and 9.7 are compared, they are seen to be very different. In the case of Fig. 9.4a, strain hardening is positive to normal strains as high as 7. In the case of Fig. 9.7, strain hardening appears to become

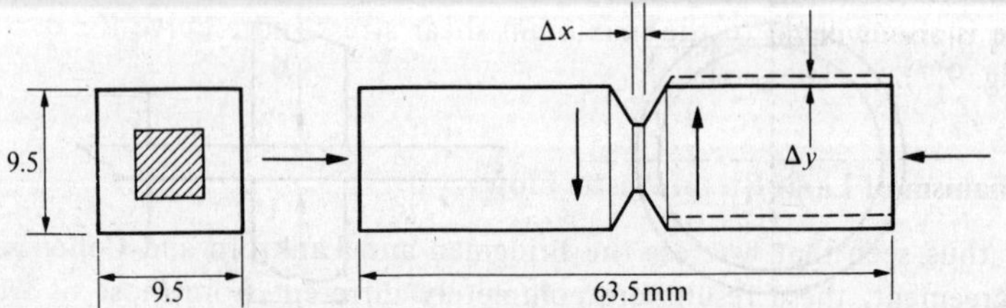

FIG. 9.6. Plane strain simple shear-compression specimen of Walker and Shaw (1969).

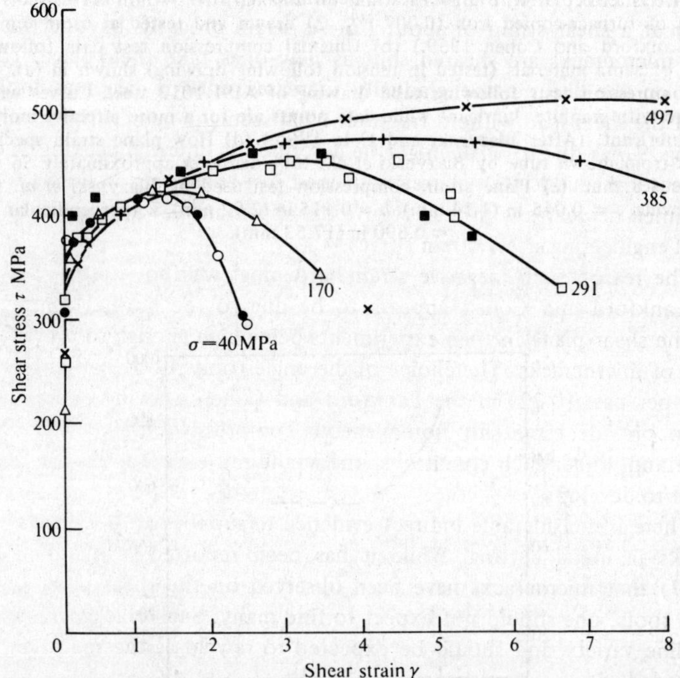

FIG. 9.7. Shear stress–shear strain results for resulfurized low carbon steel for specimen of Fig. 9.6. σ = normal stress on shear plane. (After Walker and Shaw 1969.)

negative above a particular shear strain that increases with normal stress on the shear plane.

From Fig. 9.7, it is seen that for a low value of normal stress on the shear plane of 40 MPa, strain hardening appears to go negative at a shear strain (γ) of about 1.5; that is, when the normal stress on the shear plane is about 10 percent of the maximum shear stress reached, negative strain hardening sets in at a shear strain of about 1.5. On the other hand, strain hardening remains positive to a normal strain of about 8 when the normal stress on the shear plane is about equal to the maximum shear stress (note curve for $\sigma = 497$ MPa in Fig. 9.7).

Mechanism of Large Strain Plastic Flow

It is thus seen that whereas the Bridgman and Lankford and Cohen results are in agreement, these results are completely different from those of Walker and Shaw. A proposed mechanism of large strain plastic flow (Shaw 1980) has been developed in which it is suggested that at moderate values of normal stress on the shear plane, microcracks begin to appear in a plane of concentrated shear at a shear strain of about 1.5. As strain proceeds beyond this point, the first microcracks are sheared shut as new ones take their place. The unsound area on the shear plane gradually increases until it becomes insufficient to resist the shear load without fracture. What seems to be negative strain-hardening in Fig. 9.7 is due to what might be described as 'internal necking' (i.e. a gradual decrease in sound internal area with load just as the area in the neck of a tensile specimen decreases with load to give the appearance of negative strain-hardening in an engineering stress–strain curve).

The reason such 'negative strain-hardening' was not observed by Bridgman or Lankford and Cohen appears to be due to the normal compressive stress on the shear plane in their experiments being high enough to prevent the formation of microcracks. The choice of die angle (only 1.50° half-angle) and reduction per pass (0.22) in the Lankford and Cohen drawing experiments is such as to provide essentially homogeneous compressive strain in the deformation zone and, under such conditions, one would not expect microcracks or a central burst to develop.

There is considerable indirect evidence to support of the formation of microcracks in metal cutting. While it has been reported (Komanduri and Brown 1967) that microcracks have been observed on the shear plane of quick-stop chip roots, one should not expect to find many. Such cracks will be of the thin, hairline variety and should be expected to reweld as the specimen is suddenly unloaded.

The new theory of plastic flow discussed here is an add-on to dislocation theory. As long as microcracks do not occur in appreciable number, a material may be deformed to very large strains with a continuous increase in dislocation density and strain hardening. This is consistent with the experimental results of Bridgman and Lankford and Cohen. However, at a particular value of shear strain, depending upon the ductility of the material and the normal stress on the shear plane, microcracks will begin to form. If the normal stress on the shear plane is tensile, these cracks will spread rapidly over the shear surface leading to gross fracture. If, however, a moderate compressive stress ($\sim \frac{1}{2}$ the shear flow stress for a ductile metal) is present on the shear plane, the new mechanism will pertain. When this occurs there will be an extended stress-strain region exhibiting a decrease in flow stress with strain as the ratio of real to apparent area on the shear plane (A_R/A) decreases from one to the critical value at which gross fracture occurs.

It is suggested that in metal cutting the shear stress on the shear plane is not in general independent of normal stress on the shear plane. That part of the shear plane that involves microcracks should show an increase in shear stress with normal stress as Piispanen and Merchant suggested. However, the part that does not involve microcracks should have a shear stress that is independent of normal stress in keeping with the experimental results of Bridgman and Lankford and Cohen.

It should not be inferred that the foregoing discussion has brought us any nearer to a solution of the shear angle problem. It should, however, serve to further explain why it is unlikely that a simple solution to this persistent problem is apt to be found.

Inhomogeneous Strain

When metallic single crystals are plastically deformed, it is found that slip does not occur uniformly on every atomic plane but that the active slip planes are relatively far apart (Fig. 9.8). It is further found that polycrystalline metals also strain blockwise rather than uniformly. Crystal imperfections are responsible for this inhomogeneous behavior.

When the volume of material deformed at one time is relatively large, there will be a uniform density of imperfections and for all practical purposes, strain (and strain hardening) may be considered to be uniform. However, as the volume deformed approaches the small volume associated with an imperfection, the

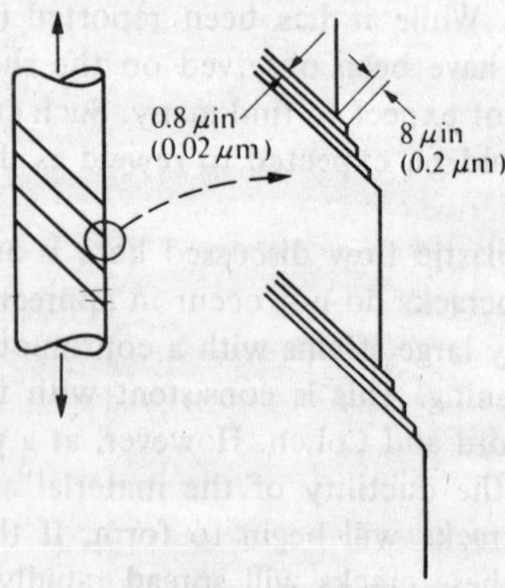

FIG. 9.8. The spacing of adjacent slip planes. Pure aluminum single crystal. (After Heidenreich and Shockley 1948.)

material will show obvious signs of the basic inhomogeneous character of strain. The mean flow stress will rise and the ends of the active shear planes will be evident in a free surface. This is called the size effect.

In metal cutting, the undeformed chip thickness (t) is small, the width of shear zone (Δy) is very small but the width of cut (b) is relatively large. It would thus appear that the volume deformed at one time would be ($bt\Delta y/\sin\phi$). However, when the back of a very continuous chip is observed under the microscope, it is found (Fig. 9.9) that the edges of the slip bands that are observed are not continuous across the width of the chip but have an extent characteristic of the imperfection spacing (a). Thus, the volume deformed at any one time should be taken to be $[a(t)\Delta y/\sin\phi]$ where $a \ll b$. In metal cutting this volume will aproach a^3 (mean imperfection volume) and there will be a size effect. This is the main reason specific energy (u) increases with decrease in undeformed chip thickness (t) as suggested by eqn (3.36).

Under ordinary conditions, the shear planes will be very closely spaced corresponding to the closeness of spacing of the weak spots in the metal. It may be assumed that slip planes are so spaced that a single weak spot is present on each plane. Drucker (1949) has employed a random array of weak points to qualitatively demonstrate the increase in unit cutting energy with decrease in depth of cut. However, inasmuch as the spacing of weak points is very small

FIG. 9.9. Back (free surface) of chip showing regions of concentrated strain.

compared with usual depths of cut, an orderly array of weak spots seems justified. The dots shown in Fig. 9.10 represent such an orderly arrangement of weak points to an exaggerated scale. These points have a uniform spacing of a units in each direction.

Let P_1 and P_2 be two shear planes making an angle ϕ with the direction of cut and passing through adjacent points in the first row below the surface. If the depth of cut is t, then t/a planes may be placed between those at P_1 and P_2 such that a single plane passes through each weak spot in the layer in the process of being cut. The number of planes per unit distance in a direction perpendicular to the shear plane will be

$$n = \frac{t}{a(a \sin \phi)} \tag{9.14}$$

or the spacing of successive planes is

$$\Delta y = \frac{a^2 \sin \phi}{t} \tag{9.15}$$

The total slip on a given shear plane will be

$$x = \frac{a^2 \sin \phi}{t} \gamma \tag{9.16}$$

where γ is the unit uniform strain.

Assuming a normal stress on the shear plane that is sufficient to suppress microcrack formation as in the Lankford and Cohen experiments, the flow stress should vary with strain as given by eqn (9.11). In veiw of the origin of strain hardening according to dislocation theory and the Lankford and Cohen

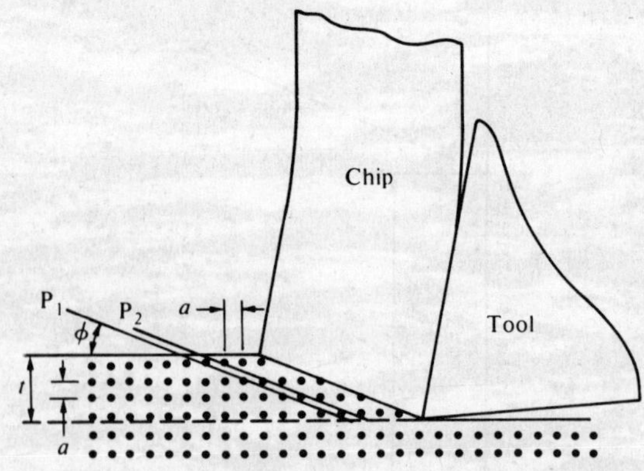

FIG. 9.10. Specimen with uniform distribution of points of stress concentration (weak points). (After Shaw 1950.)

strain hardening results for large strains, it appears reasonable to assume that strain hardening will increase linearly with the extent of the slip that occurs on a given shear plane (x given by eqn 9.16). Fig. 9.11 shows the assumed variation of shear stress (τ') with x where B' is the linear rate of strain hardening relative to x. The mean shear stress on the shear plane during slip may thus be expressed as follows (substituting for γ from eqn 3.16).

$$\tau' = A' + B'x = A' + \frac{B' a^2 \sin\phi}{2t} [\cot\phi + \tan(\phi - \alpha)] \tag{9.17}$$

This equation suggests that the shear stress on the shear plane is not constant as for an ideal plastic but depends upon the rate of strain hardening, the strain in the chip, the density of imperfections (as measured by a), and the undeformed chip thickness relative to the imperfection spacing (t/a).

The only convenient method of evaluating the constants in eqn (9.17) (A' and B') is from metal cutting. To illustrate the application of eqn (9.17), the orthogonal data of Table 3.1 may be considered. When the shear stress on the shear plane (τ) is plotted against the quantity $\gamma \sin\phi/2t$ as in Fig. 9.12 and the best straight line drawn through the points, values for A' and a^2B' may be readily determined:

$$A' = 87\,000 \text{ p.s.i. } (601 \text{ MN m}^{-2})$$

$$a^2B' = 34.55 \text{ lb in}^{-1} (60.59 \text{ N cm}^{-1})$$

The maximum difference in shear stress on the shear plane due to the difference in 'specimen size' and its influence on strain hardening is 14 000 p.s.i. (96.5 MPa) or 15 percent of the minimum observed value in this particular case.

When the orthogonal cutting data of Table 3.2 are similarly plotted Fig. 9.13

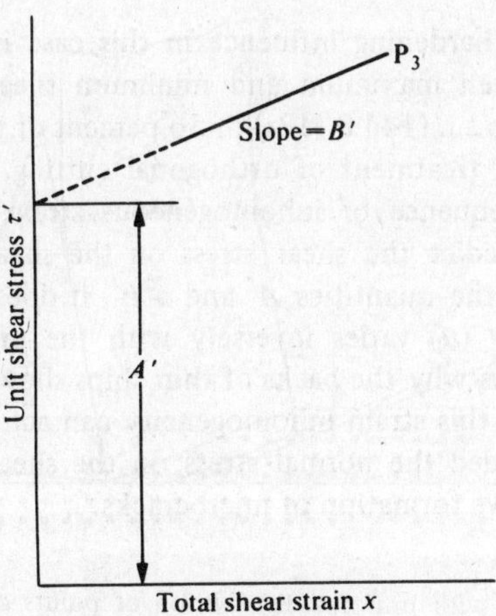

FIG. 9.11. Assumed variation of unit shear stress (τ') vs. total shear strain (x) diagram.

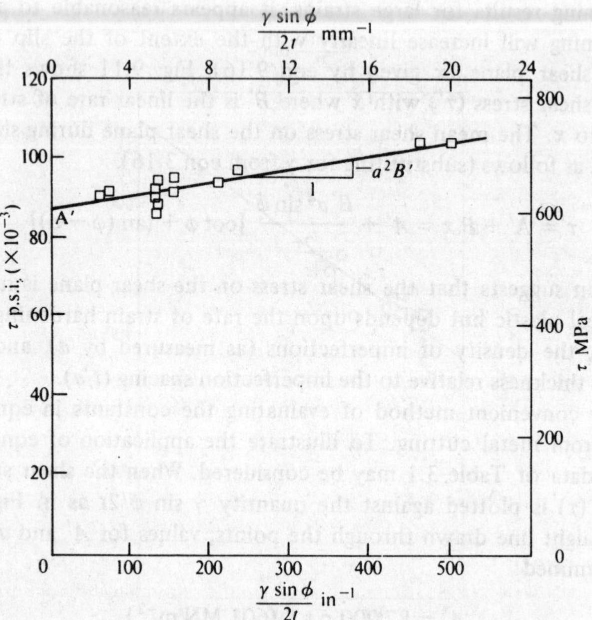

FIG. 9.12. Variation of shear stress on shear plane τ with $(\gamma \sin \phi)/2t$ for data of Table 3.1.

is obtained. For this case

$$A' = 52\,000 \text{ p.s.i. } (359 \text{ MN m}^{-2})$$

$$a^2B' = 153 \text{ lb in}^{-1} \ (268 \text{ N cm}^{-1})$$

The size effect strain hardening influence in this case is larger corresponding to a difference between maximum and minimum shear stress on the shear plane of about 21 000 p.s.i. (144.8 MPa) or 36 percent of the minimum value.

While the foregoing treatment of orthogonal cutting, which involves a size effect that is a consequence of inhomogeneous strain and strain hardening cannot be used to predict the shear stress on the shear plane since there is no way of estimating the quantities A' and a^2B', it does serve to explain why specific cutting energy (u) varies inversely with the undeformed chip thickness (t). It also explains why the backs of thin chips show evidence of inhomogeneous strain and how this strain inhomogeneity can act to intensify the strain hardening effect provided the normal stress on the shear plane is sufficiently high to prevent extensive formation of microcracks.

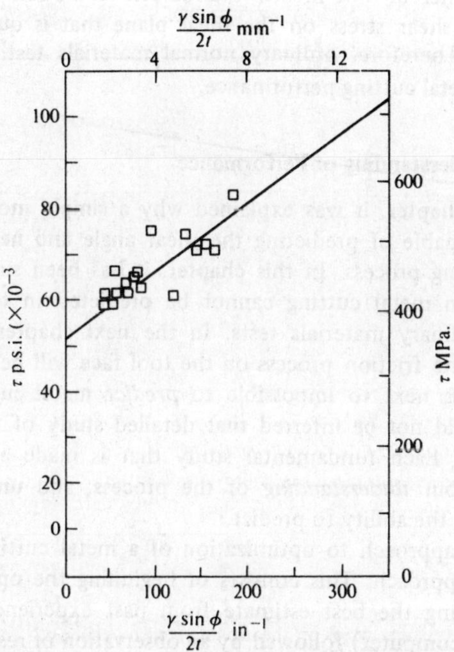

FIG. 9.13. Variation of shear stress on shear plane τ with $(\gamma \sin \phi)/2t$ for data of Table 3.2.

Interaction Between Shear Plane and Tool Face

In most materials tests things get very complicated when friction on a die face in close proximity to the plastic flow zone plays a significant role. An example is the barreling that occurs in the uniaxial compression test due to friction on the die faces. While this complication may be handled by an extension of the Bridgman correction to situations with negative radius of curvature at the free surface, the best solution is to reduce the friction to zero or to periodically remove the barrel shape by machining.

In metal cutting, action on the shear plane and on the tool face are closely associated and connected by a single stress field that must be compatible with both processes. Flow will not occur at maximum shear stress on the shear plane unless it can simultaneously occur on a plane of maximum shear stress on the tool face. This is possible only when the shear angle (ϕ) equals the rake angle (α) which is not normally the case.

In the foregoing discussion, it has been suggested that normal stress on the shear plane can play an important role in influencing the resulting flow stress on the shear plane in cutting due to the formation of microcracks. Ordinary materials tests either do not involve microcrack formation or have a ratio of normal stress to shear stress on the shear plane that is quite different from that in cutting. Therefore, ordinary normal materials test results cannot be used to predict metal cutting performance.

Prediction vs. Understanding of Performance

In the previous chapter, it was explained why a simple model of the cutting process is not capable of predicting the shear angle and hence the strain pertaining in a cutting process. In this chapter, it has been similarly argued that the flow stress in metal cutting cannot be predicted in terms of properties derived from ordinary materials tests. In the next chapter, the complexities associated with the friction process on the tool face will be discussed. All this suggests that it is next to impossible to *predict* metal cutting performance. However, it should not be inferred that detailed study of the cutting process is without value. Each fundamental study that is made and properly interpreted adds to our *understanding* of the process, and understanding is the next best thing to the ability to predict.

The practical approach to optimization of a metal cutting process involves an incremental approach. This consists of beginning the operation under conditions representing the best estimate from past experience (a handbook or data stored in a computer) followed by an observation of resulting performance relative to the optimization variable of interest. A small change in operating variables is then made and the effect on the optimization variable noted. Optimum performance may be gradually approached as useful parts are produced. The role of understanding of course lies in helping decide which variables to alter and by how much so that optimum performance may be reached in the shortest time and recognized when achieved.

References

Becker, H. (1926). *Zeit. Tech. Phys.* 7, 547.

Bridgman, P. W. (1952). *Studies in large plastic flow and fracture.* McGraw-Hill, New York.

Blazynski, T. Z. and Cole, I. M. (1960). *Proc. Instn mech. Engrs* **174**, 797.

Drucker, D. C. (1949). *J. appl. phys.* **20**, 1013.

Heidenreich, R. O. and Shockley, W. (1948). *Report on strength of solids.* Physics Society, London, p. 57.

Komanduri, R. and Brown, R. H. (1967). *Metals Mater.* **95**, 308.

Lankford, G. and Cohen, M. (1969). *Trans. Am. Soc. Metals* **62**, 623.

MacGregor, C. W. (1944). *J. Franklin Inst.* **238**, 111, 159.

MacGregor, C. W. and Fisher, J. C. (1946). *J. appl. Mech.* **68**, A-11.

Orowan, E. (1934). *Zeit. Phys.* **89**, 605, 614, 634.

Orowan, E. (1935). *Zeit. Phys.* **98**, 382.

Shaw, M. C. (1954). *Metal cutting principles* (3rd edn). MIT Press, Cambridge.

Shaw, M. C. (1950). *J. appl. Phys.* **21**, 599.
Shaw, M. C. (1980). *Int. J. mech. Sci.* **22**, 673.
Shaw, M. C. and Finnie, I. (1954). *Trans. Am. Soc. mech. Engrs* **77**, 115.
Sproull, R. L. (1956). *Modern physics.* John Wiley and Sons, New York.
Walker, T. J. (1967). Dissertation, Carnegie Mellon University, Pittsburgh, PA, USA.
Walker, T. J. and Shaw, M. C. (1969). *Proc. 10th Int. Mach. Tool Des. and Res. Conf.* Pergamon Press, Oxford, p. 291.

10
FRICTION

Friction which is the resisting force one surface experiences when it slides over another is always oppositely directed to the relative velocity vector for the two surfaces. The entire field of friction may be divided into two general regimes—lightly loaded sliders and heavily loaded sliders. Each of these will be briefly discussed followed by a consideration of the special friction conditions that pertain in metal cutting. What constitutes a lightly loaded slider will be defined in a later section. Tribology is the term used to cover the performance of surfaces relative to lubrication, friction, and wear. The complete tribological characteristics of lightly loaded sliding surfaces is treated in detail in the two-volume classical work by Bowden and Tabor (1950, 1964).

Lightly Loaded Sliders

Types of Sliding Contact

There are five types of sliding contact:

1. Fluid film lubrication—this is the ideal type of lubrication in which there is no metal-to-metal contact. The load is supported by a pressurized fluid film. Viscosity is the physical property of the system of major interest, and the principles of fluid mechanics enable the load capacity and friction characteristics of surfaces that are so lubricated to be computed. Well-designed journal and slider bearings are of this type.
2. Elastohydrodynamic lubrication—this is the type of lubrication in which the localized stresses in the bearing surfaces are sufficiently high to cause a significant change in surface geometry and the fluid pressures are sufficiently high to alter the effective viscosity of the fluid. The lubrication of gear teeth, cam surfaces, and rolling contact bearings is of this type. This is a special version of hydrodynamic lubrication.
3. Boundary lubrication—here the film is not complete and the surfaces come together close enough so that the resulting frictional resistance is not entirely due to viscous drag. The surface finish and the physical and chemical properties of the surfaces and lubricant are of major interest. Effective boundary lubricants are usually additives consisting of long-chain polar molecules that are physically adsorbed on the high points of mating surfaces and tend to prevent metal-to-metal contact.

4. Extreme boundary lubrication—this is the type of lubrication that obtains under the most severe sliding conditions (highly loaded bearing surfaces operating at relatively low speed). The additive in this case reacts chemically with one of the bearing metals to form a low-shear-strength solid layer that prevents metal-to-metal contact, welding, and metal transfer (wear). The solid films that are formed are usually inorganic, have a high melting point, and hence are suitable for use at elevated temperatures without melting. Hypoid gear lubricants and heavy-duty type cutting fluids are examples of extreme boundary lubricants.
5. Clean metal surfaces—this represents the extreme in adverse conditions of sliding contact. Hardness, shear strength, and surface finish are the main characteristics of interest in this type of action.

Real Area of Contact

All finished surfaces are found to have irregularities that are very large in comparison with atomic dimensions (i.e. large compared with 10^{-8} inch or 3×10^{-10} m).

As a consequence of these micro irregularities, the real area of contact A_R is much less than the apparent area of contact (A). For example, if two carefully ground blocks, each of one inch square cross-sectional area, are placed in contact and loaded with 10 lb, the real area of contact will be less than 10^{-4} square inches (< 0.1 mm^2), while of course the apparent area (A) is actually one square inch (645 mm^2). When such surfaces first make contact A_R is zero. As load is applied the relatively few high points that make contact are plastically deformed and A_R increases. Thus, the real area of contact will be independent of the apparent area of contact or the surface finish but will be determined solely by the applied load and the flow stress or hardness of the protuberances. The area A_R developed must be just sufficient to equal the applied load P divided by the hardness H of the metal asperities, or

$$P = A_R H \tag{10.1}$$

The characteristic roughness found on most surfaces consists of irregularities with a pitch that is at least an order of magnitude greater than the peak-to-valley distance.

Freshly generated surfaces will be clean but will quickly oxidize or be covered by an adsorbed layer of water vapor or other material depending on the environment and properties of the metal. A freshly generated surface will have a density of electrons and an atom spacing corresponding to the metal in bulk. However, electrons will leave the surface (these are called exoelectrons) as the atom spacing increases to the equilibrium value. The energy associated with the formation of a new surface after equilibrium has been established is called surface energy (T), $[FL/L^2]$. Freshly generated surfaces are extremely active chemically before equilibrium is established.

Beilby (1921) showed that the upper layer of a polished surface is different from the underlying material. His microscopic observations led him to believe that, in polishing, surface metal is caused to move from point to point as though it were molten. The net effect was to smooth the surface by transferring metal from peaks to valleys. Beilby further observed that the resulting surface layer had a glasslike or amorphous structure. The existence of such an amorphous Beilby layer on polished metals has been disputed for many years, electron diffraction data being used successively to prove and disprove the existence of such a layer. More recently, a high-energy laser beam has been used to traverse surfaces following a rastor pattern to produce a surface texture that resembles the Beilby layer. In this case (laser glazing), a very thin layer of metal is melted and then solidifies so rapidly by heat flowing into the cold subsurface that either an amorphous surface is produced or one that has a crystal size so small as to be unresolvable.

Bowden (1945) has presented the most logical arguments in favor of the existence of the Beilby layer. He points out that, in the polishing process, the polishing agent is generally embedded in a relatively soft lap, which is rubbed against the specimen in the presence of some vehicle such as water. Hot spots will develop at the points of contact between the abrasive and the specimen, and local surface temperatures up to the melting point of the metal may be encountered if the polishing speed is sufficiently high. The molten or softened metal will be smeared over the surface and will quickly solidify to form the layer which Beilby observed. Probably the molten surface layer solidifies so rapidly that the crystals produced are extremely small and the material appears amorphous by present methods of study. Bowden states that this hypothesis may be tested by the following procedure: if polishing is due primarily to mechanical abrasion, then the relative hardness of specimen and polisher is the property of importance; but if the action is caused by surface melting, then the relative melting points are of major concern. In a series of simple experiments, it was observed that camphor, which melts at 350 °F (177 °C), readily polishes metals that melt at a lower temperature, such as Woods metal, but has no effect on tin, lead, white metal, or zinc, all of which melt at higher temperatures. That camphor is softer than Woods metal indicates that relative hardness is unimportant. Similarly, zinc oxide, which is comparatively soft, readily polishes quartz, which is harder, because the melting point of zinc oxide is higher than that of quartz. Many other cases support the view that the relative melting point is the important property, rather than hardness, when choosing a polishing agent.

Origins of Friction

The friction of a lightly loaded slider is due to one or a combination of the following causes (ignoring viscous effects);

1. Adhesion (fracture of microwelds formed at the tips of contacting asperities.

2. Plowing or burnishing (displacement of metal without generation of wear particles).
3. Abrasion (microcutting action producing fine chips i.e. wear particles).
4. Transverse displacement (movement of one surface transversely to the other against the applied load due to interference of hard strong asperities).

Rules of Dry Sliding

When two dry clean surfaces slide together under relatively light load, the following rules pertain over a fairly wide range:

1. The coefficient of friction is independent of the applied load.
2. The coefficient of friction is independent of the sliding speed.
3. The coefficient of friction is independent of surface finish.
4. The coefficient of friction is independent of the apparent area of contact (A).
5. The coefficient of friction is independent of the temperature of the sliding surfaces.

The first of these rules is known as Amontons' law (Amontons, 1699). All of these rules fail to hold at extreme values. For example, the coefficient of friction at zero speed (static friction) is normally greater than the dynamic value. Also, the coefficient of friction will normally decrease with speed above that required to cause thermal softening of the less refractory surface. However, for speeds between these two extremes, the coefficient of friction will be approximately independent of speed.

A lightly loaded slider is one for which Amontons' law represents a good approximation (μ independent of P).

Adhesive Friction

For two surfaces in adhesive sliding contact, the friction force (F) will be

$$F = \alpha\tau A_R \tag{10.2}$$

where τ is the shear strength of the weaker of the two sliding materials and α is the fraction of the real area that involves clean adhering asperities. For an unlubricated case, α will approach unity.

For dry surfaces sliding in air, the coefficient of friction will be from eqns (10.1) and (10.2)

$$\mu = \frac{F}{P} = \frac{\tau A_R}{HA_R} = \frac{\tau}{H} \tag{10.3}$$

For low friction, one of the surfaces should have low shear strength and high hardness. Since this is a combination not found in nature, most bearing metals are designed to consist of a composite in which a low-shear-strength material is smeared over a harder substrate. The copper–lead bearing (small particles of insoluble lead in copper) is one of many examples of bearing materials based on this principle.

When two metal surfaces are forced together the high points that make contact will weld by establishing metallic bonds, provided the surfaces are clean. The surfaces need not be heated, for a *pressure weld* may be established at room temperature if sufficient pressure is applied. However, when such a weld is produced between materials of high stiffness (i.e. materials of large Young's modulus, E) the junctions at the tops of the asperities frequently rupture when the load is released. Elastic recovery in such cases causes the junctions to rupture one by one as the load is removed. The larger the A_R/A ratio and the smaller the stiffness of the metals united, the smaller will be the tendency for a pressure weld to rupture when load is released.

The blacksmith normally heats two pieces of steel to be united in order to decrease E and increase A_R/A for the limited force that can be applied. Similarly, it is possible to produce a pressure weld between clean indium and silver surfaces at room temperature with very light pressure applied, due to the softness and low E of indium. However, such a weld cannot be formed between steel and silver due to the relatively low A_R/A ratio and high values of E involved.

The importance of a large value of A_R/A is further illustrated by the relative ease with which a fine, freshly drawn glass filament may be made to adhere to a similar filament, compared with larger more rigid glass rods. The greater conformability of thin specimens is also demonstrated by the relative ease of welding freshly cut sheets of mica of thick and thin sections. A large real area of contact may be obtained with little applied pressure by previously depositing films in the liquid state on the surfaces to be joined. When the adhesive is transformed into a solid film by cooling, by evaporation of a solvent, or by polymerization or other chemical change a strong joint may be obtained.

In addition to the magnitude of the load, the cleanness of the surfaces is also important in pressure welding. If an oxide or other film is present it may be broken through only with difficulty when a purely normal load is applied between the surfaces. If, however, there is some tangential motion between two contaminated mating surfaces, it is easier to penetrate the film and establish a weld. A brittle oxide or other surface film is more easily penetrated and the blacksmith frequently uses a flux to react with oxide films on surfaces to be joined to render them more brittle and hence more easily pierced.

Other Sources of Friction

Cocks (1962, 1964, 1965) has studied the detailed mechanism involved when one dry surface slides over another. Asperities do not simply slide over each other but what Cocks calls 'prows' are produced. These are wedges of plastic metal that tend to force the surfaces apart. By using a flexible system that enabled normal separation of the surfaces by as much as 0.25 mm (0.010 in), large wedges (0.3 mm wide by 0.8 mm (0.012 by 0.032 in) in direction of motion) were produced with copper sliding on copper. Figure 10.1 shows the mechanism of prow formation observed when a moving soft asperity encounters a harder one. At (a) the two asperities are about to make contact while (b) shows the asperities shortly after contact. The soft (lower) metal is deformed causing the prow to rise up thus tending to force the surfaces apart. The surfaces deform locally allowing the prow to flow past the hard asperity. With a stiff system, the shear angle (ϕ) will be small and the horizontal prow generating force will be high. The prow generating force will contribute to the friction force and occasionally the strength of the softer material will be exceeded during prow formation giving rise to a wear particle coming from the moving surface. Alternatively, the strength of the harder asperity may be exceeded giving rise to a wear particle coming from the harder (upper) region.

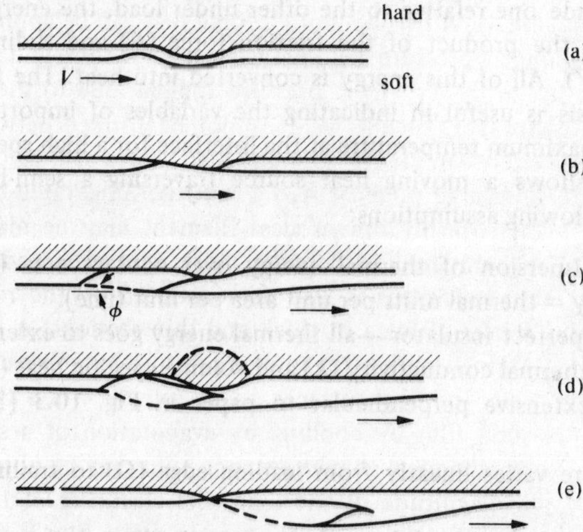

FIG. 10.1. Mechanism of prow formation.

Feng (1952) has suggested that the tips of asperities will interpenetrate (Fig. 10.2) due to the inhomogeneous nature of strain associated with the weak points present in all real materials. Subsequent motion will give rise to a friction component of force associated with:

1. The transverse displacement of one roughened surface over the other (in the case of hard strong shear plane ends).
2. The shearing-off of the interfering shear plane ends.
3. Plastic flow of the shear plane ends (burnishing action).

The friction component of force associated with Fig. 10.2 may be due to any combination of the above causes depending upon ductility.

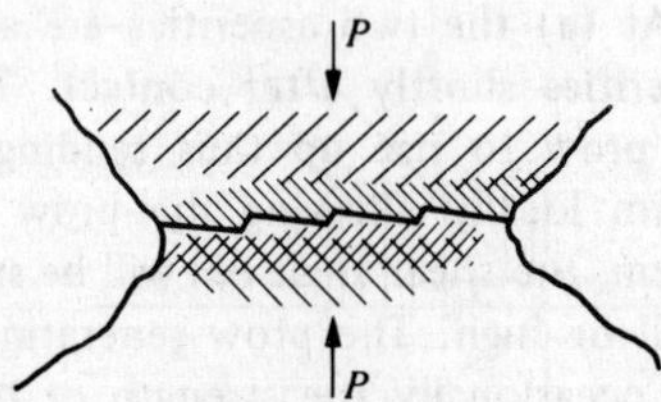

FIG. 10.2. Interface of pair of contacting asperities roughened by plastic deformation. (After Feng 1952.)

Surface Temperature

When surfaces slide one relative to the other under load, the energy dissipated per unit time is the product of the friction force (F) and sliding speed (V) ($U = FV = \mu PV$). All of this energy is converted into heat. The following approximate analysis is useful in indicating the variables of importance relative to the mean or maximum temperature at the interface for a high speed slider.

Figure 10.3 shows a moving heat source traversing a semi-infinite body subject to the following assumptions:

1. Uniform dispersion of thermal energy over contact area (q = thermal flux density = thermal units per unit area per unit time).
2. Slider is a perfect insulator—all thermal energy goes to extensive member which has thermal conductivity (k) and volume specific heat (ρC).
3. Slider is extensive perpendicular to paper in Fig. 10.3 (2-dimensional heat flow).
4. Temperature varies linearly from leading edge (O) to trailing edge (θ_m) of slider.
5. Depth of penetration of thermal energy varies linearly from leading edge (O) to trailing edge (h).

Equating the two expressions that may be written for total heat flux, Q,

$$Q = \rho C\left(\frac{\theta_m}{2}\right)(bhV) = k(bl)\frac{\theta_m}{h} \tag{10.4}$$

and hence

$$h = \left[\frac{2kl}{\rho CV}\right]^{\frac{1}{2}} = \left[\frac{2lK}{V}\right]^{\frac{1}{2}} \tag{10.5}$$

where $K = k/\rho C$ = diffusivity, b is slider width; and l is slider length. Substituting into (10.4)

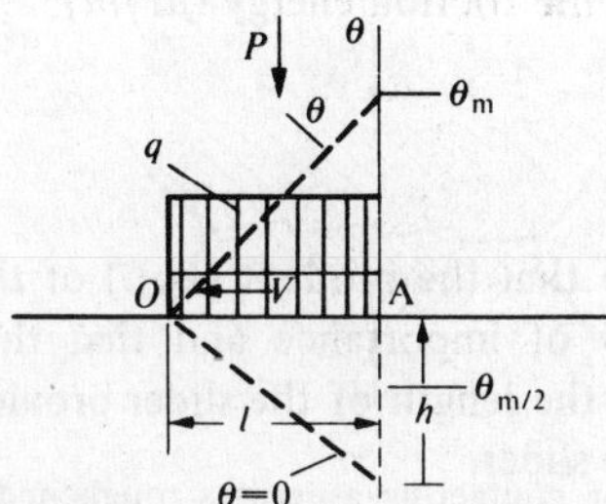

FIG. 10.3. Moving heat source corresponding to a friction slider.

$$Q = \rho C \frac{\theta_m}{2}(bV)\left[\frac{2lK}{V}\right]^{\frac{1}{2}}$$

or

$$\frac{\theta_m}{2} = 0.707 \frac{Q}{b\,[lV(k\rho C)]^{\frac{1}{2}}}. \qquad (10.6)$$

The more exact Jaeger (1942) solution leads to the same result except that coefficient (0.707) is (0.754). However, this agreement is purely fortuitous. What is significant, however, is that the somewhat overly simplified analysis leads to the same functional relationships for $(\theta_m/2)$ as the more exact Jaeger approach. Therefore, the coefficient in eqn (10.6) should be omitted and the mean surface temperature $(\theta_m/2)$ merely considered to be proportional to the quantity

$$\left(\frac{Q}{b\,[lV(k\rho C)]^{\frac{1}{2}}}\right).$$

For a friction slider q will be constant along the slider if μ is constant and

$$Q = \frac{\mu PV}{J} \qquad (10.7)$$

where μ is the mean coefficient of friction, P is the normal load and J is the mechanical equivalent of heat.

From eqns (10.6) and (10.7)

$$\bar{\theta} = \frac{\theta_m}{2} \sim \frac{\mu}{J}\left(\frac{P}{bl}\right)\bigg/\sqrt{\left(\frac{Vl}{k\rho C}\right)} \qquad (10.8)$$

For a high-speed slider, practically all of the thermal energy will be convected away by the extensive member (slider—a perfect insulator in effect) and eqn (10.8) will represent a good approximation. This equation states that the mean interface temperature ($\bar{\theta}$) varies

1. Directly with the specific friction energy ($\mu P/bl$);
2. Directly as $V^{\frac{1}{2}}$
3. Directly as $l^{\frac{1}{2}}$
4. Inversely as $(k\rho C)^{\frac{1}{2}}$.

It is of interest to note that the product ($k\rho C$) of the extensive member is the only thermal property of importance and that the mean and maximum temperatures increase with the length of the slider provided the specific friction energy is constant along the slider.

Heavily Loaded Sliders

Moore Effect

As the load on a slider increases, the size of the plastic zone associated with each asperity will increase as indicated in Fig. 10.4. At a critical value of load the plastic zones will join up and then for further increase in load there will be no further flattening of asperities. Instead, the subsurface will flow plastically. Moore (1948) drew attention to this phenomenon by pressing a hard cylinder into a soft copper surface having small parallel grooves (0.025 mm [0.001 in] apart × 100° included angle) machined in the surface. He found that even though the subsurface flowed extensively, the grooves were still substantially present (Fig. 10.5).

For an array of flattened asperities, as shown in Fig. 10.6, the mean normal stress on each of the flat areas will be $3Y$, assuming the areas to be far enough apart so there is no interaction between asperities. The asperities should flatten until the total load on all asperities will be just sufficient to cause the subsurface to flow plastically (i.e. until the mean stress in the subsurface = Y).

If asperity interaction is ignored, the real area (A_R) should increase in accordance with eqn (10.1) until $A_R/A = Y/3Y = \frac{1}{3}$. This represents a relatively crude first approximation since there will be interaction between asperities, and this will cause the limiting A_R/A to be $> \frac{1}{3}$.

Williamson (1968) has performed tests to more exactly identify the transition from a lightly loaded situation to a heavily loaded one. An aluminum specimen with bead-blasted surface was encased in a steel container and loaded by a polished flat piston that closely fitted the steel container. As load was applied, the surface asperities were flattened but did not disappear. Figure 10.7a shows the change in the number of plateaus with load. At first the number increases linearly with load and then remains constant with further increase in load. Figure 10.7b shows the variation in area of contact with load and this curve is seen to have a knee at about the same load as that where the previous one ceased to rise. Figure 10.7c shows the variation in separation of two surfaces versus load. Again, there is an abrupt change in the slope of this curve at a load

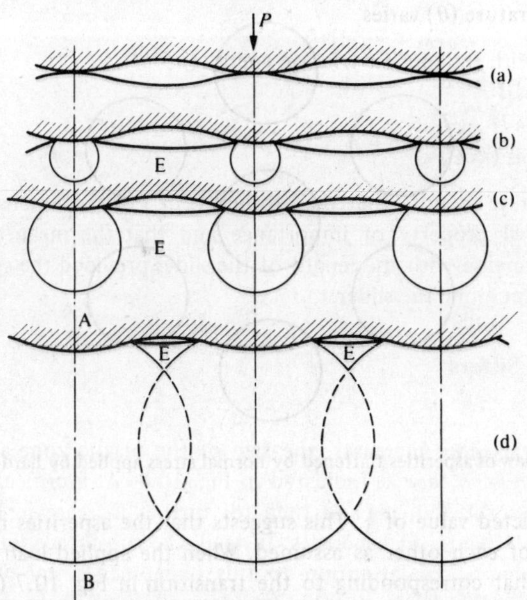

FIG. 10.4. Plane slider under progressively increasing load (P). Upper cross hatched surface is hard; lower surface, soft. Region E is elastic. (After Shaw 1963.)

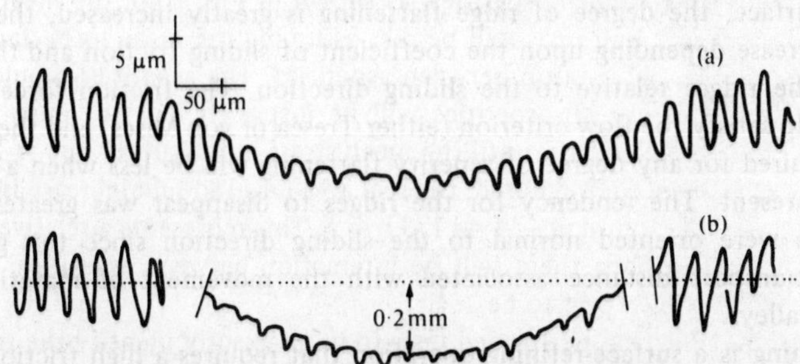

FIG. 10.5. Tallysurf traces of work-hardened copper (Vickers hardness = 84) having grooves 0.025 mm (0.001 in) apart machined in surface before indentation by 6 mm diameter × 0.18 mm (0.25 × 0.007 in) cylinder. (a) Applied load = 200 kg. (b) Applied load = 2500 kg. (After Moore 1948.)

of about 900 pounds (4009 N) which corresponds to the force required for bulk plastic flow of the specimen. It therefore appears as though the abrupt change in behavior in the curves of Fig. 10.4c corresponds to a transition from a lightly loaded situation to a heavily loaded one.

From Fig. 10.7b departure from Amontons' law occurs at A_R/A close to the

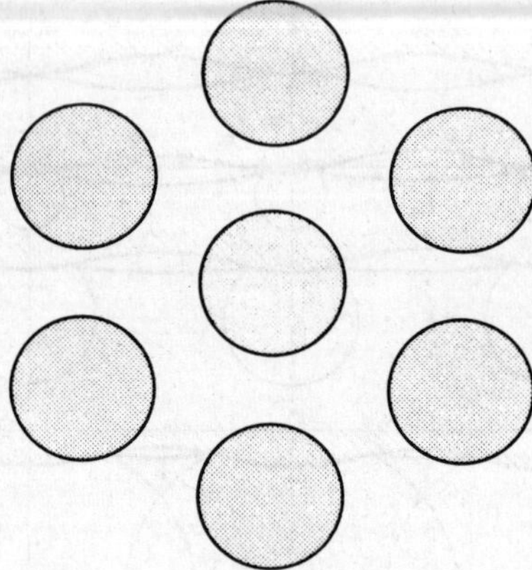

FIG. 10.6. Plan view of asperities flattened by normal stress applied by hard smooth surface.

previously predicted value of $\frac{1}{3}$. This suggests that the asperities may in fact act independently of each other as assumed. When the applied load was increased to four times that corresponding to the transition in Fig. 10.7 (4×900 lb $\cong$ 16 000 N), the real area of contact was only 50 percent of the apparent area.

Junction Growth

Moore (1948) found that when a hard smooth indentor slides across a grooved copper surface, the degree of ridge flattening is greatly increased, the extent of the increase depending upon the coefficient of sliding friction and the direction of the ridges relative to the sliding direction. The friction force will of course help satisfy the flow criterion (either Tresca or von Mises) and the normal stress required for any degree of asperity flattening will be less when a friction force is present. The tendency for the ridges to disappear was greatest when the ridges were oriented normal to the sliding direction since this gave the shortest transport distance associated with the movement of material from peaks to valleys.

Burnishing is a surface-refining operation that requires a high friction force. In wire drawing, surface roughness will be excessive if the friction force is not sufficient to iron out the roughness due to strain inhomogeneity (Fig. 10.2). The challenge is to have sufficient frictional force to provide burnishing without galling. The difference in texture of the front and back of a continuous metal cutting chip is due to removal of shear plane roughness due to burnishing on the tool face but not on the free surface of the chip.

The increase in real area of contact that accompanies sliding (friction) is called junction growth in tribology and this is what Moore found when a hard smooth indentor was slid across a grooved surface.

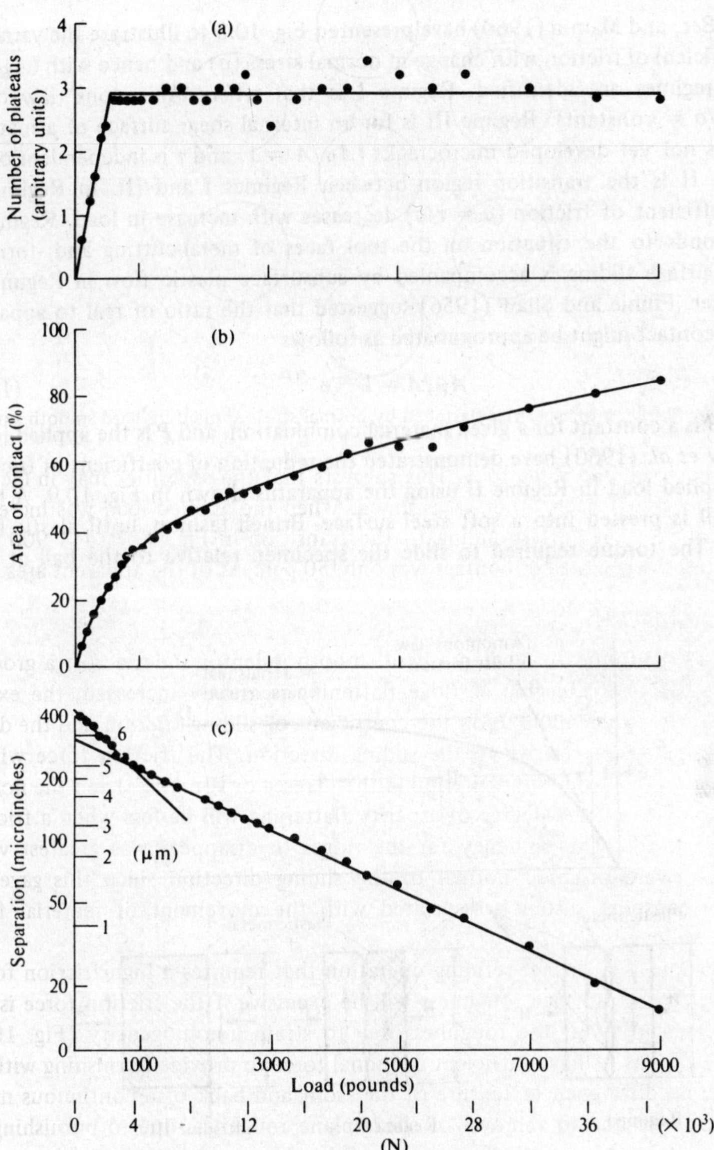

FIG. 10.7. Behavior of contacting surfaces under high normal stress. Bead-blasted aluminum surface encased in a steel container and loaded by a hard-polished steel piston closely fitting the steel container. (After Williamson 1968.)

Heavily Loaded Slider Friction

Shaw, Ber, and Mamin (1960) have presented Fig. 10.8 to illustrate the variation in coefficient of friction with change in normal stress (σ) and hence with (A_R/A). Three regimes are identified. Regime I is that where Amontons' law holds ($\mu = \tau/\sigma$ = constant). Regime III is for an internal shear surface of a material that has not yet developed microcracks ($A_R/A = 1$, and τ is independent of σ). Regime II is the transition region between Regimes I and III. In Regime II, the coefficient of friction ($\mu = \tau/\sigma$) decreases with increase in load. Regime II corresponds to the situation on the tool faces of metal-cutting and -forming tools. Surface sliding is accompanied by subsurface plastic flow in Regime II.

Earlier, Finnie and Shaw (1956) suggested that the ratio of real to apparent area of contact might be approximated as follows

$$A_R/A = 1 - e^{-BP} \tag{10.9}$$

where B is a constant for a given material combination, and P is the applied load.

Shaw *et al.* (1960) have demonstrated the reduction of coefficient of friction with applied load in Regime II using the apparatus shown in Fig. 10.9. A hard steel ball is pressed into a soft steel surface, Brinell fashion, until plastic flow occurs. The torque required to slide the specimen relative to the ball at low

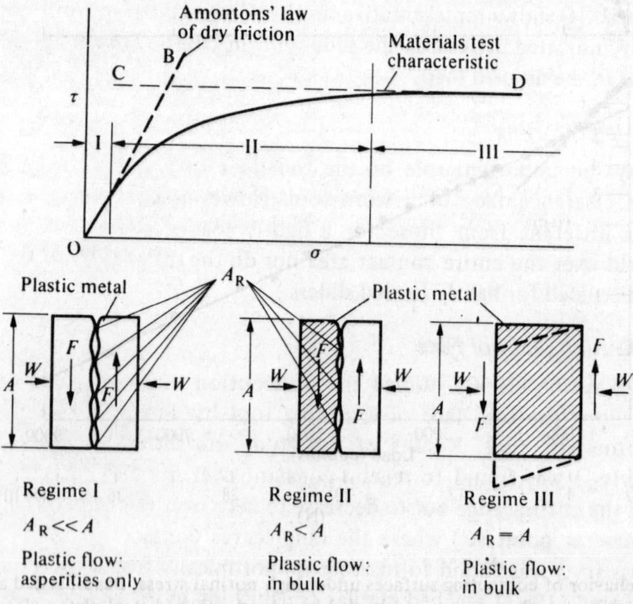

FIG. 10.8. Three regimes of solid friction. (After Shaw, Ber, and Mamin 1960.)

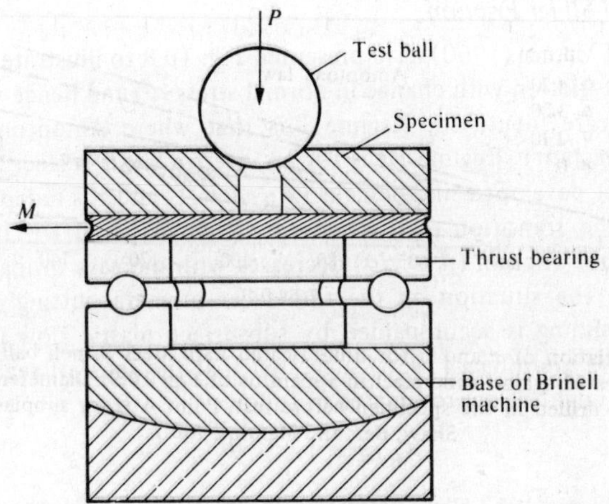

FIG. 10.9. Test apparatus for measuring friction with subsurface plastic flow. (After Shaw, Ber, and Mamin 1960.)

velocity is then measured. A small hole is provided at the center to eliminate the singularity that would otherwise exist there. The mean shear stress (τ) and normal stress (σ) may then be estimated for different values of applied load. Figure 10.10 shows representative results where deviation from Amontons' law is clearly indicated as well as the reduction in coefficient of sliding friction with increase in the applied load.

Friction in Metal Cutting

Friction plays an important role on the tool face of a sharp cutting tool and also on the clearance face of a worn tool. However, conditions on the tool face are far different from those for a lightly loaded slider. Amontons' law does not hold over the entire contact area nor do the other rules of dry friction previously discussed for lightly loaded sliders.

Stress Distribution on Tool Face

Usui and Takeyama (1960) studied the distribution of shear (τ_C) and normal (σ_C) stress along the tool face of a cutting tool by cutting lead at low speed with a photoelastic tool. Results of this study are shown in Fig. 10.11. the shear stress (τ_C) was found to remain constant over the half of tool-chip contact nearest the cutting edge but to decrease to zero over the other half, reaching zero of course at point (C) where the chip leaves contact with the tool. The normal stress (σ_C) was found to increase monotonically from (C) to the cutting edge (A). Zorev (1963) reached similar conclusions by observing that grinding scratches parallel to the cutting edge were transferred to the fully plastic surface of the chip from A to D but that these markings were replaced by an orthogonal

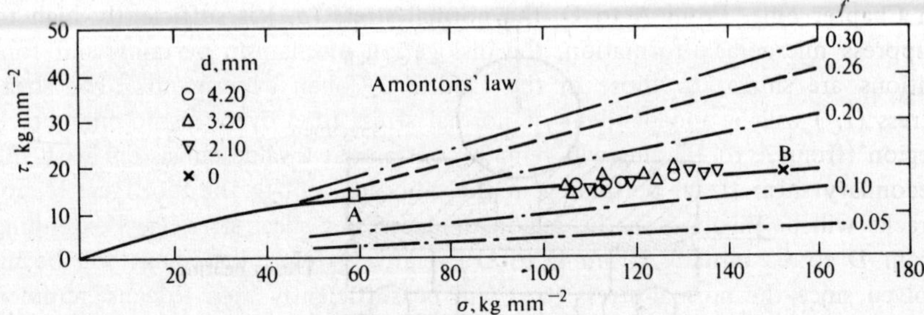

FIG. 10.10. Variation of τ and σ for unlubricated hard steel Brinell ball of $\frac{1}{2}$ inch (12.7 mm) diameter tested against mild steel in apparatus of Fig. 10.9. Diameter d is the size of the central hole drilled in the specimen before test. Point A is for subplastic load. (After Shaw, Ber, and Mamin 1960.)

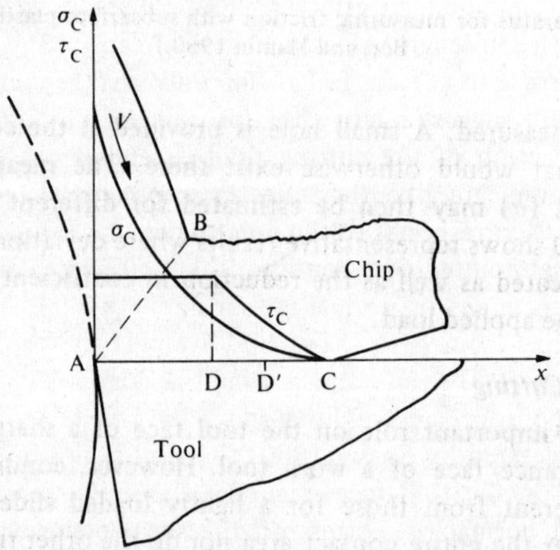

FIG. 10.11. Variation of shear and normal stresses on tool face of tool. (After Usui and Takeyama 1960.)

set of scratches running in the direction of chip motion over the second region of contact extending from D to C where sliding actually occurred. Zorev concluded that chip flow was completely within the chip (subsurface) from A to D but occurred at the interface from D to C. Kato and Yamaguchi (1972) used a special tool dynamometer with divided rake face to verify these results. They found stress distributions similar to those of Fig. 10.11 and that the region of constant shear stress decreased relative to total contact length as the rake angle increased.

Figure 10.11 is consistent with the microcrack theory of plastic flow presented in Chapter nine. From A to D, the normal stress (σ_C) is sufficiently high to suppress microcrack formation, the dislocation mechanism pertains, and conditions are similar to those in the Lankford-Cohen experiments. The shear stress (τ_C) will be independent of normal stress (σ_C) over this no-microcrack region (from A to D) and will remain constant at a value consistent with the secondary shear strain pertaining in the chip adjacent to the tool face. Microcracks will be involved in the region of decreasing shear stress (τ_C) extending from D to C. Initially (from D to D′) relatively few microcracks will be involved since the normal stress (σ_C) will be sufficiently high to cause appreciable rewelding. However, as (σ_C) decreases, there will be less and less rewelding relative to fracture and conditions will approach those for a lightly loaded slider where $A_R/A \ll 1$ and the 'microcracks' communicate. The picture of tool-face stresses shown in Fig. 10.11 has been further verified and elaborated recently by other workers (Amini 1968; Trent 1977).

It has been suggested by De Chiffre (1977) that only region DC is influenced by a cutting fluid and that an effective fluid causes a decrease in length DC and hence a decrease in length AC which in turn causes an increase in the shear angle ϕ and a decrease in cutting force. The only change the microcrack theory of Chapter 9 would suggest relative to this concept of cutting fluid action is that the primary action of the cutting fluid extends from D′ to C where the microcracks communicate like the holes in a sponge. The results of Fig. 9.7 suggest that point D corresponds to the point where $\sigma_C \cong \tau_C/2$.

As Zorev (1963) pointed out, Amontons' law of sliding friction (coefficient of friction is independent of normal stress) should not be expected to hold in the region extending from A to D but only from D to C. There will obviously be a transition region (D to D′) and it would now appear to be more accurate to suggest that Amontons' law will hold only from D′ to C.

It has been suggested by Shaw (1980) that the situation on the shear plane is similar to that on the tool face. Figure 10.12 shows the proposed variation in shear (τ_S) and normal (σ_S) stress on the shear plane with distance across the shear plane. In region AE (nearest the cutting edge), τ_S will be independent of normal stress on the shear plane (σ_S), while in region E′B (nearest the free surface) τ_S/σ_S will be approximately constant corresponding to Amontons' law for a friction slider. These two regions should be connected by a transition region (EE′ in Fig. 10.12) where the density of microcracks is increasing to the point of intercommunication. Figure 10.12 explains why the microcracks mechanism of plastic flow of Chapter 9 involving the rewelding of microcracks should be expected to pertain over part of the shear plane in metal cutting even though the *mean* normal stress on the shear plane will generally be $\geqslant \bar{\tau}_S$.

Variation of Coefficient of Tool-Face Friction with Rake Angle

One of the paradoxes associated with metal cutting involves the variation of tool-face friction coefficient (μ) with rake angle (α). Table 10.1 presents typical

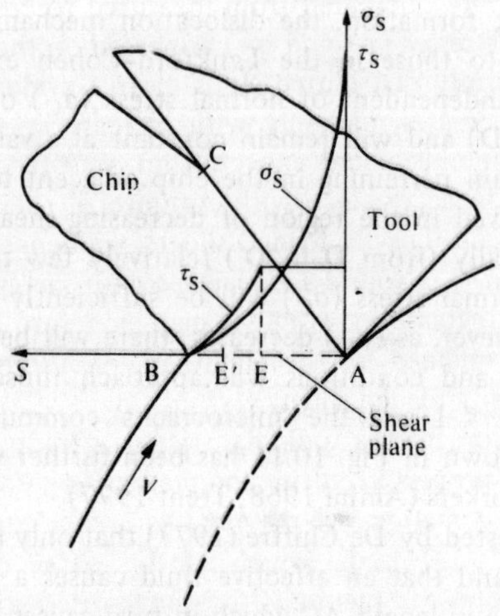

FIG. 10.12. Proposed variation of shear and normal stress across shear plane. (After Shaw 1980.)

TABLE 10.1
Representative low-speed cutting data

Work material	Rake angle α (deg)	Average stresses at tool–chip interface (p.s.i.) Shear stress (p.s.i. †) τ_C	Normal stress (p.s.i.) σ_C	Friction coefficient $\mu = \tau_C/\sigma_C$
Electrolytic tough pitch copper; cutting speed (V), 0.3 f.p.m. (0.0016 m s⁻¹); depth of cut (t), 0.002 in (0.05 mm)	30	35 600	43 300	0.82
	45	34 700	30 000	1.15
	60	36 700	16 000	2.30
SAE B1112 steel; cutting speed (V), 0.04 f.p.m. (0.0002 m s⁻¹) depth of cut (t), 0.003 in (0.076 mm)	16	46 000	70 000	0.66
	30	49 300	57 000	0.86
	45	51 700	43 200	1.20

† 1 p.s.i. = 6905 N m⁻²

orthogonal cutting data for copper and steel cut at low speed with HSS tools in air. The stresses were calculated from measured forces and contact areas on the tool face. These stresses (τ_C and σ_C) are based on the apparent area and are assumed to be uniformly distributed. It is clearly evident that the coefficient of cutting friction increases markedly with increase in rake angle. This is just the opposite behavior one would expect from experience with lightly loaded sliders that perform in accordance with Amontons' rule. However, it is what should be expected from heavily loaded slider experience where the coefficient of friction ($\mu = \tau_C/\sigma_C$) is found to decrease with increase in normal stress as a consequence of an increase in A_R/A. It may therefore be concluded that the paradox is in part explained by the fact that conditions at the tool face are largely those of a heavily loaded slider.

An additional reason for the tool-face friction paradox appears to lie in the fact that the friction process in cutting interacts strongly with the shear process. As a first crude approximation, we may characterize the shear process in terms of the Lee-Shaffer (1951) model which appears to be the least objectionable simple model yet proposed. Figure 10.13 shows the slip line fields for sharp high- and low-rake-angle tools cutting the same materials. Assuming the shear plane to be a plane of maximum shear stress and the resultant cutting force (R) to be displaced 45° from the shear plane in accordance with the Lee–Shaffer solution, angle β ($\beta = \tan^{-1} \mu$) will be smaller for small rake angles.

Indentation Force at Tool Tip

Masuko (1953) suggested that a sharp cutting tool will deform elastically to give a radius at the tip of even a sharp cutting tool. Figure 10.14 shows a cutting tool with the assumed radius at the tip. Masuko assumed that in addition to the chip forming force (R_c) on the tool face there would be an indentation force at the tool tip (R_n in Fig. 10.14). He assumed that the indentation force R_n bisects the included angle of the tool (δ). The components of indentation force R_n were substracted from the measured force components F_P and F_Q and the resulting values used to compute stresses and strains as outlined in Chapter 3.

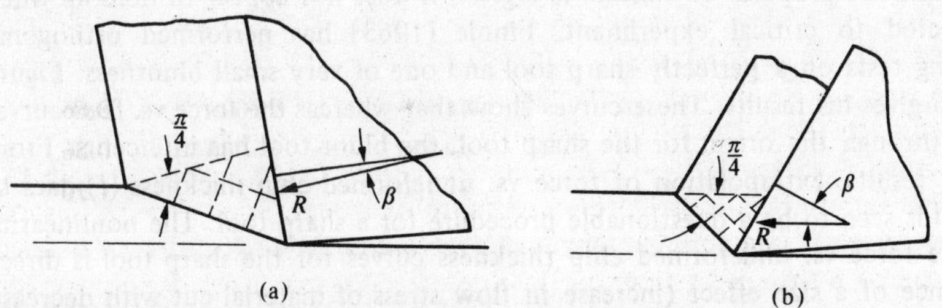

FIG. 10.13. Variation in magnitude and direction of resultant cutting force with rake angle.

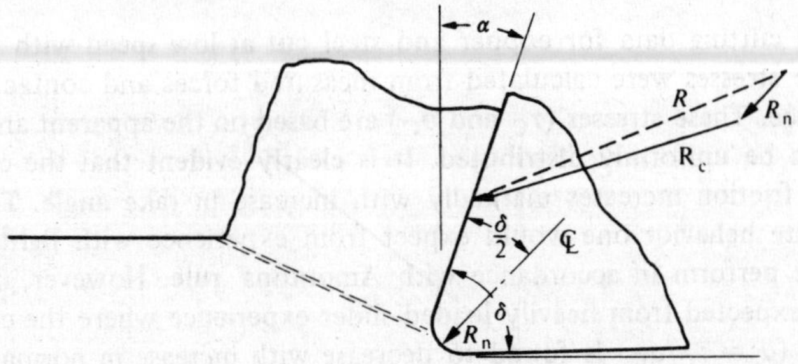

FIG. 10.14. Influence of tool tip radius on cutting forces. R_n = indenting force due to tool radius and R_c = resultant cutting force on remainder of tool face. (After Masuko 1953.)

The concept of an indenting component of force was probably suggested to Masuko by the fact that when measured values of F_P and F_Q are plotted against undeformed chip thickness (t), straight lines with an intercept at $t = 0$ are obtained as shown diagrammatically in Fig. 10.15. The cutting (R_{nP}) and feed (R_{nQ}) components of indentation force R_n were assumed to correspond to the intercepts of the F_P and F_Q verus t curves in Fig. 10.15.

The actual values of R_{nP} and R_{nQ} were obtained by an iterative approximate elastic analysis. When the above procedure was applied to data for a duralumin specimen cut dry using HSS tools having rake angles ranging from 0 to 44°, the coefficient of friction for cutting on the tool face was found to be constant relative to α and t. At the same time when the shear stress on the shear plane (τ) was compared with extrapolated values of τ vs. γ for torsion tests on the same material, no size effect was found. However, in extrapolating the torsion data to the larger strains pertaining in cutting, eqn (5.8) was used instead of eqn (9.11). From these results, Masuko concluded that there is no size effect in cutting and that the unusually high values of shear stress on the shear plane in cutting are due to strain hardening alone.

Results practically identical to those of Masuko have been reported by Albrecht (1960, 1961, 1963).

While the proposal of Masuko is logical, it does not appear to hold up when subjected to critical experiment. Finnie (1963) has performed orthogonal cutting tests on a perfectly sharp tool and one of very small bluntness. Figure 10.16 gives his results. These curves show that whereas the force vs. feed curves pass through the origin for the sharp tool, the blunt tool has intercepts. From these results, extrapolation of force vs. undeformed chip thickness (t) data to $t = 0$ is seen to be a questionable procedure for a sharp tool. The nonlinearity of the force vs. undeformed chip thickness curves for the sharp tool is direct evidence of a size effect (increase in flow stress of material cut with decrease

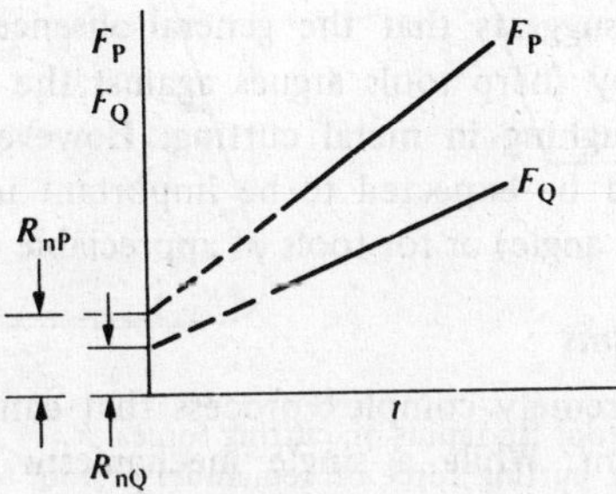

FIG. 10.15. Variation of power (F_P) and feed (F_Q) components of cutting force with undeformed chip thickness (t). R_{Pn} and R_{Qn} are the horizontal (power and vertical components of R_n in Fig. 10.14.

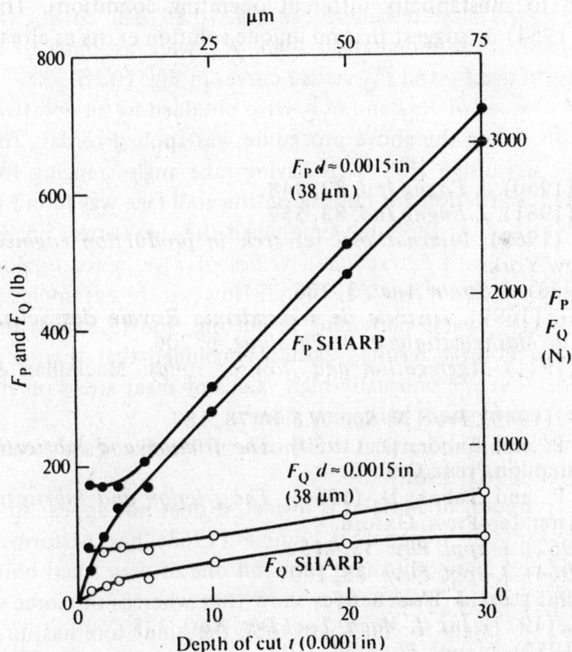

FIG. 10.16. Variation of cutting force components in velocity (F_P) and feed (F_Q) directions with undeformed chip thickness (t) for sharp and dull tools. Dimension d is extent of tool bluntness measured in feed direction. (After Finnie 1963.)

of t). Finnie's results indicate that the height of the rounded nose of a blunt tool must exceed about 20 percent of the undeformed chip thickness (t) before its influence becomes noticeable.

Finnie (1963) also suggests that the general absence of a deformed layer on surfaces produced by sharp tools argues against the general importance of an indentation or ploughing in metal cutting. However, the indentation or ploughing effect should be expected to be important in grinding (very small t and large negative rake angle) or for tools of appreciable dullness.

Multiplicity of Mechanisms

Metal cutting is an extremely complex process that cannot be described by a single simple mechanism. While a single mechanicsm may be predominant over a limited range of operating conditions, experience teaches that no single mechanism holds in general. This accounts for the extremely wide range of views that have appeared in the literature to explain cutting results most of which are supported by sound experimental data. In most cases, the relatively simple mechanism suggested by the author to explain his results fails to hold when applied to substantially different operating conditions. This condition has led Hill (1954) to suggest that no unique solution exists as already discussed in Chapter 8.

References

Albrecht, P. (1960). *J. Engng Ind.* **82**, 348.
Albrecht, P. (1961). *J. Engng Ind.* **83**, 557.
Albrecht, P. (1963). *International research in production engineering*, p. 32. ASME, New York.
Amini, E. (1968). *J. Strain Anal.* **3**, 206.
Amontons, G. (1699). *Historie de L'Academie Royale des Sciences avec les Memories de Mathematique et de Physique*, p. 206.
Beilby, G. (1921). *Aggregation and flow of solids.* Macmillan & Co., Ltd., London.
Bowden, F. P. (1945). *Proc. R. Soc. N.S.W.* **78**, 187.
Bowden, F. P. and Tabor, D. (1950). *The friction and lubrication of solids* Vol. I. Clarendon Press, Oxford.
Bowden, F. P. and Tabor, D. (1964). *The friction and lubrication of solids* Vol. II. Clarendon Press, Oxford.
Cocks, M. (1962). *J. appl. Phy.* **33**, 2152.
Cocks, M. (1964). *J. appl. Phys.* **35**, 1807.
Cocks, M. (1965). *Int. J. Wear* **8**, 85.
De Chiffre, L. (1977). *Int. J. Mach. Tool Des. Res.*, 225.
Feng, I. M. (1952). *J. appl. Phys.* **22**, 1011.
Finnie, I. (1963). In *International research in production engineering,* p. 76. ASME, New York.
Finnie, I. and Shaw, M. C. (1956). *J. Engng Ind.* **78**, 1649.
Hill, R. (1954). *The mathematical theory of plasticity*. Clarendon Press, Oxford. (Reissued 1983.)

Jaeger, J. C. (1942). *Proc. R. Soc. N.S.W.* **76**, 203.
Kato, H. and Yamaguchi, K. (1972). *Trans. Am. Soc. mech. Engrs* **94**, 683.
Lee, E. H. and Shaffer, B. W. (1951). *J. appl. Mech.* **18**, 405.
Masuko, M. (1953). *Trans. Jap. Soc. mech. Engrs* **19**, 32.
Moore, A. J. W. (1948). *Proc. R. Soc.* **A195**, 231.
Shaw, M. C. (1963). *Int. J. Wear* **6**, 140.
Shaw, M. C. (1980). *Int. J. mech. Sci.* **22**, 673.
Shaw, M. C., Ber, A., and Mamin, P. A. (1960). *J. Engng Ind.* **82**, 341.
Trent, E. M. (1977). *Metal cutting.* Butterworths, London.
Usui, E. and Takayama, H. (1960). *J. Engng Ind.* **B82**, 303.
Williamson, J. B. P. (1968). In *Interdisciplinary approach to friction and wear,* p. 85. Office of Technology Utilization, Washington, DC.
Zorev, N. N. (1963). In *International research in production engineering,* p. 42. ASME, New York.

11

WEAR AND TOOL LIFE

Wear is usually undesirable and to be minimized. This is certainly the case with tool wear or when machine surfaces rub together and a loss of material from one or both surfaces results in a change in the desired geometry of the system. In some cases wear is useful:

1. Writing with a pencil on paper.
2. Erasing unwanted pencil lines.
3. Metal removal by honing (fixed abrasive), lapping (free abrasive), or ultrasonic machining.
4. Liquid honing of surfaces.
5. Run-in of a bearing surface to improve its load capacity and conformity.
6. Comminution.
7. The self-sharpening action of grinding wheels.

Types of Wear

Wear may be classified into several types as follows:

1. Attritious (small particle) wear associated with adhesion, prow formation, and shear plane ends (Fig. 10.2).
2. Abrasive wear (due to the cutting action of hard particles).
3. Erosive wear (cutting action of particles in a fluid).
4. Diffusion wear at high surface temperatures.
5. Corrosive wear (due to chemical attack of a surface).
6. Fracture wear (chipping of brittle surfaces).

Practical wear situations rarely involve only one of these types of wear and there are important interactions. For example, adhesive wear is often accompanied by oxidation of the wear debris which may in turn act as a solid lubricant to reduce further adhesive wear or as a hard particle which causes abrasive wear. Fretting corrosion is a well-known composite of several actions. In this case two surfaces have a relative motion of small amplitude so that wear particles generated cannot escape the system. The wear particles first formed oxidize and produce abrasive particles that generate further wear particles by a cutting action. These larger wear particles in turn oxidize and produce further abrasive wear particles, etc. The result is an appreciable number of loose oxidized wear particles (Fe_2O_3 or rouge) between steel surfaces which are evident as a red powder when they are separated. Possible solutions for fretting corrosion are

1. Prevent motion;
2. Exclude oxygen (grease);
3. Use metals that either do not oxidize (gold) or form soft oxides (silver).

Wear is always accompanied by a great deal of friction and in some cases, subsurface plastic flow. In general, the milder the sliding conditions, the smaller the wear particles and the lower the wear rate but the greater the specific energy required to generate the wear particles that are formed.

Diffusion wear is a special type of considerable importance in metal cutting. In this case, the tool surface is decomposed and the decomposition products diffuse into the surface of the chip. This type of wear is most important at high surface temperatures and with systems of high solubility.

In addition to these types of wear, tools may fail due to gross fracture or due to plastic flow of the tool tip in the case of high tool tip temperature. The ratio A_R/A discussed in Chapter 10 plays a very important role relative to all types of tool wear.

Adhesive Wear

In the case of adhesive wear the mating surfaces come close enough together to form strong bonds. If the bonds so established are stronger than the local strength of the material a particle may transfer from one surface to the other. After this has occurred several times a loose fragment may be formed and leave the system as a wear particle.

If the particles so removed are very small (submicroscopic) we refer to the process as attritious wear. If the particles are visible under the microscope the process is referred to as galling. In all cases the mechanism is the same except for the size of particle generated. Figure 11.1 shows a lightly loaded situation where Amontons' law holds.

If a weak point of stress concentration is found close to a point of contact, a particle will transfer if the strength of the asperity at this weak point is less than the strength of the bond between asperities. As a real area of contact increases the size of the particle transferred increases and so does the volume rate of wear (B). It is empirically found that

$$\frac{B}{L} = KA_R \tag{11.1}$$

where B = volume worn away
L = sliding length
A_R = real area of contact
K = probability that a real contact will result in a wear particle.

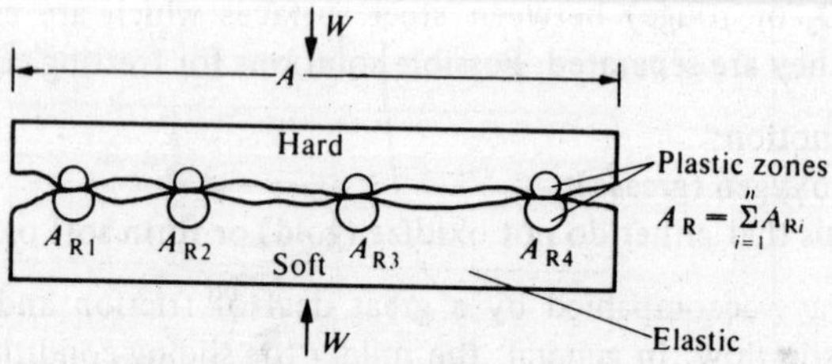

FIG. 11.1. Asperities of lightly loaded surfaces. A_R is the real area of contact and A is the apparent area of contact.

Combining eqns (10.1) and (11.1) we have

$$\frac{B}{L} = K\frac{P}{H} \tag{11.2}$$

or

$$\frac{B}{LP} = \frac{K}{H}. \tag{11.3}$$

The quantity K/H should be constant for a given pair of lightly loaded surfaces for which $A_R \ll A$. The quantity K will be constant when the hardness is changed as long as the chemistry of the system is not changed.

When the applied load over the apparent area of contact (P/A) becomes equal to the uniaxial flow stress of the material, which is approximately equal to $(H/3)$, bulk subsurface flow occurs instead of localized flow at the asperities. The model of Fig. 11.1 then no longer holds. In such a case there is a sudden increase in the real area of contact (A_R) and hence in the volume rate of wear (eqn 11.2).

Burwell and Strang (1952) have performed pin-on-disc experiments in which the wear volume B was measured for different sliding lengths (L) and for the different loads (P). Figure 11.2 shows their results plotted in accordance with eqn (11.3). The quantity K/H should be a constant as long as eqn (10.1) is operative (i.e. until gross subsurface flow occurs). The quantity B/LP should therefore be a constant until $P/A = H/3$ at which time subsurface flow occurs and A_R rises abruptly as does the wear rate.

It is thus evident that low rates of adhesive wear will result as long as there is no bulk subsurface flow but that the wear rate will increase abruptly when

$$\frac{P}{A} > \frac{H}{3} = Y \tag{11.4}$$

In some of the experiments of Burwell and Strang the pins were cylindrical in shape and softer than the moving plate. For this situation it is evident that

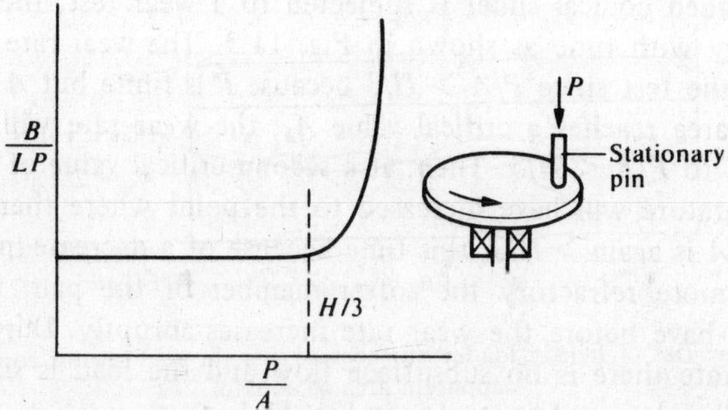

FIG. 11.2. Results obtained by Burwell and Strang (1952) with soft pin sliding on hard disc.

the subsurface of the pin will flow when $P/A = Y$ (where Y is the uniaxial flow stress) if friction is negligible. However, if the coefficient of friction is 0.3, then the value of P/A (where P is the normal load component) required for bulk subsurface flow of the pin will be less than Y since the friction component of force helps make the pin go plastic. By the maximum shear theory this is found to be $0.86Y$ for = 0.3.

When the softer member of the sliding pair is the disc and the coefficient of friction is 0.3, Shaw (1971) has estimated the critical value of P/A for rapid increase in A_R/A to be Y, the uniaxial flow stress of the disc material. To a first approximation, therefore, we may assume that subsurface flow commences when $P/A = Y$ for the softer member in all cases.

Archard (1953) has presented a derivation leading to eqn (11.3) based upon a relatively simple idealized model. Equation (11.3) may also be obtained by dimensional analysis for a given pair of metals operating with a given lubricant (Shaw, 1977). For a lightly loaded slider, the volume (B) worn away is assumed to depend upon sliding distance (L), applied load (P), the sliding velocity (V) and the hardness of the softer member of the sliding pair (H):

$$B = \psi_1(L, P. H. V).$$

After performing the dimensional analysis in the usual way

$$\frac{BH}{LP} = N_W \text{ (a constant)}$$

The velocity (V) is seen to drop out and there is finally only one nondimensional quantity which will be called the wear number (N_W)—a constant for a given sliding system. It has been suggested by Shaw (1977) that the wear number (N_W) is proportional to the ratio of volume of metal worn away to the volume plastically deformed. The American Society of Mechanical Engineers (1980) has published a Wear Control Handbook in which values of N_W are to be found for a wide range of tribological systems.

When a loaded conical slider is subjected to a wear test, the rate of wear (B/L) will vary with time as shown in Fig. 11.3. The wear rate is high at the beginning of the test since $P/A > H/3$ because P is finite but A is zero. After the apparent area reaches a critical value A_1, the wear rate will drop to that corresponding to $P/A < H/3$. Then, at a second critical value of apparent area A_2 the temperature will have increased to the point where thermal softening occurs and P/A is again $> H/3$, this time because of a decrease in H with temperature. The more refractory the softer member of the pair, the larger the value A_2 may have before the wear rate increases abruptly. During the period of low wear rate there is no subsurface flow and the load is supported on a relatively small real area of contact as in Fig. 11.1.

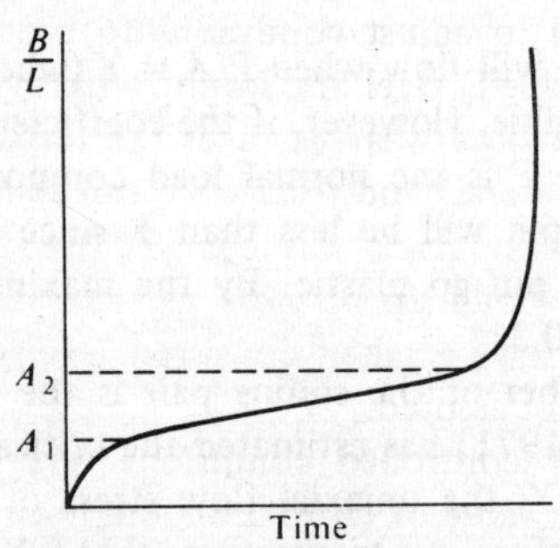

FIG. 11.3 Variation of B/L with time for lightly loaded conical pin in pin-on-disc test (Fig. 11.2).

In the foregoing discussion clean sliding surfaces are visualized. These surfaces may, however, be of different hardnesses or strengths. The stronger of the two surfaces will then wear at a lower rate than the weaker one and this is reflected in a lower value of K or N_W for the stronger member of the pair. The wear rate of the stronger surface will not be zero however, since occasionally a weak point in the generally stronger surface may occur opposite a strong point in the generally weaker surface. Transfer will then involve a particle leaving the stronger surface. This explains why the wear rate of the stronger member of a metal pair is lower than that for the weaker element but not zero.

In the presence of a contaminating environment the value of K will generally be reduced to (αK) where α is a factor less than one. The contaminating action may involve:

1. Physical adsorption of a low shear strength film (example is oleic acid).
2. Formation of an inorganic film by reaction (example is oxidation in air).

The first of these actions is often called boundary lubrication while the second is referred to as extreme boundary lubrication.

When eqn (11.4) pertains, and there is subsurface flow, then a third type of environmental action known as the Rehbinder effect (1947) may occur. In this case a contaminating vapour is adsorbed into microcracks in the deforming subsurface which in turn alters the equilibrium between the generation and rewelding of microcracks. The layer of metal directly beneath the surface then behaves as though it were weaker, and in Russia this is sometimes referred to as 'hardness reduction'. An example of this type of action is when carbon tetrachloride is applied to a heavily loaded surface subjected to very large plastic strains (> 2). If the subsurface strains are not large enough microcracks will not be formed and the Rehbinder effect will not be operative.

It is relatively difficult to achieve pure adhesive wear, for the material transferred during adhesive wear will often cause some abrasive wear. Burwell and Strang were very careful to adjust conditions to provide adhesive wear. They used a hardened (550 Brinell) steel disc having a finish of 1 μin (R_a). To prevent corrosion and to remove loose wear particles the surfaces were flooded with an inert liquid (pure hexadecane) and tests were run at relative humidities of less than 20 percent. The wear track was changed frequently to avoid grooving or abrasive wear due to build up of wear particles on the surface of the disc. Unless such precaustions were taken consisten results were not obtained and pure adhesive wear was not assured.

Under these carefully controlled conditions very low wear rates were obtained. For example, the value of (B/PL) for the horizontal region of Fig. 11.2 was found to be approximately 10^{-12} in^2 lb^{-1} (1.45×10^{-12} cm^2 N^{-1}) when the Brinell hardness of the 120° cone was either 223 or 430 kg mm^{-2}. The value of $K = N_W$ may be estimated from eqn (11.2). Thus, for the softer pin

$$N_W = \frac{B}{PL} H = 10^{-12}(223)(1420) = 0.317 \times 10^{-6}.$$

This suggests that only about one in a million encounters between asperities resulted in a wear particle.

Physical Significance of B/PL

The quantity (B/PL) has a special physical significance. The quantity P may be replaced by F/μ, where F is the friction force and μ is the coefficient of friction. Hence,

$$\frac{B}{PL} = \frac{\mu B}{FL}. \tag{11.5}$$

The quantity (FL) is the energy dissipated in sliding a distance L and hence the energy per unit volume for material removal or specific energy will be

$$u = \frac{FL}{B}. \tag{11.6}$$

From this it follows that

$$\frac{B}{PL} = \frac{\mu}{u}. \tag{11.7}$$

If we assume μ to be 0.5 in the experiments of Burwell and Strang for which $B/PL = 10^{-12}$ in^2 lb^{-1}, 1.45×10^{-12} cm^2 N^{-1}) then from eqn (11.7)

$$u = \frac{\mu}{B/PL} = 0.5 \times 10^{12} \text{ in lb in}^{-3} \ (3450 \times 10^6 \text{ MPa}).$$

This extremely high value of specific energy is due to the fact that very many bonds between asperities had to be broken before a single wear particle was generated. If we assume that when each such bond is broken the material is stressed to the point of generating a wear particle, then the energy per unit volume required to produce a wear particle in the absence of all the unsuccessful attempts may be estimated to be

$$Ku = (0.317 \times 10^{-6})\,(0.5)\,(10^{12})$$

$$= 0.158 \times 10^6 \text{ in lb in}^{-3} \ (1.09 \times 10^6 \text{ MPa})$$

The actual value of specific energy required to produce the one wear particle in the absence of the approximately one million unsuccessful attempts will be somewhat greater than this since the unsuccessful attempts will involve a lower specific energy than the one actually producing a wear particle. If we assume that on the average the bonds formed that do not result in a wear particle are only 1 percent as strong as those that do, due to the contaminating influence of the lubricant used, then the specific energy associated with the formation of a wear particle would be 15.8×10^6 in lb in^{-3} (109×10^6 MPa).

Abrasive Wear

Abrasive wear involves the loss of material by the formation of chips as in abrasive machining. To have such wear it is necessary that one material be harder (or have harder constituents) than the other member of the sliding pair or that hard particles be formed by chemical reaction of the wear debris, by oxidation for example.

A convenient way of studying abrasive machining or abrasive wear is in terms of specific energy (u in lb in^{-3} or MPa) which is the energy required to remove a unit volume of material. Specific energy is experimentally found to be a strong function of the undeformed chip size (t)† as shown in Fig. 11.4.

† Values of t in Fig. 11.4 are estimates based on mean grinding wheel properties and may be in error by a factor of 2 or more. For our purposes here the relative values are of the greatest concern and these are undoubtedly more accurate than the absolute values.

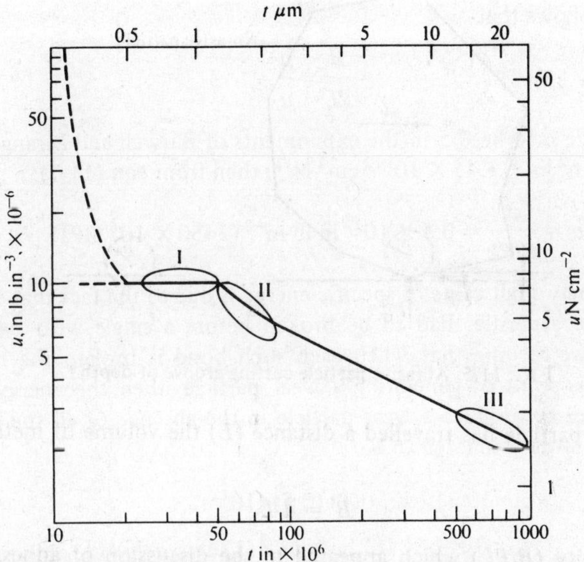

FIG. 11. 4. Variation of specific energy (u) with undeformed chip thickness (t) when grinding mild steel. (After Backer, Marshall, and Shaw 1952.)

The dotted curve represents the transition from abrasive action to adhesive action. For each abrasive particle there is a load below which it will not cut, which depends upon its fine geometry (sharpness). As we move upward on the dotted curve there is more rubbing and adhesive action and less abrasive action.

We should expect the performance of an abrasive particle giving rise to abrasive wear to resemble that of abrasive grains in a grinding wheel. Part of the action in each case will be associated with material removal and part will be associated with rubbing and friction.

Figure 11.5 shows an abrasive particle with radius ρ at its tip cutting a groove of depth t.

The cross-section involved is A' and the tangential force F'_P will be

$$F'_P = uA' \tag{11.8}$$

where u is the specific energy to produce a 'chip'.

In grinding, the radial force is approximately twice the tangential force and hence

$$F'_Q = 2uA' \tag{11.9}$$

This corresponds to a 'coefficient' of friction' $\mu = F'_P/F'_Q$ of 0.5 for the grinding process.

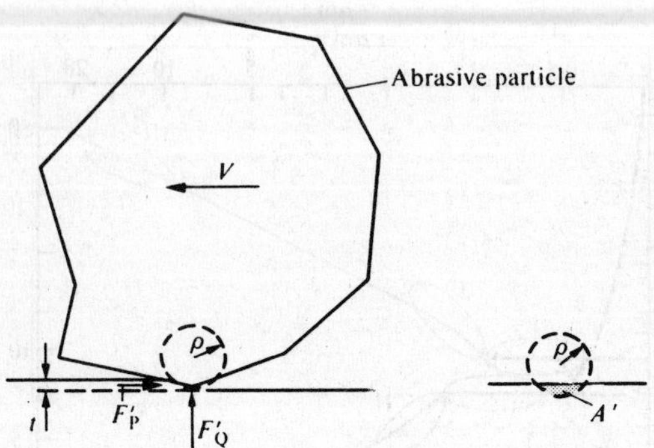

FIG. 11.5. Abrasive particle cutting groove of depth *t*.

When the particle has travelled a distance (L) the volume of metal removed (B') will be

$$B' = A'L \tag{11.10}$$

The quantity (B/PL) which appeared in the discussion of adhesive wear is a convenient parameter for wear studies and may be evaluated for abrasive as follows (with the aid of eqns (11.9) and (11.10)):

$$\frac{B}{PL} = \frac{\Sigma B'}{\Sigma F'_Q L} = \frac{B'}{F'_Q L} = \frac{A'L}{2uA'L} = \frac{1}{2u} = \frac{\mu}{u} \tag{11.11}$$

Substituting values of u from Fig. 11.4 into eqn (11.11) gives Fig. 11.6. The dotted curve represents the transition from abrasive to adhesive wear.

When the wear rate (B/PL) for fine abrasion (5×10^{-8} in^2 lb^{-1} (7.25×10^{-8} cm^2 N^{-1}) is compared with that for adhesive wear (10^{-12} in^2 lb^{-1} (1.45×10^{-12} cm^2 N^{-1}) (Burwell and Strang) it is evident that abrasive wear rates are very much larger. This is because there is much less rubbing and non-removal work done in the abrasive case. Where the total specific energy for adhesive wear was found to be 0.5×10^{12} in lb in^{-3} (0.34×10^{12} N cm^{-2}) it is seen to be about 10^7 in lb in^{-3} (0.69×10^7 N cm^{-2}) for fine abrasive action (region I, Fig. 11.4). However, when the estimated non-removal action in adhesive wear is extracted (15.8×10^6 in lb in^{-3} or 11×10^6 N cm^{-2}), it is found to correspond closely with the specific energy for fine abrasive action (10×10^6 in lb in^{-3} or 6.9×10^6 N cm^{-2}).

It is evident that many wear mechanisms exist and frequently several play an important role in a given test situation. This makes the study of wear phenomena one of the most challenging engineering activities.

The specific wear parameter (B/PL) is proportional to the ratio of coefficient of friction to specific energy (μ/u), and will generally increase when the size of

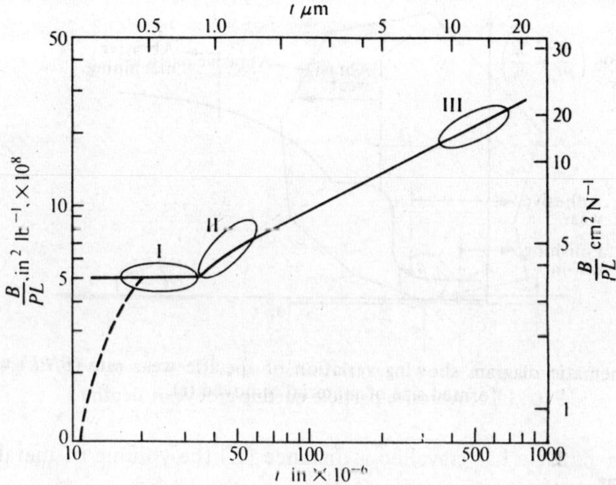

FIG. 11.6. Variation of specific wear parameter (B/PL) with undeformed chip size (t). (After Shaw 1977.)

the wear particles increase. Figure 11.7 shows a composite specific wear diagram, where t may be considered to be either the depth of scratches produced or the size of wear particles. The relationship between the polishing or run-in, adhesive wear, abrasive wear, and abrasive machining regimes is clearly evident in this diagram. The specific energy varies over about eight orders of magnitude in going from one extreme (polishing) to the other (abrasive machining), decreasing of course as the size of particle removed increases.

To minimize wear the hardness of the sliding surfaces should be as high as possible, and the materials should be chosen to be compatible and hence to give a low K. The surfaces should be well lubricated so as to further reduce the effective value of K and provide a low coefficient of friction. The surface roughness should be as low as possible and the asperities that remain should be large in number and be as blunt as possible (large ρ). Polishing or run-in will tend to provide the desired surface texture. The atmosphere should be controlled to decrease wear (sometimes oxygen is beneficial and sometimes detrimental to wear) and means should be provided to remove abrasive wear particles from the system when they tend to form (grooving and copious oil supply with filtering). Finally, the load on the sliding surfaces should be as low as possible.

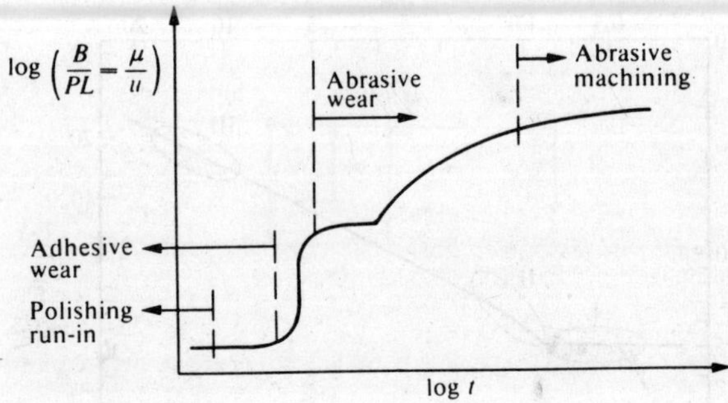

FIG. 11.7. Schematic diagram showing variation of specific wear rate (B/PL) with undeformed size of material removed (t).

Diffusion Wear

When surface temperatures become very high and surface velocities are low, it is possible for solid state diffusion to play a role in the wear process. This unusual combination (high surface temperature and low surface velocity) is found only with heavily loaded sliders involving appreciable subsurface flow. In such a case there is a boundary layer of material across which there is a large velocity gradient (Fig. 11.8). High temperatures are then possible due to the high velocity pertaining at the upper surface of the boundary layer (A) while long contact times are available for diffusion along the lower surface of the boundary layer (B).

When diffusion wear is important, the relative solubilities of the two sliding materials become an important consideration as discussed by Roach, Goodzeit, and Hunnicutt (1956).

At very high speeds the boundary layer may actually become molten. Under such conditions the thickness of the boundary layer will increase in the direction of sliding motion (Fig. 11.8) and this is capable of developing hydrodynamic pressure within the molten layer in the same way that load supporting pressure is developed in a journal bearing (DeSalvo and Shaw 1968). When the boundary layer is liquid the real area of contact will equal the apparent area ($A_R = A$) and diffusion rates will be very large. These conditions will give rise to very high wear rates if the system tends to wear by diffusion. Hence, a liquid boundary layer is to be avoided in such cases even though it gives rise to a load-supporting hydrodynamic film.

Diffusion wear is often accompanied by the decomposition of a component of one of the sliding surfaces. For example, in cutting a ferrous alloy with a tungsten carbide tool at high speed (temperature), Opitz (1963) presents evidence in support of a transformation from α-iron to γ-iron on the surface of the

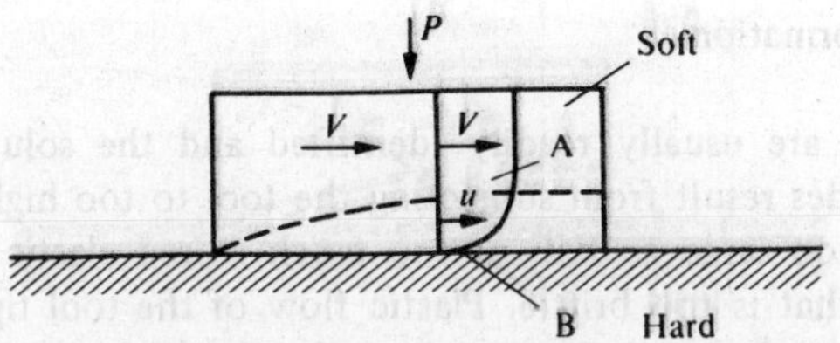

FIG. 11.8. Heavily loaded slider showing boundary layer of subsurface plastic flow.

chip. The γ-iron has a strong affinity for carbon—the WC crystals in the surface decompose and the carbon released diffuses into the surface of the chip. According to Opitz, the increased carbon concentration strengthens the surface of the chip which in turn increases the wear rate.

The fact that the wear rate of cutting tool constituents increases with increased solubility in the chip (Kramer and Suh 1980) suggests that formation of a solid solution precedes decomposition. Transport by diffusion depends on the chemistry and structure of the materials, temperature, time, and the concentration gradient of the diffusing species. Secondary shear flow on the tool face plays an important role relative to all aspects of diffusion transport. Research on this type of tool wear is very difficult to carry out since it is not possible to duplicate all conditions except by use of a cutting test and this will involve other types of wear in addition to that due to diffusion.

Static diffusion studies of pieces of tungsten carbide in contact with various tool materials held at constant temperature in a furnace for a long time (1 or 2 hours) have been somewhat useful in screening the compatability (solubility) of different pairs of materials (Dawihl, 1940a, b, 1941). However, such tests involve conditions far different from those in cutting (no freshly generated surface, no high local deformation strains, and temperatures that are too low, times that are too long, and the absence of realistic temperature and concentration gradients).

Types of Cutting Tool Wear

Tool wear may be classified as follows:

1. Adhesive wear.
2. Abrasive wear.
3. Diffusion wear.
4. Fatigue.

All of these are generally present in combination, the predominant wear mechanism depending upon cutting conditions. In addition to the above sources of tool failure, the following also pertain:

5. Microchipping.
6. Gross fracture.
7. Plastic deformation.

These, however, are usually readily identified and the solution is apparent. The fracture modes result from subjecting the tool to too high a cutting force, operating with too large a BUE or too much stored elastic energy, or using a tool material that is too brittle. Plastic flow of the tool tip arises when the temperature is too high relative to the softening point of the tool material. When plastic flow occurs at the tool tip, tool clearance is lost, the temperature rises abruptly, and total tool failure occurs rather rapidly. The obvious solution to the latter difficulty is to use a lower cutting speed or a tool material that is more refractory. While cemented tungsten carbide is a very refractory material, even this material will suffer from plastic flow at the tool tip if the temperature is too high (Trent, 1967).

Cutting tools wear predominantly in different ways, depending on cutting conditions (principally cutting speed V and undeformed chip thickness t). Opitz (1956) has identified the principal types of tool wear, together with the approximate ranges of values of V and t where each type of wear is predominant (Fig. 11.9). At a low values of the product ($Vt^{0.6}$) tool wear consists predominantly of a rounding of the cutting point and a loss of tool sharpness. This is the main type of tool wear in a broaching operation where both V and t are small. As the product ($Vt^{0.6}$) increases, the predominant type of tool wear shifts as shown in Fig. 11.9. The most important types of tool wear for a single point turning tool is wear-land formation (Fig. 11.9c) and crater formation (Fig. 11.9d). Crater formation is apt to be more important than wear-land formation in situations where cutting temperatures are very high.

Wear-Land Wear

Wear-land wear results in a loss of relief angle on the clearance face of the tool. This gives rise to increased frictional resistance. The wear land is normally a lightly loaded slider with maximum temperature at the trailing edge which increases with increase in wear-land length (w) in accordance with eqn (10.8). Wear-land wear is usually followed by observing the change in w with time, w being measured by a measuring microscope.

Figure 11.10 shows a typical set of curves of w vs. cutting time for different values of cutting speed (V). These curves are generally similar to Fig. 11.3, the extent of the low wear region decreasing with increased cutting speed. The wear rate rises abruptly when the temperature at the trailing edge of the wear-land reaches the thermal softening point of the work material. The A_R/A ratio then increases rapidly with time along with the wear rate.

The tool life (T min) for a given cutting speed is either the cutting time at

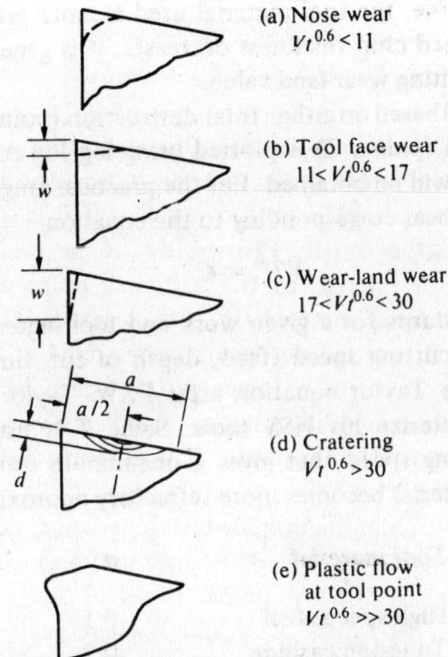

FIG. 11.9. Types of predominant tool wear depending on product ($Vt^{0.6}$). (After Opitz 1956).

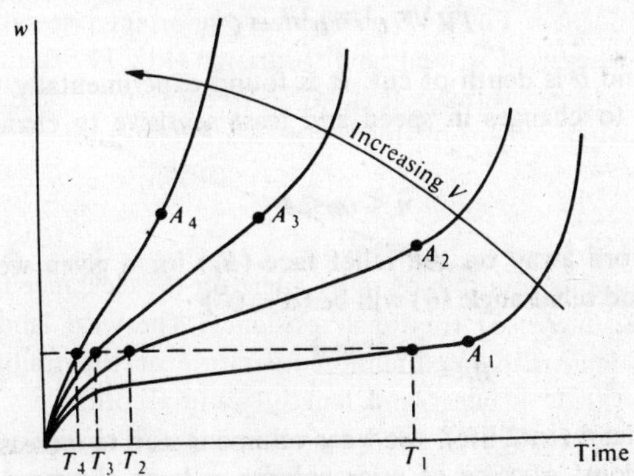

FIG. 11.10. Variation of wear land (*w*) with time for different cutting speeds (*V*).

the limit of the low wear rate region (time corresponding to points A in Fig. 11.10) typically for a high-speed steel (HSS) tool or the time to reach a given wear-land value (time corresponding to T_1, T_2, T_3 in Fig. 11.10) typically for a tungsten carbide tool. A common limiting wear-land value for tungsten carbide cutting mild steel is 0.030 in (0.75 mm). However, as the work material becomes more difficult to machine, the tool material used is more refractory and hence brittle or the undeformed chip thickness decreases, it is generally advantageous to employ a smaller limiting wear-land value.

When the tool life (T) based on either total destruction (points A in Fig. 11.10) or a limiting wear land (points T) is plotted using log–log coordinates, a result as shown in Fig. 11.11 will be obtained. For the practical range of cutting speeds this curve is generally linear corresponding to the equation

$$VT^n = C \tag{11.12}$$

where n and C are constants for a given work and tool material and machining conditions other than cutting speed (feed, depth of cut, fluid, tool geometry, etc). This is called the Taylor equation after F. W. Taylor (1907) who first employed it to characterize his HSS tools. Since T is usually measured in minutes, C is the cutting speed that gives a one-minute tool-life. The value n increases as the tool material becomes more refractory approximately as follows:

Tool material	n
High speed steel	0.1
Tungsten carbide	0.2
Al_2O_3 (ceramic)	0.4

The Taylor equation may be rearranged and generalized as follows

$$TV^{1/n}\, t^{1/m} b^{1/l} = C' \tag{11.13}$$

where t is feed and b is depth of cut. It is found experimentally that tool life is most sensitive to changes in speed and least sensitive to changes in depth of cut. Hence

$$n < m < l$$

The volume worn away on the relief face (B_r) for a given wear land (w), depth of cut (b) and relief angle (θ) will be ($\alpha = 0°$)

$$B_r = \frac{bw^2 \tan\theta}{2} \tag{11.14}$$

For a given wear land (tool life), the wear volume is seen to increase with $\tan\theta$. Thus, from the point of view of wear volume, a large clearance angle is advantageous. However, the strength of the cutting edge, capacity to absorb thermal energy, and ability to maintain dimensional accuracy decreases with an increase in relief angle. Therefore, an optimum relief angle exists in the range of 5 to 10 degrees.

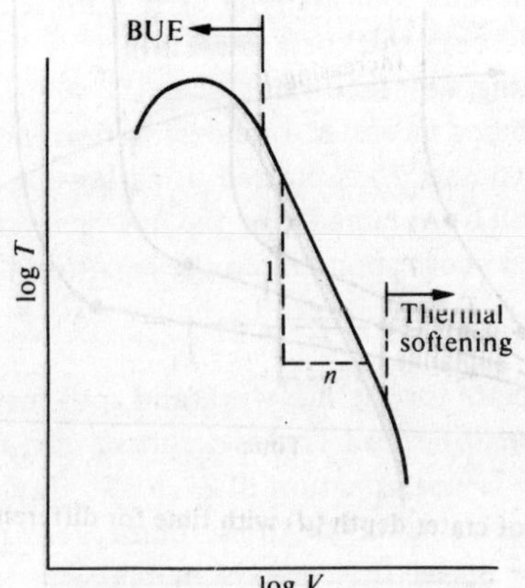

FIG. 11.11. Typical tool life curves based on wear-land wear.

Crater Wear

Conditions on the tool face are much more severe than those on the clearance face. Temperatures, pressures, and A_R/A are much higher and the tool face constitutes a very heavily loaded slider. Diffusion wear is predominant at high cutting speeds and the characteristic wear pattern is a crater (Fig. 11.9*d*). The point of greatest crater depth usually occurs near the midpoint of contact length (*a*) since this is where the tool face temperature is normally maximum.

Crater wear is followed by plotting crater depth versus time and results similar to those of Fig. 11.12 are obtained when this is done.

When the crater approaches the cutting edge the tool temperature increases as does the rate of increase of crater depth (*d*). Crater depth vs. time curves for different cutting speeds (Fig. 11.12) are very similar to wear-land curves (Fig. 11.10). Crater tool-life may be taken just short of total destruction when the crater approaches the cutting edge and edge chipping occurs (points A in Fig. 11.12) or when *d* reaches a critical value (points T_1, T_2, T_3). The tool life curve for crater wear will have the same appearance as Fig. 11.11. A Taylor type equation with different values of *n* and *C* also holds for tool life based on crater wear.

Fortunately, the volume available to be worn away before total destruction is much greater for crater wear than for wear-land wear to compensate for the much greater rate of wear on the tool face. As a consequence, tool life due to crater wear and due to wear-land development are approximately the same despite the very large difference in volume rates of wear.

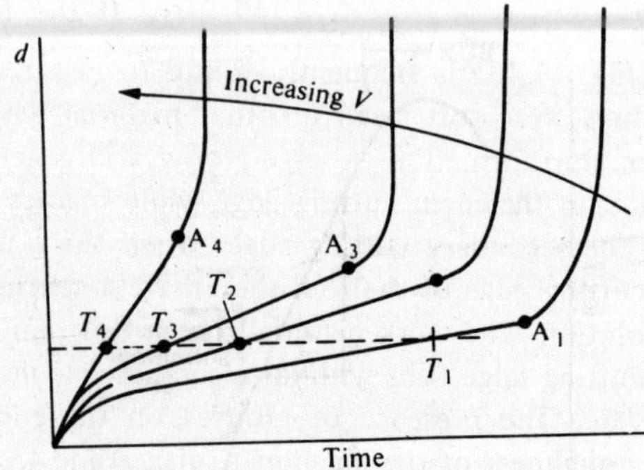

FIG. 11.12. Variation of crater depth (d) with time for different cutting speeds (V).

Tool Life Testing

Unfortunately, due to the complexity of the cutting process, it is not possible to use noncutting tests to evaluate cutting tools, work material structures, cutting fluids, etc. or to appreciably accelerate or truncate tool life tests. This is because cutting temperatures and pressures play such an important and complex role.

Soon after radioactive tracers were first used in wear studies (Burwell, 1947), Shaw and Strang (1950) cut AISI 4027 steel that had been made radioactive by neutron irradiation at the Oak Ridge Laboratory. It was found by the auto-radiograph technique that even under the most ideal cutting fluid conditions, a small amount of work material was transferred to the cutting edges of an end-milling cutter.

Merchant, Ernst, and Krabacher (1953) first used radioactive tools to study the rate of tool wear. Cutting tests of very short duration (~ 10 s) were made and the γ-radiation of tool material transferred to the chips was measured to obtain *relative* wear rates under different cutting conditions.

Since that time, there have been a number of major radioactive tool life studies but the technique has failed to provide a satisfactory substitute for conventional cutting tests for the following reasons:

1. Cost and inconvenience associated with safety precautions.
2. Tool wear is not uniform with time but frequently 'cascades' making it unreliable to extrapolate a very short time radioactive determination to predict the life of a tool.
3. Wear is not uniform across a cutting edge or with regard to the relative amounts of different wear patterns. Grooving (presently to be discussed) and the nonuniform development of the wear land along the cutting edge are examples of why it is difficult to interpret radioactive data in terms of tool life.

Groove Formation

Deep groove wear (Fig. 11.13) is frequently found to occur when machining high temperature alloys, very soft steel, or other material with a strong tendency to strain harden chip formation. A large groove will generally form at the free edge of the chip on the main cutting edge while smaller grooves form at the free surface on the secondary cutting edge. When there is more than one groove on the end-cutting edge of the tool these have a spacing corresponding to the feed per revolution. The work material that is not removed by the first groove on the end-cutting edge rubs without clearance on the tool, generating the second groove, etc. The presence of grooves on the end-cutting edge is a source of surface roughness of the machined surface and frequently dictates tool life in finish machining.

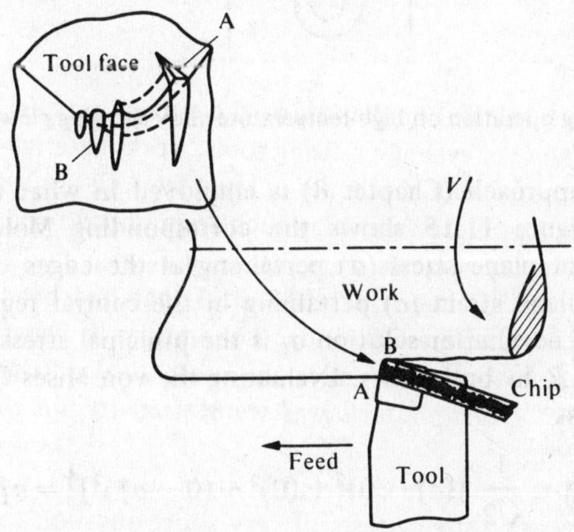

FIG. 11.13. Turning operation showing wear grooves formed at free edges of chip (A and B).

While groove formation is a complex problem, it is believed to be primarily a temperature-related phenomenon. This was suggested when a high-temperature alloy (Waspaloy) was machined in total darkness and the radiation from the chip used to make a photograph on high-speed infrared film. The photograph revealed at the point of chip generation two red-hot chip edges with a dark cooler central region (Fig. 11.14). Since one should expect a greater rate of heat transfer from the edges of the chip than from the central region, the observed high-temperature edges can only mean that more energy was expended in chip formation in the vicinity of the edges of the chip than elsewhere.

Shaw, Thurman, and Ahlgren (1966) have explained the high temperatures observed at the free edges of the chip by noting that whereas the edges of the chip will deform in plane stress, the central region will deform in plane strain.

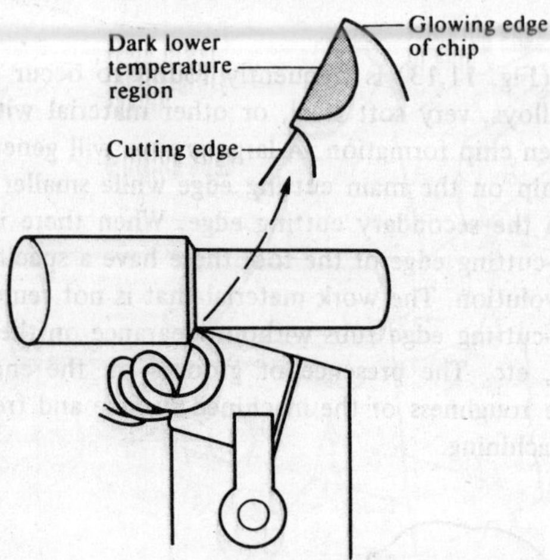

FIG. 11.14. Turning operation on high-temperature alloy showing glowing edges of chip.

The Lee–Shaffer approach (Chapter 8) is employed in what follows as a first approximation. Figure 11.15 shows the corresponding Mohr's stress circles for region ABC for plane stress (*d*) pertaining at the edges of the chip (Fig. 11.15a) and for plane strain (*c*) pertaining in the central region of the chip. According to the Lee–Shaffer solution σ_1 is the principal stress in the direction of resultant force *R* in both cases. Evaluating the von Mises flow criterion we have for plane stress

$$\sigma_e^* = \frac{1}{\sqrt{2}}\,[(\sigma_1^* - 0)^2 + (0)^2 + (0 - \sigma_1^*)^2]^{\frac{1}{2}} = \sigma_1^* \qquad (11.15)$$

where σ_1^* is plus when compressive and the * identifies the effective stress for plane stress. For plane strain (no *)

$$\sigma_e = \frac{1}{\sqrt{2}}\left[\left(\sigma_1 - \frac{\sigma_1}{2}\right)^2 + \left(\frac{\sigma_1}{2} - 0\right)^2 + (0 - \sigma_1)^2\right]^{\frac{1}{2}} = \frac{\sqrt{3}}{2}\sigma_1. \qquad (11.16)$$

For the same material, flow should occur when $\sigma_e^* = \sigma_e$ or when

$$\sigma_1^* = \frac{\sqrt{3}}{2}\sigma_1 = 0.87\,\sigma_1. \qquad (11.17)$$

This means the material at the edges of the chip will be sheared twice—once parallel to the cutting edge before the main shear plane is reached and again on crossing the shear plane when the main body of the chip is also sheared in direction AB.

The additional energy expended in double shear at the edges of the chip will

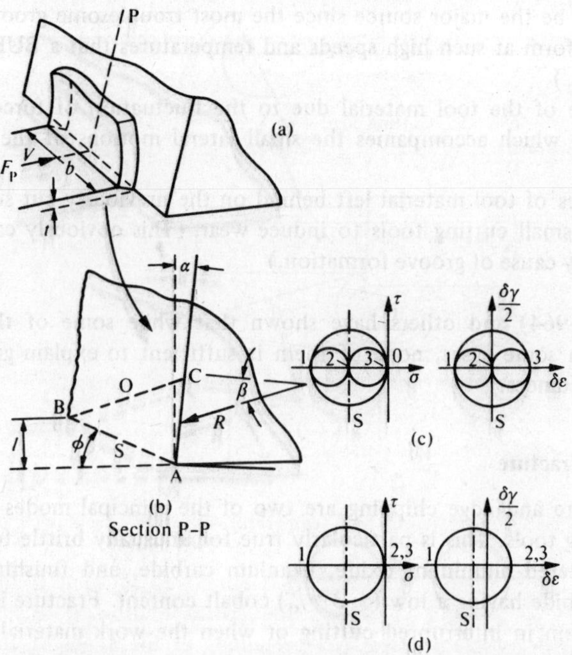

FIG. 11.15. (a) Orthogonal cutting operation. (b) Corresponding Lee–Shaffer slip line field. (c) Mohr's circle for plane strain. (d) Mohr's circle for plane stress.

be only slightly greater for a material that does not strain harden significantly (such as steel) but will be significantly greater for a material having a strong tendency to strain harden at high strain rate (such as Waspaloy). This explains why hot chip edges are not observed with relatively non-strain-hardening steel but are observed with Waspaloy.

A number of additional reasons for groove formation have been proposed (Albrecht 1956, Solaja 1958, Leyensetter 1956, Lambert 1962).

1. Presence of a work-hardened layer on the previously cut surface.
2. Stress concentration due to the stress gradient at the free surface.
3. Formation of thermal cracks due to the steep temperature gradient at the free surface.
4. Presence of a burr at the edge of the freshly machined surface.
5. A higher velocity at the outside diameter. (This obviously cannot be the answer since a groove also forms at the point of lowest speed, B in Fig. 11.13.)
6. Presence of an abrasive oxide layer on the previously cut surface.
7. Flow of built-up edge (BUE) material parallel to the cutting edge. (This

cannot be the major source since the most troublesome grooves are those which form at such high speeds and temperatures that a BUE will not be present.)

8. Fatigue of the tool material due to the fluctuation of force at the free surface which accompanies the small lateral motions of the edge of the chip.
9. Particles of tool material left behind on the previously cut surface which act as small cutting tools to induce wear. (This obviously cannot be the primary cause of groove formation.)

Moltrecht (1964) and others have shown that while some of these may be significant in some cases, none of them is sufficient to explain groove formation in all instances.

Gross Tool Fracture

Gross fracture and edge chipping are two of the principal modes of failure of metal cutting tools. This is particularly true for unusually brittle tool materials such as sintered aluminum oxide, titanium carbide, and finishing grades of tungsten carbide having a low (~ 3 $^{w}/_{o}$) cobalt content. Fracture is also a more severe problem in interrupted cutting or when the work material is unusually hard and strong.

Pekelharing (1978) presented conclusive proof that carbide tools are more apt to chip when leaving a cut than when entering. By use of plane stress, elastic, finite-element analysis, it was shown that upon approaching the end of a cut, the shear 'plane' rotates from its usual positive direction to a negative one and as a consequence, the maximum tensile stress on the tool face increases rapidly and moves closer to the cutting edge. Pekelharing also showed that, as the crack develops, the pattern of stress in the tool shifts so as to cause the crack to bend toward the clearance face. Figure 11.16 shows diagrammatically the situation pertaining at the tip of a cutting tool operating under equilibrium conditions when a considerable distance from the end of a cut. The tensile stress on the rake face will have its maximum value at a point **A** some distance from the tool tip and the state of stress at the tool tip will be highly compressive. Under such conditions there is relatively little tendency toward tool tip chipping.

Pekelharing (1978) cited experimental evidence, based on dynamometer measurements, that there is no 'overshoot' of stress in a tool as it enters a cut. This is to be expected on theoretical grounds. When a zero-degree rake-angle tool first contacts the work, the mean stress on the tool face will correspond to the indentation hardness of the work. However, after cutting equilibrium is established, this stress will correspond to the specific cutting energy. This latter quantity is considerably greater than the hardness since it includes a

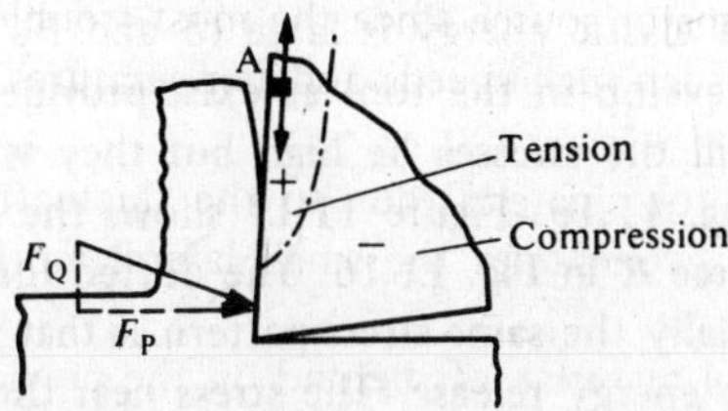

FIG. 11.16. Schematic diagram showing stress pattern for tool cutting under equilibrium conditions.

substantial friction component not involved with hardness and is associated with a much higher plastic strain than in the case of hardness. Thus, we should expect tool stresses to grow gradually toward an equilibrium value when *entering* a cut. This of course assumes a work material that is homogeneous and which does not have a hard surface layer. There will also be some elastic deflection of the tool as it enters the work and this will tend to 'cushion' the impact at entrance.

The elastic energy stored in the system will be approximately ($F_P\delta/2$) where F_P is the cutting component of force and δ is the total tool deflection in the same direction.

It is thus expected that the tendency for fracture is no greater when the tool enters the work than it is under equilibrium cutting conditions unless a hard surface is involved.

Consider the classical problem of a weight slowly and then suddenly applied to a body. If the weight (W) is slowly applied, the energy absorbed by the body will be $W\delta/2$ (where δ is the final elastic deformation). If, however, the weight is suddenly applied as by cutting a string supporting the weight just contacting the body, the energy initially absorbed will be ($W\delta$) and the maximum deformation and hence the maximum stress will be twice as great as in the quasi-static case. The body will oscillate about the equilibrium deformation position ($\delta/2$) with an initial amplitude of $\delta/2$ until the additional energy ($W\delta/2$) is converted to thermal energy by damping.

In the case of a suddenly removed load (W), we should expect the body to overshoot its equilibrium unloaded position by an amount δ if the equilibrium deformation is δ. The resulting stresses should then be the same as though a load were slowly applied in the opposite direction until the deformation were $-\delta$. Thus, a suddenly removed load should be expected to produce stresses equal to equilibrium values but in the opposite direction. This action is somewhat akin to the 'whip lash' that frequently does damage to the back of the neck in an auto crash. The source of the whip lash is different in the two cases of course, being due to the inertia of the head in the case of the crash and to the sudden release of stored elastic energy in the case of the tool.

If the total stored elastic energy is large (δ and F_P large) then very large reverse stresses can develop in the tool at exit, provided the load is suddenly released. Not only will the stresses be high but they will be opposite in sign to those shown in Fig. 11.16. Figure 11.17 shows the situation following the sudden removal of force R in Fig. 11.16. The dotted force vector is an equivalent one giving essentially the same stress pattern as that produced by overshoot due to sudden elastic energy release. The stress near the cutting edge which is normally highly compressive will now be tensile just as though the direction of the resultant cutting force (R) had reversed.

Pekelharing (1978) has shown that as the cutting tool approaches the end of a cut, the shear plane rotates rapidly in the direction of a decrease in shear angle, assuming a negative value before fracture occurs. This shear plane rotation is important in providing the rapid rate of unloading required by the foregoing mechanism of tool fracture. The rate of unloading should increase with cutting speed and hence the tendency for exit failure to occur should increase with increased cutting speed, provided the accompanying increase in tool tip temperature does not cause an increase in tool strength.

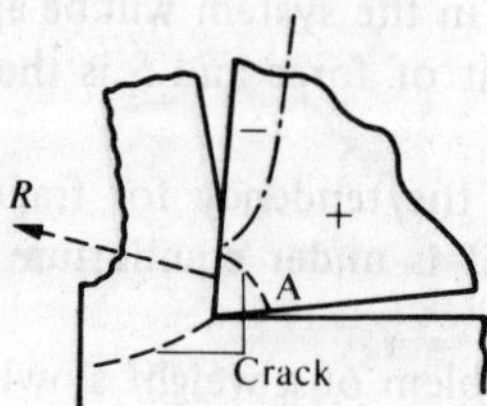

FIG. 11.17. Schematic diagram showing stress pattern for tool subjected to sudden release of stored elastic energy.

Dimensional Analysis

When the input to a brittle body is a static force the appropriate fracture criterion is the maximum tensile stress criterion provided a uniaxial tensile stress is involved at the critical point of fracture. However, when the input is an energy, the fracture criterion is the maximum tensile specific strain energy criterion (Shaw 1977). This means that for an energy input, fracture will occur when (σ_A^2/E) reaches a critical value where σ_A is the maximum uniaxial tensile stress in the loaded body. For a tool of given geometry and material, operating at a given cutting speed, we should expect

$$\sigma_A = \psi_1(E, F_P, \delta, b, t) \tag{11.18}$$

or

$$\frac{\sigma_A^2}{E} = \psi_2\left(\frac{F_P\delta}{b}, t\right) \tag{11.19}$$

F_P = cutting force component in direction of cutting speed
δ = total deflection of tool relative to work in the same direction
t = undeformed chip thickness
b = width of cut

and after performing a dimensional analysis, we have

$$\frac{\sigma_A^2 bt^2}{EF_P\delta} = \text{constant} = N_F \tag{11.20}$$

where N_F is the nondimensional fracture number. But, we may write

$$\delta = \frac{F_P}{K} \tag{11.21}$$

and

$$F_P = ubt \tag{11.22}$$

where

K = spring constant of system in the F_P direction $[FL^{-1}]$
u = specific cutting energy

Substituting into eqn (11.20)

$$N_F = \frac{\sigma_A^2(K)}{Eu^2(b)} \tag{11.23}$$

The nondimensional fracture number (N_F) will be a constant at fracture for a given tool material and geometry when the tool is operating at a given speed and with a given cutting fluid. Just as the Reynolds number in fluid mechanics is proportional to the ratio of the inertia to viscous forces acting on a fluid particle, N_F is proportional to the ratio of the specific elastic energy at the critical point at fracture (σ_A^2/E) to the specific elastic energy stored in the cutting system $F_P\delta/2 = u^2b/2K$. For a given system stiffness (K) and width of cut (b), the maximum allowable specific cutting energy (u) will vary as $(\sigma_A^2/E)^{\frac{1}{2}}$ where σ_A is the critical value of uniaxial tensile stress at fracture as obtained from a 'static' tensile test. As harder materials are cut (higher u), a stronger work material ($>\sigma_A$) and/or a stiffer system (greater K) is called for if fracture at exit is to be avoided.

Exoelectrons

Electrons (exoelectrons) are emitted from a freshly scratched cut or fractured surface as Kramer demonstrated at the PTB in Braunschweig about 1950. This emission appears to be due to the slightly greater equilibrium atom spacing at a surface than internally. Associated with the denser internal atom spacing is a greater density of free electrons. When a new surface is generated there will be an excess of electrons in the surface initially which must leave as exoelectrons in order that the atom spacing in the surface expand to its equilibrium value. The energy involved in this adjustment is surface energy.

A freshly generated surface will have an excess of free electrons and hence will be in a highly active chemical state relative to oxidation reactions (reactions involving a loss of electrons by the surface as for example: $Fe \rightarrow Fe^{++} + 2e^{-}$).

It would appear that the flow of electrons coming from a freshly scratched asperity could have an important influence on bond formation between asperities making contact as well as on surface chemical reactions. This subject is therefore of importance to those interested in wear and is particularly applicable to the wear of heavily loaded sliders where more new surface will be generated as a result of an increased wear rate.

The enormous influence that the generation of fresh surfaces has on surface reactions is dramatically illustrated by the mechanical activation process (Shaw 1948) in which organometallic reactions that will not go under ordinary laboratory conditions may be readily carried out by cutting the metal under the surface of liquid reactants. For example, when magnesium is cut in the presence of phenyl chloride and ethyl ether (solvent) the corresponding Grignard reagent (phenylmagnesium chloride) is produced even though this reaction cannot be induced in the ordinary glass apparatus of the organic chemist. The electrons emitted from the freshly cut surface undoubtedly figure in these reactions, although the increased electron mobility at the freshly cut surface is augmented by local autoclave conditions in the zone of reaction (pressures up to the flow stress of the deforming magnesium and high local temperatures on the surface of the deforming metal).

Over the years there have been many claims and counter-claims concerning the role of the heavy thermoelectric currents that are generated when a tungsten carbide tool cuts steel (Opitz and Axer 1956). The current flow in these cases is from chip to tool and it is claimed that wear may be reduced if this current flow may be prevented by insulating the tool from the system (Fig. 11.18a) or by introducing a counter e.m.f. that reduces the current flow to zero (Fig. 11.18b).

Great claims have been made for this method of preventing tool wear but these have generally not been substantiated. It would appear that the explanation for this anomoly lies in the fact that in general the heavy thermoelectric currents are eddy currents as shown in Fig. 11.19. Insulating the tool or applying

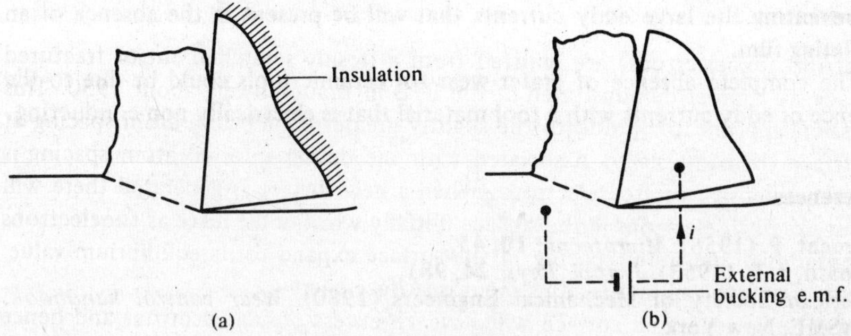

FIG. 11.18. Prevention of current flow between chip and tool. (a) By insulating tool. (b) By use of a bucking potential.

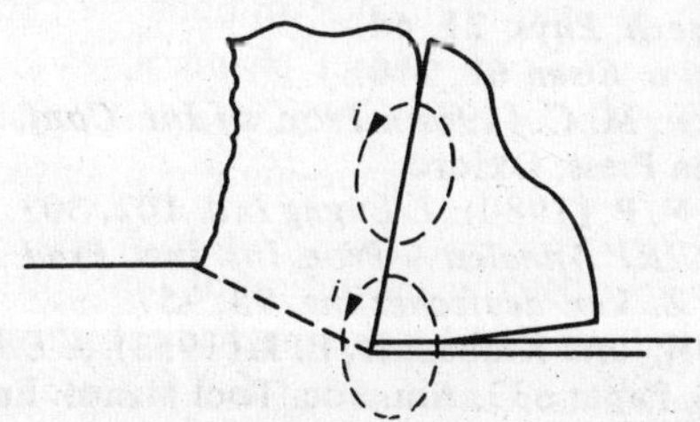

FIG. 11.19. Eddy current flow from tool to chip.

a counter e.m.f. will have little influence on these eddy currents. Occasionally the path resistance may be less through the machine and work than for the path of an eddy current, which would be the case if a semi-conducting oxide were developed between chip and tool. In such instances it would be possible to alter the current flow by tool insulation or by use of a bucking circuit. It is probable that in such cases good results have been obtained and reported in the literature using these techniques. However, it is believed that in general the situation is as shown in Fig. 11.19. The only thing that will reduce the extremely high current flow here is the formation of an insulating oxide film that greatly increases the electrical resistance between chip and tool, or use of a nonconducting (ceramic) tool.

There have been a number of reports of greatly improved tool life (from 2 to 10 times) due to the formation of stationary oxide films on the surface of titanium-carbide-containing tools. This oxide has been shown to be due to the build up of inclusions in the steel resulting from special deoxidation techniques (ferrosilicon and calcium oxide) (Opitz, Gappisch, Koenig, and Wicker 1962). While these oxide films have generally been thought to act as a diffusion barrier to prevent the loss of carbon from the WC crystals in the tool face to the austenitic layer on the chip face, it is also possible that the oxide is preventing the large eddy currents that will be present in the absence of an insulating film.

The complete absence of crater wear for ceramic tools could be due to the absence of eddy currents with a tool material that is electrically non-conducting.

References

Albrecht, P. (1956). *Microtecnic* **10**, 45.

Archard, J. F. (1953). *J. appl. Phys.* **24**, 981.

American Society of Mechanical Engineers (1980). *Wear control handbook.* ASME, New York.

Backer, W. R., Marshall, E. R., and Shaw, M. C. (1952). *J. Engng Ind.* **74**, 61.

Burwell, J. T (1947). *Nucleonics* **(Dec)**, 38.

Burwell, J. T. and Strang, C. D. (1952). *J. appl. Phys.* **23**, 18.

Dawihl, W. (1940a). *Z. Metallkunde* **32**, 320.

Dawihl, W. (1940b). *Z. tech. Phys.* **21**, 44.

Dawihl, W. (1941). *Stahl u. Eisen* **61**, 210.

DeSalvo, G. J. and Shaw, M. C. (1968). *Proc. of Int. Conf. on Mach. Des. and Res.* p. 961. Pergamon Press, Oxford.

Kramer, B. M. and Suh, N. P. (1980). *J. Engng Ind.* **102**, 303.

Lambert, H. J. (1961). *CIRP Annalen—Proc. Int. Inst. Prod. Eng. Res.* **10**, 246.

Leyensetter, W. (1956). *Z. Ver. deafscher Ing.* **98**, 957.

Merchant, M. E., Ernst, H., and Krabacher, E. J. (1953). *J. Engng Ind.* **75**, 549.

Moltrecht, K. H. (1964). Paper 637. Am. Soc. Tool Manuf. Eng. Dearborn.

Opitz, H. (1956). *Werkstattstch. u. Maschinenbau* **46**, 210.

Opitz, H. (1963). In *International Research in Production Engineering*, p. 107. ASME, New York.

Opitz, H. and Axer, A. (1956). *Forschungs Berichte des Wirtschafts und Verkehrsministeriums Nordrhein-Westfallen*, Nr. 271.

Opitz, H., Gappisch, M., Koenig, W., and Wicker, A. (1962). *Archiv. fur Eisenhuttenwesen* **33**, 841.

Pekelharing, J. (1978). *Annals of CIRP—Proc. Int. Prod. Eng. Res.* **27-1**, 8.

Rehbinder, P. A. (1947). *Nature* **159**, 866.

Roach, A. E., Goodzeit, C. L., and Hunnicutt, R. P. (1956). *Trans. Am. Soc, mech. Engrs* **78**, 1639 and 1669.

Shaw, M. C. (1948). *J. appl. Mech.* **70**, A137.

Shaw, M. C. (1971). *Annals of CIRP* **19**, 533.

Shaw, M. C. (1977). *Int. J. Wear* **43**, 263. Elsevier Sequoia SA, Lausanne.

Shaw, M. C. and Strang, C. D. (1950). *J. appl. Phys.* **21**, 349.

Shaw, M. C., Thurman, A. L., and Ahlgren, H. J. (1966). *J. Engng Ind.* **88**, 142.

Solaja, V. (1958). *Int. J. Wear* **2**, 40.

Taylor, F. W. (1907). *Trans. Am. Soc. mech. Engrs* **28**, 31.

Trent, E. M. 1963). *Proc Iron and Steel Inst.* **1001**.

Trent, E. M. (1967). *Proc. Eighth Mach. Tool Des. and Res. Conf.*, p. 629. Pergamon Press, Oxford.

12

CUTTING TEMPERATURES

There are several temperatures of importance in metal cutting. The shear plane temperature (θ_S) is of importance for its influence on flow stress and since it has a major influence on temperatures on the tool face (θ_T) and on the relief surface (θ_R). The latter two temperatures (θ_T and θ_R) are very important to crater wear rate and rate of wear-land development respectively. The temperature on the tool face also plays a major role relative to the size and stability of the BUE. The ambient temperature of the work approaching the cutting zone (θ_o) is also important since it directly affects θ_S, θ_T and θ_R.

Thermal Energy in Cutting

Practically all of the mechanical energy associated with chip formation ends up as thermal energy. One of the first measurements of the mechanical equivalent of heat (*J*) was made by Benjamin Thomson (better known as Count Rumford). Rumford (1799) measured the heat evolved during the boring of brass cannon in Bavaria. He immersed the work, tool, and chips in a known quantity of water and measured the temperature rise corresponding to a measured input of mechanical energy. These experiments not only provided a good approximation to the mechanical equivalent of heat that stood as the accepted value for several decades but also provided new insights into the nature of thermal energy at a time when most people believed that heat was a special form of fluid called 'caloric'. It is well known that some of the energy associated with plastic deformation remains in the deformed material. Taylor and Quinney (1934, 1937) using a very accurate calorimetric technique measured the residual energy involved when metal bars were deformed in torsion. It was found that the percentage of deformation energy retained by the bars decreased with increase in strain energy involved. When these results are extrapolated to strain energy levels in chip formation, it is estimated that the energy that is not converted to thermal energy is only between 1 and 3 percent of the total cutting energy. Bever, Marshall, and Titchener (1953) have directly measured the residual energy stored in metal cutting chips and Bever, Holt, and Titchener (1974) have discussed the stored energy in plastically deformed bodies from a broad point of view.

All of these results suggest it is safe to assume as a first approximation that all of the energy associated with chip formation is converted to thermal energy.

The energy retained in the chips and that associated with the generation of new surface area is negligible relative to the total energy expended in chip formation.

Experimental Techniques

Tool temperatures could not be treated quantitatively until the 1920s when means were devised by Shore (1924) in USA and almost simultaneously by Gottwein (1925) in Germany and Herbert (1926) in Great Britain for measuring the mean temperature along the face of a cutting tool by the tool–work thermocouple technique. This method is based on the fact that an e.m.f. is generated at the interface of two dissimilar metals when the temperature of the interface changes.

The laws of thermoelectric circuits that are applicable here may be summarized as follows:

1. The e.m.f. in a thermoelectric circuit depends only on the difference in temperature between the hot and cold junctions, and is independent of the gradients in the parts making up the system.
2. The e.m.f. generated is independent of the size and resistance of the conductors.
3. If the junction of two metals is at uniform temperature, the e.m.f. generated is not affected if a third metal, which is at the same temperature, is used to make the junction between the first two.

Just how these principles are applied in practice is illustrated in Fig. 12.1, which shows a lathe equipped for measuring tool face temperatures. The chip and tool junction H constitutes the hot junction, while A and B, the cold junctions, remain at room temperature. Should these points be heated appreciably above room temperature during tests, special procedures are available to compensate for this. The mercury contact is used to make electrical contact with

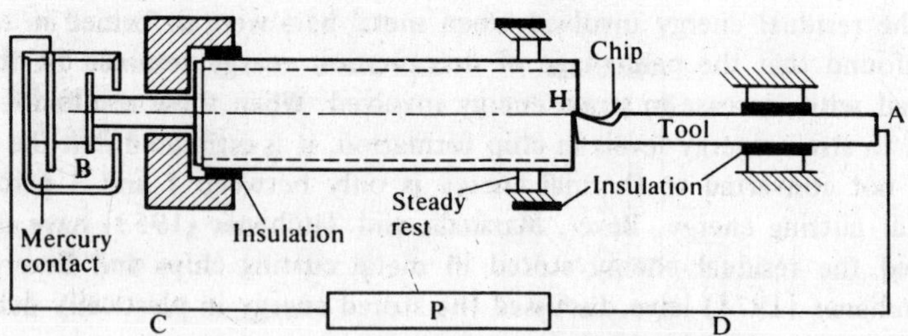

FIG. 12.1. Schematic diagram showing the manner in which tool–chip interface temperatures are measured by the thermoelectric technique.

the rotating member without introducing extraneous voltages that frequently accompany the use of slip rings. The wires C and D may be ordinary copper wires since both ends are at the same temperature and thus have no influence on the e.m.f. generated at H (rule 3), as measured with the potentiometer P.

The simplest and possibly the most accurate method for calibrating the tool–work thermocouple is shown in Fig. 12.2. The most convenient form of workpiece is a long chip. The end of the tool is best ground to a $\frac{1}{8}$ in (3 mm) diameter, both to ensure uniform temperature and to limit the amount of heat transferred to the 'cold' end of the tool. To keep the 'cold' end at room temperature it is advisable to use a long tool, or possibly several similar carbide inserts brazed together, otherwise corrections may be necessary. If the workpiece does not form a long chip it must be prepared similar to the tool. As long as the lead bath is at uniform temperature its presence will have no effect on the e.m.f. measured, according to the third law of thermoelectricity. Its temperature is measured with a standard chromel–alumel couple. In case a large number of calibrations are to be made with various tool and work materials it is often more convenient to measure the e.m.f.s of the tool and work materials separately, both against alumel wire and then obtain the tool–work calibration by difference. That this will yield the desired calibration is also evident from the foregoing third law of thermoelectricity. Two typical calibration curves obtained in the manner described are shown in Fig. 12.3. They are for the same low-carbon, free machining steel (0.32% sulphur), but one is for 18-4-1 high-speed steel while the other is for K2S cemented carbide.

Representative cutting temperatures measured with the tools and work material of Fig. 12.3, over a range of cutting speeds and at several depths of cut, are given in Fig. 12.4. The cutting temperatures are seen to be noticeably higher with the HSS tool, than with the carbide at the same values of speed and feed. The carbide tool had a rake angle of 20°, while the HSS tool had a rake angle of 15°. Both tools had clearance angles of 5°, and all cuts were made under orthogonal conditions.

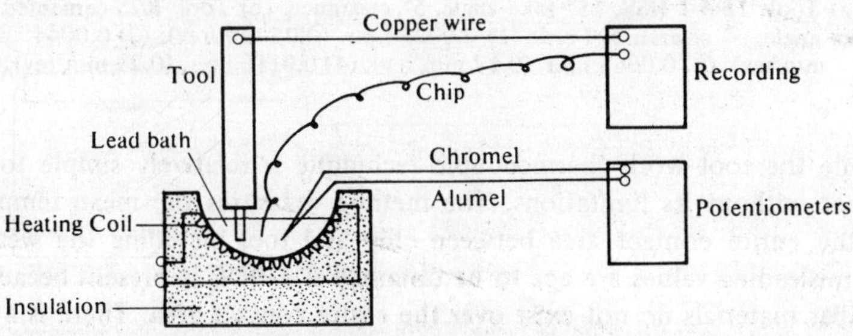

FIG. 12.2. Arrangement for calibration of tool–work thermocouple.

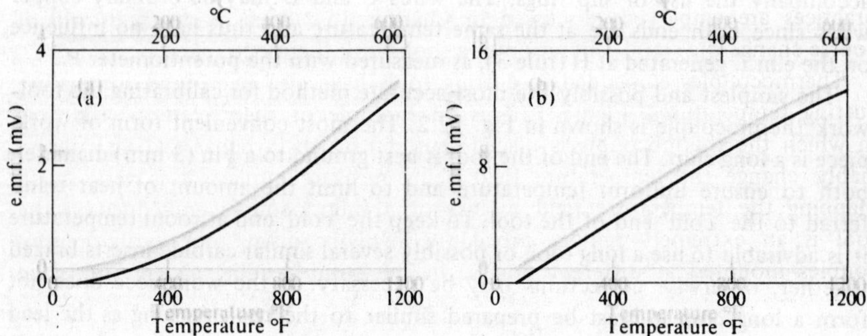

FIG. 12.3. Temperature calibration curves of AISI 1113 steel against (a) 18-4-1 HSS; and (b) K2S cemented carbide.

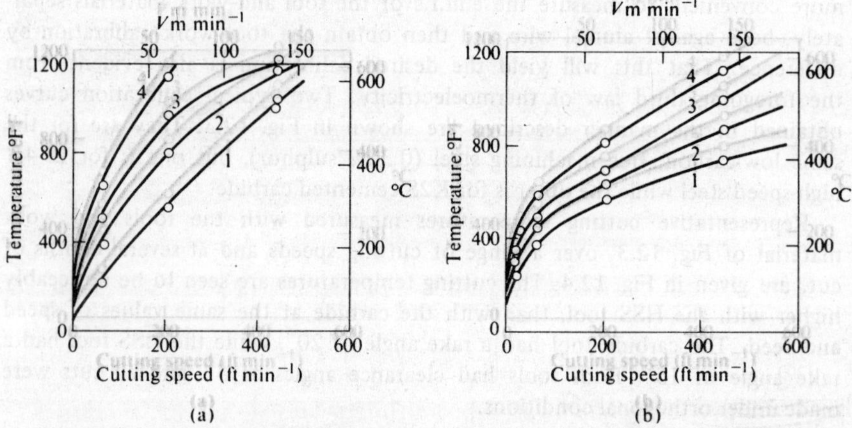

FIG. 12.4. Representative temperature data for two-dimensional dry cutting of AISI 1113 steel. (a) Tool: 18-4-1 HSS, 15° rake angle, 5° clearance. (b) Tool: K2S cemented carbide, 20° rake angle, 5° clearance. Feed: (1) 0.0023 i.p.r. (0.05 mm/rev); (2) 0.0044 i.p.r. (0.11 mm/rev); (3) 0.0065 i.p.r. (0.17 mm/rev); (4) 0.0116 i.p.r. (0.29 mm/rev).

While the tool-work thermocouple technique is relatively simple to apply, it is not without its limitations. The method measures the mean temperature over the entire contact area between chip and tool including the wear land. Also, misleading values are apt to be obtained if a BUE is present because then dissimilar materials do not exist over the entire contact area. There is a further question as to whether a static calibration is satisfactory for the dynamic cutting situation involving exoelectrons. Also, oxide layers tend to form on carbide tools when low-melting nonmetallic inclusions resulting from certain deoxidation practices are employed (discussed in Chapter 15). Such oxide layers will of course change the static calibration of a tool-work thermocouple if present.

Thermocouples embedded in cutting tools have also been used to measure cutting temperatures. Figure 12.5 is an example of such a study. The hole in which the thermocouple is placed represents a disturbance and may appreciably change the temperature field being measured unless it is very small. Mapping the temperature field by use of a thermocouple embedded in the tool is an extremely tedious procedure involving many tools with thermocouples mounted at different points. In the study leading to Fig. 12.5, the

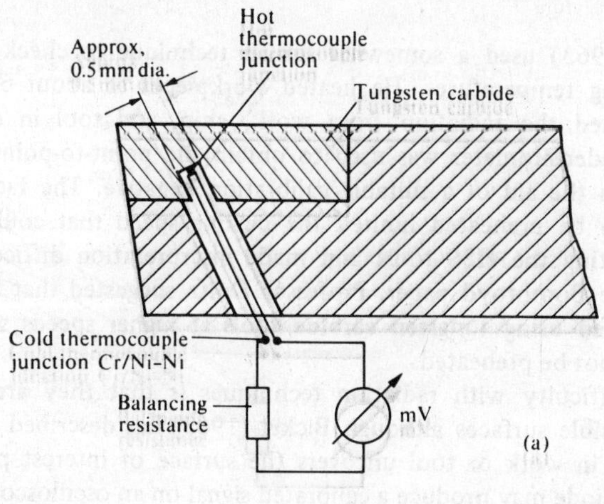

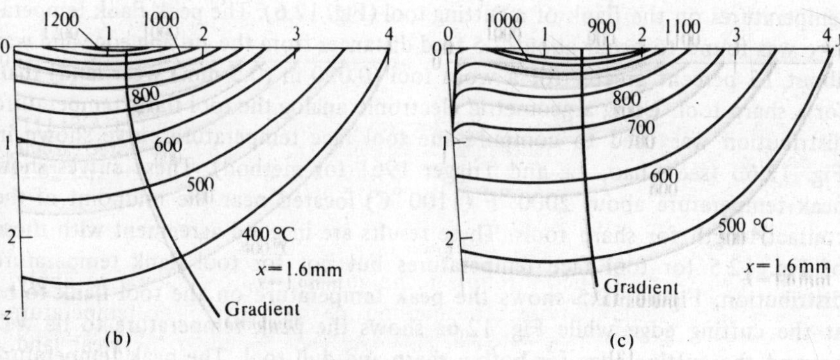

FIG. 12.5. Embedded thermocouple technique for mapping temperatures in cutting tool used in turning operation. (a) Experimental arrangement. (b) Sharp tool. (c) Worn tool. Cutting conditions: work, 0.3 w/o C, 1 w/o Mn steel; cutting speed, 180 m min^{-1} (600 f.p.m.); depth of cut, 3 mm (0.12 in); feed, 0.74 mm/rev (0.03 i.p.r.); rake angle, 0°; relief angle 4°; side cutting edge angle, 45°; fluid, none. (After Küster 1956.)

very small holes were made by electroerosion and temperatures within 0.2 mm of the surface could be measured. Figure 12.5 shows the maximum temperature in the tool surface to occur an appreciable distance from the cutting edge. Figure 12.5b is for a new tool and Fig. 12.5c is for a worn tool. The center of the crater that develops is seen to occur very near the point of maximum surface temperature.

Schwerd (1937) measured the infrared radiation from the tool, work, and chip in orthogonal cutting to establish the temperature field on the outside surfaces of these regions. While this method did not reveal internal temperatures, it did provide a valuable early insight into the temperature distribution on shear plane, tool, and chip. The field covered by each measurement was 0.2 mm.

Boothroyd (1963) used a somewhat similar technique to check analytical studies of cutting temperatures. He heated workpieces to about 600 °C and then photographed the radiation from work, chip, and tool in orthogonal cutting. A microdensitometer was used to obtain the point-to-point intensity of radiation with the aid of a suitable calibration negative. The fact that the specimen had to be preheated limited the cutting speed that could be used without overheating the HSS tools and made interpretation difficult. In the discussion of the Boothroyd paper, Professor Opitz suggested that the experiments be repeated using tungsten carbide tools at higher speeds so that the workpiece need not be preheated.

The main difficulty with radiation techniques is that they are generally limited to accessible surfaces although Bickel (1963) has described techniques in which a hole in work or tool uncovers the surface of interest periodically so that a photo diode may produce a calibrated signal on an oscilloscope.

Trigger (1963) has similarly used a lead sulfide radiation sensor to map temperatures on the flank of a cutting tool (Fig. 12.6). The peak flank temperature was found to occur about 2.5 feed distances from the cutting edge and was about 15 percent greater for a worn tool (0.020 in (0.5 mm) wear land) than for a sharp tool. Using a geometric electronic analog the tool flank temperature distribution was used to compute the tool face temperature curve shown in Fig. 12.6b (see Chao, Li, and Trigger 1961 for method). These curves show peak temperature about 2000 °F (1100 °C) located near the midpoint of the contact length for sharp tools. These results are in good agreement with those of Fig. 12.5 for tool face temperatures but not for tool flank temperature distribution. Figure 12.5 shows the peak temperature on the tool flank to be at the cutting edge while Fig. 12.6a shows the peak temperature to lie well beyond the cutting edge for both a sharp and dull tool. The peak temperature on the clearance face should be expected to occur at the tool point for a sharp tool but at the end of the wear land for a worn tool.

Wright and Trent (1973) have mapped the softening of a HSS tool when operated at high speed for a short time (30 s). When an M-34 HSS tool

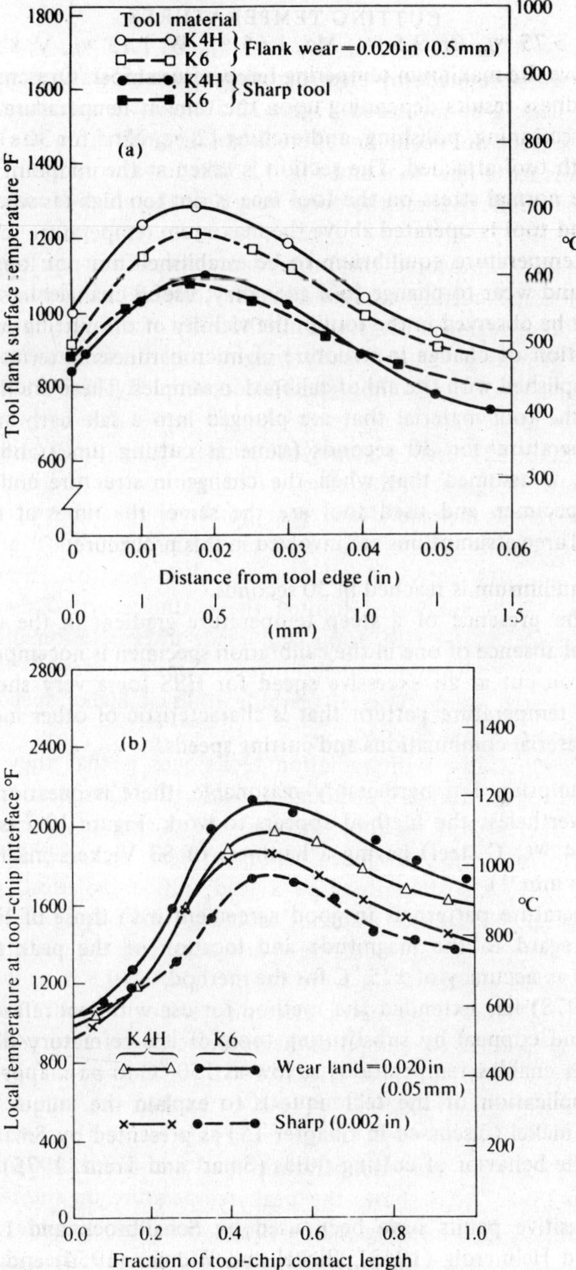

FIG. 12.6. Temperature distribution on (a) tool flank; and (b) tool face as determined by radiation measurement (flank) and computation (face) when turning AISI 52100 steel (380 Brinell hardness) with sharp and dull carbide tools. Machining conditions: Tool geometry, 0, 5, 5, 5, 5, 5, 0.045 in (1.1 mm); cutting speed, (a) 150 f.p.m. (47 m min^{-1}); (b) 200 f.p.m. (63 m min^{-1}); feed, 0.0074 i.p.r. (0.19 mm/rev). (After Trigger 1963.)

(0.87 w/o C, 3.75 w/o Cr, 9.5 w/o Mo, 1.65 w/o W, 1.15 w/o V, 8.25 w/o Co) is operated above the maximum tempering temperature (550 °C) a change in structure and hardness results depending upon the time at temperature. Structure is revealed by sectioning, polishing, and etching (2 w/o Nital for 30 s) a quick-stop chip root with tool attached. The section is taken at the midpoint of the depth of cut. If the normal stress on the tool face is not too high (as when machining pure iron) and tool is operated above the maximum temperature for a time long enough for temperature equilibrium to be established but not long enough for plastic flow and wear to change tool geometry, useful changes in structure and hardness may be observed in the tool in the vicinity of the cutting edge.

Interpretation of change in structure or microhardness in terms of temperature is accomplished with the aid of calibration samples. These are thin (1.5 mm) platelets of the tool material that are plunged into a salt bath and held at a known temperature for 30 seconds (same as cutting time) and then water quenched. It is assumed that when the change in structure and hardness in calibration specimen and used tool are the same, the times at temperatures correspond. Three assumptions are involved in this procedure:

1. That equilibrium is reached in 30 seconds.
2. That the presence of a steep temperature gradient in the tool and the essential absence of one in the calibration specimen is not important.
3. That iron cut at an excessive speed for HSS for a very short time will yield a temperature pattern that is characteristic of other more practical tool–material combinations and cutting speeds.

While assumption 1 is perfectably reasonable, there is question concerning 2 and 3. Nevertheless the method appears to work. Figure 12.7 is an example for iron (0.04 w/o C steel) having a hardness of 83 Vickers machined at 600 f.p.m. (183 m min^{-1}).

This temperature pattern is in good agreement with those of Fig. 12.5 and 12.6b with regard to the magnitude and location of the peak temperature. Wright claims an accuracy of ±25 °C for the method.

Wright (1978) has extended the method for use with less refractory metals (aluminum and copper) by substituting tools of less refractory die steel (01) for HSS. This enables temperatures as low as 150 °C to be mapped. The most important application of the technique is to explain the unique cutting performance of nickel (discussed in Chapter 15) as presented by Smart and Trent (1974) and the behavior of cutting fluids (Smart and Trent, 1975) as discussed in Chapter 13.

Thermosensitive paints have been used by Schallbrock and Lang (1943), Pahlitzsch and Helmerdig (1943), Bickel and Widmer (1954) and others. This technique is limited to accessible surfaces under steady state conditions. This method is of limited value to estimate relative tool temperatures, and is not capable of giving accurate temperatures at actual wear surfaces.

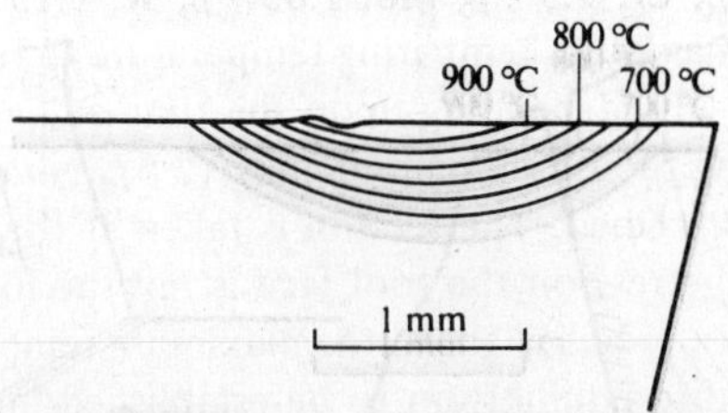

FIG. 12.7. Isotherms for machining of iron (0.04 w/o C) at 183 m min^{-1} (600 f.p.m.), feed of 0.25 mm/rev (0.010 i.p.r.), depth of cut of 1.25 mm (0.050 in) and cutting time of 30 s. (After Wright and Trent 1973.)

A somewhat related method of assessing cutting temperatures is to note the temper color of a chip. The temper color of a chip is due to optical interference that depends on the thickness of oxide produced on the surfaces of the chip. This thickness depends on time at temperature and oxygen concentration as well as temperature which makes interpretation difficult. For example, a silver white chip produced with a flood of cutting fluid may actually have involved a higher cutting temperature than a blue chip produced in air. The fluid would exclude oxygen from the surface of the chip in this case and hence prevent the temper color from developing.

There are no simple reliable methods of measuring the temperature field in the chip–tool and workpiece for even steady state orthogonal metal-cutting. Therefore, analytical approaches, even though often more complex than we would like, must be relied on to obtain a broad insight into the all important thermal aspects of metal cutting.

Analytical Approach

There are two major sources of thermal energy in metal cutting with a sharp tool—in the shear zone (1) and along the tool face (2) as shown in Fig. 12.8a. As a first approximation, it is assumed that:

1. All of the energy expended at (1) and (2) is converted to thermal energy.
2. The energy at (1) and (2) is concentrated on a plane surface (Fig. 12.8*b*).
3. The energy at (1) and (2) is uniformly distributed (i.e. the heat flux along (1) and (2) is constant).

Even with these assumptions, the problem of estimating the mean temperatures on the shear plane ($\bar{\theta}_S$) and tool face ($\bar{\theta}_T$) is complex. This is because part of the energy at (1) will be convected away by the chip and part will flow into the work. Also, part of the energy at (2) will usually go to the chip and part to the tool. There are thus two partition coefficients to be evaluated. R_1 is the fraction of energy (1) going to the chip; R_2 is the fraction of energy at (2) going

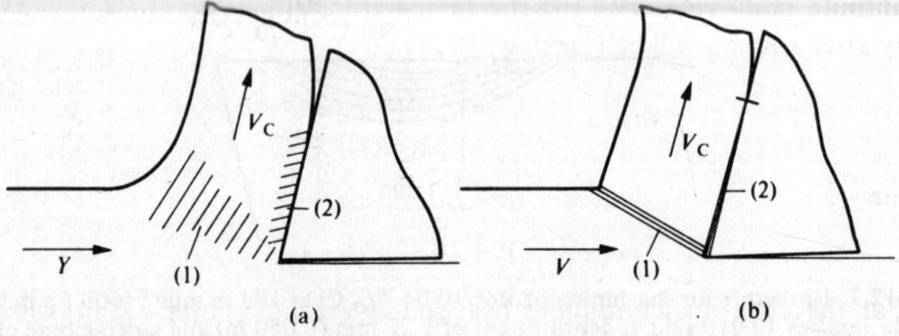

FIG. 12.8. (a) Actual and (b) idealized heat sources in orthogonal metal-cutting.

to the chip. A fourth assumption is that none of the thermal energy goes to the environment during the chip forming process. This means that the energy per unit volume going to the chip on the shear plane (1) is $u_{C1} = R_1 u_S$ while the energy per unit volume going to the chip at (2) is $u_{C2} = R_2 u_F$ where u_S and u_F are the specific energies involved in shear and friction respectively.

The method of estimating the partition coefficients is due to Blok (1938). This method evaluates the mean interface temperature from two points of view:

1. Assuming all of the heat to flow into a stationary member.
2. Assuming all of the heat to flow into the extensive member. This latter calculation involves the moving heat source solution of Jaeger (1942) for an insulated slider moving across a semi-infinite solid.

When the mean temperature from calculation (1) using a heat flux of $(1-R)q$ (q = total thermal units per unit area per unit time) is equated to that of calculation (2) using a heat flux of Rq, an equation with R as the only unknown is obtained which may be solved for R.

Friction Slider

Before treating shear plane and tool face temperatures the aforementioned method of obtaining the heat flow partition coefficient (R) will be illustrated by a simpler application—that of a friction slider corresponding to a moving rectangular heat source of dimensions $2l \times 2m$ of uniform strength q $[FL/L^2T = FL^{-1}T^{-1}]$ moving over a semi-infinite body insulated everywhere except across the interface (Fig. 12.11a with $V = 0$).

The partial differential equation that relates the temperature and energy input is

$$\frac{k}{\rho C}\left(\frac{\partial^2 \theta}{\partial x^2} + \frac{\partial^2 \theta}{\partial y^2} + \frac{\partial^2 \theta}{\partial z^2}\right) + \frac{q}{\rho C} = \frac{d\theta}{dt} \tag{12.1}$$

When a quantity of heat Q is liberated instantaneously at a point x', y', z' in an infinite body (Fig. 12.11b) the temperature at any point $P\,(x, y, z)$ in the body after a time t will be (Carslaw, 1921).

$$\theta(x, y, z, t) = \left[\frac{QK}{8k(\pi Kt)^{\frac{3}{2}}}\right] e^{-r^2/4Kt} \tag{12.2}$$

where

$$r^2 = (x - x')^2 + (y - y')^2 + (z - z')^2. \tag{12.3}$$

For the case of a continuous heat source extending over a finite area, both time and area integration are required. If the time integration is performed first, for the semi-infinite body, and if we concentrate our interest on the steady-state solution (i.e. let $t \to \infty$), then the equation for the steady-state temperature rise anywhere in the semi-infinite body with a uniform heat source, of strength $q\,[FL^{-1}T^{-1}]$ extending over $-l < x' < l$ and $-m < y' < m$ is given by

$$\Delta\theta = \frac{q}{2\pi k}\int_{-l}^{l} dx' \int_{-m}^{m} \frac{dy'}{[(x - x')^2 + (y - y')^2 + z^2]^{\frac{1}{2}}}. \tag{12.4}$$

The main interest lies in the temperature rise in the plane containing the source. Then, upon integrating eqn (12.4) and letting $z = 0$, we have

$$\begin{aligned}\Delta\theta = \frac{q}{2\pi k}\Bigg[& |x + l|\left\{\sinh^{-1}\left(\frac{y + m}{x + l}\right) - \sinh^{-1}\left(\frac{y - m}{x + l}\right)\right\} \\ & + |x - l|\left\{\sinh^{-1}\left(\frac{y - m}{x - l}\right) - \sinh^{-1}\left(\frac{y + m}{x - l}\right)\right\} \\ & + |y + m|\left\{\sinh^{-1}\left(\frac{x + l}{y + m}\right) - \sinh^{-1}\left(\frac{x - l}{y + m}\right)\right\} \\ & + |y - m|\left\{\sinh^{-1}\left(\frac{x - l}{y - m}\right) - \sinh^{-1}\left(\frac{x + l}{y - m}\right)\right\}\Bigg].\end{aligned} \tag{12.5}$$

The mean surface-temperature rise over the area of the source may be obtained by integrating eqn (12.5) over the area of the source and dividing the result by the area, thus

$$\Delta\bar{\theta} = \frac{\int_{-l}^{l}\int_{-m}^{m} (\Delta\theta)\,dx\,dy}{4lm} \tag{12.6}$$

$$\Delta\bar{\theta} = \frac{ql}{k}\,\frac{2}{\pi}\left[\sinh^{-1}\left(\frac{m}{l}\right) + \left(\frac{m}{l}\right)\sinh^{-1}\left(\frac{l}{m}\right) - \tfrac{1}{3}\left(\frac{m}{l}\right)^2 + \tfrac{1}{3}\left(\frac{l}{m}\right) - \tfrac{1}{3}\left\{\left(\frac{l}{m}\right) + \left(\frac{m}{l}\right)\right\}\left\{1 + \left(\frac{m}{l}\right)^2\right\}^{\frac{1}{2}}\right]. \tag{12.7}$$

This equation may be written

$$\Delta\bar{\theta} = \frac{ql}{k}\bar{A} \tag{12.8}$$

where $\bar{A}$ is the area factor (a function of the aspect ratio of the surface area m/l only), plotted in Fig. 12.9. The function A_m, also shown in Fig. 12.9, has a similar significance for the maximum temperature rise in the surface ($\Delta\theta_m$) and is defined as

$$\Delta\theta_m = \frac{ql}{k}A_m. \tag{12.9}$$

For values of (m/l) greater than 20, the quantities $\bar{A}$ and A_m may be found to a good approximation from the equations

$$\bar{A} = \frac{2}{\pi}\left(\ln\frac{2m}{l} + \tfrac{1}{3}\frac{l}{m} + \tfrac{1}{2}\right) \tag{12.10}$$

$$A_m = \frac{2}{\pi}\left(\ln\frac{2m}{l} + 1\right). \tag{12.11}$$

The nature of the temperature distribution over the surface as well as below it is shown in Fig. 12.10 for a stationary heat source.

Jaeger (1942) has presented a more exact solution to that given in Chapter 10 for the problem where a perfect insulator slides over the surface of a semi-infinite body with velocity V (Fig. 12.11) and heat flux q at the interface. For $m/l > 2$,

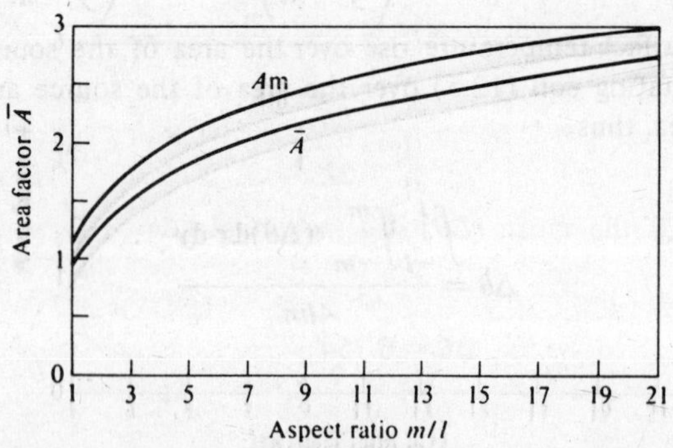

FIG. 12.9. Variation of area factors $\bar{A}$ and A_m with aspect ratio m/l.

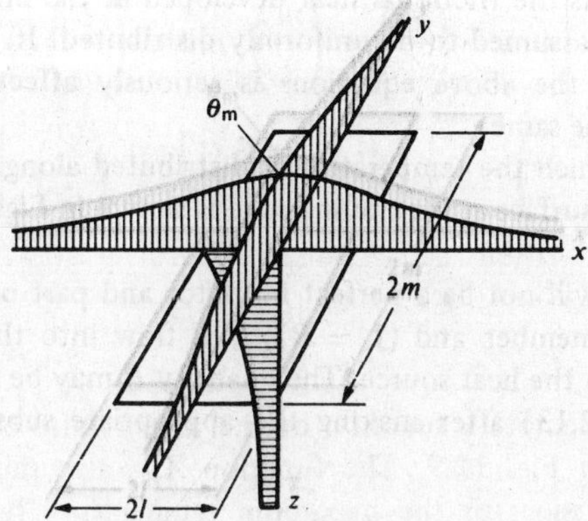

FIG. 12.10. Temperature distribution across and below a stationary heat source on a semi-infinite body.

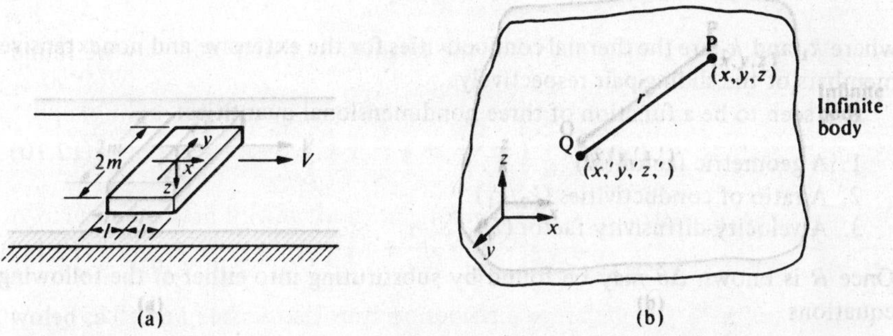

FIG. 12.11. (a) Insulated slider on conducting surface. (b) Infinite body with sudden energy release at Q at time zero.

the influence of the changes in m will be negligible over the velocity range of interest here and results will involve a single dimensionless velocity parameter L defined as follows:

$$L = \frac{Vl}{2K}. \tag{12.12}$$

Provided $L > 5.0$ the mean ($\Delta\bar{\theta}$) and maximum ($\Delta\theta_m$) surface temperature rises will be

$$\Delta\bar{\theta} = 0.754 \frac{ql}{k\sqrt{L}} \tag{12.13}$$

$$\Delta\theta_m = 1.130 \frac{ql}{k\sqrt{L}} \tag{12.14}$$

where $q[FL^{-1}T^{-1}]$ is the frictional heat developed at the interface in the case of a slider which is assumed to be uniformly distributed. If, however, q is not uniform, neither of the above equations is seriously affected as long as the heat input remains the same.

The manner in which the temperature is distributed along the surface of the slider and below the surface at the trailing edge is shown in Figs. 12.12 and 12.13 respectively.

An actual slider will not be a perfect insulator and part of q (Rq) will flow into the extensive member and $(1 - R)q$ will flow into the member that is stationary relative to the heat source. The quantity R may be found by equating eqns (12.8) and (12.13) after making the appropriate substitutions for q in each case.

$$R = \frac{1}{1 + \dfrac{0.754\,(k_2/k_1)}{\bar{A}\sqrt{L}}} \tag{12.15}$$

where k_1 and k_2 are the thermal conductivities for the extensive and nonextensive members of the sliding pair respectively.

R is seen to be a function of three nondimensional quantities:

1. A geometric factor ($\bar{A}$)
2. A ratio of conductivities (k_2/k_1)
3. A velocity-diffusivity factor (L)

Once R is known $\Delta\bar{\theta}$ may be found by substituting into either of the following equations

$$\Delta\bar{\theta} = \frac{0.754\,Rql}{k_1\sqrt{L}} \tag{12.16}$$

$$\Delta\bar{\theta} = \frac{(1-R)ql}{k_2}\,\bar{A} \tag{12.17}$$

where $\Delta\bar{\theta}$ is the temperature rise above ambient (θ_0).

It should be noted that equating eqns (12.16) and (12.17) is not exact. From Figs. 12.10 and 12.12, it is evident that the temperature distribution for the stationary and moving heat source situations are quite different. The procedure of equating average temperatures is an approximation to a very complicated situation. However, experimental results are found to be in reasonably good agreement with calculated mean temperature values and hence the procedure is useful.

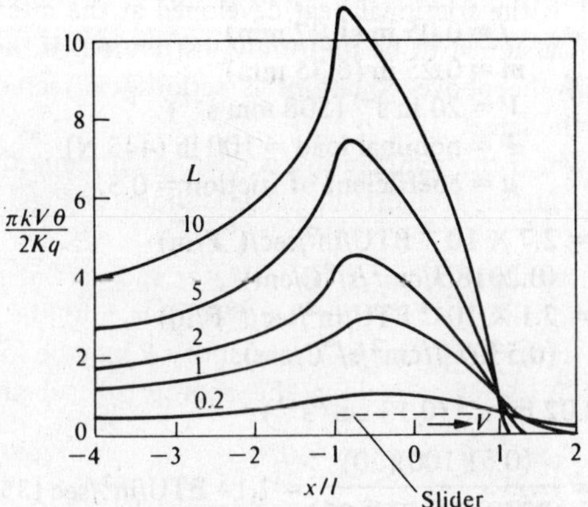

FIG. 12.12. Temperature distribution curves for a non-conducting friction slider of infinite width sliding on a plane conducting surface.

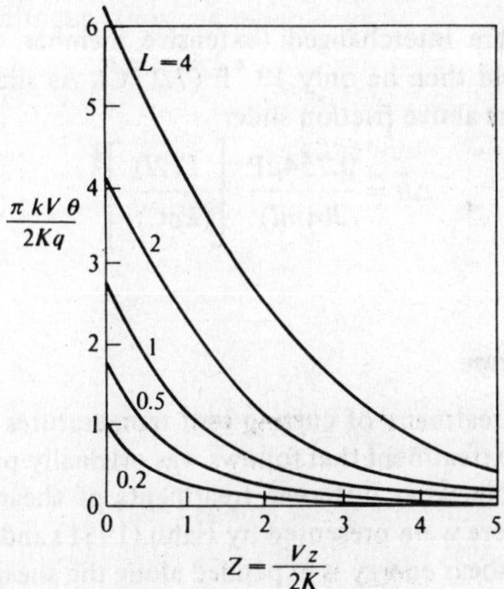

FIG. 12.13. Temperature distribution perpendicular to the plane of sliding at the trailing edge of slider.

Example—Friction Slider

The following example illustrates the application of the foregoing analysis to estimate the mean temperature rise for a stainless steel slider moving across a plane carbon steel infinite surface.

$l = 0.05$ in (1.27 mm)
$m = 0.25$ in (6.35 mm)
$V = 20$ in s^{-1} (508 mm s^{-1})
P = nominal load = 100 lb (445 N)
μ = coefficient of friction = 0.5.

Conductivity $k_1 = 2.7 \times 10^{-4}$ BTU/in²/sec/(°F/in)
(0.2018 J/cm²/s/°C/cm)
$k_2 = 7.1 \times 10^{-4}$ BTU/in²/sec/(°F/in)
(0.5307 J/cm²/s/°C/cm).

Diffusivity $K_1 = 0.02$ in² s^{-1} (0.13 cm² s^{-1})

$$q = \frac{\mu PV}{J(4lm)} = \frac{(0.5)(100)(20)}{(9340)(4)(.05)(.25)} = 2.14 \text{ BTU/in}^2\text{/sec } (350 \text{ J/cm}^2\text{/s}).$$

From Fig. 12.9: $\bar{A} = 1.80$
Eqn (12.12): $L = 25$
Eqn (12.15): $R = 0.82$
Eqn (12.16): $\Delta\bar{\theta} = 49$ °F (27.2 °C)

If the materials are interchanged (extensive member, stainless steel), the temperature rise would then be only 13 °F (7.2 °C). As sliding speed increases, R increases and for the above friction slider

$$\Delta\bar{\theta} = \frac{0.754\mu P}{J(4ml)}\left[\frac{V(2l)}{(k\rho C)_1}\right]^{\frac{1}{2}} \tag{12.18}$$

when $R = 1$.

Shear Plane Temperature

The first analytical treatment of cutting tool temperatures was that of Trigger and Chao (1951). The treatment that follows was originally presented by Loewen and Shaw (1954). Somewhat different treatments of shear plane temperature than that presented here were presented by Hahn (1951) and Leone (1954).

The rate at which shear energy is expended along the shear plane will be

$$U_S = F_S V_S \tag{12.19}$$

where F_S is the component of force directed along the shear plane, and V_S is the velocity of the chip relative to the workpiece, which also is directed along the shear plane. The rate at which energy is expended per unit area on the shear plane for orthogonal cutting conditions will be

$$U_S{}' = \frac{F_S V_S}{tb \csc \phi} \tag{12.20}$$

where t is the undeformed chip thickness, b is the width of the workpiece, and ϕ is the shear angle.

To a good approximation, it may be assumed that all of the mechanical energy associated with the shearing process is converted into thermal energy, and the heat which flows from the shear zone per unit time per unit area will be

$$q_1 = \frac{F_S V_S}{Jtb \csc \phi} = \frac{u_S V \sin \phi}{J} \tag{12.21}$$

where J is the mechanical equivalent of heat, and u_S is the shear energy per unit volume of metal cut, and V the cutting speed.

If $R_1 q_1$ is the heat per unit time per unit area which leaves the shear zone with the chip, then $(1 - R_1)q_1$ is the heat per unit time per unit area that flows into the workpiece. The mean temperature of the metal in the chip, in the vicinity of the shear plane, will be

$$\bar{\theta}_S = \frac{R_1 q_1 (bt \csc \phi)}{C_1 \rho_1 (Vbt)} + \theta_0 = \frac{R_1 u_S}{JC_1 \rho_1} + \theta_0 \tag{12.22}$$

where θ_0 is the ambient workpiece temperature, $C_1 \rho_1$ is the volume specific heat at the mean temperature between $\bar{\theta}_S$ and θ_0.

Equation 12.13 may be used to compute the shear-plane temperature rise as shown in Fig. 12.14. Here the chip may be considered to be a perfect insulator if the total heat flowing from the interface is $(1 - R_1)q_1$. The velocity of sliding is taken as V_S and not V in accordance with the generally adopted Piispanen picture of metal-cutting shown in Fig. 12.15. By this picture, metal when cut behaves as a deck of cards, one card at a time sliding a finite distance across its neighbor. Heat is generated only when sliding occurs in direction V_S and we might consider that one plate at a time is always in motion. An approximation involved in the use of Fig. 12.14 is that the metal which is assumed present in wedge A is not there, while that in wedge B is ignored. Since the heat flows associated with wedges A and B will tend to cancel each other, it is considered that the idealized picture of Fig. 12.14 should provide a good approximation.

When eqn (12.13) is used to compute the shear-plane temperature, it is found that

$$\bar{\theta}_S = 0.754 \frac{(1 - R_1)q_1 \left(\dfrac{t \csc \phi}{2}\right)}{k_2 \sqrt{L_1}} + \theta_0 \tag{12.23}$$

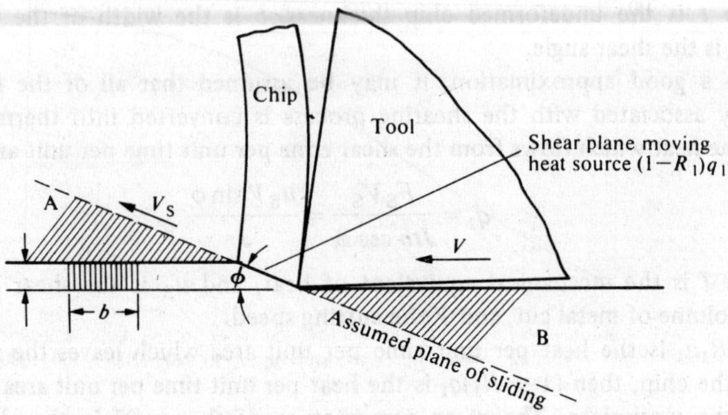

FIG. 12.14. Idealized diagram of shear plane moving heat source.

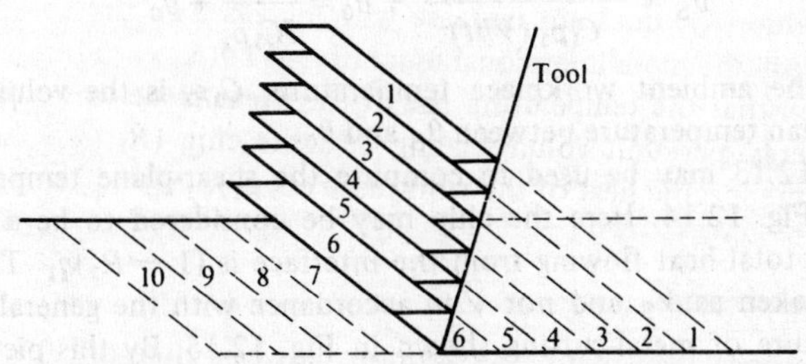

FIG. 12.15. Piispanen's mechanism of chip formation.

and

$$L_1 = \frac{V_S\left(\dfrac{t \csc \phi}{2}\right)}{2K_1} = \frac{V\gamma t}{4K_1} \tag{12.24}$$

where k_1 is the conductivity, and K_1 the diffusivity of the workpiece material at temperature $\bar{\theta}_S$, and γ is the strain in the chip. Then, equating eqns (12.22) and (12.23), and solving for R_1, it is found that

$$R_1 = \frac{1}{1 + 1.328\left[\dfrac{K_1\gamma}{Vt}\right]^{\frac{1}{2}}}. \tag{12.25}$$

Once R_1 is known, $\bar{\theta}_S$ may be calculated from eqn (12.22), Fig. 12.16 is convenient for us in determining R_1. It is evident that the percentage of energy

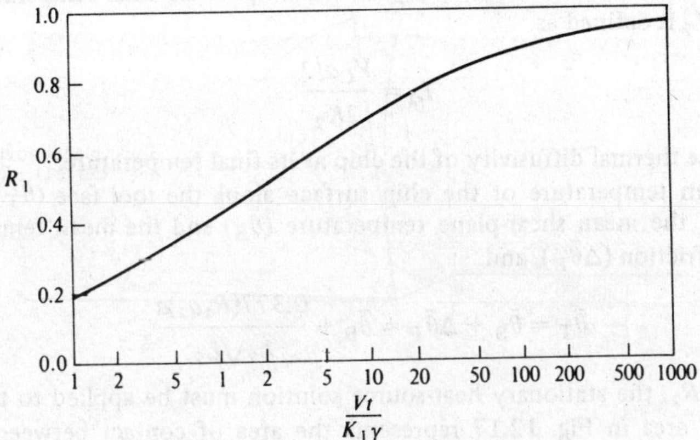

FIG. 12.16. Curve for use in computing R_1.

going to the chip (R_1) does not increase with increased cutting speed (V) alone, but rather with the nondimensional quantity $(Vt)/(K_1\gamma)$. From eqn (12.22) it may be seen that the temperature rise at the shear plane varies directly with the shear energy per unit volume going into the chip ($R_1\, u_S$), and inversely with the volume specific heat of the workpiece ($C_1\, \rho_1$).

Tool Face Temperature

From an analytical viewpoint, the friction between chip and tool can be regarded as a heat source that is moving in relation to the tool. As before, it is a double treatment that allows calculation of the partition of friction energy between chip and tool, and hence the resulting temperature.

The total friction energy $q_2\,[FL^{-1}T^{-1}]$ that will be dissipated at the chip–tool interface per unit time per unit area will be

$$q_2 = \frac{F_C V_C}{Jab} = \frac{u_F Vt}{Ja} \tag{12.26}$$

where F_C is the friction force along the tool face, V_C is the velocity of the chip relative to the tool, a is the length of contact between chip and tool in the direction of motion, b is the width of the chip, and u_F is the energy per unit volume of metal cut that is consumed by friction on the tool face.

If R_2 is the fraction of q_2 that flows into the chip, then the average temperature rise in the surface of the chip due to friction $\Delta\bar{\theta}_F$ is from eqn (12.13).

$$\Delta\bar{\theta}_F = \frac{0.754(R_2 q_2)a/2}{k_2\sqrt{L_2}} \tag{12.27}$$

where k_2 is the thermal conductivity of the chip at its final temperature, the parameter L_2 is defined as

$$L_2 = \frac{V_C a/2}{2K_2} \tag{12.28}$$

and K_2 is the thermal diffusivity of the chip at its final temperature.

The mean temperature of the chip surface along the tool face ($\bar{\theta}_T$) will be the sum of the mean shear-plane temperature ($\bar{\theta}_S$) and the mean temperature rise due to friction ($\Delta\bar{\theta}_F$), and

$$\bar{\theta}_T = \bar{\theta}_S + \Delta\bar{\theta}_F = \bar{\theta}_S + \frac{0.377(R_2 q_2)a}{k_2\sqrt{L_2}}. \tag{12.29}$$

To find R_2, the stationary heat-source solution must be applied to the tool. The shaded area in Fig. 12.17 represents the area of contact between a chip and a two-dimensional cutting tool, the cutting edge being along the y-axis and the tool face in the xy-plane. The tool represented by the solid lines may be considered a quarter-infinite body relative to the shaded area of contact. This is seen to represent a good approximation of an orthogonal cutting tool, provided the rake and clearance angles are not too large. By symmetry it is evident that the temperature at any point in the surface of the quarter-infinite body subjected to the uniform heat source represented by the shaded area would be the same as the temperature of the corresponding point in the semi-infinite body when subjected to a uniform heat source extending over A–B–C–D, provided that the yz-plane is a perfect insulator, just as the xy-plane, which is a valid assumption. Thus, the aspect ratio to be used in obtaining the mean tool face-temperature rise is

$$\frac{m}{l} = \frac{b}{2a} \quad \text{(orthogonal tool)} \tag{12.30}$$

By similar reasoning the aspect ratio for a lathe tool which cuts at the corner of the tool would be

$$\frac{m}{l} = \frac{b}{a} \quad \text{(lathe tool)} \tag{12.31}$$

With the aspect ratio thus defined for any tool, the shape factor $\bar{A}$ may be determined from Fig. 12.9 and the temperature $\bar{\theta}_T$ solved for as a point in the tool from eqn (12.8)

$$\bar{\theta}_T = \frac{(1-R_2)q_2 a}{k_3}\bar{A} + \theta_0' \tag{12.32}$$

where k_3 is the thermal conductivity of the tool material at the final tool temperature $\bar{\theta}_T$, and θ_0' is the ambient temperature of the tool.

By equating the two values of $\bar{\theta}_T$, given in eqns (12.29) and (12.32), it is

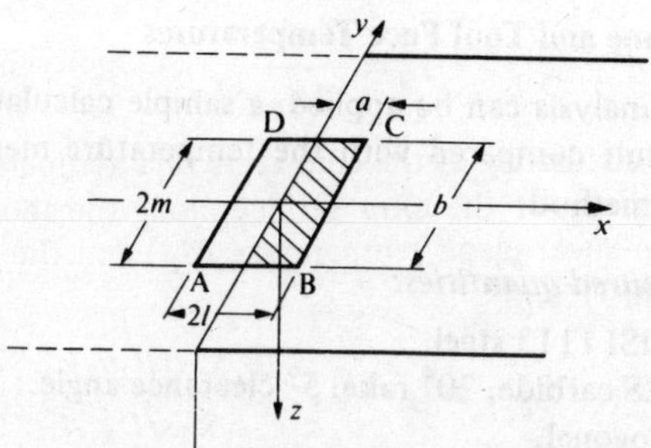

FIG. 12.17. Relation between two-dimensional tool and semi-infinite body.

possible to solve for R_2.

$$R_2 = \frac{q_2(a\bar{A}/k_3) - \bar{\theta}_S + \theta_0'}{q_2(a\bar{A}/k_3) + q_2(0.377a/k_2\sqrt{L_2})} \tag{12.33}$$

or, substituting from eqns (12.26) and (12.28), and noting that $V_C = rV$, where r is the chip-thickness ratio (i.e. the ratio of the depth of cut t to chip thickness when taking an orthogonal cut), we have

$$R_2 = \frac{(u_F Vt\bar{A}/Jk_3) - \bar{\theta}_S + \theta_0'}{\{(u_F Vt\bar{A}/Jk_3) + (0.754u_F/J\rho_2 C_2)\}\{Vt^2/arK_2\}^{\frac{1}{2}}}. \tag{12.34}$$

When R_2 is determined, $\bar{\theta}_T$ may be found readily from either eqn (12.29) or eqn (12.32).

As usual, it has been necessary in the foregoing derivations to consider all thermal quantities to be constants. Actually the significant thermal quantities are functions of temperature, so that the question arises as to just what sort of mean values should be substituted when evaluating the final equations. In the foregoing equations the following quantities occur: ρ_1, C_1, K_1, ρ_2, C_2, K_2, k_2, and k_3. The thermal quantities ρ_1, C_1, and K_1 refer to the shear plane, and, to obtain a good approximation, should be evaluated at a temperature halfway between the shear-plane temperature $\bar{\theta}_S$ and the ambient temperature θ_0. The thermal quantities having the subscript 2 refer to the workpiece material in the vicinity of the tool face. Inasmuch as the heat flow in the chip by conduction and convection will be controlled predominantly by the properties very near the tool face, because of the large temperature gradient that exists in the surface, it would appear that the best temperature to use in connection with ρ_2, C_2, K_2, and k_2 would be the mean temperature at the surface ($\bar{\theta}_T$). By similar reasoning, it appears that the value of the thermal conductivity of the tool k_3 also should be taken as that corresponding to temperature $\bar{\theta}_T$. In view of the uncertainties involved, it is fortunate that the numerical solutions obtained are not sensitive to the particular mean temperature that are used. This is confirmed in the calculations that follow.

Example—Shear Plane and Tool Face Temperatures

To show how this analysis can be applied, a sample calculation will be carried through and the result compared with the temperature measured by the tool-work thermocouple method.

Experimentally measured quantities:

Work material, AISI 1113 steel.
Tool material, K2S carbide, 20° rake, 5° clearance angle.
Type of cut, orthogonal.
Cutting speed, $V = 445$ f.p.m. $= 139$ m min^{-1}.
Undeformed chip thickness, $t = 0.0023$ in $= 0.06$ mm.
Width of cut, $b = 0.151$ in $= 3.84$ mm.
Chip contact length†, $a = 0.009$ in $= 0.23$ mm.
Cutting force, $(F_P) = 80$ lb $= 356$ N.
Feed Force, $F_Q = 28$ lb $= 125$ N.
Chip-thickness ratio, $r = 0.51$.
Thermal properties of steel, given in Fig. 12.18.

Calculated Values:

Average shear strain in chip, $\gamma = 1.91$.
Shear energy per unit volume, $u_S = 153\,000$ in lb in^{-3} $= 1056$ N mm^{-2}.
Shear angle, $\phi = 30°$.
Friction energy per unit volume, $u_F = 78\,000$ in lb in^{-3} $= 544$ N mm^{-2}.

Outline of Solution:

(a) Estimate the value of $\bar{\theta}_S$, e.g. 500 °F (260 °C) and determine corresponding values of K_1 and $\rho_1 C_1$ at $(500 + 75)/2 = 288$ °F (142 °C). From Fig. 12.18, $K_1 = 0.023$ in^2 sec^{-1} (14.84 mm^2 s^{-1}) and $\rho_1 C_1 = 0.034$ B.T.U. in^{-3} °F^{-1} (3.94 J cm^{-3} °C^{-1}).
(b) Calculate quantity $Vt/K_1\gamma = 4.65$.
(c) Obtain R_1 from Fig. 12.16: $R_1 = 0.62$.
(d) Calculate $(\bar{\theta}_S - \theta_0)$ from eqn (12.22), $\bar{\theta}_S - \theta_0 = 299$ °F (156 °C) and if $\theta_0 = 75$ °F (24 °C) then $\bar{\theta}_S = 374$ °F (190 °C).
(e) Repeat steps (a) to (d) if original estimate of $\bar{\theta}_S$ is in error by more than 25 °F (14 °C). In this case K_1 remains at 0.023 in^2 sec^{-1} (14.84 mm^2 s^{-1}) and $\rho_1 C_1$ is 0.033 B.T.U. in^{-3} °F^{-1} (3.82 J cm^{-3} °C^{-1}). The $\bar{\theta}_S$ changes to 383 °F (195 °C).
(f) Estimate value of $\bar{\theta}_T$, e.g. 800 °F (444 °C) and determine K_2 and $\rho_2 C_2$ from Fig. 12.18 at this temperature: $K_2 = 0.016$ in^2 sec^{-1} (0.103 cm^2 s^{-1}) and $\rho_2 C_2 = 0.042$ B.T.U. in^{-3} °F^{-1} (4.87 J cm^{-3} °C^{-1}).

† Obtained by measuring a thin layer of metal deposited on tool face with a microscope.

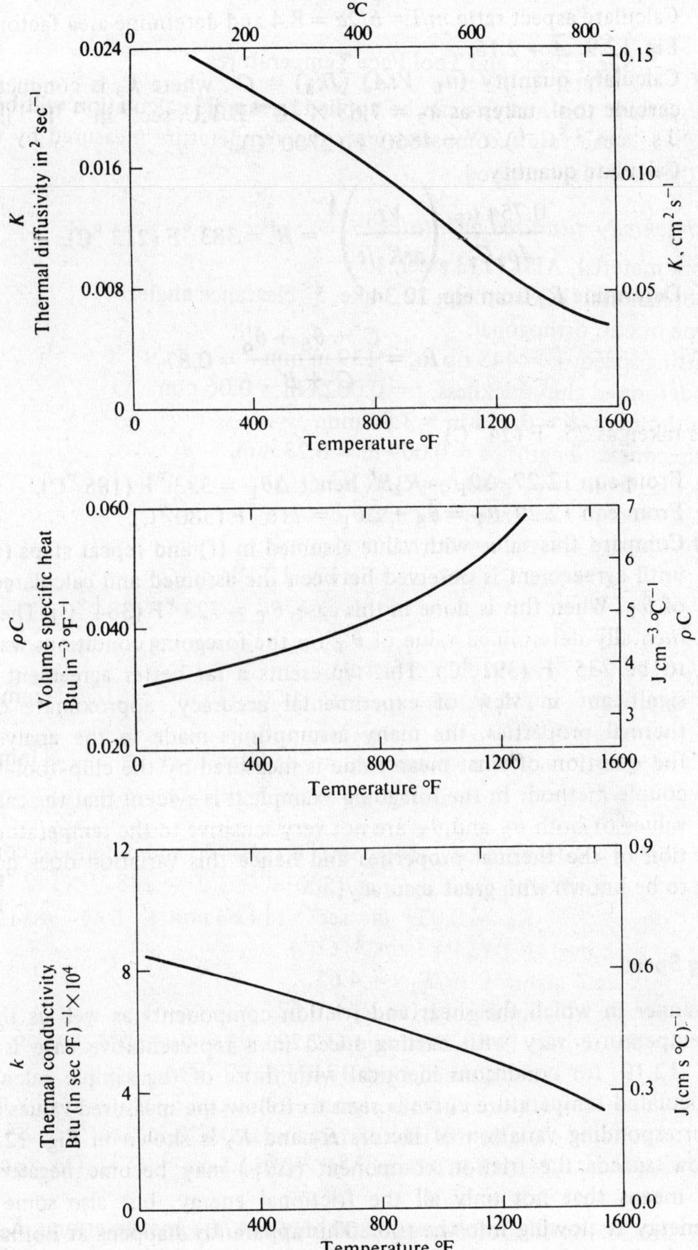

FIG. 12.18. Thermal properties of a low-carbon steel with 0.39 percent Mn. (Anon, 1939, 1946.)

(g) Calculate aspect ratio $m/l = b/2a = 8.4$ and determine area factor $\bar{A}$ from Fig. 12.9: $\bar{A} = 2.15$.

(h) Calculate quantity $(u_F\ VtA)\ (Jk_3) = C'$, where k_3 is conductvitiy of carbide tool, taken as $k_3 = 7.63 \times 10^{-4}$ B.T.U. sec^{-1} in^{-2} °F^{-1} in (0.570 J s^{-1} cm^{-2} °C^{-1}): $C' = 4860$ °F (2700 °C).

(i) Calculate quantity

$$\frac{0.754\, u_F}{J\rho_2 C_2}\left(\frac{Vt}{arK_2/t}\right)^{\frac{1}{2}} = B' = 383\ \text{°F}\ (213\ \text{°C}).$$

(j) Determine R_2 from eqn 12.34

$$R_2 = \frac{C' - \theta_S + \theta_o'}{C' + B'} = 0.87$$

If θ_o' is taken as 75 °F (24 °C)

(k) From eqn 12.27, $\Delta\bar{\theta}_F = R_2 B'$, hence $\Delta\theta_F = 333$ °F (185 °C).

(l) From eqn 12.29, $\bar{\theta}_T = \bar{\theta}_S + \Delta\bar{\theta}_F = 716$ °F (380 °C).

(m) Compare this value with value assumed in (f) and repeat steps (f) to (l) until agreeement is observed between the assumed and calculated values of $\bar{\theta}_T$. When this is done in this case, $\theta_T = 723$ °F (384 °C). The experimentally determined value of $\bar{\theta}_T$ for the foregoing conditions was found to be 735 °F (391 °C). This represents a far better agreement than is significant in view of experimental accuracy, approximate data on thermal properties, the many assumptions made in the analysis, and the question of what mean value is measured by the chip–tool–thermocouple method. In the foregoing example it is evident that the calculated values of both $\bar{\theta}_S$ and $\bar{\theta}_T$ are not very sensitive to the temperature variation of the thermal properties and hence this variation does not need to be known with great accuracy.

Cutting Speed

The manner in which the shear and friction components as well as the total tool temperature vary with cutting speed in a representative case is shown in Fig. 12.19, for conditions identical with those of the sample calculations. The calculated temperature curve is seen to follow the measured values closely. The corresponding variation of factors R_1 and R_2 is shown in Fig. 12.20. At very low speeds the friction component $(\Delta\bar{\theta}_F)$ may become negative. This merely means that not only all the frictional energy, but also some of the shear energy is flowing into the tool. This apparently happens at normal turning speeds for nickel as discussed in Chapter 15. Both the observed and calculated resultant temperatures $\bar{\theta}_T$ are seen to vary approximately parabolically with cutting speed (i.e. as $V^{\frac{1}{2}}$).

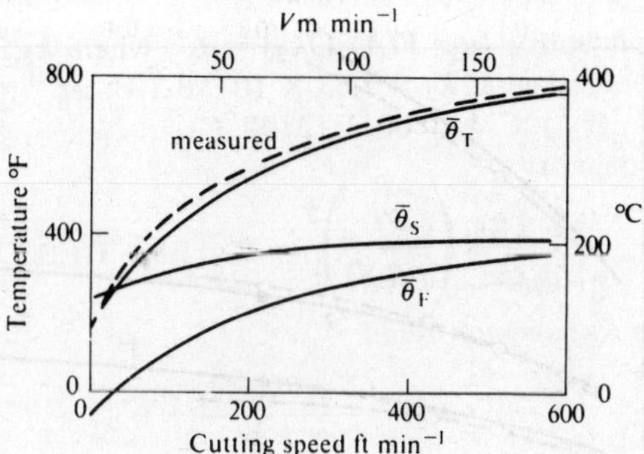

FIG. 12.19. Comparison between observed and computed cutting temperatures for AISI B1113 steel when cut with a K2S cemented carbide tool having 20° rake angle, $t = 0.0023$ in (0.06 mm), $b = 0.151$ in (3.84 mm).

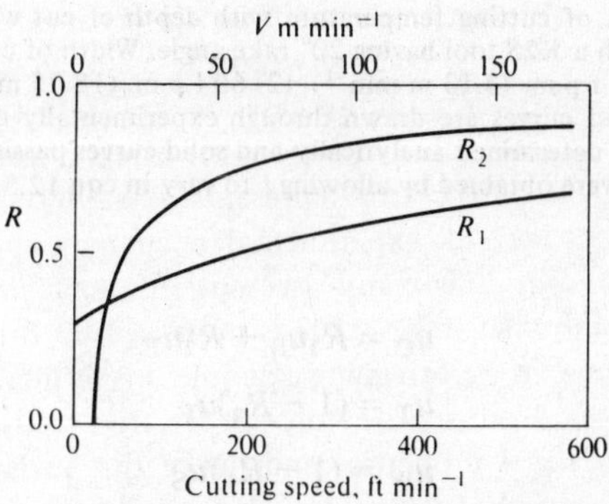

FIG. 12.20. Variation of R_1 and R_2 for data of Fig. 12.19.

Cutting Geometry

Experimental results of tests in which the undeformed chip thickness t and width of cut b were varied are shown in Fig. 12.21 and 12.22 respectively. The cutting temperature is very nearly independent of the width of cut b.

Energy Balance

The energies going to the chip (u_C), tool (u_T), and workpiece (u_W) per unit volume of metal removed are functions of R_1 and R_2, and the specific energies of the cutting process (u_S = shear energy per unit volume, u_F = friction energy per unit unit volume). Thus

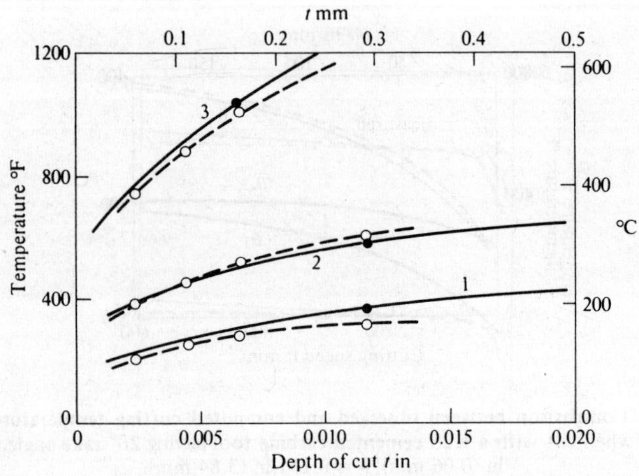

FIG. 12.21. Variation of cutting temperature with depth of cut when cutting AISI 1113 steel orthogonally with a K2S tool having 20° rake angle. Width of cut 0.151 in (3.84 mm): cutting speeds: (1) 10 f.p.m. (3.13 m min⁻¹), (2) 60 f.p.m. (18.75 m min⁻¹), (3) 450 f.p.m. (141 m min⁻¹). Dotted curves are drawn through experimentally measured temperatures. Points marked • were determined analytically and solid curves passing through these points were obtained by allowing *t* to vary in eqn 12.32.

$$u_C = R_1 u_S + R_2 u_F \tag{12.35}$$

$$u_T = (1 - R_2) u_F \tag{12.36}$$

$$u_W = (1 - R_1) u_S \tag{12.37}$$

When the approximate values are substituted into these equations for the tests in Fig. 12.19 and 12.20, the distribution of energy shown in Fig. 12.23 is obtained. The percentage of the total energy going to the chip increases with increased speed although the percentages going to the tool and workpiece decrease. At very high cutting speeds, practically all of the energy is carried away in the chip, a small amount going into the workpiece and a still smaller amount to the tool.

Calorimetric measurements of the three energies would be an ideal method for checking the validity of the analysis were it not for the experimental difficulties involved.

Schmidt and Roubik (1949) have drilled metal in a calorimeter mounted in a drill press as shown in Fig. 12.24. Total energy was measured by noting the temperature rise of the water as $\frac{3}{8}$ in (9.53 mm) diameter rods of magnesium alloy with 0.110 in (2.79 mm) bore were drilled using a 0.440 in (11.18 mm)

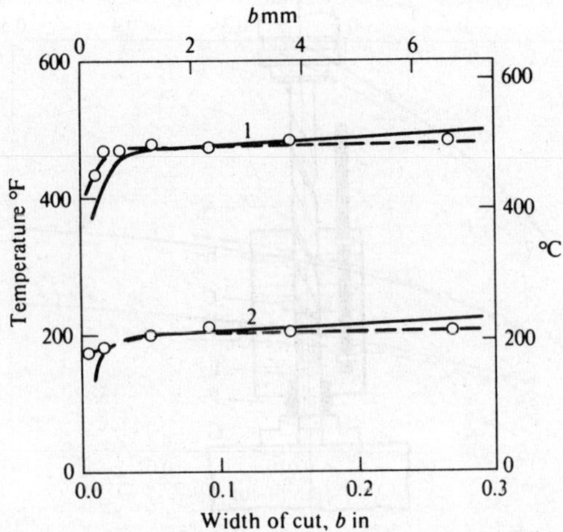

FIG. 12.22. Variation of cutting temperature with width of cut when cutting AISI 1113 steel orthogonally with K2S carbide tool having 20° rake angle. Cutting speeds: (1) 226 f.p.m. (70.6 m min^{-1}) and (2) 15 f.p.m. (4.7 m min^{-1}). Depth of cut 0.0023 in (0.06 mm). Dotted curves are drawn through experimentally measured temperatures. Temperatures for b = 0.151 in (3.84 mm) were determined analytically in each case and solid curves passing through these points were obtained by allowing only b to vary in eqn 12.32.

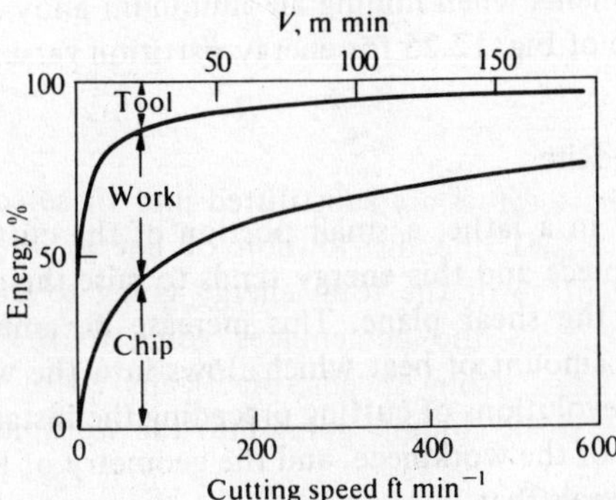

FIG. 12.23. Calculated variation of energy distribution with cutting speed for cutting conditions of Fig. 12.21.

drill. The energy to the chips was obtained by drilling in air, letting the chips fall into the water, again noting the temperature rise. Finally, the same small quantity of metal was again cut in air and the drill quickly plunged into water

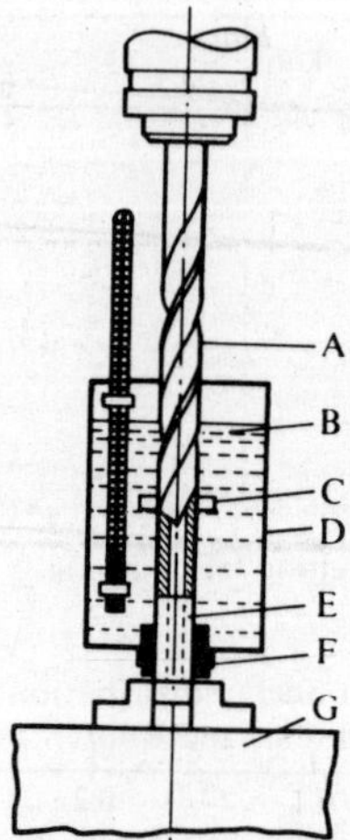

FIG. 12.24. Calorimetric apparatus for determining energy distribution in drilling. A, drill; B, distilled water; C, agitator; D, container, E, $\frac{3}{8}$ in (9.5 mm) specimen; F, rubber grommet; G, chuck. (After Schmidt and Roubik 1949.)

to obtain the energy going into the drill. The results obtained for a feed rate of 0.0091 in sec^{-1} (0.23 mm s^{-1}) are given in Fig. 12.25. Although absolute values differ since different materials are involved, the trends of Fig. 12.25 are similar to those of Fig. 12.23. Schmidt and Roubik (1954) have repeated the calorometric drilling experiments when milling an aluminum alloy and have obtained similar results to those of Fig. 12.25 for energy partition versus cutting speed.

Ambient Temperature Rise

When a bar is turned in a lathe, a small portion of the cutting energy is conducted into the workpiece and this energy tends to raise the temperature of the material approaching the shear plane. This increase in 'ambient' temperature will depend upon the amount of heat which flows into the workpiece per unit time, the number of revolutions of cutting preceding the instant under question, the thermal properties of the workpiece, and the geometry of the system.

Detailed analysis reveals that

1. $\Delta\theta_1$ increases proportionately with the shear energy per unit volume (u_S), which is normally approximately proportional to the Brinell hardness of the work material and will generally increase with tool dulling.
2. $\Delta\theta_1$ varies inversely with the workpiece diameter (D) as has been found experimentally on many occasions.
3. $\Delta\theta_1$ varies linearly with the depth of cut (b).

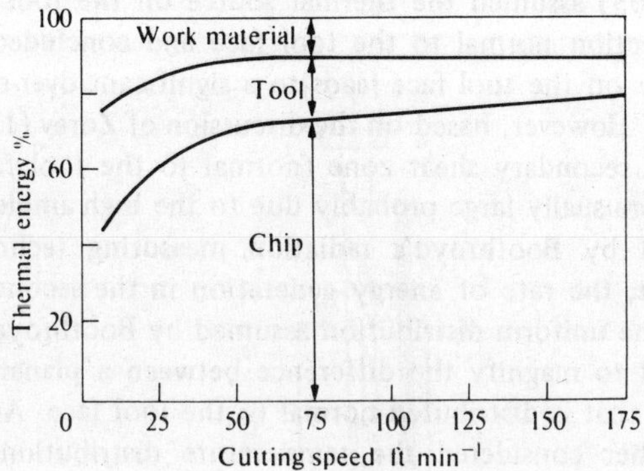

FIG. 12.25. Variation of energy distribution with drilling speed. (After Schmidt and Roubik 1949.)

4. $\Delta\theta_1$ varies inversely with the thermal conductivity of the workpiece material (k).
5. $\Delta\theta_1$ varies approximately as the 0.8-power of the feed (i.e. $\Delta\theta_1 \sim t^{0.8}$).
6. The influence of cutting speed is complex resulting in a relatively small change in ($\Delta\theta_1$) with (V) which may be either positive or negative.
7. A coolant should decrease $\Delta\theta_1$ and this is observed experimentally.

While the ambient workpiece temperature rise is the lowest of those considered, it will directly influence the temperatures of importance to tool wear on the tool face and tool flank. It is frequently reported, for example, that tool life decreases when all cutting conditions are maintained the same except the workpiece diameter which is decreased.

Analytical Refinements

Since the mid 1950s there have been a number of analytical refinements in the approach to cutting-temperature analysis first introduced by Trigger and Chao (1951). However, these refinements have generally been accompanied by increased complexity.

Chao, Trigger, and Zylstra (1952) and Chao and Trigger (1953) improved on the original 1951 analysis placing considerable emphasis upon a nondimensional thermal number (Vt/K, where K is the diffusivity of the work material). Chao and Trigger (1958) and Chao, Li and Trigger (1961) discussed the temperature distributions along the chip–tool and chip–work interfaces. Rapier (1954) Weiner (1955) and Nakayama (1956) have also considered the temperature distribution in orthogonal cutting.

Boothroyd (1963) assumed the thermal source on the tool face to be distributed in a direction normal to the tool face and concluded that use of a planar heat source on the tool face leads to a significant over-estimate of tool face temperatures. However, based on the discussion of Zorev (1963), it appears that the depth of secondary shear zone (normal to the tool face) considered by Boothroyd is unusually large probably due to the high ambient temperature (600 °C) required by Boothroyd's radiation measuring technique. Also, as Zorev pointed out, the rate of energy generation in the secondary shear zone will be far from the uniform distribution assumed by Boothroyd. Both of these discrepancies tend to magnify the difference between a planar source on the tool face and one that is distributed normal to the tool face. Arndt and Brown (1967) have further considered the temperature distribution in orthogonal machining.

Finite Element Analysis

Tay, Stevenson, and Davis (1974) have applied the finite element method to obtain temperature fields for three specific cases. While this technique is a very powerful one enabling variable material properties to be included in the analysis, the results will be only as good as the properties fed in. For example, it was assumed throughout the Tay *et al.* analysis that the flow stress of the material corresponds to eqn 5.8 ($\sigma = \sigma_0\epsilon^n$) and that uniaxial stress and strain are related to shear stress and shear strain in cutting through the concept of equivalent stress and equivalent strain. As discussed in Chapter 9 these are very questionable assumptions for the large plastic strains involved in cutting. Nevertheless, this analysis is one of the most comprehensive available and provides a valuable insight into the thermal fields associated with metal cutting.

However, the theory presented is not predictive since it requires inputs that cannot be arrived at from material properties but which must be obtained from metal-cutting tests for the specific combination of tool and work materials, tool geometry, and cutting conditions pertaining. For example, the extent of secondary shear was estimated from the distortion of lines on a gridded specimen and the magnitude of the internal secondary shear stress was estimated from measured values of tool face friction and chip–tool contact length.

The secondary shear zone that occurs on the tool face plays a very important role relative to the magnitude and distribution of thermal energy and hence with regard to tool face temperatures as Boothroyd (1963) has indicated. This shear zone may be of three types:

1. No subsurface shear flow in which case there is sliding and friction along the entire chip tool contact plane.
2. Subsurface plastic flow in which case the surface velocity of the chip will be less than V_C at the cutting edge but will increase as the chip

moves up the tool reaching a value of V_C at about half the contact length.

3. Subsurface fluid flow in situations where a molten layer actually forms on the tool face.

An important example of the latter type will be discussed in Chapter 15 in connection with the performance of specially deoxidized steels.

Tay *et al.* studied the thermal fields produced when a free-machining (0.13 $^w/_o$ C, 1.45 $^w/_o$ Mn, 0.25 $^w/_o$ S and 0.19 $^w/_o$ P) steel was machined with a brazed tungsten carbide tool having a rake angle of 20° and a relief angle of 6°. Orthogonal cuts of 9.50 mm width (b) were made at a feed rate (t) of 0.274 mm/rev. Three cutting speeds (V) were employed: 29.6, 78.0, and 155.4 m min^{-1}. The specimens were provided with a grid that could be analyzed in the chip root after suddenly interrupting the cut using an explosive device. This enabled strain rates in the primary and secondary shear zones to be determined from which the rate of energy dissipation and hence the heat flux could be found.

Interpretation of the deformed grid pattern on the shear plane was relatively easy but much more difficult for the secondary shear zone. This was due to the thinness of the secondary shear zone (~ 0.10 mm) and the fact that the grid on the free chip surface had been sheared three times in different directions (first in plane stress, then in plane strain in the primary shear zone, and finally in the secondary shear zone). When cutting at the lowest speed (29.6 m min^{-1}) there was no subsurface flow in the chip but there was at the two higher speeds.

Figure 12.26 shows deformed grid lines in the chip root obtained at the highest cutting speed (155.4 m min^{-1}). The velocity in the main body of the chip (V_C = 68.9 m min^{-1}) is seen to be uniform except in the secondary shear zone of width δ. The velocity at the cutting edge in the direction of V_C was assumed to be $V_C/3$ which gradually accelerated to V_C as the chip moved over about half of the chip-tool contact length (a = chip tool contact length = 0.91 mm and δ = width of the secondary shear zone = 0.10 mm). The region over which secondary internal shear occurs is triangle ABC. As the chip passes up the tool face, the extent of the internal shear decreases to zero as the velocity of the surface of the chip accelerates. Beyond point C the velocity of the surface of the chip equals V_C and the heat source then lies in the tool face.

This interpretation follows that of Nakayama (1959) and is in agreement with the stress distribution on the tool face discussed in Chapter 10 where the shear stress was found to be constant over the first half of the chip tool contact and to fall to zero over the second half of the contact area. The value of the internal shear stress (τ_i) was obtained by Tay *et al.* as follows by assuming τ_i to decrease linearly over the second half of the contact area:

$$F_C = \tau_i(a/2\,b) + \tau_i/2\,(a/2\,b) \tag{12.38}$$

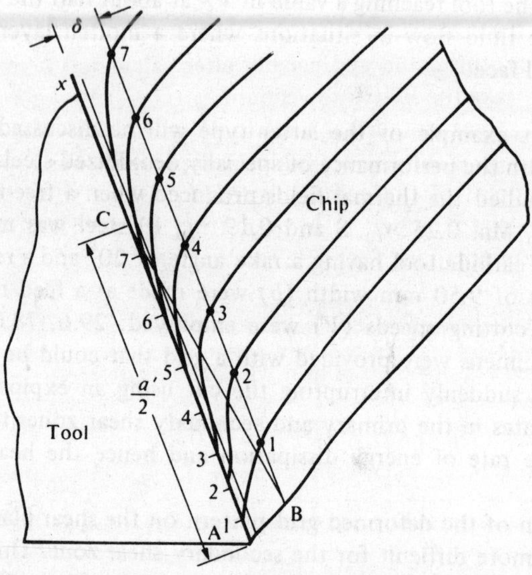

FIG. 12.26. Secondary shear zone for dry orthogonal cutting of free machining steel with a carbide tool ($\alpha = 20°$) at 155.5 m min^{-1} (497 f.p.m.) and a feed of 0.274 mm/rev (0.01 i.p.r.). (After Tay, Stevenson, and Davis 1974.)

or

$$\tau_i = 4F_C/3ab \tag{12.39}$$

where F_C is the measured amount of force along the tool face, b is the width of cut, and a is the total chip tool contact length obtained by noting the extent of the scar on a polished tool face.

It should be noted that the foregoing interpretation of secondary flow of Tay *et al.* is not consistent with the observation of Zorev (1963) that grinding scratches parallel to the cutting edge were not removed over the first half of the contact zone but were replaced by scratches in the cutting direction over the second half of contact. It is also not in agreement with the experimentally based views of Trent (1976). The foregoing interpretation of flow lines in a chip root assumes there is some sliding of the chip over the tool face all the way to the cutting edge (merely lower sliding velocities over the first half of the contact area). Zorev's observation would require a thin layer of completely stationary chip material between the main body of the chip and the tool face. This thin layer of metal would be strongly bonded to the tool face, would remain on the tool in a quick-stop test and would not be evident if only the chip and not the tool were subsequently examined at high power. Such built-up layers (BUL) have been observed on the face of a carbide tool after cutting

(Schaller 1964). If a stationary BUL is present between the main body of the chip and the tool face, this could appreciably alter calculated temperature distributions that do not take the BUL into account.

Figure 12.27 shows isothermal lines for $V = 155.4$ m min^{-1} and $t = 0.274$ mm/rev. due to Tay *et al.* The maximum temperature was found to occur just beyond the midpoint of the contact length (at 0.68a from the cutting edge) but was closer to the midpoint at lower cutting speeds. The following measured values were used in obtaining Fig. 12.27:

Cutting force (F_P) = 3750 N (842 lb).
Feed force (F_Q) = 1190 N (267 lb).
Bulk chip speed (V_C) = 68.9 m min^{-1} (220 f.p.m.)
Shear angle (ϕ) = 27.8°.
Chip tool contact length (a) = 0.91 mm (0.0358 in).
Maximum width of secondary shear zone (δ) = 0.10 mm (0.004 in).

The following calculated values are obtained:

Average shear plane temperature = 192 °C (378 °F).
Maximum tool face temperature = 1016 °C (1861 °F).

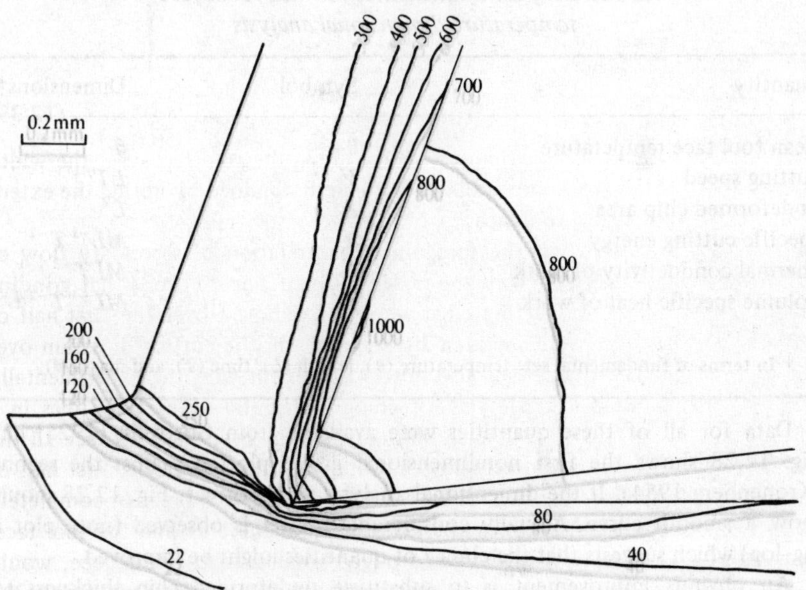

FIG. 12.27. Isothermal lines for dry orthogonal cutting of free maching steel with a carbide tool (α = 20°) at 155.4 m min^{-1} (497 f.p.m.) and a feed of 0.274 mm/rev (0.01 i.p.r.). (After Tay, Stevenson, and Davis 1974.)

The temperature of the shear plane in the vicinity of the tool tip is seen to be considerably higher than the mean temperature on the shear plane.

Dimensional Analysis

It is obvious that a simpler approach to cutting temperatures, although more approximate, is needed for practical application in the workshop. Dimensional analysis satisfies this need.

Kronenberg (1949) was apparently the first to apply dimensional analysis to the tool face temperature problem. Based on his broad experience and intuitive skill, he reasoned that a first approximation to the magnitude of mean tool face temperature ($\bar{\theta}_T$) could be obtained by considering the variables listed in Table 12.1. There are four dimensionally independent quantities in this list which may be chosen to be V, u, k, and ρC. When these four quantities are combined with each of the remaining two ($\bar{\theta}_T$ and A for area), two non-dimensional quantities are obtained and hence by the principle of dimensional homogeneity

$$\frac{\bar{\theta}_T(\rho C)}{u} = \psi_1\left(\frac{AV^2(\rho C)^2}{k^2}\right). \tag{12.40}$$

TABLE 12.1
Kronenberg list of quantities for mean tool face temperature dimensional analysis

Quantity	Symbol	Dimensions†
Mean tool face temperature	$\bar{\theta}_T$	θ
Cutting speed	V	LT^{-1}
Undeformed chip area	A	L^2
Specific cutting energy	u	$ML^{-1}T^{-2}$
Thermal conductivity of work	k	$MLT^{-3}\theta^{-1}$
Volume specific heat of work	ρC	$ML^{-1}T^{-2}\theta^{-1}$

† In terms of fundamental set: temperature (θ), length (L), time (T), and mass (M).

Data for all of these quantities were available from Gottwein (1925) and Fig. 12.28 shows the first nondimensional group plotted against the second (Kronenberg 1954). If the dimensional analysis were correct, Fig. 12.28 should show a smooth curve. Actually considerable scatter is observed (note plot is log–log) which suggests that the choice of quantities might be improved.

An obvious improvement is to substitute undeformed chip thickness (t) for undeformed chip area (bt) since b has a negligible influence on $\bar{\theta}_T$ while t has an appreciable influence. Another improvement is to consider the product

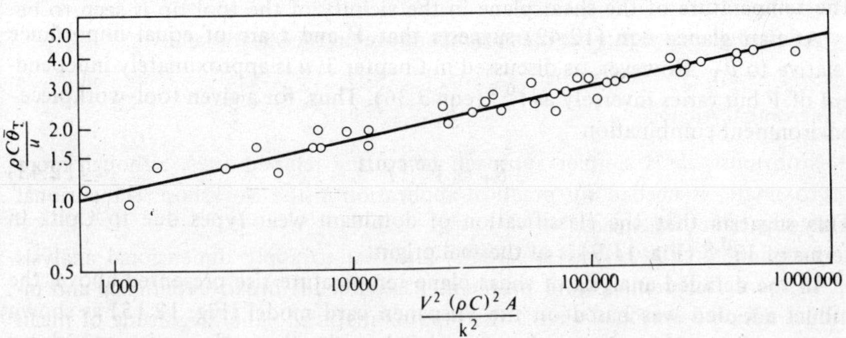

FIG. 12.28. Nondimensional plot of Gottwein data. (After Kronenberg 1954.)

($k\rho C$) as a single variable instead of including both k and ρC. All of Kronenberg's other variables appear to be good choices.

Table 12.2 gives an improved list of quantities which leads to a single non-dimensional quantity and hence (Shaw 1958)

$$\frac{\bar{\theta}_T}{u}\left[\frac{k\rho C}{Vt}\right]^{\frac{1}{2}} = \text{constant} \tag{12.41}$$

or

$$\bar{\theta}_T \sim u\left[\frac{Vt}{k\rho C}\right]^{\frac{1}{2}} \tag{12.42}$$

where the constant is for a given tool–environment (fluid) combination.

TABLE 12.2

Improved list of quantities for mean tool face temperature dimensional analysis

Quantity	Symbol	Dimensions†
Mean tool face temperature	$\bar{\theta}_T$	θ
Cutting speed	V	LT^{-1}
Undeformed chip thickness	t	L
Specific cutting energy	u	FL^{-2}
$k\rho C$ of work	$k\rho C$	$F^2L^{-2}T^{-1}\theta^{-2}$

† In terms of fundamental set: temperature (θ), length (L), time (T) and force (F).

At first glance eqn (12.42) suggests that V and t are of equal importance relative to $\bar{\theta}_T$. However, as discussed in Chapter 3, u is approximately independent of V but varies inversely as $t^{0.2}$ (eqn 3.36). Thus, for a given tool–workpiece–environment combination

$$\bar{\theta}_T \sim V^{0.5} t^{0.3} \tag{12.43}$$

This suggests that the classification of dominant wear types due to Opitz in terms of $Vt^{0.6}$ (Fig. 11.9) is of thermal origin.

In the detailed analysis of shear plane temperature rise presented above, the model adopted was based on the Piispanen card model (Fig. 12.15) as shown in Fig. 12.14. Here the work material below the shear plane was considered to be the extensive member while the shear plane itself was considered to be the moving heat source. When q_1 from eqn (12.21) and L_1 from eqn (12.24) are substituted into eqn (12.23), the following expression is obtained:

$$\bar{\theta}_S = 0.754 \frac{R_1 u_S}{J} \sqrt{\left[\frac{Vt/\gamma}{(k\rho C)_1}\right]} + \theta_0. \tag{12.44}$$

An equally valid interpretation of Fig. 12.14 is one in which the *chip* is the extensive member and then the following equation for θ_S is obtained directly from eqn 12.13:

$$\bar{\theta}_S = 0.754 \frac{R_1' u_S}{J} \sqrt{\left[\frac{V_S l_S}{(k\rho C)_1}\right]} + \theta_0 \tag{12.45}$$

where l_S is the length of the shear plane ($t/\sin\phi$) and R_1' is not the same as R_1. Due to the convective heat transfer associated with chip velocity (V_C) (not considered in eqn 12.45) R_1' in this equation will be essentially 1 and therefore

$$\Delta\bar{\theta}_S \sim u_S \sqrt{\left[\frac{V_S l_S}{(k\rho C)_1}\right]} \tag{12.46}$$

From eqn (12.27), it may be similarly shown that

$$\Delta\bar{\theta}_F \sim u_F \sqrt{\left[\frac{V_c a}{(k\rho C)_1}\right]} \tag{12.47}$$

where a is the chip–tool contact length.

The mean temperature rise on the tool face will be

$$\Delta\bar{\theta}_T \sim \bar{\theta}_S + \Delta\bar{\theta}_F \tag{12.48}$$

if it is assumed $\Delta\bar{\theta}_S$ and $\Delta\bar{\theta}_F$ are about equal. Noting that $u_S + u_F = u$ (the total specific cutting energy) and that $V_S l_S \sim V_c a \cong Vt$, it follows from eqn (12.48) that

$$\Delta\bar{\theta}_T \sim u \sqrt{\left[\frac{Vt}{(k\rho C)_1}\right]} \tag{12.49}$$

which is the result obtained directly by dimensional analysis (eqn 12.42).

Equation (12.46) for the mean shear plane temperature appears to be more realistic than eqn (12.22) since the former indicates a substantial rise in $\Delta\bar{\theta}_S$ with speed (i.e. as $\sqrt{V_S}$ since u_S is essentially independent of speed) in agreement with experience while the latter indicates a small increase in $\Delta\theta_S$ with speed (due only to a small increase in R_1 with speed).

Hot Machining

Tests have been made in several laboratories to determine the practicability of artifically heating the surface of the workpiece just before a cut is taken. This procedue will raise the temperature of the work piece closer to or slightly above the recrystallization temperature and hence reduce the tendency to strain harden. While the method does not appear to be of general interest, it may prove to be of particular value in the machining of high temperature alloys where strain hardening is serious. The power of this method is strikingly illustrated when the machining characteristics of vitallium are compared at room temperature and at 2000 °F (1903 °C). At room temperature, this material is almost completely unmachinable while at 2000 °F (1903 °C) a good continuous chip is obtained. Although less spectacular results are obtained when machining other high temperature alloys, the method deserves careful study.

Several methods of heating the workpiece have been employed including: furnace, gas torch, induction coil, carbon arc, and electrical resistance. When a furnace is used, the part is heated throughout. This is only feasible for small parts as in the drilling of turbine buckets. In the other methods, one object is to supply enough heat to the surface to raise the temperature of a layer about equal to the depth of cut to the desired value. This heat must be supplied just in front of the cutting tool so that there is not sufficient time for a large portion of it to flow into the inner regions of the bar.

It will be recalled that the tool temperature is the sum of the shear, friction and ambient temperatures. The friction temperature ($\bar{\theta}_F$) is probably not greatly influenced by heating the workpiece. Obviously the base temperature θ_0 is raised in hot machining, since it represents the temperature of the workpiece before cutting. Increased tool life can be expected on this basis only if the shear temperature increment ($\bar{\theta}_S - \theta_0$) decreases faster with temperature than θ_0 increases. This should be expected to be the case for materials having a strong tendency to strain harden during cutting. For materials that do fall in this category there should be an optimum hot-machining ambient workpiece temperature. This was actually found by Krabacher and Merchant (1951) in milling tests on hot Timken alloy (16 w/o Cr, 25 w/o Ni, 6 w/o Mo). Milling allowed the tests to be carried on at higher temperatues than would be possible in a lathe.

It is interesting that the temperature for maximum tool life varies with the

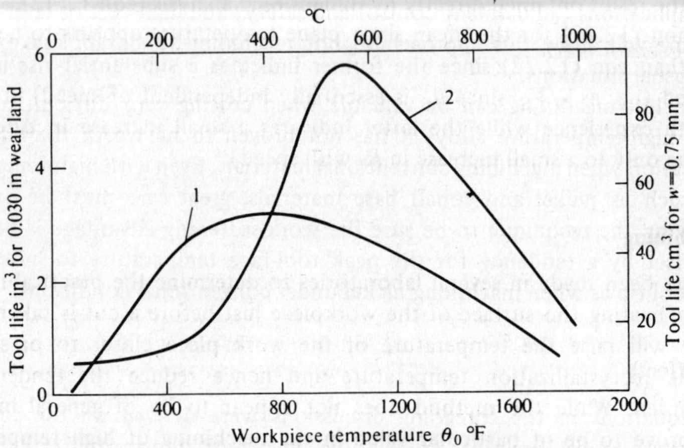

FIG. 12.29. Variations of tool life with workpiece bulk temperature when milling Cr–Ni–Mo steel at speeds of (1) 150 f.p.m. (47 m min^{-1}) and (2) 200 f.p.m. (63 m min^{-1}). Tool characteristics: fly-milling end-mill of KM carbide; rake angle, – 14°; inclination, 0°; relief angle, 7°; corner angle, 30°; feed per tooth, 0.01 in (0.25 mm); depth of cut, 0.05 in (1.27 mm). (After Krabacher and Merchant 1951.)

cutting speed as shown in Fig. 12.29. The higher the cutting speed the higher the optimum ambient temperature.

The data of Fig. 12.29 illustrate the nature of the improvement in tool life that may be achieved by hot machining. The best cutting conditions were found to be $\theta_0 = 1000$ °F (538 °C) and a cutting speed of about 200 f.p.m. (69 m min^{-1}) for this particular material (16-25-6). Under these conditions the tool life was found to be over ten times the corresponding value when θ_0 was at room temperature. All of the tool life values in Fig. 12.29 are considerably lower than would be obtained under similar conditions in turning. These low values are probably due in part to the discontinuous nature of the cutting operation studied which involves shock loading and thermal fatigue.

Laser-assisted machining employs a laser beam focused on the work material just in front of the cutting tool. The laser beam extends over the entire depth of cut and is positioned and adjusted to soften the material on the shear plane without allowing appreciable laser energy to flow into the tool. This constitutes a difficult control problem and requires a relatively expensive (and potentially dangerous) high-power (15 kW) continuous wave or pulsed laser. The high power required is due in part to the fact that the absorptivity of a high-temperature alloy is only about 10 percent. A sprayed-on coating of material such as potassium silicate may be used to increase the absorptivity somewhat and thus decrease the power of the laser required (Rajagopal 1982). However, this represents an added complication. All things considered, it does not appear likely that the complications (adjustment control and safety) and high cost of laser-assisted machining will allow this approach to hot machining to be applied except in very unusual situations.

While hot machining can be valuable when cutting very difficult materials such as high-temperature alloys it has not proven to be worth the additional complication when machining conventional materials. Even with high-temperature alloys such as nickel and cobalt base materials, great care must be exercised in applying the technique to be sure the work-softening advantage is not more than offset by a tendency for the peak tool-face temperature to move closer to the tool tip as when machining nickel under conventional conditions.

Time Effects

While equations in the foregoing discussions were derived on the basis of a steady state, most practical applications involve a nonsteady state. When a slider starts from rest, a certain time must elapse before a steady equilibrium temperature distribution obtains. The steady state is approached asymptotically and theoretically the absolute steady state is obtained at time $T = \infty$. However, the temperature distribution is indistinguishable from the absolute distribution after an elapsed time of but

$$T = \frac{6LK}{(V)^2} \tag{12.50}$$

When values are substituted into this equation for a typical machining case, T is found to be about 10^{-3} sec. This corresponds to a tool movement of but 0.040 in (1 mm) at a cutting speed of 200 f.p.m. (63 m min^{-1}). Therefore the degree of approximation involved in applying steady-state theory to nonsteady state machining problems should be relatively insignificant.

The foregoing conclusions has been experimentally verified by Opitz who found, by use of oscillograms of thermoelectric voltages generated with a milling cutter, that equilibrium temperatures are reached in the tool surface in about 10^{-4} s.

Even though tool temperatures obtained in intermittent and continuous cutting will be the same, the effects of these temperatures upon tool life will not be the same. The rest period that occurs between cuts in an intermittent operation such as milling is very significant for the temperature will drop very rapidly when cutting ceases by heat flow from the surface of the tool into its interior. For this reason, milling cutter teeth can operate at significantly higher peak temperatures than can a lathe tool. This point is well illustrated by some thermoelectric measurements made by Herbert (1926). He found that there was no softening of the teeth of a saw when subjected to a temperature of 1830 °F (1000 °C) for $\frac{1}{3}$ second during an actual sawing operation whereas the teeth were completely softened when held in a lead pot at 1650 °F (900 °C) for 3 seconds.

References

Anon. (1939). Sp. Rep. **24**, p. 215. *The Iron and Steel Institute.*

Anon. (1946). *Proc. Iron and Steel Inst.* **154**, 83.

Arndt, G. and Brown, R. H. (1967). *Int. J. Mach. Tool Des. Res.* **7**, 39.

Bever, M. B., Holt, R., and Tichener, L. B. (1974). The stored energy of cold work. *Prog. Mater. Sci.* **17**.

Bever, M. B., Marshall, E. R., and Tichener, L. B. (1953). *J. Appl. Phys.* **24**, 117.

Bickel, E. (1963). In *International research in production engineering,* p. 89. ASME, New York.

Bickel, E. and Widmer, W. (1951). The temperature on the cutting edge. *Industrial Organization* **8**.

Blok, H. (1938). In *Proc. General Discussion in Lubrication and Lubricants,* p. 222. Instn Mech. Eng., London.

Boothroyd, G. (1963). *Proc. Instn Mech. Engrs* **177**, 789.

Carslaw, H. S. (1921). *Introduction to the mathematical theory of the conduction of heat in solids.* The Macmillan Co. Ltd., London.

Chao, B. T., Li, H. L., and Trigger, K. J. (1961). *J. Eng. Ind. Trans. Am. Soc. mech. Engrs* **83**, 496.

Chao, B. T. and Trigger, K. J. (1953). *Trans. Am. Soc. mech. Engrs* **75**, 109.

Chao, B. T. and Trigger, K. J. (1955). *Trans. Am. Soc. mech. Engrs* **77**, 1107.

Chao, B. T. and Trigger, K. J. (1958). *Trans. Am. Soc. mech. Engrs* **80**, 311.

Chao, B. T., Trigger, K. J. and Zylstra L. B. (1952). *Trans. Am. Soc. mech. Engrs* **74**, 1039.

Gottwein, K. (1925). *Maschinenbau* **4**, 1129.

Hahn, R. S. (1951). *Proc. First U.S. Nat. Cong. Appl. Mech.*, p. 661.

Herbert, E. G. (1926). *Proc. Instn mech. Engrs* **1**, 289.

Jaeger, J. C. (1942). *Proc. R. Soc. N.S.W.* **78**, 203.

Krabacher, E. J. and Merchant, M. E. (1951). *Trans. Am. Soc. mech. Engrs* **73**, 761.

Kronenberg, M. (1949). *Am. Mach.* **93**, 104.

Kronenberg, M. (1954). In discussion to Loewen and Shaw.

Kusters, K. J. (1956). Dissertation, T. H. Aachen.

Leone, W. C. (1954). *Trans. Am. Soc. mech. Engrs.* **76**, 121.

Loewen, E. G. and Shaw, M. C. (1954). *Trans. Am. Soc. mech. Engrs* **76**, 217.

Nakayama, K. (1956). *Bull Fac. Engng Yokohama nat. Univ.* **5**, 1.

Nakayama, K. (1959). *Bull Fac. Engng Yokohama nat. Univ.* **8**, 1.

Pahlitzsch, G. and Helmerdig, H. (1943). *A des Ver. deutscher Ing.* **81**, 564. and 691.

Rajogopal. S. (1982). *Laser Focus* **18**, 49.

Rapier, A. C. (1954). *Br. J. appl. Phys.* **5**, 400.

Rumford, Count (1799). In *Essays political, economical and philosophical.* David West, Boston.

Schallbrock, H. and Lang, M. (1943). *Z. Ver. deutscher Ing.* **87**, 15.

Schaller, E. (1964). Beitrag Zur Untersuchung von Spannungen und Dynamischer Vorgangen in der Grenzschict Zwischen Werkzeug und Span bei der Stahlzerspannung mit Hartmetalwerkzeugen, Dissertation, T. H. Aachen.

Schmidt, A. O. and Roubik, J. R. (1949). *Trans. Am. Soc. mech. Engrs* **71**, 245.

Schmidt, A. O. and Roubik, J. R. (1954). In discussion to Loewen and Shaw (1954), p. 229.

Schwerd, F. (1937). *Z. des Ver. deutscher Ing.* **77**, 211.

Shaw, M. C. (1958). *Technische Mitteilungen* (Essen) **5**, 211.

Shore, H. (1924). Tool and Chip Temperatures in Machine Shop Practice. M.S. Dissertation, M. I. T. Also *J. Wash. Acad. Sci.* **15**, 85 (1925).
Tay, A. O., Stevenson, M. G., and Davis, G. da V. (1974). *Proc. Instn mech. Engrs* **188**, 627.
Taylor, G. I. and Quinney, H. (1934). *Proc. R. Soc.* **A143**, 307.
Taylor, G. I. and Quinney, H. (1937). *Proc. R. Soc.* **A163**, 157.
Trent, E. M. (1976). *Metal cutting.* Butterworths, London.
Trigger, K. (1963). In *International research in production engineering,* p. 95, ASME, NY.
Trigger, K. J. and Chao, B. T. (1951). *Trans. Am. Soc. mech. Engrs* **73**, 57.
Weiner, J. (1955). *Trans. Am. Soc. mech. Engrs* **77**, 1331.
Wright, P. K. (1978). *J. Engng Ind.* **100**, 131.
Wright, P. K. and Trent, E. M. (1973). *J. Iron Steel Inst.* **211**, 361.
Zorev, N. N. (1963). In *International research in production engineering,* p. 42. ASME, New York.

13

CUTTING FLUIDS

The two main functions of cutting fluids are lubrication at relatively low cutting speeds and cooling at relatively high cutting speeds. At high cutting speeds there is not time for fluid to penetrate the chip–tool interface, the wear land, or microcracks on the back of the chip to provide lubrication; and at low cutting speeds cooling is relatively unimportant. Typical low-speed cutting operations where lubrication is important are broaching and tapping while typical high-speed machining operations where cooling is important are turning and milling. Water-base cutting fluids are more likely to be used in the high-speed cooling regime while oil-base fluids are more apt to be used in the low-speed lubrication regime.

There are many other secondary considerations associated with cutting fluid selection (Shaw 1970) such as

Chip disposal.
Corrosion.
Health, safety, and aesthetic considerations.
Cost.

The main objective in use of a cutting fluid is either reduction in total cost per part or an increase in the rate of production.

Low-Speed Cutting Fluid Action

The data presented in Table 13.1 illustrates the very large difference a cutting fluid can make when used in an orthogonal cutting operation performed at a very low cutting speed. In these tests the workpiece was moved past the stationary tool (planer fashion) at very low speed ($V = 1$ i.p.m. $= 0.025$ m min^{-1}). The undeformed chip thickness was 0.003 in (0.08 mm) and the width of cut 1.000 in (25.4 mm). Forces were measured using a two-component dynamometer and fluid was applied dropwise at the chip–tool interface. Only those values are recorded that correspond to a continuous chip and no BUE was evident in any of the tests recorded in Table 13.1.

Figure 13.1 shows values of shear stress (τ) and normal stress (σ) on the tool face plotted with shear stress positive (clockwise) and with shear stress negative (counter-clockwise) for points on the shear plane. The numbers refer to the test numbers in Table 13.1. Single Mohr's circles passing through the

TABLE 13.1

Cutting data for annealed AISI 1112 in orthogonal machining tests at low cutting speed. $V = 1$ i.p.m. (0.025 m min^{-1}); $t = 0.003$ i.p.r. (0.08 mm/rev); $b = 1$ in (25.4 mm). (After Shaw, Cook, and Finnie 1953)

Test No.	Fluid	α deg	F_P lb	F_Q lb	r	ϕ deg	γ	$\tau(\times 10^{-3})$ p.s.i.	$\sigma(\times 10^{-3})$ p.s.i.	μ	$\beta = \tan^{-1} \mu$ deg
1	Benzene	45	699	84	0.602	36.5	1.20	85.6	84.0	1.33	53.0
2	Air	45	504	46	0.629	38.6	1.14	70.1	72.7	1.20	50.2
3	Ethanol	45	374	−6	0.680	42.7	1.04	63.1	56.5	0.97	44.1
4	CCl_4	45	329	−41	0.739	47.4	0.96	62.1	52.5	0.78	37.9
5	Benzene	30	912	251	0.380	22.2	2.31	94.4	72.7	1.01	45.4
6	Air	30	740	142	0.454	27.0	1.91	89.9	69.8	0.86	40.9
7	Ethanol	30	541	40	0.521	31.4	1.66	76.4	54.8	0.68	34.2
8	CCl_4	30	437	−30	0.625	38.2	1.41	74.9	50.7	0.49	26.1
9	Air	16	1115	347	0.301	17.5	3.20	96.5	66.9	0.66	33.3
10	Ethanol	16	850	168	0.362	21.1	2.68	87.9	55.6	0.51	27.2
11	CCl_4	16	576	19	0.485	28.1	2.07	78.9	46.2	0.32	17.9
12	CCl_4	0	816	108	0.340	18.8	2.28	79.2	39.2	0.13	7.6

1 lb = 4.45 N
1 p.s.i. = 0.69 N cm^{-2}

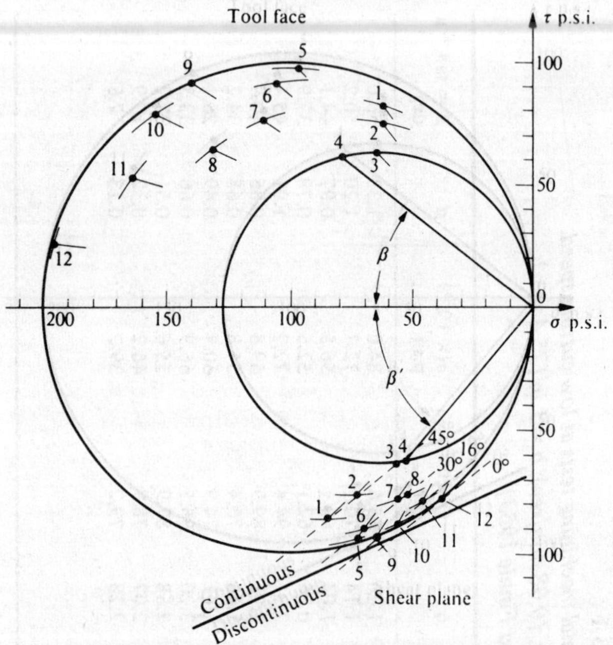

FIG. 13.1. Mohr's circle diagram for data of Table 13.1.

origin and corresponding points (same test number) for tool face and shear plane are shown in each case, only the smallest and largest circles being shown completely. From Fig. 13.1, it is evident that for the same Mohr's circle to hold for shear plane and tool face (no strain hardening) neither the shear plane nor the tool face can be a plane of maximum shear stress. Fig. 13.2a shows a slip line field corresponding schematically to any of the points in Fig. 13.1 while Fig. 13.2b is the corresponding Mohr's circle. Here the maximum shear direction deviates by an angle η' from the shear plane and an angle η from the tool face.

Several observations may be made concerning the data of Table 13.1.

1. Benzene which is a very stable compound does not contaminate a clean metal surface by reacting with it to prevent adhesion but merely excludes air which is a useful (friction-reducing) contaminant. Thus benzene is a negative 'lubricant' relative to dry cutting in air.
2. Ethanol and Carbon tetrachloride are positive lubricants relative to air since they lower cutting forces and increase shear angle.
3. The coefficient of tool face friction (μ) decreases as the effectiveness of the cutting fluid increases and also as the rake angle decreases.
4. The force component F_Q tends to be negative with increased effectiveness

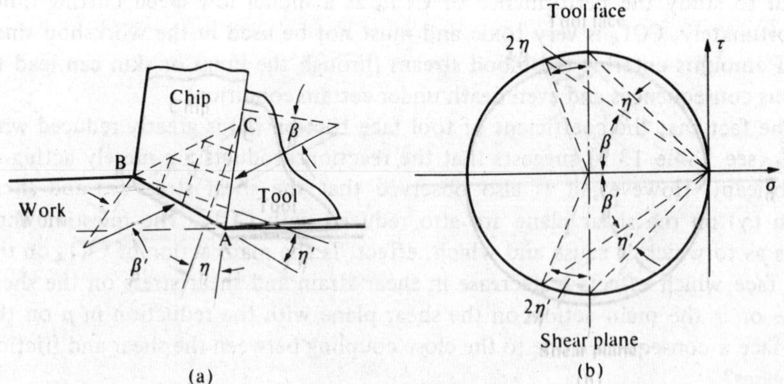

FIG. 13.2. (a) Slip line field for typical Mohr's circle diagram of Fig. 13.1. (b) Corresponding Mohr's circle diagram.

of cutting fluid and increased rake angle. A negative F_Q represents an unstable condition in that tool deflection will tend to cause an increase in $-F_Q$, a further increase in t, etc. Thus, when F_Q is negative the tool will tend to cut deeper with time until its stiffness becomes nonlinear or the tool fractures.

5. The tendency to form a discontinuous chip increases with decrease in rake angle but decreases with effectiveness of the cutting fluid. The only fluid that was capable of producing a continuous chip with a rake angle of zero degrees under the conditions of Table 3.1 was CCl_4 (carbon tetrachloride).
6. A BUE will not form at very low cutting speeds such as that for Table 13.1. The tendency to form a BUE at first increases and then decreases with cutting speed reaching a maximum at a speed of about 10 f.m.p. (3 m min^{-1}), when cutting in air. In the speed range where a BUE forms (say 5 f.p.m. (1.5 m min^{-1})) it will be suppressed by effective cutting fluids such as CCl_4. A BUE will not form if η is negative. For test nos. 7–12, η is positive and a BUE could form were it not for the very low cutting speed.
7. The shear and normal stresses on the tool face are far from constant.
8. The shear stress on the shear plane increases with normal stress on the shear plane.
9. All values of η are negative; that is, the material approaching the shear plane appears to shear before reaching the maximum value of shear stress which is within the chip.

Carbon Tetrachloride

Carbon tetrachloride (CCl_4) is one of the most effective cutting fluids known for a wide range of work materials at low cutting speeds. Other effective low-speed fluids perform in a similar way to CCl_4 but less decisively. It is therefore useful to study the performance of CCl_4 as a model low-speed cutting fluid. Unfortunately, CCl_4 is very toxic and must not be used in the workshop since small amounts entering the blood stream through the lungs or skin can lead to serious consequences and even death under certain conditions.

The fact that the coefficient of tool face friction (μ) is greatly reduced with CCl_4 (see Table 13.1) suggests that the reaction products are merely acting as a lubricant. However, it is also observed that the shear stress (τ) and shear strain (γ) on the shear plane are also reduced with CCl_4. The question then arises as to which is cause and which, effect. Is the main action of CCl_4 on the tool face which effects a decrease in shear strain and shear stress on the shear plane or is the main action on the shear plane with the reduction in μ on the tool-face a consequence due to the close coupling between the shear and friction processes?

Carbon Tetrachloride as a Lubricant

There appears to be no evidence that either carbon tetrachloride or its reaction product with steel ($FeCl_3$) exhibits good lubrication action in the sense that these materials give low sliding friction. For example, when CCl_4 is used as a wire-drawing lubricant for steel wire, very high friction results whereas materials such as oleic acid give low die friction.

A number of friction tests have shown that both CCl_4 and $FeCl_3$ may give much higher friction than oleic acid between sliding steel surfaces and in fact even higher friction than that pertaining for dry surfaces sliding in air. These tests were run by first making a planing type cut of $\frac{1}{4}$ in (6.35 mm) width on a specimen 3 in (76.2 mm) long. The clearance face of the high-speed steel tool used to make the cut was then immediately loaded against the freshly cut surface and drawn back after first applying the appropriate lubricant. The coefficient of friction recorded using a cutting dynamometer was found to be independent of the normal load. Values given in Table 13.2 are averages of five or more tests each and are representative of a large number of such sliding tests.

As long as the cut surface is smooth (i.e. when cut using CCl_4 or oleic acid) the coefficient of sliding friction in air is small and much below that pertaining for carbon tetrachloride. When, however, the surface is rough (as when cut in air), the oxide breaks down and the friction is very high. In some tests the surface cut with CCl_4 was first washed with water before testing with air, or oleic acid, but this made no difference in the results obtained. In no case was the friction coefficient obtained with CCl_4 in the same class with such good lubricants as oleic acid, graphite, or molybdenum disulfide.

In Table 13.3 friction values are given for a steel ball rubbing on a steel turntable. Before each tests a new spot on the ball was selected and both sliding surfaces were washed with benzene. The steel plate was then abraded with 4/O aluminum oxide paper immediately before the test lubricant was applied. The plate was somewhat rougher than the surfaces mentioned in Table 13.3

TABLE 13.2

Friction experiments on freshly cut surfaces

Surface treatment	Fluid	Mean coefficient of friction μ
Cut with 0.010 in (0.25 mm) feed using CCl_4; allowed to dry	CCl_4	0.23
	Air	0.18
	Oleic Acid	0.14
Cut with 0.010 in (0.25 mm) feed using oleic acid; washed with CCl_4 and allowed to dry	CCl_4	0.32
	Air	0.22
	Oleic Acid	0.19
Cut with 0.005 in (0.125 mm) feed in air	CCl_4	0.29
	Air	0.65
	Oleic Acid	0.17
Cut with 0.002 in (0.05 mm) feed using water; allowed to dry	Water	0.39
	CCl_4	0.38
	Air	0.38
	Graphite	0.15
	Molybdenum disulfide	0.10

TABLE 13.3

Friction tests with $\frac{1}{4}$ in (6.35 mm) diameter AISI 51100 hardened steel ball sliding on an AISI 1020 steel turntable at 6 f.p.m. (1.88 m min^{-1})

Fluid	Mean coefficient of friction μ
Air	0.52
$FeCl_3$	0.49
CCl_4	0.29
Mineral oil	0.18
Oleic acid	0.13
Graphite	0.11
Molybdenum disulfide	0.08

and hence the friction results in air were high. In this test the CCl_4 values were of intermediate magnitude but distinctly higher than those for mineral oil or the better lubricants: oleic acid, graphite, or molybdenum disulfide.

From these considerations it would appear that CCl_4 is not a lubricant in the sense that oleic acid or graphite is or in fact even in the sense that a thin oxide film present on a smooth metal surface is.

The apparatus of Fig. 10.9 in which a smooth steel ball is rotated in an AISI 1020 steel socket was used with different lubricants applied to the socket before the test (Fig. 13.3). CCl_4 was again found to be poor lubricant relative to graphite, MoS_2, or even air. The results shown in Fig. 13.3 were initial values corresponding to a fraction of 1 revolution of the ball in the socket.

A new subsurface plastic flow friction apparatus was constructed in which the ball could be rotated continuously at low speed. When running a test in air (AISI 52100 ball against AISI 1020 steel) it was found that the low initial friction torque rose gradually to a steady equilibrium value that was about six times as high as the initial value (Fig. 13.4). During the period of rising torque the soft steel surface was severely galled. Figure 13.5 shows mean frictional stress (τ) versus mean normal stress (σ) when sliding in air under initial (smooth) and steady-state (galled) conditions.

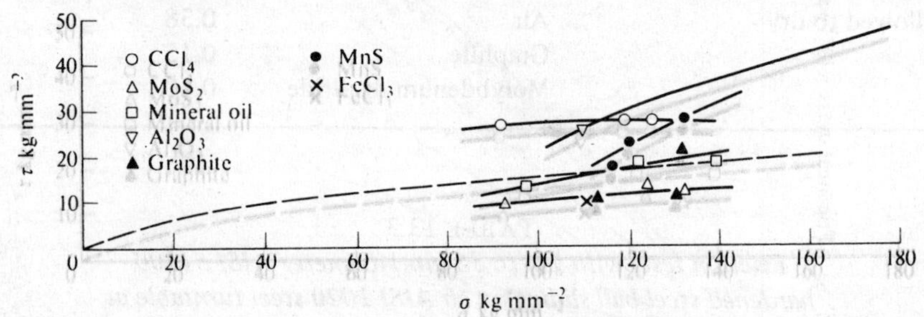

FIG. 13.3. Variation of τ and σ for friction tests involving subsurface plastic flow with different lubricants present. The dotted curve is the curve for tests in dry air given previously in Fig. 10.10. (After Shaw, Ber, and Mamin 1960.)

Figure 13.6a shows results similar to those of Fig. 13.4 while Fig. 13.6b shows results similar to Fig. 13.5 for CCl_4. It is evident that with CCl_4 the steady-state value is reached immediately. It should also be noted that whereas the steady-state value of coefficient of friction (μ) for air is about 0.5, that for CCl_4 is now only about 0.28. CCl_4 now looks like a lubricant relative to air.

Additional tests were run in which sliding in air was continued until equilibrium (galling) was obtained in air followed by sliding with CCl_4. Figure 13.7a

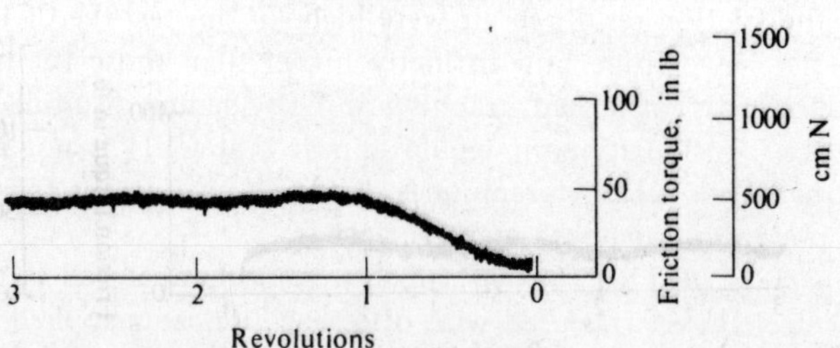

FIG. 13.4. Friction torque versus rotational displacement for sliding in air. Load on sphere = 1000 kg (2200 lb); mean sliding speed = 0.05 m min^{-1} (0.16 f.p.m.); sphere diam. = 12.7 mm (0.500 in); hole diam. = 3.8 mm (0.125 in). (After Shirakashi, Komanduri, and Shaw 1978.)

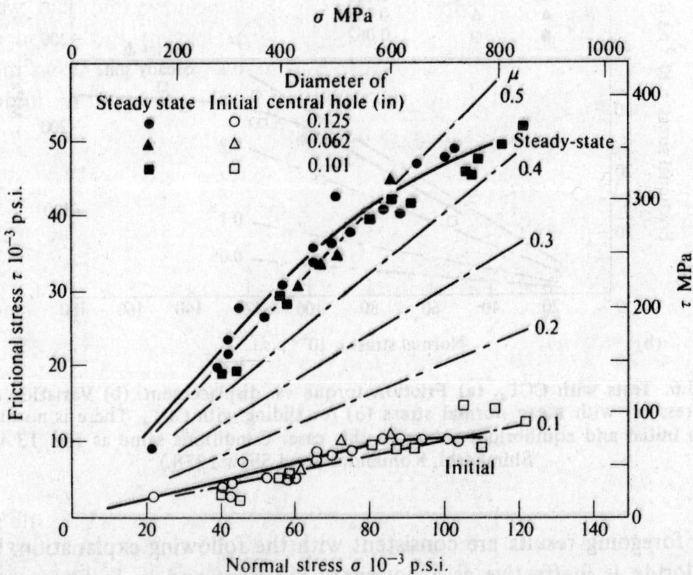

FIG. 13.5. Variation of mean shear stress (τ) with mean normal stress σ for sliding in air under initial (open points) and steady state (solid points). Conditions same as Fig. 13.4. (After Shirakashi, Komanduri, and Shaw 1978.)

shows the torque versus displacement curves for a typical test while Fig. 13.7b shows that the coefficient of friction (μ) for $CC1_4$ is very low (~ 0.05) when sliding on a very rough surface. The $CC1_4$ might be said to be a super lubricant under these conditions.

Figure 13.8 shows values of τ and σ for surfaces of difference initial roughness with $CC1_4$ as lubricant.

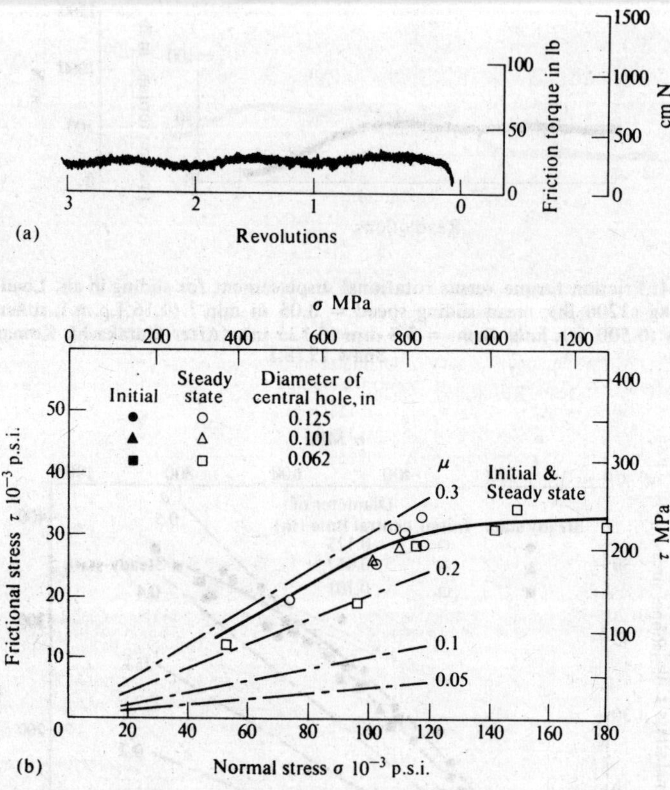

FIG. 13.6. Tests with CCl_4. (a) Friction torque vs. displacement. (b) Variation of mean shear stress (τ) with mean normal stress (σ) for sliding with CCl_4. There is no difference between initial and equilibrium values in this case. Conditions same as Fig. 13.4. (After Shirakashi, Komanduri, and Shaw 1978.)

The foregoing results are consistent with the following explanation. Carbon tetrachloride is ineffective as a boundary lubricant and is, in fact, a negative boundary lubricant if it does not react chemically with the surface being lubricated to form a brittle compound ($FeCl_3$ in the case of steel). The negative action comes from the fact that it excludes oxygen in air. Since metal oxides are somewhat beneficial in lowering frictional resistance, the presence of air is important.

Surface asperities are merely flattened by an indentor but not removed completely unless the indentor is slid across the surface (Chapter 10). This is because the plastic zones beneath each asperity grow as they are flattened under the increased load, but only until adjacent plastic zones interact. After that, there is gross plastic flow beneath the indentor instead of localized flow

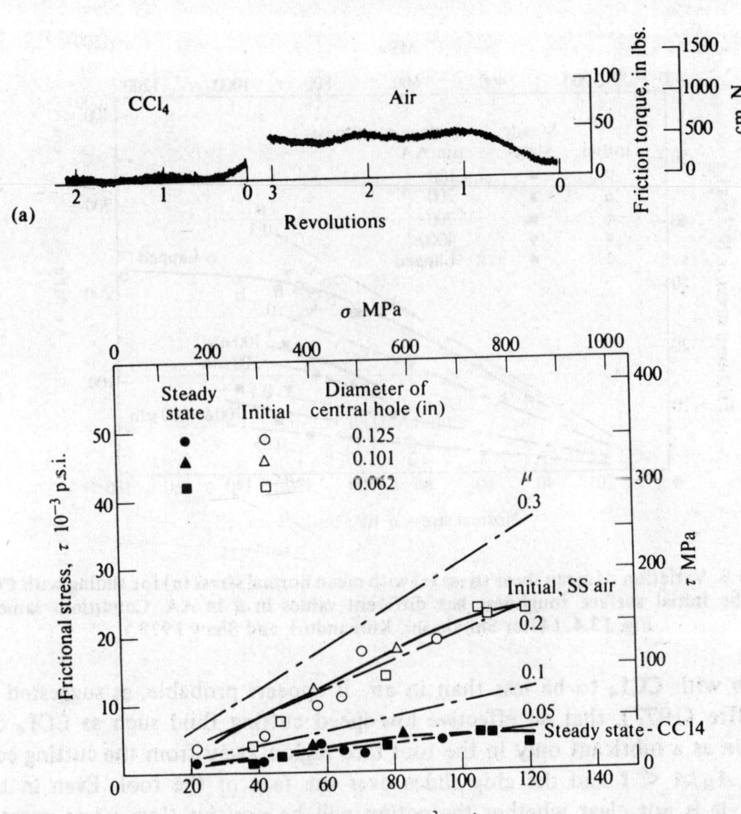

FIG. 13.7. Test to equilibrium surface condition in air followed by CCl_4. (a) Friction torque vs. displacement (b) Variation of mean shear stress (τ) with mean normal stress (σ). Conditions same as Fig. 13.4. (After Shirakashi, Komanduri, and Shaw 1978.)

beneath individual asperities. The tendency for CCl_4 to react with iron will increase with pressure. For a smooth surface, the pressure will be insufficient to cause CCl_4 to react to form $FeCl_3$ and the role of CCl_4 will be to exclude oxygen which in turn results in high friction. This explains why the friction was so high when a tool used to cut steel or aluminum using CCl_4 was slid back over a freshly cut surface. The surface produced was so smooth that the CCl_4 did not react to form a metal chloride on the backward pass. The foregoing explanation is also consistent with the results of Fig. 13.8. The pressure is too low on asperities of a smooth surface to cause CCl_4 to react and hence its role is the negative one of excluding oxygen.

When local pressure is sufficiently high, CCl_4 will react with the metal and the reaction product will prevent galling and junction growth. This then causes

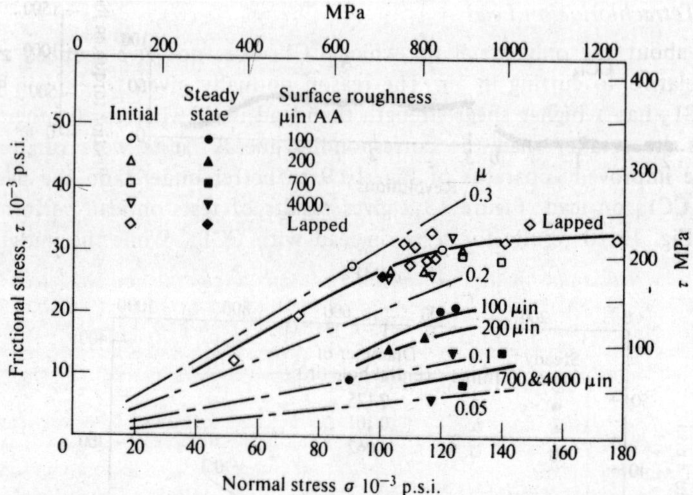

FIG. 13.8. Variation of mean shear stress (τ) with mean normal stress (σ) for sliding with CCl_4 when the initial surface roughness has different values in μ in AA. Conditions same as Fig. 13.4. (After Shirakashi, Komanduri, and Shaw 1978.)

friction with CCl_4 to be less than in air. It appears probable, as suggested by DeChiffre (1977), that an effective low-speed cutting fluid such as CCl_4 can function as a lubricant only in the tool face region away from the cutting edge where $A_R/A < 1$ and the chip slides over the face of the tool. Even in this region, it is not clear whether the action will be positive (low μ) or negative (high μ) because of the decreasing normal stress on the tool face with distance form the cutting edge.

It is necessary for CCl_4 to penetrate between chip and tool before it can react with the chip surface to prevent junction growth and galling. Vapor would have a greater chance of penetrating the intercommunicating voids in the region of contact further from the cutting edge than liquid. It has been demonstrated that CCl_4 vapor is just as effective as the liquid in slow-speed cutting (Shaw 1951). This was shown by cutting metal at low speed in a heated box into which superheated CCl_4 vapor was introduced. By maintaining the temperature of the box and tool above the boiling point of CCl_4 (80 °C) it was certain that only vapor reached the cutting zone. When pure aluminum was cut at low speed in the presence of CCl_4 vapor, the same dramatic improvement of Table 13.1 was obtained and the following reaction was found to occur on the freshly cut and highly active surfaces contacting the vapor.

$$2\,Al + 6\,CCl_4 \rightarrow 2\,AlCl_3 + 3\,C_2Cl_6. \tag{13.1}$$

The reaction products on the right-hand side of this equation were identified.

Carbon Tetrachloride on Lead

Lead is about the only metal for which CCl_4 does not give reduced cutting forces relative to cutting in air. The reason normally given to explain this is that $PbCl_3$ has a higher shear strength than lead itself whereas all other metal chlorides are weaker than the corresponding metal. Tests were run on lead using the improved apparatus of Fig. 10.9 to better understand the performance of CCl_4 on lead. Figure 13.9 gives results of tests on lead performed in air and Fig. 13.10 results for tests on lead with CCl_4. While the initial value

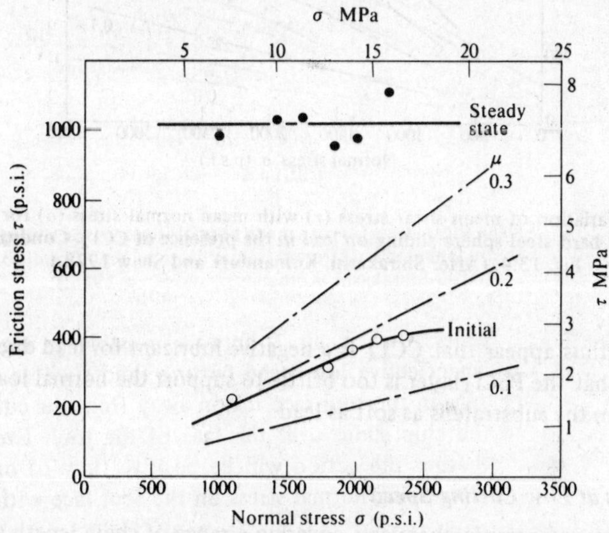

FIG. 13.9. Variation of mean shear stress (τ) with mean normal stress (σ) for a $\frac{1}{2}$ in (12.5 mm) diameter hard steel sphere sliding on lead in air. Conditions same as Fig. 13.4. (After Shirakashi, Komanduri, and Shaw 1978.)

of μ is slightly greater for CCl_4 than for air, the steady-state values are seen to be about the same. The explanation for this is that the air-lubricated specimen gives low friction until the oxide film initially present on the asperities is worn away. The friction in air is then that of clean metal resulting in high friction and galling. In the case of CCl_4, the $PbCl_3$ formed is too brittle on a soft substrate (lead) and fractures exposing clean lead as in the equilibrium air case. To check this, tests were performed using a hydrogen peroxide solution (H_2O_2) as a source of oxygen to replace the oxide worn away in the tests of Fig. 13.9. Figure 13.11 shows that in this case the steady-state value is the same as the initial value.

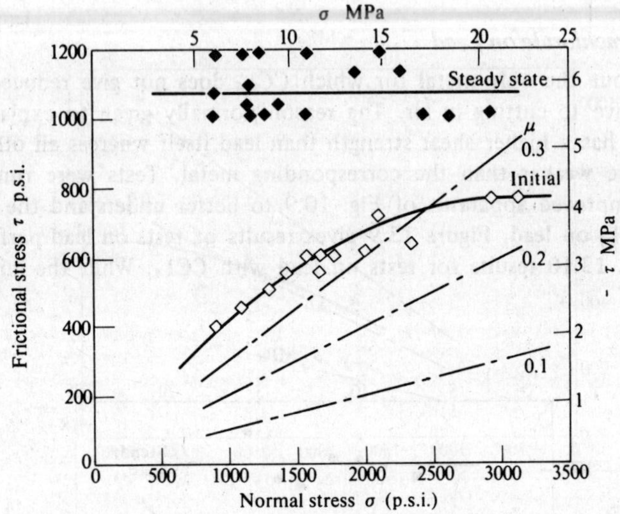

FIG. 13.10. Variation of mean shear stress (τ) with mean normal stress (σ) for a $\frac{1}{2}$ in (12.5 mm) diameter hard steel sphere sliding on lead in the presence of CCl_4. Conditions same as Fig. 13.4. (After Shirakashi, Komanduri, and Shaw 1978.)

It would thus appear that CCl_4 is a negative lubricant for lead due primarily to the fact that the $PbCl_3$ layer is too brittle to support the normal load without fracture when the substrate is as soft as lead.

Other Fluids at Low Cutting Speed

A wide variety of organic chemicals covering a range of chain length and several homologous series have been found to give low coefficient of tool face friction and a high cutting ratio relative to air as shown in Table 13.4. All fluids were put through a small laboratory still (nine theoretical plates, 5 ml holdup). Low-speed (5.5 in m^{-1} = 0.14 m min^{-1}) planer type cuts of 0.005 in (0.125 mm) undeformed chip thickness and 0.250 in (6.35 mm) width were made on commerically pure aluminum, the kneenness of the cutting edge being checked periodically using CCl_4 as a standard. The first three cuts with each new fluid were ignored being considered as part of the tool cleaning process. The values in Table 13.4 are average values for five cuts.

The disulfides, mercaptans, and long chain length esters were found to be outstanding in their ability to reduce tool face friction. The lowest coefficient of friction obtained was 0.237. A general decrease in μ with increased chain length was observed for all homologous series. The cutting ratio (r) and corresponding shear angle (ϕ) are shown plotted against the coefficient of tool face

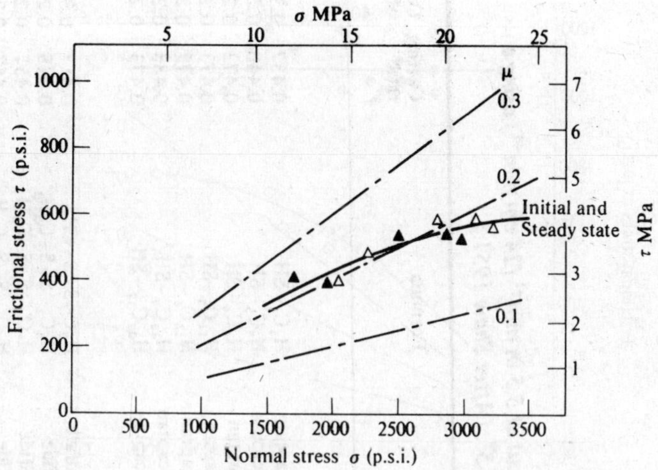

FIG. 13.11. Variation of mean shear stress (τ) with mean normal stress (σ) for a $\frac{1}{2}$ in (12.5 mm) diameter hard steel sphere sliding on lead in the presence of hydrogen peroxide. Conditions same as Fig. 13.4. (After Shirakashi, Komanduri, and Shaw 1978.)

friction (μ) in Fig. 13.12. All points except those for water and pentachlorethane fall on a smooth curve which suggests that in general a relation between r and μ exists.

It is evident from the data in Table 13.4 for the several chlorinated hydrocarbons that there is no correlation between the chlorine content of the molecule and the observed coefficient of friction; nor is there any indication that a long chain length or highly polar molecule is essential for low friction, since carbon tetrachloride gave the lowest coefficient of friction of all of the chlorinated hydrocarbons tested. The unsaturated chlorinated hydrocarbons gave particularly high values for the coefficient of friction.

A further series of tests was performed upon the chlorinated hydrocarbons listed in Table 13.4 in order to determine the relative reactivities of these compounds toward aluminum, at their boiling points under a pressure near 1 atmosphere. Each fluid (20 mls) was refluxed in a glass apparatus with six discs of aluminum, measuring 0.25 in diameter (6.4 mm) by 0.06 in (1.5 mm) thick, in contact with the boiling liquid. The condenser was attached to the flask with a ground glass joint and moisture was kept from entering the top of the condenser by means of a drying tube. When the fluid was boiling, the aluminum discs were introduced and the time noted. The fluid was refluxed until visible signs of a reaction were noted or for 24 hours in those cases where no reaction occurred. The reactions were all autocatalytic (some violently so) and the reaction time, therefore, could be easily recognized and reproduced.

TABLE 13.4

Friction data for a variety of organic compounds when aluminum is cut at 5.5 in min^{-1} (14 cm min^{-1}) using a high-speed steel tool of rake angle, 15°, relief angle, 5°. (After Shaw 1951.)

Fluid	Formula	Cutting ratio r	Coefficient of friction μ	Fluid	Formula	Cutting ratio r	Coefficient of friction μ
Chlorinated Hydrocarbons:				Mercaptans:			
Carbon tetrachloride	CCl_4	0.427	0.288	Ethyl mercaptan	H_5C_2-SH	0.437	0.308
Chloroform	$CHCl_3$	0.401	0.312	Propyl mercaptan	H_7C_3-SH	0.440	0.276
Dichloromethane	CH_2Cl_2	0.265	0.477	Butyl mercaptan	H_9C_4-SH	0.473	0.267
Pentachlorethane	$Cl_2HC-CCl_3$	0.230	0.648	Amyl mercaptan	$H_{11}C_5-SH$	0.473	0.265
S. tetrachlorethane	$Cl_2HC-CHCl_2$	0.320	0.419	Hexyl mercaptan	$H_{13}C_6-SH$	0.477	0.262
β Trichlorethane	$ClH_2C-CHCl_2$	0.344	0.407	Heptyl mercaptan	$H_{15}C_7-SH$	0.483	0.255
Methyl chloroform	Cl_3C-CH_3	0.340	0.422	Lauryl mercaptan	$H_{25}C_{12}-SH$	0.473	0.237
Ethylene dichloride	ClH_2C-CH_2Cl	0.250	0.481				
Ethylidene dichloride	Cl_2HC-CH_3	0.330	0.406	Disulfides:			
Ethyl chloride	H_3C-CH_2Cl	0.255	0.511	Methyl disulfide	$H_3C-S-S-CH_3$	0.417	0.302
Tetrachloro-ethylene	$Cl_2C=CCl_2$	0.195	0.603	Ethyl disulfide	$H_5C_2-S-S-C_2H_5$	0.448	0.271
Trichloroethylene	$Cl_2C=CHCl$	0.188	0.623	Propyl disulfide	$H_7C_3-S-S-C_3H_7$	0.457	0.256
T. dichloroethylene	$ClHC=CHCl$	0.185	0.612	Butyl disulfide	$H_9C_4-S-S-C_4H_9$	0.457	0.257
Hexyl choride	$C_6H_{13}Cl$	0.308	0.440	Amyl disulfide	$H_{11}C_5-S-S-C_5H_{11}$	0.462	0.250
Lauryl chloride	$C_{12}H_{25}Cl$	0.385	0.369				

Esters:			
Ethyl acetate	$H_3C-C(=O)-O-C_2H_5$	0.227	0.518
Hexyl acetate	$H_{11}C_5-C(=O)-O-C_2H_5$	0.403	0.299
Lauryl acetate	$H_{23}C_{11}-C(=O)-O-C_2H_5$	0.440	0.240
Ethyl caproate	$H_3C-C(=O)-O-C_6H_{13}$	0.391	0.353
Ethyl laurate	$H_3C-C(=O)-O-C_{12}H_{25}$	0.447	0.244
Alcohols			
Ethyl alcohol	H_5C_2-OH	0.327	0.425
Hexyl alcohol	$H_{13}C_6-OH$	0.396	0.320
Decyl alcohol	$H_{21}C_{10}-OH$	0.400	0.325
Undecyl alcohol	$H_{23}C_{11}-OH$	0.412	0.325
Lauryl alcohol	$H_{25}C_{12}-OH$	0.385	0.331
Hydrocarbons:			
Hexane	C_6H_{14}	0.163	0.639
Dodecane	$C_{12}H_{26}$	0.156	0.703
Miscellaneous:			
Benzene	C_6H_6	0.123	0.892
Acetic acid	$H_3C-C(=O)-OH$	0.247	0.496
Hexanoic acid	$H_{11}C_5-C(=O)-OH$	0.349	0.353
Heptaldehyde	$H_{13}C_6-C(=O)-OH$	0.430	0.298
Water	H_2O	0.187	0.690
Air		0.140	0.785

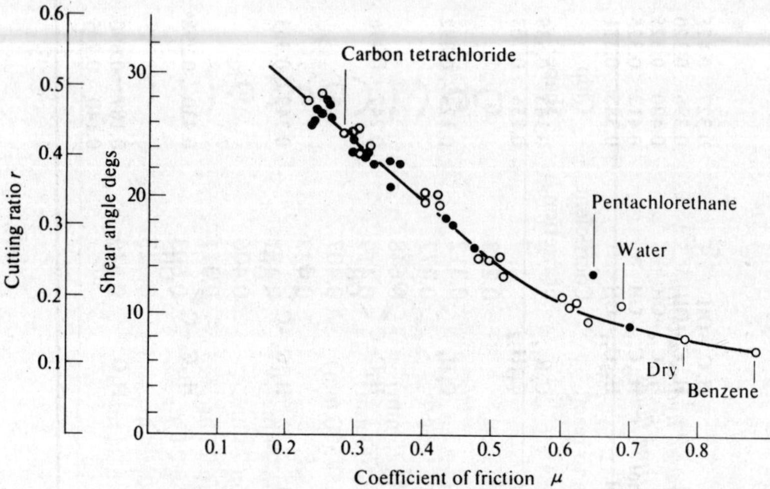

FIG. 13.12. Variation of cutting ratio and shear angle with coefficient of friction for data of Table 13.4.

Since three of the compounds decomposed on boiling at atmospheric pressure, it was necessary to test these compounds at reduced pressure. The pressures used in these cases are so stated with the corresponding times in Table 13.5 in which the results of the reflux tests are summarized.

It is evident that those fluids which did not react are the poorer cutting fluids and those which reacted readily are the better cutting fluids. It is not possible, however, to correlate the reaction time directly with the coefficient of friction. This is not surprising, inasmuch as the conditions of temperature, pressure, residual stress, and cleanness of surface are quite different along the tool face from those on the surface of the aluminum discs in the reflux experiments. The observation that the unsaturated chlorinated hydrocarbons are relatively stable toward aluminum in both the cutting and reflux experiments is consistent with the known reactivity of these compounds. In general, it is found that a halogen atom attached to a carbon atom holding a double bond is very stable, relative to one in a saturated compound, whereas a halogen atom attached to a carbon atom one removed from a carbon atom holding a double bond, is relatively active.

Further evidence of the physico-chemical nature of cutting fluid action at low cutting speeds is presented in Fig. 13.13. Here values of cutting force (F_P) are shown plotted against the chain length of a series of pure normal primary monohydric alcohols when machining pure aluminum at a speed of 5.5 i.p.m. (0.14 m min^{-1}), rake angle of 15°, and undeformed chip thickness of 0.005 in (0.125 mm). A systematic variation of cutting properties is noted with chain length and a significant difference in performance occurs for compounds having even and odd numbers of carbon atoms.

TABLE 13.5

Comparisons of the relative reactivities of several chlorinated hydrocarbons toward aluminum with their performance as cutting fluids

Fluid	Formula	Boiling point, °C	Time to react, min.	Coefficient of friction μ	Chip shape
Carbon tetrachloride	CCl_4	76.7	8.5	0.288	
Chloroform	$CHCl_3$	61.2	135	0.312	
Dichloromethane	CH_2Cl_2	39.8	360	0.477	
Pentachlorethane	$Cl_3C{-}CCl_2H$	161.9 (180 mm)	6 (125 mm)	0.648	
S. tetrachlorethane	$HCl_2C{-}CCl_2H$	146.5 (360 mm)	2 (200 mm)	0.419	
β. Trichlorethane	$H_2ClC{-}CCl_2H$	113.5 (700 mm)	1.5 (500 mm)	0.407	
Methyl chloroform	$H_2C{-}CCl_3$	74.1	2	0.422	
Ethylene dichloride	$H_2ClC{-}CClH_2$	83.7	not in 24 hr.	0.481	
Ethylidene dichloride	$H_3C{-}CCl_2H$	57.3	not in 24 hr.	0.406	
Ethyl chloride	$H_3C{-}CClH_2$	12.5	not in 24 hr.	0.511	
Tetrachloroethylene	$Cl_2C{=}CCl_2$	120.8	not in 24 hr.	0.603	
Trichloroethylene	$HClC{=}CCl_2$	86.7	not in 24 hr.	0.623	
T. dichloroethylene	$HClC{=}CClH$	48.4	not in 24 hr.	0.612	

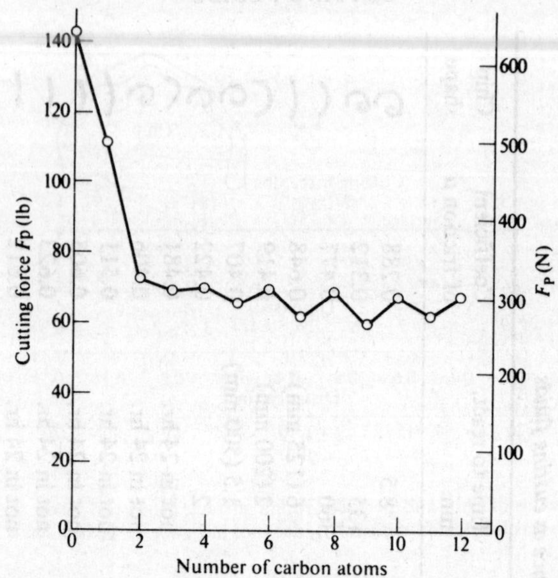

FIG. 13.13. Variation of cutting force (F_P) with chain length when machining aluminum with pure normal primary monohydric alcohols. (After Shaw 1944.)

An interesting phenomen was observed when aluminum was cut at 0.5 i.p.m. (0.01 m min^{-1}) using a 15° rake tool and a 3 percent mixture of *n*-decanol in benzene as the fluid. The chip thickness and the forces were observed to vary periodically as shown in Fig. 13.14. The finished surface was found to be alternately rough and smooth with a period corresponding to the variation in thickness of the chip. From the time scale that is shown it is evident that the fluctuations observed have too high a period (6 sec) for the variation to be due to chatter or any other periodic variation of the mechanical system.

This periodic cutting action can be explained in terms of the chemico-physical action of cutting fluids as follows. It is reasonable to expect that a certain minimum quantity of reaction product will be required to produce an appreciable reduction in the coefficient of friction between chip and tool and consequently a noticeable improvement in finish and a decrease in cutting force. At the beginning of a cut the surface temperature and the force increase gradually. At 3 percent decanol, the concentration of the active ingredient is just sufficient at the higher surface temperature and force to produce enough reaction product for effective cutting fluid action, thus decreasing the frictional resistance and the cutting force. At this reduced cutting force the surface temperature and area of contact are decreased, and thus the quantity of the chemical product formed is insufficient for effective cutting fluid action; therefore, the force and surface temperature once more increase. This cycle is repeated periodically.

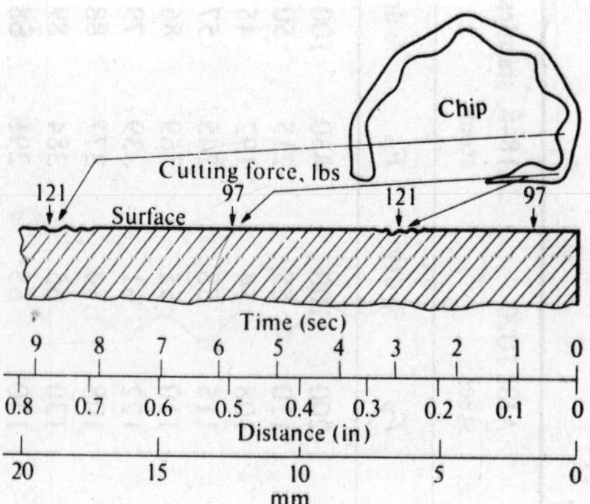

FIG. 13.14. Drawing of chip and periodic rough surface produced when tool cutting pure aluminum was lubricated with 3 $^{w}/_{o}$ decanol in benzene. (After Shaw 1942.)

Specificity

The fact that a cutting fluid in the vapor phase may be as effective as the fluid in the liquid phase supports the view that cutting fluid action is at least in part chemical rather than totally physical in nature. The fact that the same cutting fluid does not give equally good results on all metals offers further evidence. The data shown in Table 13.6 illustrates the variety of results that may be obtained when different fluids are used with different metals. Benzene is seen to be uniformly poor with all metals except copper. Turpentine is very good with copper but poor with stainless steel. Diamyl sulphide and carbon tetrachloride are seen to be good in all cases (lead not included here) and acetic acid is unusually poor with aluminum.

Shear Plane Action

In addition to lubrication action on the tool face there is evidence that cutting fluids may also play an important role on the shear plane at low cutting speeds by preventing microcracks from rewelding as the material passes through the shear zone. Russian workers appear to favor such an action to explain low-speed cutting behavior (Rehbinder, 1948a, 1948b, 1949, 1951, 1954a, 1954b). A large number of microcracks is assumed to be produced in the chip at the cutting edge when a new surface is generated. The absorbed fluid is believed to keep the microcracks from welding under load. The increased number of points of stress concentration corresponding to the unhealed cracks is believed to yield a closer spacing of active slip planes. This action has been termed 'hardness reduction' in the USSR.

TABLE 13.6

Values of cutting force (F_P) in pounds for several metal–fluid combinations at slow cutting speeds. Cutting conditions: undeformed chip thickness, 0.002 in (0.05 mm); width of cut, 0.25 in (6.35 mm); cutting speed, 5.5 in min^{-1} (0.14 m min^{-1}); rake angle, 15°.

Fluid	Pure Iron		Pure aluminum		Copper		AISI 1020 steel		18–8 stainless steel	
	F_P	% dry	F_P	% dry	F_P	% dry	F_P	% dry	F_P	% dry
Dry	370	100	35	100	292	100	200	100	430	100
n-Diamyl Sulphide	153	41	12	34	31	11	110	55	215	50
Carbon tetrachloride	119	32	18	51	28	10	108	54	192	45
Oleic Acid	122	33	16	46	111	38	115	58	245	57
n-Propyl Alcohol	192	52	18	51	200	68	110	55	369	86
Carbon Disulphide	253	68	29	83	80	27	122	61	339	79
Acetic Acid	160	43	32	91	130	45	108	54	377	88
Turpentine	245	66	14	40	46	16	130	65	384	89
Paraffin Oil	208	56	23	66	153	52	130	65	291	68
Kerosene	340	92	31	89	139	48	165	83	384	89
Water	200	54	18	51	123	42	139	70	284	66
Benzene	340	92	37	106	122	42	240	120	407	95

Epifanov, Soloshko, and Rehbinder (1955) also suggest that an active cutting fluid makes the metal more brittle in a manner similar to hydrogen embrittlement.

A shear test (shown in Fig. 13.15) was devised to study the influence of a fluid on the shearing action in low speed cutting in the absence of sliding. The protruding element of dimensions *b* and *d* is made $\frac{1}{8}$ in (3.18 mm) high and resembles a partly formed chip. However, this is not a chip but a piece of undeformed work material rising out of the surface. The specimen is clamped horizontally in a dynamometer and the tool is fed downward, shaving the front of the specimen to ensure a perfectly flat surface against which to push. The tool is then set exactly at the level of the rear surface with the aid of a microscope. When all is ready, the horizontal feed is engaged and the protruding member is pushed to the right as the horizontal (F_P) and vertical (F_Q) components of force are recorded.

The shearing test resembles a two-dimensional cutting operation performed by a tool with zero rake angle making a cut with zero shear angle and with negligible motion of the chip along the tool face. The dimension *b* is comparable to a typical width of cut while dimension *d* is made to correspond to a representative shear plane length (about 0.05 in, or 1.27 mm).

Before these tests were performed it was expected:

1. That the 'chip' could be broken free from the specimen after a very short displacement (of the order of about 0.01*d*).
2. That CCl_4 would be without influence on the force displacement curves.

Figure 13.16, representative of a large number of similar tests, reveals that both

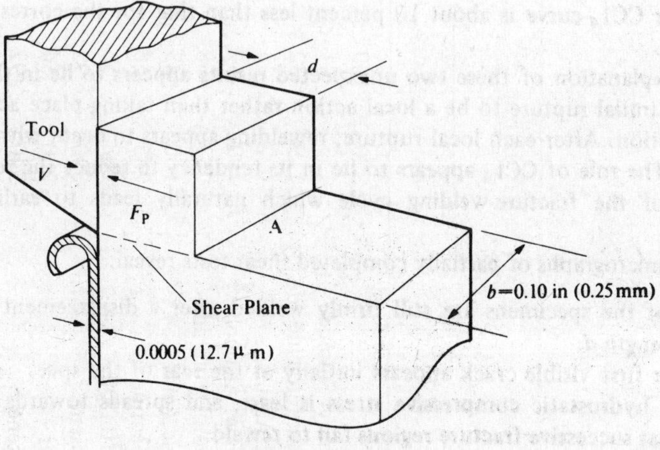

Fig. 13.15. Shear test arrangement (after Usui, Gajral, and Shaw 1961).

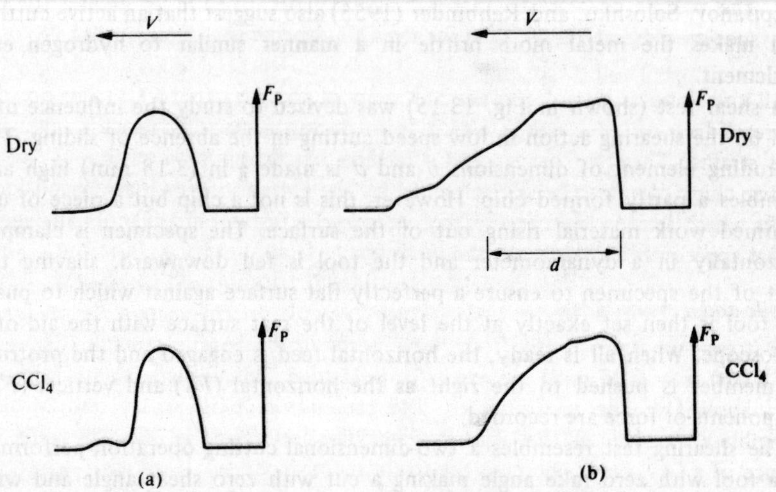

FIG. 13.16. Variation of shearing force (F_P) with tool displacement with and without drop of CCl_4 placed at A in Fig. 13.15. (a) Low carbon steel; (b) Soft copper. Speed = 0.4 i.p.m. (10.2 mm min^{-1}); d = 0.05 in (1.25 mm); b = 0.1 in (2.5 mm); relief angle = 20°. (After Usui *et al.* 1961.)

of these expectations are false. The force does not reach zero until the 'chip' has been displaced a distance in excess of d.

The curves obtained with CCl_4 are seen to fall to zero more quickly than those obtained in dry air. While the presence of the CCl_4 does not influence the maximum force or the displacement corresponding to this force, it has a distinct influence on the shape of the falling portion of the curve. The area under the CCl_4-curve is about 10 percent less than that for the corresponding test in air.

The explanation of these two unexpected results appears to lie in the tendency for initial rupture to be a local action rather than taking place across the entire section. After each local rupture, rewelding appears to occur with further motion. The role of CCl_4 appears to lie in its tendency to reduce the rewelding portion of the fracture-welding cycle which naturally leads to earlier total fracture.

Photomicrographs of partially completed shear tests reveal:

1. That the specimens are still firmly welded after a displacement as large as length d.
2. The first visible crack appears initially at the rear of the specimen where the hydrostatic compressive stress is least, and spreads towards the tool tip as successive fracture regions fail to reweld.
3. The extent of the region failing to reweld is greater when CCl_4 is present than in air.

4. The volume of material that is plastically deformed is far greater for the specimen sheared in air than for the CCl_4-specimen.

All the foregoing results appear to support the idea that CCl_4 can prevent the rewelding of local cracks which form during a shearing process involving large strains. They further suggest that the formation of 'continuous' metal cutting chips may also be accompanied by the formations of many small local cracks which tend to weld and refracture in cyclic fashion in a manner analagous to that apparently pertaining in the shear test.

Tensile and Torsion Tests

True-stress–true-strain tensile tests were performed upon very small (diameter $d = 0.010$ in $= 0.25$ mm) specimens of mild steel and copper with and without CCl_4 applied to the test surfaces. The presence of CCl_4 was found to have negligible influence on the shapes of the flow curves. The only observed influence of CCl_4 was to increase the strain at fracture a small amount. A mineral oil gave tensile results that were identical to the tests in air.

Miniature torsion tests were next performed upon cylindrical specimens to provide further materials test data on specimens of large area to volume ratio. These tests were unsuccessful since the small specimens strained in erratic ways due to the large influence of grain boundaries and other crystal imperfections. It was not until a specimen of the shape shown in Fig. 13.17a was adopted that reproducible results were obtained.

There was no difference in the torsional curves with and without a thin film of CCl_4 on the copper surface as long as the axial load was zero or tensile. However, when an axial compressive load was present there was a considerable difference in the strain at fracture as shown in Fig. 13.17b. The fracture strain with CCl_4 corresponded to 800 degrees of twist while that without CCl_4 corresponded to about 1100 degrees of twist.

These tests seem to indicate that although CCl_4 has no important influence in conventional tensile or torsion tests, it does have a significant influence upon the combined torsion-compression characteristics at large strains. Apparently a torsion specimen develops small surface cracks which subsequently weld and refracture before total failure if a compressive stress is also present on the shear plane to induce rewelding. When CCl_4 is present, rewelding is largely prevented and total failure comes earlier.

Microcracks in Machined Surface

A number of orthogonal cutting tests were performed at low speed (1 i.p.m. $= 0.025$ m min^{-1}) with a 20° rake angle HSS tool. In these tests several cuts ($t = 0.005$ in $= 0.127$ mm) were first made in air. Then a small amount of CCl_4 was placed on the second half of the last surface produced. A final cut was then taken and the forces recorded. The first half of this cut was produced with

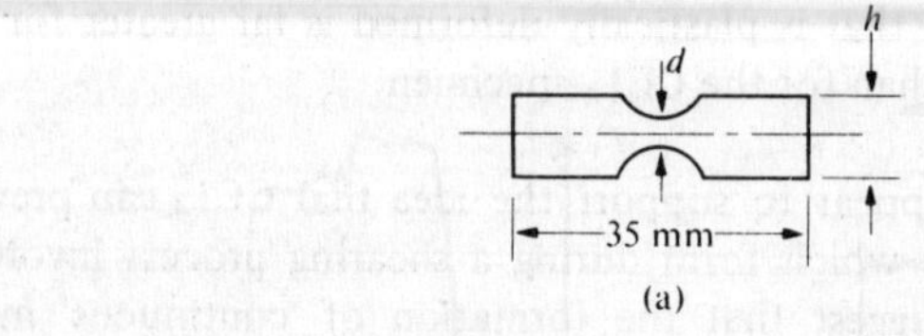

(a)

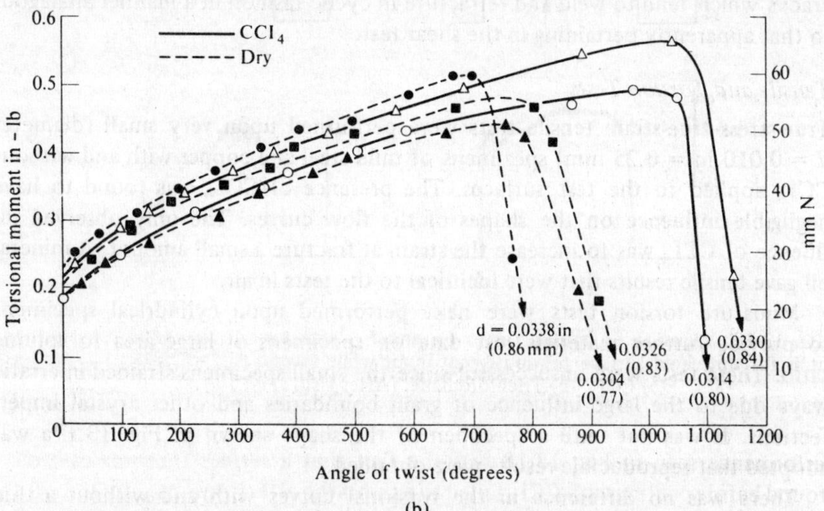

(b)

FIG. 13.17. Torsion test for copper with axial compressive load of 10 lb (44.5 N). (a) Specimen. (b) Torsional moment vs. angle of twist with and without CCl_4 on surface. (After Usui *et al.* 1961.)

a dry tool face while the second half of the cut still had a dry tool face but what became the back surface of the chip had been coated with CCl_4. Fig. 13.18 shows the suprising result representative of many others. At the first point where CCl_4 had been placed, the chip curled abruptly and force (F_P) dropped about 10 percent. This occurred whether the CCl_4 was allowed to evaporate or not. The obvious explanation of this unexpected result is that the machined surface contained microcracks which absorb CCl_4 which in turn keeps the microcracks from rewelding as the material passes through the shear zone.

When the surface was heated to the boiling point of the fluid after evaporation (80 °C in the case of CCl_4) the same result was obtained. Only when the surface was heated about 20 °C above the boiling point of CCl_4 did the sudden change in cutting performance at the midpoint of the cut fail to develop. Apparently, the boiling point of the fluid in small metal capillaries is 20° higher than in bulk at atmospheric pressure. Another observation in agreement with the microcrack contamination explanation for the unexpected change in cutting

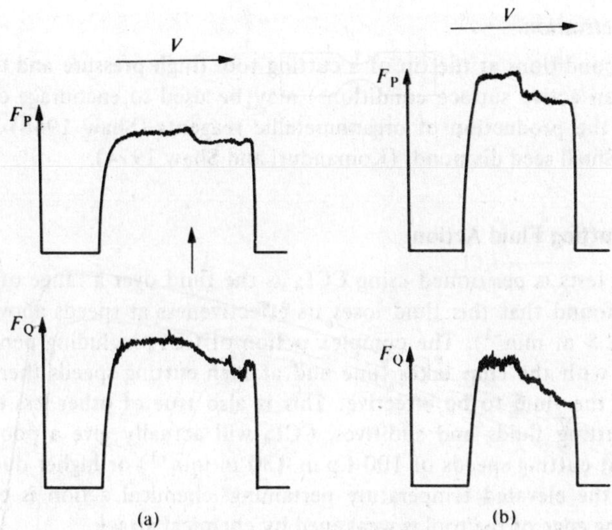

FIG. 13.18. Variation of cutting forces F_P and F_Q with tool displacement. First half of cut is for untreated surface previously cut in air while second half was coated with fluid. (a) CCl_4. (b) Trichlorethane. (After Usui *et al.* 1961.)

performance seen in Fig. 13.18 was the fact that a surface that was carefully ground before coating with CCl_4 failed to exhibit the effect. Apparently, the grinding operation removed microcracks that had developed during previous cutting.

It appears from all of the foregoing tests that at low cutting speeds important cutting-fluid action occurs on both the tool face and shear plane. On the tool face extreme boundary lubrication involving reaction between chip and tool is important over the part of the chip-tool contact area where sliding occurs. The function of the fluid here is apparently to prevent galling and to shorten the total chip-tool contact length which in turn causes an increase in the shear angle. The cutting fluid appears also to play an important role in preventing rewelding of microcracks that are formed during initial chip formation. Since the tendency for microcrack formation increases with decrease in normal stress on the shear plane, a greater density of microcracks should be expected on the free surface of the chip away from the tool face. This should promote chip curl which in turn will further decrease chip-tool contact length.

It would thus appear that the spectacular benefit observed with CCl_4 at low cutting speed is due to the combined effect of a decreased chip-tool contact area, a reduction in the flow stress on the shear plane and tool face due to microcrack contamination (hardness reduction) and a reduction in the coefficient of friction on the sliding portion of the chip-tool contact due to the prevention of junction growth and galling.

Mechanical Activation

The unusual conditions at the tip of a cutting tool (high pressure and temperature, and clean active surface conditions) may be used to encourage chemical action as in the production of organometallic reagents (Shaw 1948) and the formation of small seed diamonds (Komanduri and Shaw 1974).

High Speed Cutting Fluid Action

If a series of tests is performed using CCl_4 as the fluid over a range of cutting speeds, it is found that this fluid loses its effectiveness at speeds above about 40 f.p.m. (12.5 m min^{-1}). The complex action of CCl_4 including penetration and reaction with the chip takes time and at high cutting speeds there just is not time for the fluid to be effective. This is also true of other less effective low speed cutting fluids and additives. CCl_4 will actually give a poorer tool life than air at cutting speeds of 100 f.p.m. (30 m min^{-1}) or higher due to the fact that at the elevated temperature pertaining, chemical action is excessive and the cutting edge of the tool is weakened by chemical attack.

At cutting speeds from about 10 f.p.m. (3.0 m min^{-1}) to 100 f.p.m. (30 m min^{-1}) for steels, a BUE is apt to be troublesome from the standpoint of surface finish. Oil-base cutting fluids containing fatty acids for boundary lubrication and more active chlorine, sulfur, or phosphorus compounds for extreme boundary lubrication (chemical reaction with the surface of the chip) are effective in decreasing surface roughness due to a BUE. An increased rake angle (α) and a reduction in the undeformed chip thickness (t) are also effective ways of controlling BUE roughness at relatively low cutting speeds.

As the cutting speed increases, the tool face temperature increases rapidly. In the practical speed range for turning, the Taylor equation ($VT^n = C$) represents a good approximation. For a HSS tool, the exponent n is about 0.1 and from eqn (12.43), the tool face temperature varies as $V^{0.5}$. It thus follows from these two results that for a HSS tool the tool life (T) should vary with tool temperature as follows

$$T \cong \bar{\theta}_t^{20} \tag{13.2}$$

This result was originally obtained empirically by Shallbrock, Schaumann, and Wallichs (1938). The corresponding exponent for a carbide tool will be 10.

At speeds between about 100 f.p.m. (30 m min^{-1}) and 200 f.p.m. (60 m min^{-1}) BUE ceases to be a problem but another problem takes its place—temperature-accelerated tool-wear. Above this speed range there is not time for the low-speed cutting-fluid action on the tool face or the back of a chip to occur and the sole function of the fluid gradually becomes one of cooling. At very high speeds (> 500 f.p.m., 150 m min^{-1}) there is not even time for a fluid to be effective in cooling. When turning at 500 f.p.m. (150 m min^{-1})

and a feed rate of 0.010 i.p.r. (0.25 mm/rev.) the total time available for cooling is about 10^{-3} second.

Just as there is the possibility of negative cutting-fluid action at low cutting speeds (benzene for example) there is also the possibility of negative fluid performance at high cutting speeds. This results from the fluid sometimes causing excessive chip-curl decreasing the chip–tool contact area to the point where the maximum tool temperature is brought too close to the cutting edge. This aspect of high-speed cutting-fluid action will be presently discussed in detail.

Another negative aspect of high-speed cutting-fluid action is thermal shock and thermal fatigue. Certain cutting operations are periodically intermittent (milling for example). In such a case use of a coolant may give too great a temperature range from the cutting (sudden approach to high tool-tip temperatures) to noncutting (strong cooling) parts of the cycle with a high range of fluctuating tensile stress due to differential thermal expansion. In such cases, it is best to cut in air. In general, dry cutting in air is recommended for carbide milling cutters.

Chip–Tool Thermocouple Tests

The chip–tool thermocouple technique may be used in conjunction with a cutting fluid provided the fluid is a poor electrical conductor. Most water-base cutting fluids provide this requirement. In a series of tests on AISI 1015 steel tubing using an orthogonal tool with a 15° rake angle, the following fluids were investigated:

1. Dry tool.
2. Commercial emulsifiable water-base cutting fluid in concentration of 40:1.
3. Commerical soluble water-base cutting fluid in concentration of 40:1.
4. Water plus 0.1 percent sodium nitrite (rust inhibitor).
5. Water plus 0.1 per cent sodium nitrite plus wetting agent.

The results of these tests are shown in Fig. 13.19.

The fluids are seen to be most effective in lowering the chip–tool interface temperature at the smaller values of feed. In all cases the effectiveness of the fluid in lowering the tool temperature on a percentage basis is seen to decrease with cutting speed. At the larger feeds and cutting speeds, the fluids are seen to be completely ineffective. The cutting speed at which the fluid becomes ineffective decreases as the feed is increased. At a feed of 0.0023 i.p.r. (0.06 mm/rev), the fluid was effective at all speeds tested; at a feed of 0.0052 i.p.r. (0.13 mm/rev) the fluid was effective to 400 f.p.m. (122 m min^{-1}) and, at a feed of 0.0104 i.p.r. (0.26 mm/rev), the fluid became ineffective at about 200 f.p.m. (60 m min^{-1}).

In these tests the fluids were applied to the chip, tool, and workpiece from above (i.e. toward the back of the chip). The increase in heat transfer path with

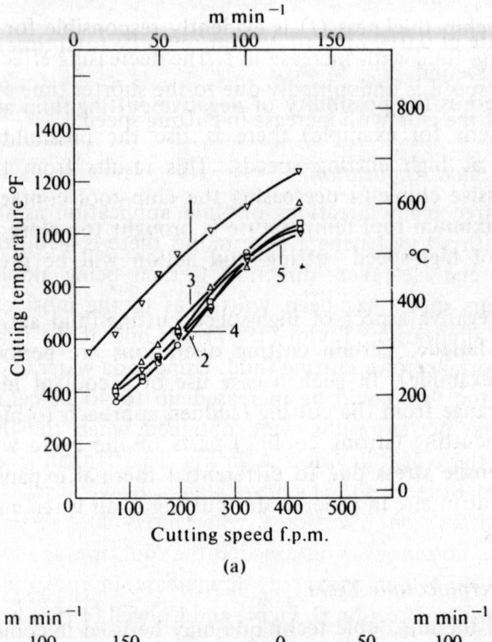

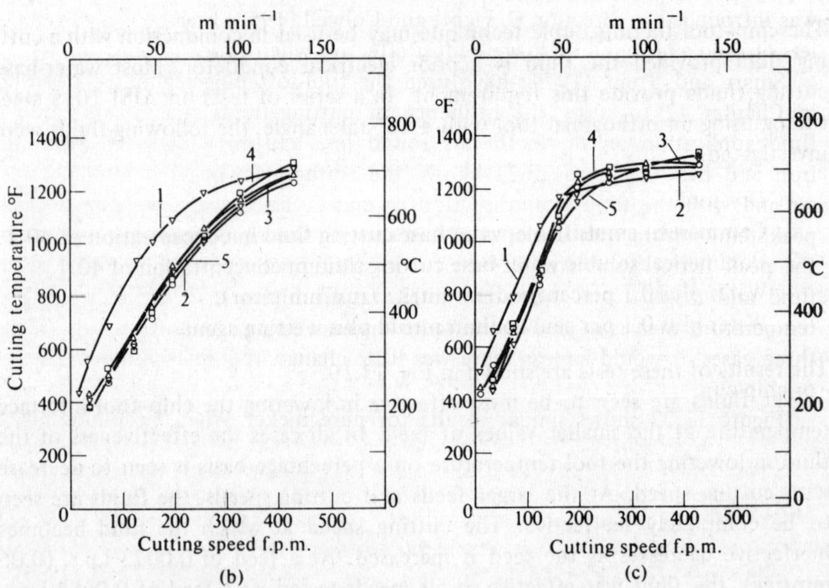

FIG. 13.19. Cutting speed versus mean tool face temperature (tool-chip thermocouple values) curves for AISI 1015 steel turned under orthogonal cutting conditions at different speeds and feeds. Numbers on curves correspond to fluids defined in text. (a) Feed = 0.0023 i.p.r. (0.06 mm/rev). (b) Feed = 0.0052 i.p.r. (0.13 mm/rev). (c) Feed = 0.0104 i.p.r. (0.26 mm/rev). (After Shaw, Pigott, and Richardson 1951.)

undeformed chip thickness (t) is evidently responsible for the decreased effectiveness of the fluid with increase in t. The decreasing effectiveness of the fluids with cutting speed is undoubtedly due to the shorter time available for meaningful cooling of the chip with increase in cutting speed.

Direction of Fluid Application

There are three main directions of fluid application as shown in Fig. 13.20. In turning with a tool having a nose radius there is a fourth direction C′ which is along the end clearance, direction C then being along the side clearance. Over the years much has been written as to the most effective direction of cutting fluid application. Taylor (1907) was one of the first engineers to recognize the importance of a cutting fluid. Using soda water as a coolant. he demonstrated that tool life could be increased up to 40 percent over cutting in air. He found that better results were obtained when the fluid was directed on the back of the chip (direction A in Fig. 13.20) than from directions B or C. Niebusch and Strieder (1951) advocated a heavy flow from direction A but also a heavy flow from direction C when heavy cuts are taken. A total flow of one gallon per horsepower consumed at the tool tip was advocated (~ 5 l kW^{-1}). Lauterbach (1952) demonstrated an increase in tool life when the lubricant was introduced in direction C. Pigott and Colwell (1952) used very high velocity jets of fluid and found these were most effective when aimed in direction C.

Smart and Trent (1974) have used the tool-softening technique described in Chapter 12 to study the influence of cutting fluids and the direction of fluid application on the isotherms found in a cutting tool when machining iron and nickel. Figure 13.21 shows the isotherms when machining iron in air and with the fluid from directions A and C. There was little change in the peak temperature with use of the fluid or with the fluid application. However, the peak temperature area is seen to be substantially reduced when using the fluid and in particular when the fluid was from direction C. Since the peak temperature area is at about the same distance from the cutting edge for all three cases, it would appear there was little change in chip–tool contact length or chip curl.

Figure 13.22 shows similar results for pure nickel. When machining nickel,

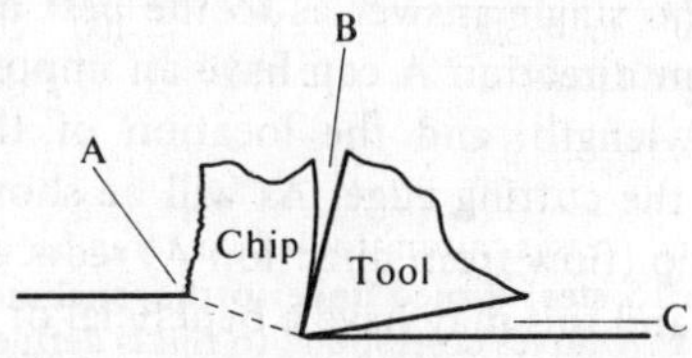

FIG. 13.20 Directions of cutting-fluid application in high-speed orthogonal machining

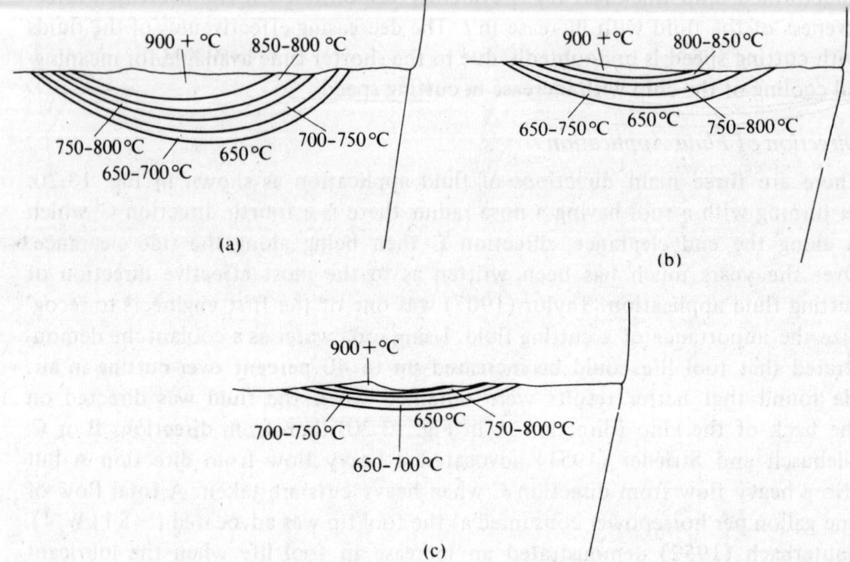

FIG. 13. 21. Isotherms for cutting iron (a) Dry; and with 30:1 soluble oil from (b) direction A and (c) direction C. Cutting conditions: cutting speed, 600 f.p.m. (183 m min^{-1}); feed, 0.010 i.p.r. (0.25 mm/rev); tool material, equivalent of M34 HSS; cutting time 30s. (After Smart and Trent 1974.)

the maximum temperature is at the tool tip (as discussed in Chapter 15) resulting in rounding of the cutting edge. With a cutting fluid, wear at the cutting edge is much less particularly when the fluid is directed along C. The peak cutting temperature is seen to be less when a fluid is used and particularly so when applied from direction C.

Many other papers concerning the optimum direction of cutting fluid application may be found in the literature—some advocating flow from direction A as in Taylor's case, some from direction C as in the case of Smart and Trent, and some advocating flow from both directions A and C as with Niebusch and Strieder. There is little support for direction B.

There appears to be no single answer as to the best method of fluid application. A strong flow from direction A can have an important influence on chip curl, tool-chip contact length, and the location of the point of maximum temperature relative to the cutting edge. As will be shown in the next section, cooling the back of a chip (flow from direction A) reduces the chip-tool contact length for cutting in air and this may have a beneficial or a detrimental influence on tool life. If application of the fluid from direction A causes overcurling of the chip bringing the maximum temperature too close to the cutting edge, then

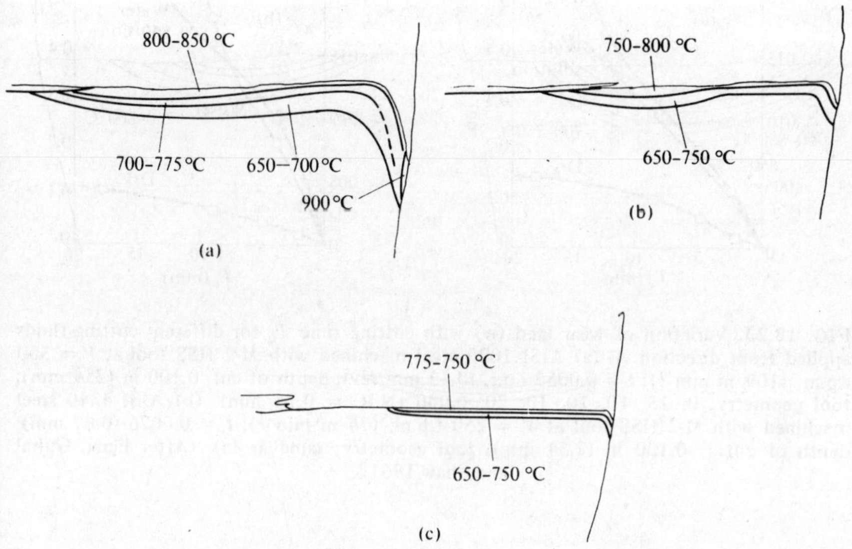

FIG. 13.22. Isotherms for cutting nickel. (a) Dry; and with 30:1 soluble oil from (b) direction A and (c) direction C. Cutting conditions: cutting speed, 150 f.p.m. (46 m min^{-1}); feed, 0.010 i.p.r. (0.25 mm/rev); tool material, equivalent of M34 HSS; cutting time, 30s. (After Smart and Trent 1974.)

application of the fluid from direction C or C′ or both will be preferable. If, however, the maximum temperature is already at the tool point in dry cutting (as when cutting nickel as discussed in Chapter 15), then direction C will give best results as Smart and Trent suggest.

Chip Curl

While the cooling function of a fluid may be beneficial relative to tool life, this is not always the case. Figure 13.23 is a case in point. Here the wear land size (w) is plotted against cutting time (T_c) when turning with a HSS tool. The rate of wear with water is considerably higher than that for a dry tool.

Figure 13.24 shows the wear land and crater development when machining AISI 1020 steel with and without a coolant (water). Each time the wear land was measured, a diamond stylus was drawn across the tool face and the shape of the build-up and crater on this face was recorded. The vertical magnification of motion is made five times the horizontal value in order to allow craters to be more conveniently represented.

The numbers next to each test-point correspond to the numbers of the crater traces. At the points indicated, the clearance face of the tool was carefully

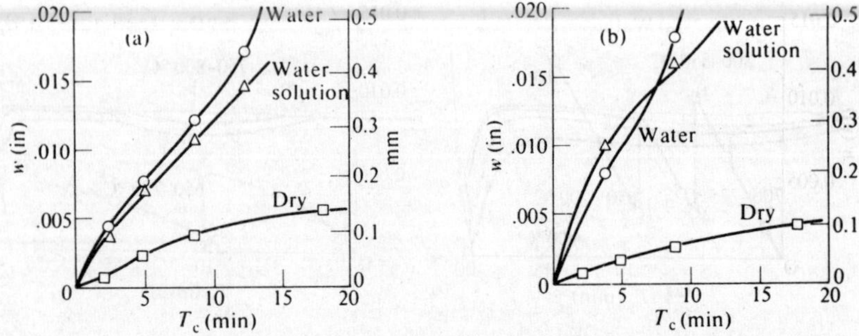

FIG. 13.23. Variation of wear land (w) with cutting time T_c for different cutting fluids applied from direction A. (a) AISI 1020 steel machined with M-2 HSS tool at $V = 350$ f.p.m. (109 m min^{-1}); $t = 0.0052$ i.p.r. (0.13 mm/rev); depth of cut, 0.100 in (254 mm); tool geometry; 0, 15, 10, 10, 10, 20, 0.030 (N.R. = 0.75 mm). (b) AISI 4340 steel machined with M-2 HSS tool at $V = 250$ f.p.m. (76 m min^{-1}); $t = 0.0026$ (0.07 mm); depth of cut = 0.100 in (2.54 mm); tool geometry; same as (a). (After Feng, Gujral and Shaw 1961.)

reground so as to decrease the extent of the wear land without disturbing the crater on the tool face. The crater produced with water is seen to form more rapidly than when cutting dry and the center of the crater is found to be closer to the tool point. This latter observation is an important one. Even though the water carries some heat away from the wear land, it simultaneously shifts the maximum temperature closer to the tool tip which results in more heat flowing into the wear land. The latter of these two effects is the predominant one in Fig. 13.24.

In order to check this thesis, the temperature of the newly developed surface was measured when cutting with and without a flood of water-base cutting fluid from direction A. In Fig. 13.25, the change in temperature of the work-piece surface in a turning operation is shown diagrammatically in developed form, from a point one revolution before the cutting edge to a point one revolution after the cut is completed. The temperature at (1) is ($\Delta\theta_1$) above the ambient temperature of the bar. Temperature (2) is the workpiece temperature at the cutting edge. The workpiece surface temperature will rise to a maximum value at (3) due to the heat which flows across the wear land from the tool. The temperatures at (4) and (5) correspond to surface temperatures at points 180 and 360 degrees from the cutting tool.

Nakayama (1956) measured the temperature at point (4) for dry, two-dimensional cutting on the end of a tube and noted a significant decrease in this temperature with increase in workpiece diameter. In making these measurements Nakayama allowed several strands of constantan wire to rub upon the freshly cut surface at a point 180° from the cutting tool. In repeating these

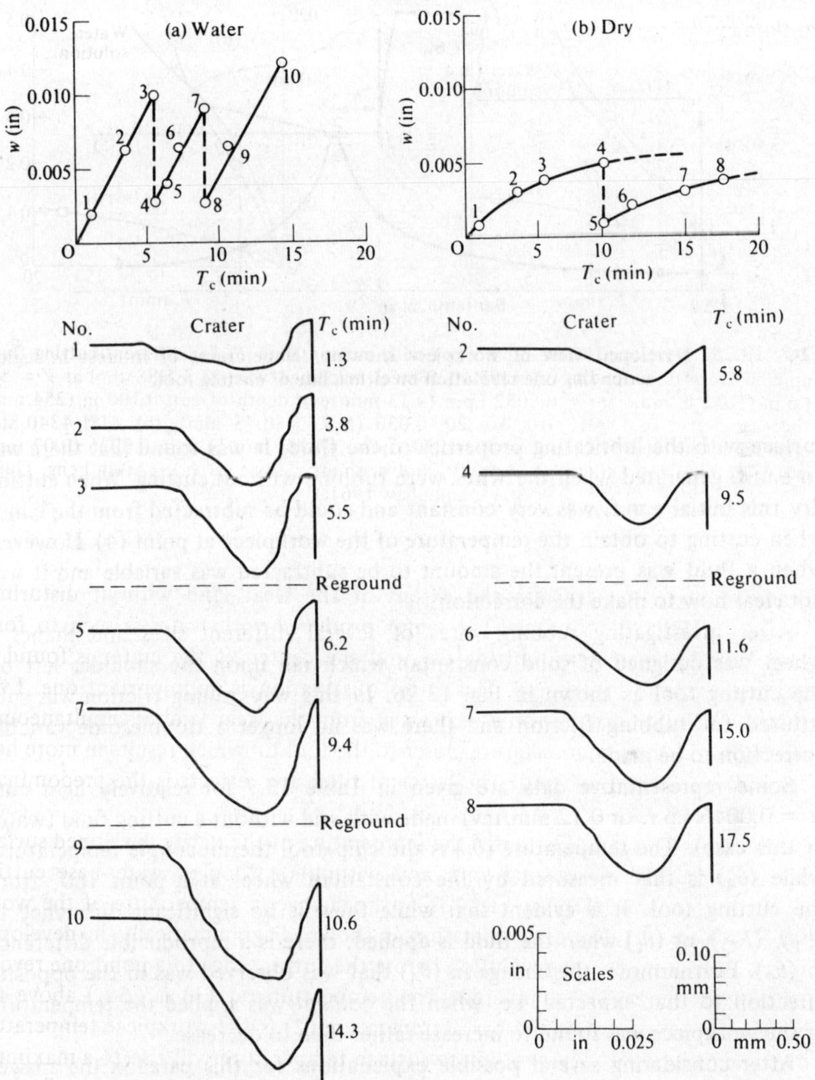

FIG. 13.24. Variation of wear land and crater with cutting time (T_c). Cutting speed = 450 f.p.m. (141 m min^{-1}); feed, 0.0052 i.p.r. (0.13 mm/rev); depth of cut = 0.100 in (2.54 mm); tool: M-2 HSS; tool geometry: 0, 15, 10, 10, 10, 20, 0.03 (N.R. = 0.75 mm). (a) Water applied from A at 2½ g.p.m. (9.5 l min^{-1}). (b) Cutting in air (dry).

tests for bar-turning it was found that this method worked quite well for dry cutting, but when a fluid was applied unsatisfactory results were obtained which were traced to a change in the rubbing friction of the wires on the workpiece

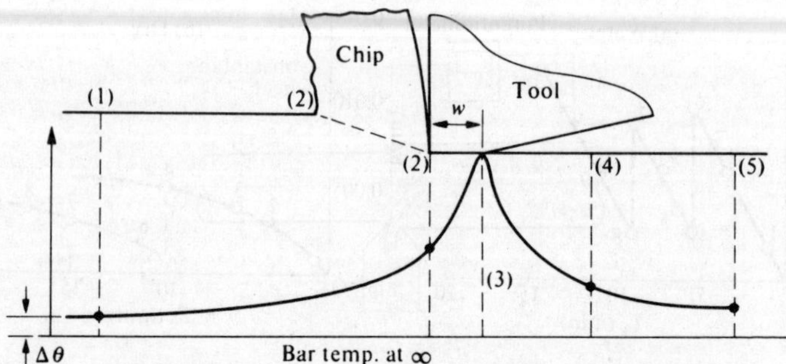

FIG. 13.25. Developed view of workpiece showing temperatures of interest in region extending one revolution on either side of cutting tool.

surface with the lubricating properties of the fluid. It was found that there was an e.m.f. generated when the wires were rubbing without cutting. When cutting dry this initial e.m.f. was very constant and could be subtracted from the e.m.f. when cutting to obtain the temperature of the workpiece at point (4). However, when a fluid was present the amount to be subtracted was variable and it was not clear how to make the correction.

After investigating rubbing wires of several different sizes and shapes a wheel was designed of solid constantan which ran upon the shoulder left by the cutting tool as shown in Fig. 13.26. In this way rolling friction was substituted for rubbing friction and there was no longer a troublesome variable correction to be made.

Some representative data are given in Table 13.7 for relatively light cuts (t = 0.0046 i.p.r. or 0.12 mm/rev) made with and without a cutting fluid (water in this case). The temperature ($\bar{\theta}_t$) is the chip-tool thermocouple temperature, while (θ_4) is that measured by the constantan wheel at a point 180° from the cutting tool. It is evident that while there is no significant difference in (F_P), (F_Q), or ($\bar{\theta}_t$) when the fluid is applied, there is a reproducible difference in (θ_4). Furthermore, the change in (θ_4) that was observed was in the opposite direction to that expected, i.e. when the coolant was applied the temperature of the workpiece was found to increase rather than to decrease.

After considering several possible explanations for this paradox the answer was found to lie in chip curl. It was found that the radius of curvature of the chip decreased appreciably when the fluid was applied, and returned to its original value when dry cutting was resumed. While there was a negligible difference in tool-chip thermocouple temperature ($\bar{\theta}_t$) with and without the fluid, the maximum temperature on the tool face was closer to the cutting edge with a tightly curled chip and hence more heat was caused to flow from the tool and into the workpiece.

All of the data of Table 13.7 are for a sharp tool. The data for Table 13.8 are

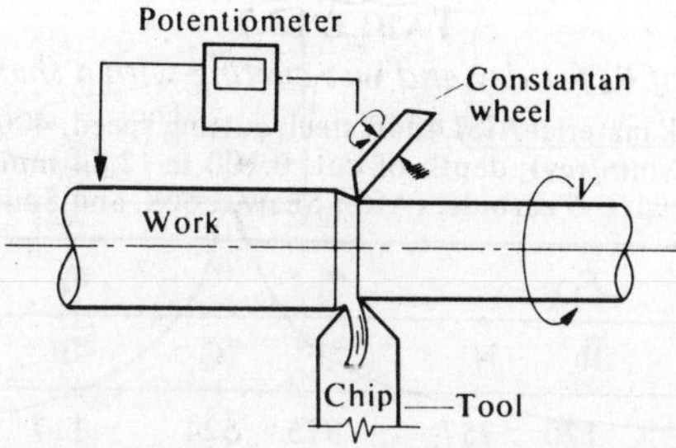

FIG. 13.26. Arrangement of apparatus for measuring cutting workpiece surface temperature one half-revolution from point of cutting.

for the same tool with a wear land present. As cutting progressed the fluid was shut off every few minutes. The values of radius of chip curvature (ρ) were the same for the worn and sharp tools. The only difference between the tests of Tables 13.7 and 13.8 lies in the wear land being either 0.008 in (0.2 mm) or 0.12 in (0.3 mm) for Table 13.8 and essentially zero for Table 13.7.

Presence of a wear land caused a reversal in the influence of the fluid upon θ_4. It is seen in Table 13.8 that θ_4 is less when the fluid is present whereas the reverse was true in Table 13.7. This result was also found in other instances with small feed, and the explanation would appear to be as follows. With a sharp tool the increased curl that accompanies chip cooling shifts the maximum tool-face temperature closer to the cutting edge, and this gives rise to more heat flow into the work as shown in Fig. 13.27a. For the dry case the maximum temperature is too far from the sharp edge for much heat to flow into the workpiece. The net effect of the direct cooling action of the fluid and of the increased flow of heat into the workpiece gives rise to a higher wear land temperature than for dry cutting. However, after an appreciable wear land has developed, heat can flow into the workpiece from the tool face in dry cutting as well (see Fig. 13.27b) and then the direct cooling action of the fluid causes the resultant wear-land temperature to be less when the fluid is used.

Sharma, Rice, and Salmon (1971) injected cutting fluids directly into the chip-tool interface through a small hole in the tool. While this gave an initial increase in chip curl due to cooling, the hole soon became blocked by chip material.

Vapors and Mist

Gaseous lubricants appear very attractive when the cutting fluid penetration problem is considered. Operations which are performed dry are actually carried out using air as a cutting fluid; and although the presence of the air is usually taken for granted, it is a very effective boundary lubricant. When air is considered as a coolant, however, it is not very attractive since all gases have relatively poor cooling capacity compared with liquids.

TABLE 13.7

Values of θ_4 for dry and wet cutting with a sharp tool.

Cutting conditions: Work material AISI 4340 steel; cutting speed, 400 f.p.m. (125 m min^{-1}); feed, 0.0046 i.p.r. (0.12 mm/rev); depth of cut, 0.100 in (2.54 mm); side rake angle, −7°; tool material, C-5 carbide. (After Shaw, Cook, and Smith 1958.)

Fluid	F_P		F_Q		θ_t		θ_4		Average
	lb	N	lb	N	°F	°C	°F	°C	
Dry	220	975	170	757	975	524	117	47	
Dry	220	975	170	757	975	524	123	51	120 °F 49 °C
Water	225	1001	175	779	975	524	139	59	
Water	230	1024	180	801	975	524	136	58	
Water	240	1068	185	823	975	524	136	58	137 °F 58 °C

TABLE 13.8

Values of θ_4 for dry and wet cutting with tool having a wear land (w). (After Shaw, Cook, and Smith 1958.)

Fluid	w		F_P		F_Q		θ_4	
	in	mm	lb	N	lb	N	°F	°C
Dry	0.008	0.20	240	1068	180	801	121	49
Water			235	1046	180	801	107	42
Dry			245	1090	180	801	125	52
Water			245	1090	180	801	115	46
Dry			255	1135	185	823	128	53
Water	0.012	0.30	255	1135	190	846	118	48

Cutting conditions; same as Table 13.7.

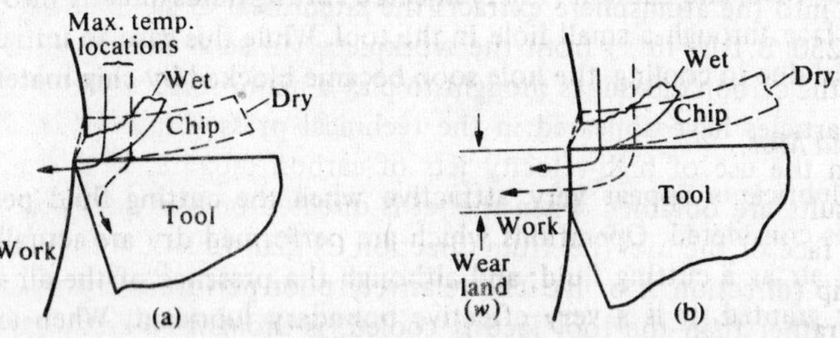

FIG. 13.27. Paths of heat flow for (a) sharp tool, (b) worn tool.

Attempts to improve the cooling capacity of air by refrigerating it have yielded different results. Olson (1948), using cooled air in a milling operation, reported a 400 percent increase in tool life over air at room temperature; while Pahlitzsch (1951) found a small difference in tool life in turning with air introduced from position B at 50 °F and −19 °F (10 °C and −28 °C).

Pahlitzsch has also presented test results using carbon dioxide gas and nitrogen gas as cutting fluids. He found an improvement in tool life over air by 150 percent using carbon dioxide and 240 percent using nitrogen. Tests using carbon dioxide gas at different temperatures gave a negligible difference in tool life and Pahlitzsch concluded from these results that the favorable action of carbon dioxide in these tests is not a cooling effect but is due to the exclusion of air. He believed that oxidation promotes friction and, therefore, the exclusion of oxygen by carbon dioxide is beneficial.

At the same time, if this is assumed to be the explanation, then it follows that nitrogen must have some sort of positive boundary lubrication effect. It has been suggested (Wister, 1936) that nitrides could form on a sliding metal surface in the presence of nitrogen which would provide lower friction as a result of the increased hardness. Further German work on this problem (Axer, 1954) confirmed Pahlitzsch's observations regarding nitrogen as a boundary lubricant. Axer used a carbide tool equipped with a small hole in the clearance surface of the tool to enable gas to be injected in direction C at high pressure. It was found that a jet of oxygen gave lower tool life than dry cutting; while a jet of nitrogen gave increased tool life. Furthermore, the tool life was observed to increase with the nitrogen pressure while it tended to show lack of improvement with increase in the oxygen pressure. Argon gave slightly better results than dry cutting, while carbon dioxide gas gave slightly inferior results to dry cutting. It should be noted that this latter observation is not in agreement with Pahlitzsch's results.

While carbon dioxide gas was used as a cutting fluid a long time ago (Reitz, 1919), it appears that it is only relatively recently that very rapid expansion of liquid carbon dioxide through a nozzle has been used. When this is done the workpiece is coated with carbon dioxide snow, which upon subsequently subliming into the atomsphere extracts the latent heat of sublimation of carbon dioxide (250 B.T.U. lb^{-1}) from the workpiece. In such instances, the cooling action of the carbon dioxide is thought to play a major role.

Many articles have appeared in the technical press (Bingham, 1954 for example) on the use of high-velocity jets of carbon dioxide. It is reported that better results are obtained when the jet is directed on the work and along the clearance face of the tool (i.e. from direction C) than when the stream is applied to the chip (direction A or B). The relatively poor performance obtained when the chip rather than the tool face is cooled, is thought to be due to the chip becoming stronger relative to the tool face which, in turn, results in an increased tendency for a crater to develop on the tool face. The widely different results that are reported with a carbon dioxide jet are undoubtedly due in part to differences in the application of the fluid.

Mist lubrication, in which very small droplets of a liquid are distributed in a large volume of air (aerosol), has been used to lubricate high-speed ball-bearings for a long time (Faust 1952). A similar scheme involving a larger concentration of liquid and coarser particle size has been adopted in workshops to combine the attractive aspects of gas and liquid cutting fluids. The very large surface-to-volume ratio for each particle of liquid provides the possibility of rapid vaporization. While the convective cooling capacity of an aerosol will be about the same as that for the gas, due to the very small amount of liquid present, rapid vaporization makes it possible to have a larger measure of evaporative cooling. For this purpose, a material of low boiling point and high latent heat of vaporization should give best results. Water is an ideal substance in this regard and oils are much less attractive.

A number of commerical mist generators are available. These units usually have filters for both air and liquid, a pressure regulator, a mixing valve, discharge nozzle, and a reservoir for the fluid. Most units can be used with either cutting oils or water-base fluids, but fluids containing solid additives in suspension are not recommended. Some manufacturers strongly recommend water-base materials over oil-base materials. The air pressure used with these devices ranges from about 1 to 5 atmospheres.

Mist lubrication has been used on essentially all machining operations and a number of articles have been written that describe industrial experience with this method of fluid application (for example: Brosheer 1953, Thuma 1954, Chamberland 1950). While the cost of the air used is significant, the liquid cutting fluid is conserved by this method. There is no drag-out problem on chips and the method may be used on operations where fluids are normally not employed, for example, on high-speed aluminum routing operations in the aircraft industry and on large planers. The contamination of the air can be a problem, depending on the nature and amount of the liquid that is employed. While there appears to be no tangible evidence that mist lubrication represents any health hazard, this problem has been raised and debated. In this connection it would seem wise to use as little liquid entrained in the air as is necessary and to install an automatic solenoid shut-off valve which closes whenever the machine is not in operation.

A water mist can be more effective than a flood of coolant with carbide tools operated at speeds in the range where the shear plane temperature exceeds the boiling point of water and an insulating steam blanket tends to form between fluid and chip. With the mist, each water droplet striking the surface of the chip will evaporate and absorb the corresponding latent heat of vaporization. If the concentration of water droplets is not too great, the steam generated has an opportunity to escape and the insulating steam blanket obtained with a flood of fluid does not develop. At cutting speeds below the temperature where a steam blanket develops, a flood of liquid provides better cooling than a mist. At very high speeds there is not time for the mist to evaporate and extract heat while the chip is still in contact with the tool.

Bench Tests

It would be convenient if cutting fluids could be evaluated in bench tests designed to more clearly characterize a fluid or fluid additive relative to a single item such as extreme boundary lubrication or cooling ability. This has been tried many times in the past but with only limited success. For example, the four-ball wear test has been used to evaluate lubrication characteristics of a fluid (Feng, Gujral, and Shaw 1961). Also Hain (1952) has measured the temperature rise of coolants when flowing through a heated capillary tube under standard conditions as a measure of their cooling capability. However, the performance of a fluid in cutting is often related to subtle secondary changes in cutting geometry such as that due to chip curl more than to direct lubrication or cooling action. These secondary actions presently cannot be predicted. While bench tests often serve a useful role as screening tests, there is unfortunately no satisfactory substitute for direct machining trials under conditions that are close to those to be used in practice. The selection and application of metal-cutting fluids is very much an art rather than a science as are so many other aspects of metal cutting. The main role of theory in metal cutting is often not to predict what will happen but to understand what is observed in order to reduce the number of trials required to reach the desired objective.

References

Axer, H. (1954). In *Fortschrittliche Fertigune und Moderne Werkzeugmaschinen.* Giradet (Essen).
Bingham, H. L. (1954). *Scient. Lubric.* **6**, 22.
Brosheer, B. C. (1953). *Am. Mach.* **97**, 137.
Chamberland, H. J. (1950). *Lubric. Engng* **6**, 21.
DeChiffre, L. (1977). *Int. J. Mach. Tool Des. Res.* **17**, 225.
Epifanov, G. I. Soloshko, E. P., and Rehbinder, P. A. (1955). *Dokl. Akad. Nauk.* **104**, 68.
Faust, D. G. (1952). *Lubric. Engng* **8**, 183.
Feng, I. M., Gujral, A., and Shaw, M. C. (1961). *Lubric. Engng* **17**, 324.
Hain, G. M. (1952). *Trans. Am. Soc. mech. Engrs* **74**, 1077.
Komanduri R. and Shaw, M. C. (1974). *Nature* **248**, 582.
Lauterbach, W. E. (1952). *Lubric. Engng* **8**, 135.
Nakayama, K. (1956). *Bull. Fac. Engng Yokohama nat. Univ.* **5**, 1.
Niebusch, R. B. and Strieder, E. H. (1951). *Mech. Eng.* **73**, 203.
Olson, G. B. (1948). *Machinery*, **43**.
Pahlitzsch, G. (1951). *Industrie Anzeiger* **69/70**, 796.
Pigott, R. J. S. and Colwell, A. T. (1952). *Trans. Soc. Automot. Engrs* **6**, 547.
Rehbinder, P. A. (1947). *Nature* **159**, 866.
Rehbinder, P. A. (1948a). *Dokl. Akad. Nauk.* **62**, 053.

Rehbinder, P. A. (1948b). *Dokl. Akad. Nauk.* **63**, 159.
Rehbinder, P. A. (1949). *Dokl. Akad. Nauk.* **64**, 874.
Rehbinder, P. A. (1951). *Dokl. Akad. Nauk.* **80**, 781.
Rehbinder, P. A. (1954a). *Dokl. Akad. Nauk.* **97**, 270.
Rehbinder, P. A. (1954b). *Dokl. Akad. Nauk.* **99**, 801.
Reitz, P. (1919). DRP (German Patent) Nr. 336960.
Schallbroch, H., Schaumann, H., and Wallichs, R. (1938). *Vortrage der Haptversammlung der Deutsche Gesellschaft fur Metallkunde,* p. 34. VDI.
Sharma, C. S., Rice, W. B., and Salmon, R. (1971). *J. Engng Ind.* **93**, 441.
Shaw, M. C. (1942). *Metal Prog.* **42**, 85.
Shaw, M. C. (1944). *J. Am. Chem. Soc.* **66**, 2057.
Shaw, M. C. (1948). *J. appl. Mech.* **15**, 37.
Shaw, M. C. (1951). *Ann. N.Y. Acad. Sci.* **53**, 962.
Shaw, M. C. (1970). *Proc. Soc. Manuf. Engrs* **MR70**, 277.
Shaw, M. C., Ber, A., and Mamin, P. A. (1960). *J. Basic Engng* **82**, 342.
Shaw, M. C., Cook, N. H., and Finnie, I. (1953). *Trans. Am. Soc. mech. Engrs* **75**, 273.
Shaw, M. C., Cook, N. H., and Smith, P. A. (1958). Research Report No. 19. Soc. Man. Eng.
Shaw, M. C., Pigott, J. D., and Richardson, L. P. (1951). *Trans. Am. Soc. mech. Engrs* **75**, 45.
Shirakashi, T., Komanduri, R., and Shaw, M. C. (1978). *J. Engng Ind. Trans. Am. Soc. mech. Engrs* **100**, 244.
Smart, E. F. and Trent, E. M. (1974). In *Proc. Int. Conf. Machine Tool Des. Res.* p. 187. Macmillan London.
Taylor, F. W. (1907). *Trans Am. Soc. mech. Engrs* **28**, 31.
Thuma, R. F. (1954). *Am. Mach.* **98**, 117.
Usui, E., Gujral, A., and Shaw, M. C. (1961). *Int. J. Mach. Tool Des. Res.* **1**, 187.
Wister, H. J. (1936). *Archiv. fur des Eisenhutten wesen Gruppe E* – **Nr. 513**, 523.

14

TOOL MATERIALS

The Machining System

Table 14.1 shows the variety of inputs and internal machine considerations involved when metals are cut. The cutting tool material is one of the important elements of the machining system. Tool material and geometry must be carefully chosen in relation to the workpiece material to be machined, the kinematics and stability of the machine tool to be employed, the amount of material to be removed, and the required accuracy and finish.

TABLE 14.1
Inputs, outputs, and internal items in the machining system when metals are cut. (After Shaw 1968.)

Inputs ⟶	Machine Internal items	⟶ Output
Workpiece	Forces	Parts/cost
Tools	Energy	
Motions	Temperatures	Parts/time
Control	Wear	
Fluids	Stationary zones	Required geometry
	Vibration	Surface integrity

The most satisfactory tool will usually be the one corresponding to the minimum total cost of performing a required operation to the specified accuracy This total cost includes:

1. Initial tool cost.
2. Tool grinding cost.
3. Tool life.
4. Labor cost as influenced by cycle time, machine, operator, and labor cost.
5. The proportion of tool changing cost assignable to one part.

Thus, the best tool material will not necessarily be the one that gives the longest life; such factors as grindability, tool material cost, and the practical levels of cutting speeds and feeds for a given tool material play important roles in the selection of the best tool material for a particular operation.

With the increased use of numerically controlled machine tools, reliability and predictability of performance are of greater significance than before and these items must be given greater weight in selecting tool materials for such applications.

Historical Background

History is a useful liberal art for explaining the present and anticipating the future relative to human events. It is similarly useful to consider the progression of developments in a given area of technology.

Figure 14.1 is a chronological list of major developments relative to tool materials from 1900 to the present. While one might question the time when each of these major developments entered our technology or whether other landmark events should have been included, it is believed that this presentation is reasonably complete and accurate. Only those items that appear to have a long (100 years) future have been included in Fig. 14.1. After making a few general observations concerning this figure, each item will be discussed in detail.

The most striking aspect of Fig. 14.1 is the increasing rate of introduction of new concepts from 1900 to the present. This was due in part to an accelerating improvement in materials technology and in part to the need for special machining characteristics.

Some of the events responsible for the acceleration of change evident in Fig. 14.1 are listed in Fig. 14.2. The new areas of technology that have appeared since 1945 have each introduced new materials to be machined to which the cutting-tool industry has had to respond. The decade of the seventies has been one of rapidly changing manufacturing conditions that is bound to be reflected in major changes in the cutting-tool industry in the 1980s and beyond.

Tool Material Requirements

The second general observation that may be made is that there is need for a wide spectrum of cutting tool materials. Three important considerations are high-temperature physical and chemical stability (HTS), abrasive wear resistance (AWR), and resistance to brittle fracture (RBF). A given tool material is generally not outstanding relative to all three of these attributes. In general, as a material is made more refractory (high HTS), it becomes more brittle (low RBF) or if it is made more abrasion resistant (high AWR), it also becomes more brittle (low RBF). This is true not only when comparing different classes of tool material such as high-speed steel (HSS), tungsten carbide (WC), and titanium carbide (TiC) but also when comparing different compositions within

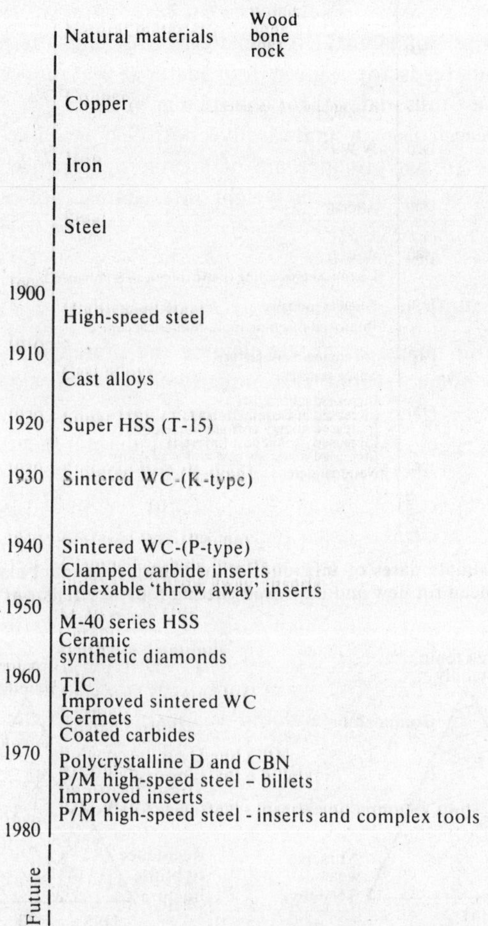

FIG. 14.1. Approximate dates of introduction of different cutting tool materials and concepts. (After Shaw 1980*a*.)

a given class such as different grades of sintered tungsten carbide. Tungsten carbide tool manufacturers frequently indicate the area of usefulness of a given product relative to other carbides on a triangular plot similar to that shown in Fig. 14.3.

A similar plot may be used to show the relative areas of usefulness of HSS, WC and TiC tools. Such a diagram can give only a rough indication of the area of usefulness of a given material since the exact location of the lines will depend upon many specific items such as

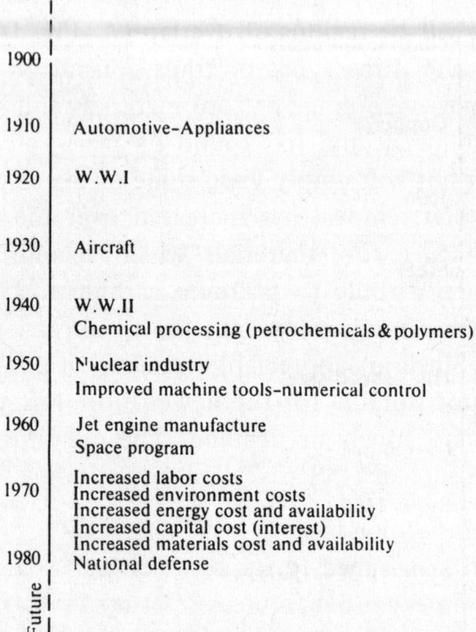

FIG. 14.2. Approximate dates of introduction of principal product classes and conditions influencing the need for new and improved cutting tool materials. (After Shaw 1980*a*.)

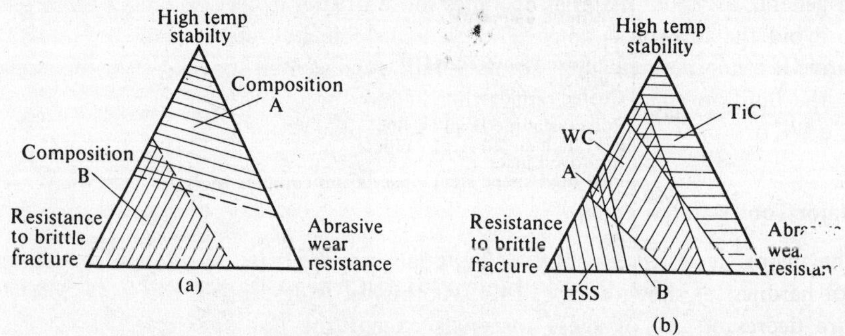

FIG. 14.3. Diagram showing (a) the relative high-temperature stability, resistance to brittle fracture, and abrasive wear resistance of a cutting tool material; (b) approximate application of (a) to high-speed steel, tungsten carbide, and titanium carbide tool materials. (After Shaw 1980*a*.)

1. Relative hardness of tool and work material (tool should be four times as hard as work).
2. Abrasive particles in work or on work surface (scale).
3. Chemical compatibility of tool and work material.
4. Tool tip temperature (speed, feed, tool geometry, coolant, etc.).
5. Condition of machine (rigidity).
6. Type of operation (continuous or interrupted cut).

In general, there will be regions of overlap (A, Fig. 14.3b) and regions that are not covered by the three types of tools considered (B, Fig. 14.3b). This explains the need for a wide range of tool materials and suggests that as a new tool material is introduced, and has found its place, an old material will not completely disappear but will simply yield some of the area it once dominated.

As the rate of metal removal has increased over the years, there has been need for more refractory tool materials which accounts for the progression from HSS to tungsten carbide to titanium carbide to ceramic tool materials. At the same time such a progression calls for an increase in machine tool horsepower and rigidity. Since an appreciable number of tools must operate under conditions that are not suitable for the newer more refractory but more brittle tools, HSS is still very much in demand. The conditions under which more ductile tool materials such as HSS are called for include:

1. Older, less rigid machine tools.
2. Under-powered low-speed machines (in such cases, temperatures do not warrant more expensive, more refractory tooling).
3. Small-diameter work that cannot be operated in the speed range to completely avoid a built-up edge.
4. Workpieces involving interrupted cuts or heavy surface scale.

In general, as a tool material becomes more brittle, it becomes more important to avoid the low-speed range where an unstable built-up edge (BUE) tends to form. If a ductile metal that forms a BUE, such as mild steel, must be machined in the built-up edge speed range, then it may be necessary to use HSS instead of a WC or ceramic tool material to avoid chipping of the cutting edge.

Major Tool Material Types

The major classes of cutting tool materials may be listed in order of increasing hot hardness as shown in Fig. 14.4. As this list is descended, the strain at fracture decreases. All of these materials except the last one have a two-phase structure—a softer continuous phase separating very hard particles. As the list is descended, the spacing between hard particles decreases. Initially (first three items) the spacing of the hard particles is sufficiently great so that Young's modulus of elasticity corresponds to that of the continuous phase (a substitutional solid solution of iron in the case of HSS). However, beginning with cemented tungsten carbide (item 4), the hard carbides are so close together that Young's modulus approaches that of the tungsten carbide particles instead of the cobalt binder-phase (i.e. 90×10^6 p.s.i. instead of about 30×10^6 p.s.i. (6.21×10^5 MPa instead of 2.07×10^5 MPa)). In the case of items 7 and 8 of Fig. 14.4, the space between hard particles is essentially that corresponding to crystal boundaries. In the case of ceramic tools, no binder material or sintering aid is normally employed, the only difference in chemistry in the crystal

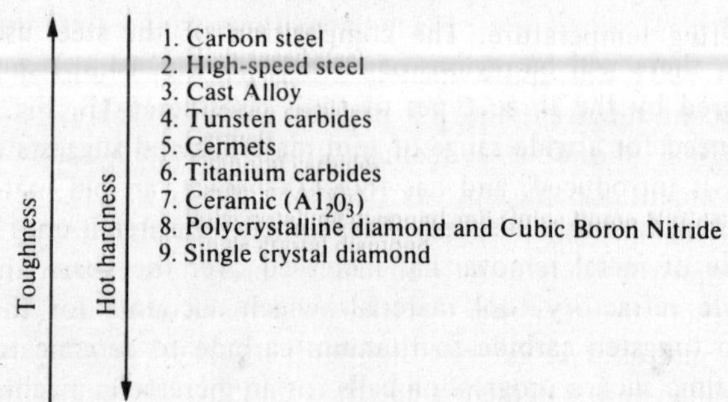

FIG. 14.4. Principal classes of cutting-tool materials and their relative toughness and hardness at elevated temperature.

boundaries being that due to the very small impurity content. In the case of item 8, a small amount of cobalt is present in the crystal boundaries that evidently diffuses there from the cemented tungsten carbide substrate on which the diamond or CBN crystals are supported. Item 9 is the only truly single-phase material. This is single crystal (natural) diamond that is extremely brittle due to the complete lack of crack-arresting mechanism as in the polycrystalline counterpart (item 8) and the presence of many structural defects.

High Speed Steel

Before 1870 all lathe tools were forged from plain carbon steel consisting of about 1 percent carbon, 0.2 percent manganese, and the remainder iron. These tools had a low hardness at elevated temperature (low hot-hardness) and could only be used to machine steel at relatively low speeds (up to about 15 f.p.m. = 0.08 m s^{-1}). This material also required a water quench in hardening which frequently resulted in quenching cracks. In 1870 Robert Mushet introduced a steel in England having approximately the composition: 2 percent carbon, 1.6 percent manganese, 5.5 percent tungsten, and 0.4 percent chromium. This steel was air-hardening and retained its hardness to higher cutting temperatures. As a result it could be used to speeds of about 25 f.p.m. (0.13 m s^{-1}). Mushet steel with modification (mainly replacement of manganese by chromium) was used for lathe tools until the turn of the century.

F. W. Taylor and M. White (1901) produced tools of greatly increased stability which allowed cutting speeds of about 60 f.p.m. (0.32 m s^{-1}) and consequently this material became known as high speed steel (HSS). Contrary to common belief this was not the development of a new steel but rather a new heat treatment for the existing material. Taylor and White found that if tools were heated quickly through the brittle temperature range of 845 to 930 °C to a temperature just short of the melting point of the steel before quenching, a steel of improved hot-hardness resulted. It was also found that such tools were improved by a higher tempering temperature. The composition of the steel used by Taylor and White in 1901 was approximately 1.9 percent carbon, 0.3 percent manganese, 8 percent tungsten, and 3.8 percent chromium. This is seen to differ from the original Mushet steel mainly in the increased amount of tungsten and the replacement of manganese by chromium.

The period from 1900 to 1906 saw the rapid development of HSS. Taylor found that improved tools were produced by using less carbon and more tungsten (the carbon decrease was necessary to make the steel forgeable when tungsten was increased to improve hot-hardness). By 1903 the carbon content of HSS had fallen to 0.7 percent while the tungsten content had risen to 14 percent. In 1904, Dr. J. A. Matthews found that the abrasive resistance of HSS could be increased by additions of vanadium and by 1906 Taylor was using about 0.3 percent vanadium in his tools. By 1910 the tungsten content had increased to 18 percent, the chromium content to 4 percent and the vanadium content to 1 percent, thus providing the well known 18–4–1 HSS which was the standard HSS for the next 40 years. This steel is designated AISI T-1†.

In the short span from 1900 to 1910 the allowable cutting speed had been increased from 25 f.p.m. (0.13 m s^{-1}) to over 100 f.p.m. (0.52 m s^{-1}) and as might be expected this had an immediate and far reaching influence on machine tool design.

The production of higher quality HSS was made possible by the introduction of electric furnace melting in 1907 but it was not until about 10 years later that such furnaces came into wide use. In 1912 it was found that the red-hardness of HSS could be improved by additions of from 3 to 5 percent cobalt. However, such materials were not extensively used because the speed and power available in lathes still had not been increased to values called for by 18–4–1 HSS.

By 1920 there were but three high speed steels in common use:

AISI symbol†	W	Cr	V	Co
T-1	18%	4%	1%	0
T-7	14%	4%	2%	0
T-4	18%	4%	1%	5%

The first of these was the standard, the second was used where greater abrasive resistance was called for, while the third was used where tool temperatures were high and greater hot-hardness was desired.

About 1923 the first of the so-called super high-speed steels appeared. This steel (T-6) had the composition

† The American Iron and Steel Institute (AISI) has introduced symbols for the two major classes of HSS: the symbol T designates a tungsten-base steel while the symbol M a molybdenum-HSS.

0.7% C, 4%Cr, 2% V, 20% W, 12% Co

and was followed by steels with even greater amounts of carbon and vanadium. However, these steels could not be made commercially due to the difficulty of forging, rolling, and heat-treating them. It was not until 1939 that Gill showed that steels of high carbon and vanadium content could be readily hot-worked without cracking if the tungsten content were decreased. This led to such super high-speed steels as T-15.

1.5% C, 4% Cr, 5% V, 12% W, 5% Co

While it was known as early as 1900 that molybdenum could be substituted for tungsten in HSS (half the weight of molybdenum need be used since the atomic weights of molybdenum and tungsten stand approximately in the ratio of 1:2), this was not done commercially until about 1923 due to difficulties associated with a narrower heat-treating range and a greater tendency for molybdenum bearing steels to decarburize. The molybdenum steels are essentially equivalent to their tungsten counterparts but were not widely used until World War II when there was a shortage of tungsten. Since this time, the molybdenum high speed steels have largely replaced the tungsten steels due to a significant cost advantage and the greater strategic availability of molybdenum. Today, M-2 HSS (composition below) is the standard in USA having displaced T-1 about 1950.

M-2: 0.8% C, 4% Cr, 2% V, 6% W, 5% Mo

Current HSS tools enable steel to be machined at speeds that are often in excess of 300 f.p.m. (1.45 m s^{-1}).

It is well known that the hardness and wear resistance of HSS depends on the composition, size, and distribution of the carbides in the steel and upon the stability of the matrix at high temperatures. The latter is increased by addition of cobalt. Harder mixed carbides are provided by increased vanadium and carbon content. A major development in the HSS area in the 1950s was the discovery that steels of increased hardness (R_c 70 instead of the R_c 65 for the more conventional HSS) and reasonable toughness may be produced by use of compositions such as

1.4% C, 4% Cr, 4% V, 9% W, 4% Mo, 12% Co

These steels have a large concentration of finely and uniformly divided carbides in a rather refractory matrix. They are rendered forgeable by substitution of Mo for W. However, like wrought T-15 HSS, they are difficult to grind.

During the 1970s there have been two additional significant developments in the HSS area. The first involved the atomization of a prealloyed HSS such as T-15, followed by consolidation into billets by hot isostatic compaction.

The main characteristics of this material are a smaller carbide size and a more uniform carbide distribution than that for the more conventionally produced wrought billets. These structural changes made it much easier to grind normally-difficult-to-grind HSS such as T-15. The atomized material has carbides that are typically less than 3μm in size. The importance of this is that carbides of this size are displaced to the side by the abrasive particles in a grinding wheel instead of resisting the motion of the abrasive through the material being ground resulting in less loss of abrasive from the grinding wheel when materials with small carbides are ground. The atomized material has less carbide segregation and produces tougher tools.

The second important development consists of a cold impact fracture process as the final comminution step before sintering. This results in individual particles having many sharp points instead of the rounded surfaces produced in ball milling. The advantage of this, of course, is that less pressure is required in preform production. This has made it possible to produce preforms in dies by cold pressing instead of by the more expensive hot isostatic pressing required for directly atomized material. The so-called cold-stream impact comminution process (Friedman, Friedman, and Zagielski 1965) has the added advantage of making it possible to use scrap HSS chips as feed material instead of 100 percent atomized material. However, the chief advantage of this latest HSS-P/M† development is that HSS *inserts* for use in standard carbide tool holders and having pressed-in chip curlers may be economically produced. Complex tools such as taps, hobs, and milling cutters may also be produced by the new HSS-P/M process in near net shape, thus greatly reducing the energy and productivity loss associated with the slitting of such a tool from a solid billet. As the cost and scarcity of energy and strategic materials such as chromium, cobalt, and tungsten increases, the HSS-P/M approach to inserts and tools of complex shape is bound to play an increasing role.

A development in the HSS area not included in Fig. 14.1 is the production of precision-cast HSS tools. This technique like the HSS-P/M approach makes it possible to use compositions of higher carbon and alloy content than would be possible if forging and rolling were employed. Krekeler (1957) has shown that a heat treatment of 3 hours at 1100 °C is beneficial in removing the embrittling microcracks that are normally closed in forging. However, cast HSS tools have in general not produced tools as shock-resistant as wrought HSS tools.

Table 14.2 gives compositions for a number of commercially available high-speed steels.

Cast Alloy Tools

About 1915 non-ferrous high temperature alloys containing significant amounts of cobalt, chromium, and tungsten were devised by Elwood Haynes. These materials were not heat-treatable but were used as cast and are, therefore, sometimes known as cast alloy tools. However, it was not until the period 1920

† P/M = powder metallurgy

TABLE 14.2

Composition (w/o) of some commerically available high-speed steels

AISI designation	Weight percent C	W	Cr	V	Mo	Co
M1	0.80	1.75	3.75	1.15	8.75	
M2, Class 1	0.85	6.25	4.00	2.00	5.00	
M3, Class 1	1.05	6.25	4.00	2.50	5.75	
M4	1.30	5.50	4.00	4.00	4.75	
M6	0.80	4.25	4.00	1.50	5.00	12.00
M7	1.02	1.75	3.75	2.00	8.75	
M8	0.80	5.00	4.00	1.50	5.00	(1.25 Cb)
M10, Class 1	0.89	0.70	4.00	2.00	8.00	
M15	1.50	6.50	4.00	5.00	8.50	5.00
M30	0.80	1.80	4.00	1.20	8.25	5.00
M34	0.90	1.75	3.75	2.10	8.75	8.25
M36	0.85	6.00	4.00	2.00	5.00	8.25
M41	1.10	6.75	4.25	2.00	3.75	5.25
M42	1.08	1.60	3.75	1.15	9.60	8.25
M43	1.20	2.70	3.75	1.60	8.00	8.20
M44	1.15	5.25	4.25	2.00	6.50	11.75
M45	1.27	8.25	4.20	1.60	5.20	5.50
M46	1.24	2.10	4.00	3.20	8.25	8.25
T1	0.73	18.00	4.00	1.00		
T2	0.85	18.00	4.00	2.00		
T3	1.05	18.00	4.00	3.00	0.60	
T4	0.75	18.00	4.00	1.00	0.60	5.00
T5	0.80	18.00	4.25	2.00	0.90	8.00
T6	0.80	20.50	4.25	1.60	0.90	12.25
T7	0.75	14.00	4.00	2.00		
T8	0.80	14.00	4.00	2.00	0.90	5.00
T9	1.20	18.00	4.00	4.00		
T15	1.55	12.50	4.50	5.00	0.60	5.00

to 1925 that cast alloy tools were commonly used in cutting. While the cast alloy tools are not quite as hard as the high-speed steels at room temperatures, they do retain their hardness to high temperatures and hence are occasionally used for such special applications as form tools.

Cast alloy tools in the past were used in the form of relatively large bars just as HSS turning tools and brazed carbides were used before clamped sintered

TABLE 14.3

Compositions of some commercially available cast alloy tool materials

Trade designation	Weight percent			
	Co	Cr	W	C
Blackalloy 525	44	24	20	2
Blackalloy T × 90	42	24	22	2
Crobalt 1	48	30	14	2
Crobalt 2	40	33	18	2.5
Crobalt 3	40	33	20	3
Haynes Stellite R Star J	43	32.5	17.5	2.5
			(1.2 B, 4 V,	4 Ni)
Haynes Stellite R3	50	31	12.5	2.4
Haynes Stellite R19	53	31	10.5	1.8
Haynes Stellite R98M2	38	30	18.5	2
Tantung G	47	30	15	3
Tantung 144	45	28	18	3

carbide inserts were introduced about 1950. The appearance of HSS-P/M inserts suggests that cast alloy inserts that do not exhibit the same casting segregation problems as HSS and have greater hot-hardness should be reexamined. The fact that such materials contain appreciable amounts of cobalt, chromium, and tungsten would be partially offset by the fact that an insert requires a much smaller volume of material than a massive tool bar. In addition, some of the work done recently in substituting nickel for cobalt in high-temperature turbine alloys might be applicable to cutting-tool technology.

Table 14.3 gives compositions of a number of cast alloy tool materials. The room temprature hardness values of these materials is generally in the range R_c 60 to R_c 65.

Cemented Tungsten Carbide

The first successful tungsten carbide tools were brought to the United States from Germany in 1928. These tools consisted of finely ground tungsten carbide particles sintered together with cobalt binder. Early work with these new tool materials was hampered by

1. Their brittleness and tendency to chip.
2. Brazing difficulties.
3. Difficulty of grinding.
4. Lack of lathe rigidity, power, and speed to use them adequately.

5. Tendency for a crater to form on the tool face at high cutting speeds, especially when machining steels.

Gradually during the thirties, tools of greater shock resistance were produced but carbide tools were used mainly to turn cast iron and non-ferrous metals due to the greater tendency for steel to cause tool face cratering. Diamond wheels for grinding carbide tools were introduced in the thirties which greatly aided tool preparation. About 1938 it was found that the tendency for a carbide tool to crater when cutting steel could be reduced by additions of titanium and tantulum carbides. During World War II the use of carbide tools expanded rapidly.

Titanium and tantalum carbides are more stable than tungsten carbide and have a greater resistance to decomposition in the presence of γ (FCC) iron. The temperature of the surface of the chip in contact with the tool in high-speed machining will generally lie above the allotropic transformation temperature for steel and the great affinity of austenite (γ-iron) for carbon will cause a loss of carbon from WC crystals in the tool surface. This results in the development of a crater having its greatest depth at the point of maximum tool-face temperature. Titanium carbide having a greater resistance to decomposition has a dramatic effect on crater resistance of carbide tools when a steel is machined at a high cutting speed (high tool face temperature). The addition of titanium carbide decreases the strength and abrasive wear resistance of a sintered tungsten carbide tool. Tantalum carbide also provides crater resistance with less loss of impact strength than titanium carbide since it gives rise to less grain growth during sintering but costs more than an equivalent amount of titanium carbide.

The net result of all of this is that there are two types of sintered tungsten carbide—one for machining gray cast iron, non-ferrous metals, and abrasive nonmetals such as fiber glass and graphite (ISO K-type) and one for machining ferrous metals (ISO P-type). The cast iron type consists of WC crystals with normally 3 to 12 percent cobalt as a binder while the steel-cutting grades of carbide have some of the WC substituted by TiC, TaC, or NbC, all of which render greater crater resistance with different degrees of loss of strength, abrasive wear, impact, and corrosion (oxidation) resistance. The strength (strain at fracture) of a carbide tool increases with increase in cobalt content and tools of higher cobalt content are required in roughing operations where the rate of material removal is higher than in finishing operations.

During the 1960s there was a gradual increase in the quality of carbide tools. Carbide tool materials were made less brittle by reduced particle size and improved binding and sintering techniques. By milling the tungsten carbide powder with the sintering material it was possible to coat each particle with the binder phase before pressing and thus provide a more homogeneous material. Tungsten carbide tools having a smaller void content and improved shock resistance were produced by subjecting the material to high hydrostatic pressures before sintering.

In recent years the U.S. Cemented Carbide Producers Association (CCPA) which was organized in 1955 to promote the effective use of cemented carbides, has been instrumental in standardizing test procedures for cemented carbides in cooperation with the American Society for Testing Materials (ASTM) and other standardizing organizations. However, there is much that remains to be done in the area of standardization. One of the most pressing needs is a system of carbide classification that could serve as a guide to the user in the selection of a carbide grade for a particular application. There are four systems of classification of cemented carbides in use in the world today:

1. The U.S. unofficial 'C' classification system which is based on performance.
2. The British Hard Metal Association (BHMA) system which is based on properties.
3. The International Standards Organization (ISO) system (R513 *Application of carbides for machining by chip removal*).
4. The Russian system which is based on composition.

The 'C' classification system now used in the USA was introduced by automotive engineers to obtain some indication of equivalence of products made by different manufacturers. This system employs eight numbers C-1 to C-8 which covers the grades of carbide manufactured for cutting-tool applications. The C-1 to C-4 grades are for machining cast iron and materials that yield short chips such as the non-ferrous alloys. In going from C-1 (roughing cuts) to C-4 (finishing cuts), the cobalt content and hence the shock resistance decreases. The C-5 to C-8 grades are steel-cutting grades which contain TiC, TaC, and NbC. In going from C-5 (roughing cuts) to C-8 (finishing cuts), the cobalt content decreases and the shock resistance decreases.

The BHMA designation consists of three numbers. The first number (1 to 9) gives the hardness category of the carbide (1 = low hardness, and 9 = high hardness) and is meant to reflect wear resistance. The second number (1 to 9) gives the transverse rupture strength (TRS) of the carbide (1 = low TRS = 150 000 p.s.i. = 1035 MPa, and 9 = high TRS = 500 000 p.s.i. = 3449 MPa) and is meant to reflect shock resistance. The third number (0-9) gives the TiC/TaC content (0 = low TiC/TaC content, and 9 = high TiC/TaC content) and is meant to reflect crater resistance. The difficulty with this system is that hardness is not a good measure of edge-wear resistance and TRS is not a good measure of shock resistance. It also fails to accommodate coated carbide tools to be discussed later.

The ISO system (Table 14.4) is somewhat like the 'C' system and classifies all cutting grade carbides into three categories: P (steel-cutting grades), K (cast iron cutting grades), and M (an intermediate category for ductile irons, hard steels, and high-temperature alloys).

TABLE 14.4
ISO carbide classification system

Symbol	Category	Color Code	Designation	Cutting conditions
P	Ferrous metals with long chips	Blue	P01 P10 P20 P30 P40 P50	Inc. Speed → Inc. Speed → Inc. Speed ↑ Inc. Feed ← Inc. Feed ← Inc. Feed ↓
M	Intermediate	Yellow	M10 M20 M30 M40	
K	Nonferrous metals and ferrous metals with short chips (gray cast iron)	Red	K01 K10 K20 K30 K40	

The Russian system is based on composition of the carbide. This does not help the relatively uninformed user choose the proper carbide and is unacceptable in the West for proprietary reasons.

It appears as though the ISO system is gaining ground and that it may someday be a world standard.

Inserts

When tungsten carbide was introduced as a tool material it was so expensive that it became necessary to braze a thin wafer of the material to a steel shank. Brazed tools of this type were used throughout World War II even though they did not generally allow the full potential of tungsten carbide to be realized. Frequently carbide tips would fail through the braze due to the abrupt change in Young's modulus and coefficient of thermal expansion in this region (sintered carbide has about one half the coefficient of expansion of steel). This difficulty was minimized by making carbide wafers thicker and using a copper shim at the brazed surface to decrease concentrations of stress. A representative brazed carbide tool is shown in Figure 14.5a.

Following World War II when improved lathe power and rigidity enabled

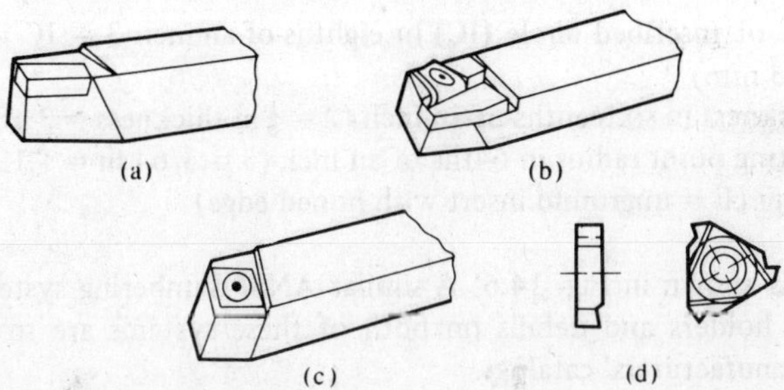

FIG. 14.5. Various turning-tool designs. (a) Brazed carbide tool. (b) Single edge clamped tool. (c) Multi-edge clamped tool. (d) Special screw thread cutting insert.

larger cuts to be taken, clamped carbide holders came into use. The first of these (Fig. 14.5b) resembled the brazed type of carbide tool having but one cutting edge and several surfaces to be ground and honed. The next development to lower tool preparation cost was a multi-edged clamped tool. The tool shown in Fig. 14.5c has four cutting edges on each end of the blank and a new cutting edge may be rotated into position without disturbing the tool setting. The decrease of grinding and tool-changing down-time costs associated with tools of this type is evident. Another feature associated with clamped carbide tools is the use of disposable inserts. In this instance an insert having several cutting edges (eight in the case of Fig. 14.5c) is ground at the factory and when all cutting edges have been used the insert is discarded rather than reconditioned. Today special inserts are even available for cutting a variety of screw thread forms (Fig. 14.5d).

Regrinding of 'throw away' carbide inserts is done to further reduce carbide tool cost and scrap carbide inserts are being recycled by mixing some finely divided used tool material with virgin powder in new tool manufacture.

Indexable carbide inserts are now available in a wide variety of shapes including; triangles, squares, circles, rectangles, hexagons, pentagons, and diamonds having a variety of angles (80°, 55°, 42°, and 35°). The American National Standards Institute (ANSI) has introduced a special numbering system for inserts and tool holders. For example, the following designation has the meaning indicated below:

Designation: T N M G 3 2 3 E

(1 2 3 4 5 6 7 8)

1 = shape of insert (T = triangle)
2 = relief angle (N = 0°)
3 = accuracy of point location, relative to pin and thickness
M = point location accuracy of ±0.002 in to ±0.004 in (±0.05 to 0.10 mm) and thickness accuracy of ±0.005 in (±0.13 mm)
4 = type of insert (G = insert with hole and molded chip breaker)

5 = size of inscribed circle (IC) in eighths-of-an-inch 3 = IC = 0.375 in = 9.53 mm)
6 = thickness in sixteenths-of-an-inch (2 = $\frac{1}{8}$ in thickness = 3.18 mm)
7 = cutting point radius in 64ths of an inch (3 = 3/64 in = 1.19 mm rad)
8 = other (E = unground insert with honed edge)

This insert is shown in Fig. 14.6. A similar ANSI numbering system is used to specify tool holders and details on both of these systems are to be found in most tool manufacturers' catalogs.

Fewer brazed tools are used today since clamped inserts eliminate brazing stresses, eliminate the need for tool grinding by the user, enable hardened tool holders of long life to be used, and decrease tool-setting time when changing tools. A carbide shim is usually employed beneath the indexable carbide for improved support and to protect the tool holder against damage. Since cemented carbides have a stiffness (Young's modulus) that is about three times that of the steel used in the tool holder, it is important to use carbide of considerable thickness to prevent a stress concentration that could cause the carbide to fracture. The carbide thickness is usually divided into two parts— that in the shim and that in the indexable cutting tool.

Chip breakers (chip curlers) may be of two types—a separate carbide plate clamped on top of the indexable carbide or a molded groove in the face of the cutting tool. Figure 14.6 shows a triangular insert with molded top surface. The geometry of the top surface of the tool into which the chip curler is molded may be as shown in Fig. 14.7a (negative rake) or Fig. 14.7b (positive rake). Both of these tools provide a limited contact on the tool face which tends to reduce cutting forces and cutting temperatures and will have improved tool life when machining low strength steels. Chip curlers of the type shown in Fig. 14.6 give good chip control over a wide range of feeds and depths of cut.

Titanium Carbides

In the 1960s efforts were made to find hard refractory materials that are more effective as tools than tungsten carbide. Titanium carbide is a material that has attracted considerable interest in this connection. This material has a specific gravity of 4.9 compared with 17.9 for tungsten carbide and in addition it is more oxidation resistant and more readily available at low cost. Best results

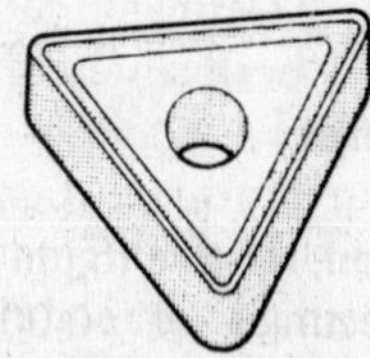

FIG. 14.6. Triangular carbide insert with center hole and molded chip curler in top surface.

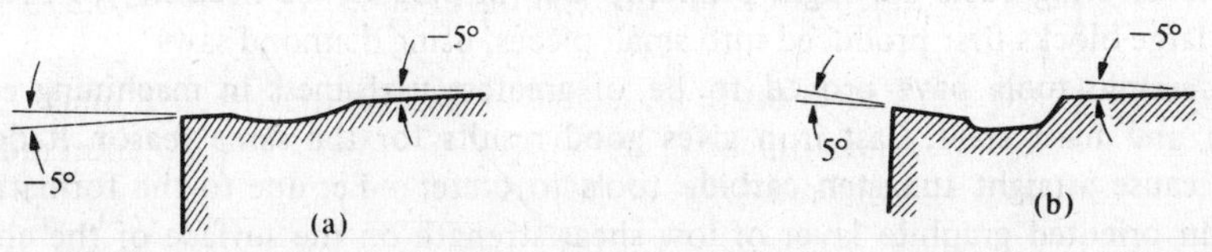

FIG. 14.7. Side elevations of inserts with molded top surface. (a) Negative rake angle. (b) Positive rake angle.

have been obtained with TiC when it is bonded with nickel. However, such materials have been non-uniform and extremely brittle. The reason for this was not known until Humanik and Parikh (1954) demonstrated the importance of *wetting* in the liquid-phase sintering of TiC. While liquid cobalt was found to have a wetting angle of zero degrees against WC (indicating perfect wetting) nickel was found to have an angle of 30° against TiC. Additives to nickel that would reduce the wetting angle against TiC were systematically studied and it was found that metals that were weak carbide formers (i.e. metals having low negative free-energy of carbide formation) had the greatest influence. When 10 percent molybdenum powder is mixed with nickel before sintering TiC, it is found that complete wetting takes place and the bonding material is thoroughly dispersed. The resulting sintered tool material is relatively tough and is now in use in industry. Details of this interesting development are discussed by Humanik *et al.* (1954).

The iron-group materials in the periodic table (Co, Ni, Fe) have been found useful as binder materials for tungsten carbide and other refractory carbides such as TiC. The solubility of WC in Co is low (~ 1 percent) while the room-temperature solubility of WC in Ni and Fe are much higher (25 percent and 5 percent respectively). The higher the carbide solubility in the binder phase at room temperature the more brittle the cemented carbide. Low solubility of WC in Co is one of the main reasons why cobalt is so widely used as the binder for tungsten carbide. The greater solubility of titanium carbide in nickel is one of the reasons TiC–Ni is more brittle than WC–Co.

Ceramic and Cermet Tools

Ceramic cutting tools were put into use in the USA in the mid 1950s following earlier use in Russia, England, and Germany. Most of these tools are made from finely divided α-alumina that has been sintered, without binders or other additives. Ceramic tools are more refractory and harder than carbide tools but also much more brittle. In order to provide reasonable toughness it is important that aluminum oxide tools have very small grain size and be sintered to essentially maximum density. Hot pressing in graphite molds yields superior tools but the dies do not last long and substantially increase the cost of the product. While the cost of the aluminum oxide used to make ceramic tools is low, manufacturing costs are high, primarily due in part to the necessity of slicing the large blocks first produced into small pieces, using diamond saws.

Ceramic tools have proved to be of greatest usefulness in machining cast iron and hard steel. Cast iron gives good results for the same reason it does not cause straight tungsten carbide tools to crater—i.e. due to the formation of an oriented graphite layer of low shear strength on the surface of the chip. Hard steel will also cause less cratering of tungsten carbide since the weld areas formed between chip and tool will be small. The brittleness of ceramic tools makes them susceptible to chipping when machining soft steels under conditions to produce a large unstable built-up edge (BUE) or when making discontinuous cuts. Ceramic tools are particularly disappointing when used to machine aluminum or titanium alloys, presumably due to the unusually high affinity of these metals for oxygen or an oxide surface. Strong bonds are then established between chip and tool and the wear rate of the tool is accelerated.

Ceramic tools have not come up to expectations. Wider use is predicted for the future, however, due to:

1. Use of composites of Al_2O_3 and TiC (the so-called cermets) that are less brittle and which have a transverse rupture stength greater than 100 000 p.s.i. (690 MPa).
2. Availability of machine tools having higher speeds (3000 f.p.m. = 914 m min^{-1}, and higher) and greater horsepower (> 50 h.p. = 37 kW for turning machines).
3. Chucks and work-holding devices that are designed for higher speed use.
4. Wider use of small grain size (< 1 μm) hot-pressed material.

It is believed that ceramics will take over a larger part of the machining field now dominated by cemented carbides as heavier cuts involving continuous chips become a possibility without tool tip fracture. In order to obtain good results with the highly refractory but brittle ceramic tools, it is important to operate at very high cutting temperatures (i.e. at high cutting speeds).

Ceramic (Al_2O_3) tools are commercially available in both cold-pressed (white) and hot-pressed (black) varieties. The cold-pressed tools are less expensive but less strong and are limited to use on cast irons having a Brinell hardness less than about 350 kg mm^{-2}. The hot-pressed materials have a smaller particle diameter (< 2 μm) and higher tensile strength.

Tools that are less brittle than those made from pure Al_2O_3 but somewhat more refractory than TiC are made from mixtures of hot-pressed Al_2O_3 and 15–30 w/o TiC. These cermet tools are of greatest usefulness in the high-speed machining of hard cast irons (H_B = 350 to 600 kg mm^{-2}) and hard steels under conditions of limited mechanical shock.

Coated Tools

The most important tool development of the 1960s was the introduction of coated carbide tools about 1969. The initial efforts in this direction involved ordinary steel-cutting grade carbide tools that were provided with a thin film of TiC, applied by vapor-phase deposition. The thickness of TiC coating was about 0.0002 inch (0.005 mm). Such coatings reduce the tendency of a cemented carbide tool to crater when machining mild steel and thus enable low-alloy steels to be machined at higher values of temperature (i.e. at higher speeds and feeds) without excessive cratering. The effect of the very thin coating exists long after a crater depth is reached corresponding to the thickness of the coating. For example, the wear-reducing effect of a 0.0002 inch (0.005 mm) coating will still be evident at crater depths of several thousandths-of-an-inch. This suggests that the coating material is carried into the crater by the moving chip.

Although other carbides and nitrides (for example, hafnium carbide and hafnium nitride) have been used as coatings to a limited extent only three coating materials are presently in wide commercial use (TiC, TiN, and Al_2O_3). While TiC and Al_2O_3 appear to provide the most chemically stable screening layer between chip and tool, TiN appears to offer the lowest tool friction. This has been attributed to a lower chip surface strengthening action of nitrogen than carbon when the coating material is decomposed in the presence of the austenitic surface of the chip (Rao, Kumar, and Shaw 1978). Although Al_2O_3 should be the most effective diffusion barrier, it is much more difficult to bond to the substrate than either TiN or TiC. Bond strength is probably the overriding attribute of a coating and, in this regard, TiC excels.

Coated tools are inherently more brittle than uncoated tools and should be used only when crater formation is a problem. Coated tools are generally not used for milling. The increased brittleness associated with a coating is in part due to the difference in properties between the substrate and the coating and it is important to provide a graded transition from one material to the other if undue brittleness is to be avoided. Some manufacturers today use a multiple coating technique. The layer next to the carbide substrate is TiC to provide a stable diffusion barrier to prevent carbon leaving the substrate or iron from entering the cobalt phase. The outer layer is a 0.0002 in (5 μm) layer of TiN for low tool-face friction with a transition layer of titanium carbonitride between the TiC and TiN coatings.

In general, coated carbides are not useful for machining high-temperature alloys (either nickel- or cobalt-base), titanium alloys, or nonmetals such as graphite, fiber glass, plastics, or nonferrous metals. Coated carbides are most useful for finishing cuts at high speed on ferrous alloys of all types including stainless steel. For such applications, few varieties of coated carbide are required to cover the range of materials normally encountered. In other words, coated carbides are generally more versatile than noncoated carbides. When first introduced, coatings were applied only to steel-cutting grades of carbide. Most manufacturers now include at least two types of coated steel-cutting grades in their catalogs and one cast iron grade of coated carbide.

While coatings may be applied by sputtering or other physical deposition techniques, better bonding is generally obtained by chemical vapor deposition where vapors such as $TiCl_4$ and methane are reacted on the hot tool surface (1200 °C or higher) to produce the coating.

In addition to chemical vapor-phase deposition of carbides, nitrides, or oxides of titanium and alumina, other methods of altering the chemistry of tool surfaces are being explored in laboratories throughout the world. However, these efforts are at present too preliminary to ascertain whether a new material can seriously challenge the present front-runners (TiC, TiN, and Al_2O_3).

Polycrystalline Diamond and Cubic Boron Nitride (CBN) Tools

Another type of tool that is giving conventional sintered carbides some competition is that which contains a thin coating of polycrystalline diamond or cubic boron nitride (CBN) on the surface of a sintered tungsten carbide substrate having a cobalt binder. These tools which were introduced in 1973 are produced by placing small grain size diamond or CBN on the carbide surface and subjecting it to high temperature and high pressure (in an apparatus similar to that used to produce synthetic diamond or CBN). No binder is used. However, a small amount of cobalt diffuses to the surface and acts as a binder phase for the superhard coating. The diamond-coated tools give poor performance on soft steel, titanium alloys, stainless steel, or high-temperature alloys and are primarily used to machine nonferrous alloys, fiberglass, silicon-aluminum alloys, graphite, and other highly abrasive materials.

The CBN-coated tools are used primarily to machine steel, high-speed steel, stainless steel, and high-temperature alloys. Polycrystalline diamond has even been used to machine sintered carbide material (Wilson, Alworth, and Ramalingam 1980).

Miscellaneous Tool Materials

Some interesting tool materials have been developed that are technically sound but either cost too much, have too limited an area of application, or have appeared at a time when another more advantageous material is also available.

One of these materials is a columbian (niobium) based material called UCON (tradename of Union Carbide Co.). These tools which consist of 50 w/o columbium, 30 w/o titanium, and 20 w/o tungsten are cast, rolled into sheets, and slit into blanks. This material is relatively soft (200 Brinell) but difficult to machine. Nitrogen is finally diffused into the surface at a temperature of 3000 °F (1650 °C) to produce a hard surface layer (about 3000 kg mm^{-2}) but a relatively ductile core having a hardness of about 500 kg mm^{-2}. These blanks thus produced are not reground or brazed but are used as clamped tools. This material is used primarily to machine steel (not useful for cast iron, titanium alloys, stainless steels, or high-temperature alloys) at large depths of cut but light and medium feeds, and high cutting speeds (2500 f.p.m. = 900 m min^{-1}). The final tools are relatively expensive (> carbide).

Castable carbide tools (Leverenz 1973) represent another type that will find difficulty competing with WC–Co tools. These materials consist of a hard carbide alloy in a high-strength refractory-metal matrix. The resulting tool material is best suited for use in the rough cutting (feeds up to 0.061 i.p.r. (1.55 mm/rev) and depths-of-cut of 1 in (25 mm), and over) of low-alloy and stainless steels at moderate and high cutting speeds. These tools are particularly resistant to crater formation and to plastic flow at high operating temperatures. Cast carbide is fabricated by consumable and nonconsumable electrode arc melting and spin-casting of the melt into graphite molds. A typical composition is 20 $^{w}/_{o}$ titanium, 58 $^{w}/_{o}$ tungsten, and 22 $^{w}/_{o}$ carbon. The resulting material contains about equimolar amounts of titanium and tungsten carbides in a tungsten-rich fine-grained (0.5 μm) metal matrix. The metal matrix is dispersion-strengthened by finely distributed interlocking carbides. The superior hot-hardness of this material at the high temperatures encountered under severe cutting conditions is responsible for its unusual performance. These cast tungsten–titanium carbide tools are not particularly hard at room temperature but are unusually resistant to thermal softening under severe machining conditions. Compared with coated sintered tungsten carbide, these materials are found to crater more rapidly at first but after equilibrium temperature is reached, the cast tungsten–titanium carbide material exhibits a slower rate of crater development. However, it does not appear that the marginal benefits over conventional WC–Co tools under an extremely limited range of conditions will warrant their adoption. For this reason, they have not been included in Fig. 14.1.

These are but two examples of tool materials that offer potential benefits over too limited a range of application to be widely adopted. Still another example is Baxtron developed by the Du Pont Co. primarily to provide a material superior to HSS for multispindle screw machining applications. In this case, a limited range of optimum field of performance was compounded by no tool marketing expertise of the developing company.

Design and Performance

Tool Material Properties

Strength, hardness, and ductility are extremely important properties of a tool material. In general, as the hardness and strength of a material increases, it becomes less ductile. The tensile strength of a tool material is usually measured in a three- or four-point bending test (Fig. 14.8) and the maximum stress at fracture is referred to as the transverse rupture strength (TRS). The TRS values for tungsten carbide tools are considerably lower than those for HSS

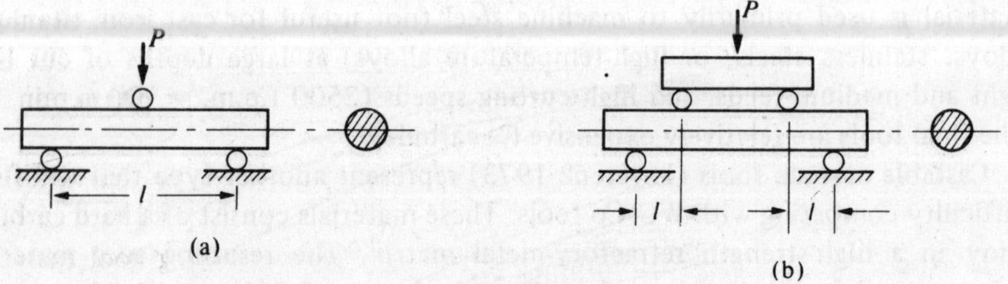

FIG. 14.8. (a) Three-point and (b) four-point bending test. Transverse rupture strength = $8\,Pl/\pi d^3$ in each case.

and the TRS of a tungsten carbide tool increases with percentage of cobalt binder used.

The hardness and Young's modulus values for different tool materials are approximately as given in Table 14.5. In general, as the material becomes more refractory (has a higher softening temperature) it has a higher value of Young's modulus (stiffness) and a higher hardness.

The hardness and strength of HSS tools falls off rapidly above the tempering temperature (~ 1000 °F or 535 °C) as shown in Fig. 14.9. It is of interest to note that the cast cobalt alloys while softer than HSS at room temperature

TABLE 14.5
Relation of Young's modulus (E) and Vickers Hardness (H_V) of tool materials

Material	*E* p.s.i. (× 10^6)	*E* MPa (× 10^6)	H_V(Kg mm $^{-2}$)
Steel	30	0.207	850
Cast alloy	30	0.207	800
Cemented carbide	60-90	0.414-0.621	1400-1800
Ceramic	60	0.414	2200
Polycrystalline diamond	>90	>0.621	7800
Single crystal diamond	>90	>0.621	8000

are harder at temperatures above 1000 °F (535 °C). The wide range of values indicated for the carbides is primarily due to the range of cobalt content (~ 3 to 12 w/o). As the cobalt content is increased, the carbide becomes more ductile and shock-resistant but less refractory.

When heat-treatable tool materials (carbon steel and HSS) are cooled to room temperature following exposure to a temperature above their tempering

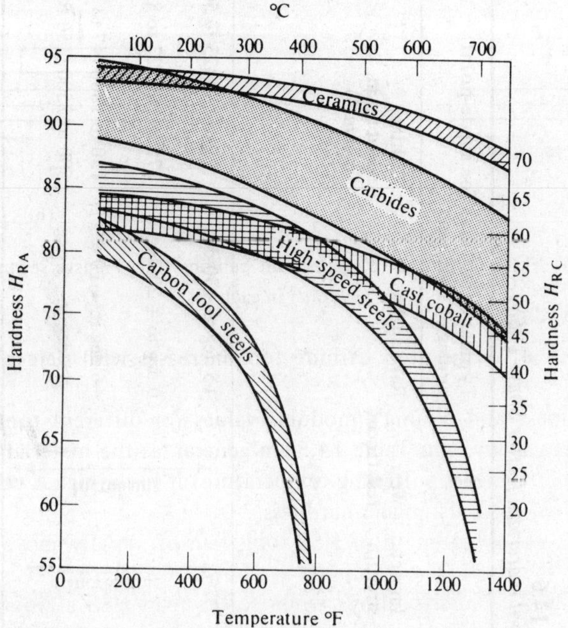

FIG. 14.9. Hot-hardness of several types of tool materials. (After Shaw 1968.)

temperature, there is a loss of hardness. Other tool materials such as tungsten carbide, titanium carbide, cast alloys, ceramics, and cermets recover their hardness when returned to room temperature following exposure to a high temperature.

Table 14.6 summarizes some of the important characteristics of different classes of tool materials. As shown in this table, the typical rake angle increases with impact strength (or ductility) while the useful range of cutting speeds increases as the tool material becomes more refractory.

Several other properties not given in Table 14.6 are of importance. These include the coefficient of expansion, (α), thermal conductivity, (k), and volume specific heat, (ρC). The coefficient of thermal expansion influences the thermal stresses and thermal shock on a material. Carbides have a coefficient of expansion about one-half that of steel, but this value increases with cobalt content. However, the carbides are so brittle that they are much more sensitive to thermal shock than high-speed steels. As a result, cutting fluids are often used when milling with high-speed steel cutters, but are rarely used with carbide cutters.

Unfortunately as the wear resistance of most cutting tool materials improves, their grindability generally decreases. For example, one of the major problems in using a highly wear-resistant high-hardness high-speed steel is the high cost of grinding. Very frequently one high-speed steel will be chosen over another

TABLE 14.6

Characteristics of several types of tool materials. (After Shaw 1968)

Cutting tool material	Hardness: Room temperature	Hardness: 1000 °F (535 °C)	Hardness: 1400 °F (760 °C)	Typical transverse rupture strength (× 10³) p.s.i. (MPa)	Impact strength	Wear resistance	Typical conditions for single-point turning of 1045 steel: Rake angles	Typical conditions for single-point turning of 1045 steel: Cutting speed f.p.m. (m min⁻¹)
High-speed steel	63 to 70 R_C (85 to 87 R_A)	50 to 58 R_C (77 to 82 R_A)	Very low	600 (4140)	↑ Increasing	↓ Increasing	+5 deg to +30 deg	75 to 180 (23 to 56)
Cast alloy	60 to 65 R_C (82 to 85 R_A)	48 to 58 R_C (75 to 82 R_A)	40 to 48 R_C (70 to 75 R_A)	300 (2070)			0 deg to + 20 deg	75 to 260 (23 to 81)
Carbide	89 to 94 R_A	80 to 87 R_A	70 to 82 R_A	250 (1724)			−6 deg to + 10 deg	150 to 1800 (47 to 560)
Ceramic	94 R_A	90 R_A	87 R_A	85 (586)			−15 deg to −5 deg	500 to 2500 (156 to 781)
Diamonds	7000 Knoop	7000 Knoop	7000 Knoop	40 (276)				

that gives somewhat better tool life, due to the differential cost of grinding finished tools from the two steels.

Special Tool Tip Geometry

Klopstock (1926) was the first to show that tool life and cutting forces could be favorably altered by restricting the contact length between chip and tool. This was done by use of a composite rake face—small positive primary rake angle of +15° followed by secondary rake surface having a greater positive rake angle of 30°. This was found to give a more stable BUE and hence better surface finish with HSS tools. Tool life was also improved. When applying this concept to carbide tools, the primary rake surface is usually made slightly negative (< 10°) in order to provide a larger included angle at the tool tip for added strength. However, when this is done, the negative primary rake angle gives high cutting-forces tending to neutralize the positive effect of the controlled contact.

A compromise solution for carbide tools is shown in Fig. 14.10a. Here the primary rake angle is positive (~ + 30°) but the cutting edge is given a large included angle for strength and thermal capacity by employing a land extending from a quarter to a half of the undeformed chip thickness in the feed direction. The rake angle of the land is usually between −5 and −10°. The contact length is controlled by providing a third negative rake surface (Fig. 14.10a).

When the rake angle of the land is −30° or more, it is called a chamfer (Fig. 14.10b). A chamfer has the added advantage of providing a very stable BUE and better surface finish.

When the cutting tool is relatively brittle and subjected to an unusually hard work material such as surface scale or to shock-loading, the very tip of the tool is frequently rounded to prevent chipping (Fig. 14.10c). This is referred to as a honed tool. Edge honing may be provided mechanically by hand stoning or by tumbling cutting inserts with shaped abrasives in a rotating basket. A lightly honed edge will have a radius of about 0.001 in (0.03 mm) while a heavily honed edge will have a radius of about 0.005 in (0.125 mm).

The purpose of the limited-contact tool is to reduce cutting forces by as much as 30 percent and hence the mean tool face temperature. In doing this, it is important to recognize there is an optimum contact length. When the contact length becomes too small, the maximum tool-face temperature will be too close to the tool tip and even though the mean cutting temperature is reduced, the tool life will then be decreased. Limited-contact cutting has been studied by Takeyama and Usui (1958), Chao and Trigger (1959), Usui and Shaw (1962), K. Hoshi and Usui (1962) and Usui, Kikuchi, and K. Hoshi (1963).

Using the Built-up Edge

Sorin (1955) demonstrated that unhardened steel in the shape of a tool was capable of producing chips in a turning operation. While tool life was short,

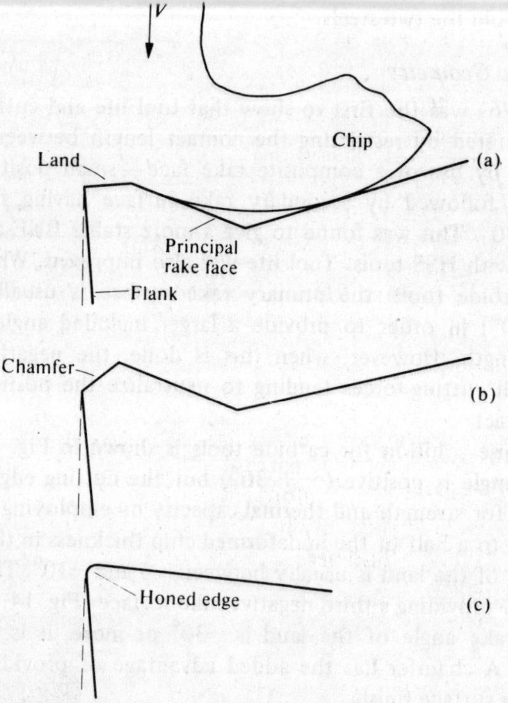

FIG. 14.10 Diagram illustrating (a) land, (b) chamfer, and (c) honed edge at the tip of a cutting tool.

the soft tool was definitely capable of producing chips. This suggests that a built-up edge (BUE) could be utilized in cutting with reduced tool wear if the BUE is continuously replaced.

K. Hoshi (1939) extensively studied the characteristics of the BUE and, in the mid-fifties, invented what he called the silver white chip (SWC) cutting technique as reported by Hitomi (1961) and T. Hoshi (1980). The SWC technique involves tool geometry that produces a BUE that is caused to flow away continuously in the form of a separate secondary chip (Fig. 14.11a). The BUE is stabilized by chamfering the cutting edge (Fig. 14.11b) and the contact length is controlled by a chip curler. The metal trapped by the chamfer acts as the cutting edge and is continuously replaced as the secondary chip flows in a direction approximately parallel to the principal cutting edge. Figure 14.12 is a photomicrograph of a partially formed chip that shows the BUE trapped by the chamfer. Figure 14.13 shows the geometry of SWC tools designed to cut low-carbon steel and Table 14.7 gives recommended tool dimensions for different feed rates in mm/rev. The SWC tool is reported to give 15 percent

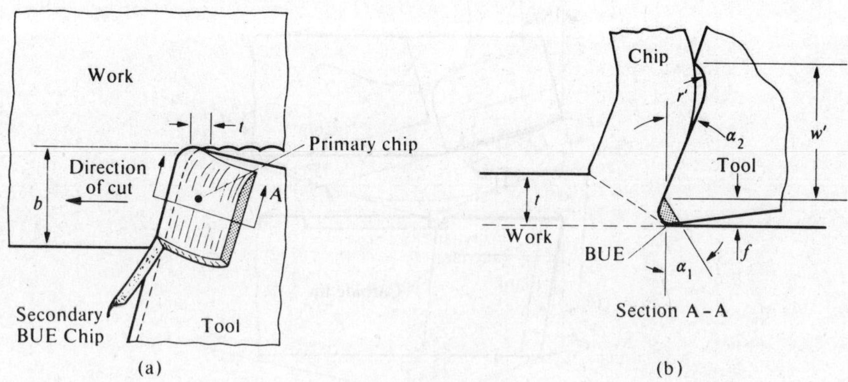

FIG. 14.11. Silver white chip (SWC) cutting mechanism. (a) Plan view. (b) Section AA normal to principal cutting edge.

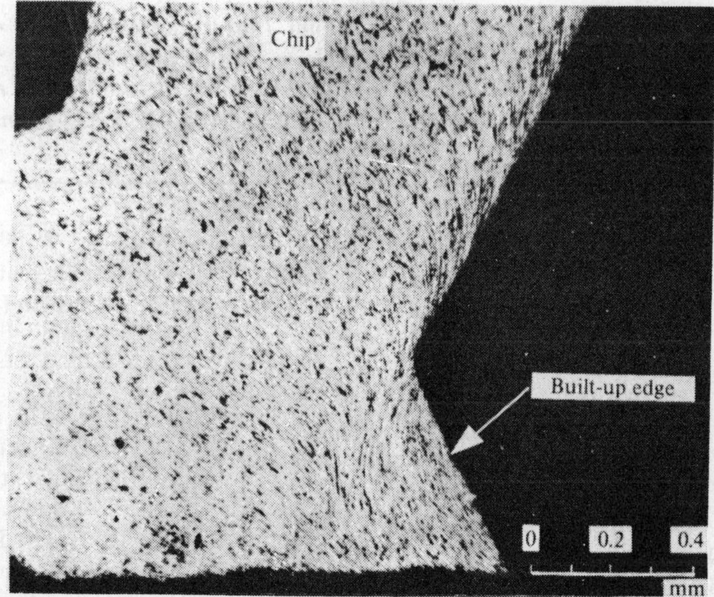

FIG. 14.12. Photomicrograph of partially formed chip produced by SWC tool. (After T. Hoshi 1980.)

lower specific energy and about 20 percent longer tool life (Hitomi 1961) than conventional tools. The lower cutting forces result in lower temperatures in the primary chip that often do not give a temper color but a shiny silver surface, hence the name 'silver white chip'.

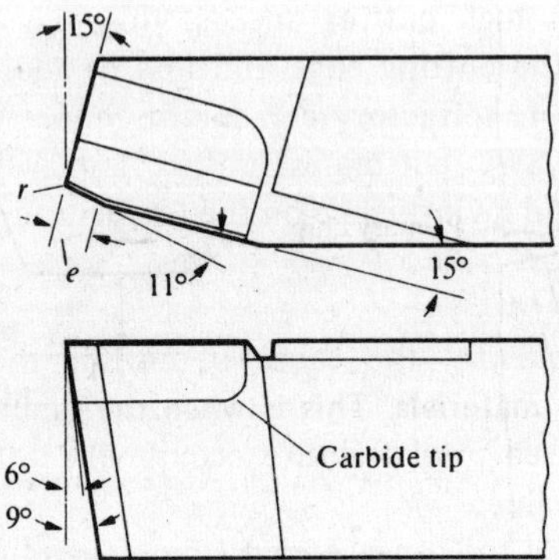

FIG. 14.13. Geometry of SWC cutting tool. (After Hitomi 1961).

TABLE 14.7

SWC cutting tool dimensions. (After Hitomi 1961.) End relief angle = 6°; end clearance angle = 9°; side relief angle = 6°; side clearance angle = 9°; first rake angle (α_1) = −30°; second rake angle (α_2) = 15° to 30°; nose radius (r) = 1 to 1.5 mm; cutting speed = 70 to 120 f.p.m. (21 to 37 m min^{-1}); fluid = none

Depth of cut mm	Feed mm/rev	Flat width *f* (mm)	Chip former (mm) Width (w′)	Chip former (mm) Radius (*r*′)	Sec. cut edge length (*e*), mm
1-3	0.2-0.4	0.1	2.0	0.5	0.0
2-5	0.3-0.5	0.2	2.5	1.0	1.0
3-6	0.4-0.7	0.3	4.0	1.5	2.0
5-8	0.5-0.9	0.4	6.0	2.0	3.0
5.5-9	0.7-1.2	0.5	8.0	4.5	4.0
6.5-10	0.8-1.4	0.6	9.0	5.0	5.5
7-15	2.0-15.0	0.7	10.0	5.5	6.0

The SWC technique has been found to be most usef[illegible] in Japan in rough-turning and planing operations at high feed rates and low to moderate cutting speeds ($V < 100$ m min^{-1}). A relatively high feed rate is important since the extent of the chamfer should be a fraction of the undeformed chip thickness (t).

A relatively large percentage of the cutting energy is convected away by the secondary chip and, at high cutting speeds, this may be red hot. This then causes a softening of the cutting edge and loss of tool geometry in the case of tungsten carbide. More refractory ceramic or cermet tools should be better. The cutting speed of SWC cutting is limited primarily by the temperature of the secondary chip and its influence on the temperature of the tool tip.

Brittle Fracture of Tools

Brittle fracture is frequently the cause of cutting tool failure, particularly for the more refractory materials. This is when the input to the body is a force that is gradually increased until fracture occurs when the stress at the critical point reaches a certain value.

In the case of cutting tool engagement, the magnitude of the contact force that develops depends on the stiffness of the material. The higher the Young's modulus, the higher the contact force, and hence the higher the stress at the critical point. Therefore, for a dynamic loading situation, a more appropriate brittle fracture criterion appears to be a maximum elastic strain energy criterion which states that fracture will occur when $\frac{1}{2}\sigma_c e = \sigma_c^2/2E$ reaches a critical value where σ_c is the stress at fracture in a static test. This is equivalent to considering the input to be a specific energy for the dynamic case just as the input is a force in the case of static loading.

Such a maximum tensile strain energy criterion has been suggested by Shaw (1980*b*) based on dimensional reasoning.

Consider a thin brittle disc with a central hole (Fig. 14.14a). If this is loaded statically as shown, fracture will occur at either A or B and in accordance with the maximum tensile stress criterion, the load at fracture (P'_f) per unit axial length before and after dimensional analysis will be

$$P'_f = \psi_1(a, b, \sigma_c) \tag{14.1}$$

$$\frac{P'_f}{\sigma_c b} = \psi_2\left(\frac{a}{b}\right) \tag{14.2}$$

If the disc is dropped from successively higher heights (h) it will fracture at A (Fig. 14.14b) at a critical drop height (h_c). The maximum unit force at O(P'_O) will depend upon the maximum deformation at O on impact and

$$P'_O \sim \gamma' h_c E \tag{14.3}$$

(where γ' and E are the specific weight and Young's modulus of the disc material) assuming completely elastic deformation, a uniform rate of deceleration and that all of the kinetic energy is absorbed on impact. Now, before and after dimensional analysis:

$$\gamma' h_c E = \psi_3(\sigma_c, a, b) \tag{14.4}$$

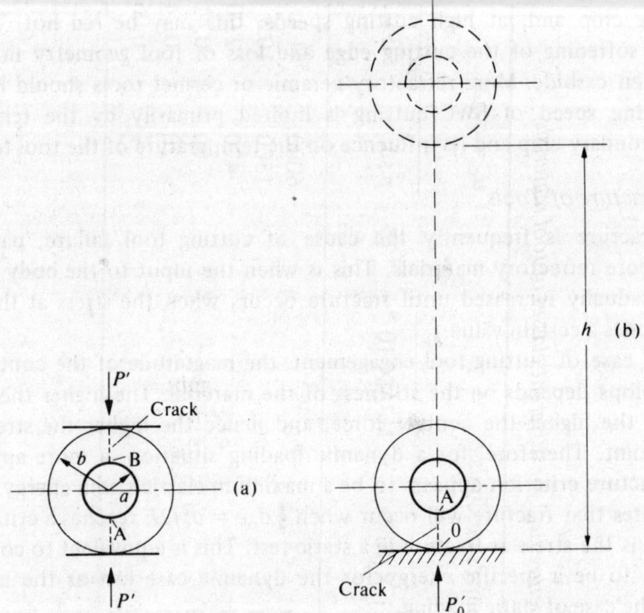

FIG. 14.14. (a) Static fracture test on disc with central hole. (b) Dynamic drop test of disc with central hole.

$$\frac{\gamma' h_c E}{\sigma_c^2} = \psi_4 \left(\frac{a}{b}\right). \tag{14.5}$$

Comparing this with the static equivalent eqn (14.2), it is seen

1. That the input is $\gamma' h_c$ instead of P'_f/b.
2. That the equivalent fracture criterion for the dynamic case will be $\sigma_c^2/E =$ constant, instead of $\sigma_c =$ constant for the static case.

Table 14.8 gives representative values of σ_c (transverse rupture stress), Young's modulus (E) and $u_c = \sigma_c^2/2E$ for different classes of cutting tool materials. From these values, it is evident why the more refractory tool materials are so much more brittle than HSS under dynamic conditions of loading. (u_c is two orders of magnitude greater for HSS than ceramic.)

The Future

Beginning about 1970 a new basic concept was introduced—the use of composite tool materials in the form of coatings or a layer of very hard particles on

TABLE 14.8

Representative values of critical specific strain energy at fracture (u_c) for different classes of tool materials

Tool material	Transverse rupture strength σ_c		Young's modulus E		$u_c = \sigma_c^2/2E$	
	p.s.i. $\times 10^{-3}$	MPa	p.s.i. $\times 10^{-6}$	MPA $\times 10^{-4}$	p.s.i.	MPa
HSS	600	4138	30	20.7	6000	41.4
Cast alloy	300	2069	30	20.7	1500	10.3
Tungsten carbide	250	1724	75	51.8	420	2.90
Ceramic	85	586	60	41.4	60	0.41
Polycrystalline diamond	60	414	90	62.1	20	0.14
Single crystal diamond	40	276	90	62.1	9	0.06

a less brittle substrate (polycrystalline diamond and CBN tools). These developments should have been anticipated and introduced much earlier because a similar concept had been in use for many years in tribology. It is well known that a good bearing surface is one that is hard (to give a small ratio of real-to-apparent area of contact to prevent galling and high friction) and which has low shear strength at the surface (to give low friction). Unfortunately shear strength increases with hardness for all materials. The solution, of course, is to provide a low-shear-strength film, that is preferably renewable from within, on a hard substrate. This is the reason cast iron, copper–lead, babbit, silver–lead–indium, etc. are such good bearing materials. High hardness is a bulk requirement while low shear strength is a surface requirement. By providing a thin film of graphite or lead of low shear strength on a harder substrate (iron, copper, or silver), the opposing requirements of high hardness and low shear strength may be met.

Similarly, there are opposing material requirements in the case of cutting tools. For machining steel at high speed, a more chemically stable carbide (TiC) is required than tungsten carbide. However, when TiC is substituted for WC, a more brittle tool results even when great care is taken in selecting the binder phase. A useful solution is to use a relatively ductile substrate on which a thin layer of the more stable carbide (TiC) is deposited. The polycrystalline diamond and CBN tools introduced about 1973 represent a second application of the same principle (composite). Single-crystal (natural diamond tools have been used for years to machine nonferrous materials but their use was limited by the extreme brittleness of a single-crystal superhard material. Cubic boron nitride does not appear in nature and only small particles previously used in grinding wheels can be synthesized. The polycrystalline tool development represents a second example of use of composites. In this case small, ultrahard particles are sintered together on a relatively more ductile tungsten carbide substrate. The composite concept is relatively new. Just as the idea of using harder and more refractory hard particles has led to a progression of new tool materials over the last eighty years, we should expect the composite concept to lead to a similar progression of new tool-material concepts over at least the next half century.

One form of technology transfer is to apply concepts that have proven useful in one engineering field to another. As already mentioned, the use of the composite concept should have appeared much earlier relative to tool materials in view of its earlier use in tribology (Bowden and Tabor, 1950).

In recent years, there have been important developments in the area of structural materials and it would appear useful to consider some of these developments in seeking clues to some of the next generation of cutting-tool materials. Argon (1980) has reviewed new structural materials with emphasis on microstructural concepts. While many of the concepts of importance in structural materials development cannot be transferred to tool material development (such as those pertaining to plastics and other nonrefractory materials), the following ideas appear to have some relevance to possible new tool material concepts:

1. Thermomechanical treatments to produce textural or orientation hardening.
2. Internal oxidation to produce very small (100 Å) hard particles that strengthen by precipitation or dispersion hardening.
3. Multiphase materials with each phase interconnected, produced by spinoidal decomposition with a phase wavelength of 100 Å or less.
4. Improved fracture-toughness of ceramics by precipitation of a crack-arresting phase.
5. Directional solidification of cast tool materials and casting techniques that eliminate the embrittling action of microshrinkage cracks and centers of tensile stress due to the dimensional change associated with the last material to solidify between dendrites.
6. Application of SPLAT cooling techniques to produce material of very fine grain size approaching an amorphous structure.
7. Continued and extended application of the composite concept.

Some of these concepts are already being applied to the development of new tool materials. For example, tool materials are now being evaluated that involve hardening by spinoidal decomposition. These materials consist of TiC + MoC hard particles in a rather brittle matrix and have cutting characteristics similar to those for sintered TiC tools. High-speed steels are being produced from powder that is rapidly cooled to yield an extremely fine carbide spacing. This is one example of item 6. Another possibility in this direction is laser-glazing of tool surfaces. In this case, the surface of a tool is scanned by a high-powered continuous wave (CW) CO_2-laser beam of 10 600 Å wavelength at a speed sufficient to melt a thin layer of material on the surface. This requires about 10^6 W cm^{-2}. The rate of heating is so high relative to the thermal diffusivity of the material that only the surface is heated and the molten layer is 'self quenched' by heat flowing into the bulk metal at a very rapid rate. The result is to produce a surface layer of very fine grain size or one that is truly amorphous. This may be considered a further extension of the composite concept.

Still another possible extension of the composite concept involves the formation of a layer of metal on the surface of a cutting tool with many diamond or CBN particles in the coating formed *in situ*. It has been demonstrated that small diamonds can be produced when a sandwich of nickel and graphite is subjected to the high temperatures and pressures associated with cutting. It is conceivable that this technique might be used to produce a thin coating of nickel with dispersed diamonds on a tungsten carbide substrate by simply coating the tool with layers of nickel and graphite and perhaps a 'protective' coating of a material such as TiC before the tool is put into use. Alternatively, thin layers of nickel and graphite on a tungsten carbide substrate might be subjected to an explosive shock-wave (another way of producing small diamond particles) before the tool is put into use.

Any one of the above-mentioned developments could lead to the next breakthrough which leads to the new idea of the decade. However, it is not possible to predict which of the many possible new developments will produce a tool material that outperforms existing tools in any corner of the huge field of cutting tool applications. It is not sufficient that a new tool material concept perform successfully to be included in Fig. 14.1 when it is updated fifty to a hundred years from now, it must also meet a particular need better and more economically than a tool material already in existence. When a new idea is introduced, such as the coated carbide, it takes a relatively long time to find out where and how it should be used, an even longer time to disseminate the information to the public, and a still longer time to overcome the inertia and resistance to change that unfortunately is a deep seated characteristic of human behavior.

In conclusion, it may be said that there are many exciting new possibilities for further improvement of cutting tools. It is unlikely that a universal tool material that meets all needs and displaces all present materials will be found. One approach to the problem of predicting the future is to ask a large group of tool-material experts for their views. This should be expected to yield a result of limited or negative value since it automatically reflects the collected bias and self-interest of the group of experts and will only extrapolate the present without sufficient regard for the fact that the future will be dominated by breakthroughs very difficult to anticipate.

References

Argon, A. S. (1980). In *Fundamentals of tribology*. MIT Press, Cambridge, Ma. USA.

Bowden, F. P. and Tabor, D. (1950). *The friction and lubrication of solids.* Clarendon Press, Oxford.

Chao, B. T. and Trigger, K. J. (1959). *Trans. Am. Soc. mech. Engrs* **81**, 139.

Friedman, L. S., Friedman, I. L., and Zagielski, K. C. (1965). U.S. Patent 3 184 169.

Hitomi, K. (1961). *J. Engng Ind.* **83**, 509.

Hoshi, K. (1939). *Trans. Jap. Soc. mech. Engrs* **5**, 137.

Hoshi, K. and Usui, E. (1962). *Proc. Int. Machine Tool Des. Res. Conf.* p. 121.

Hoshi, T. (1980). In *Proc. Int. Conf. Cutting Tool Mater.*, p. 199. Am. Soc. for Metals.

Humanik, M. and Parikh, N. H. (1954). *J. Am. Ceram. Soc.* **37**, 18.

Klopstock, H. (1926). In *Berichte des Versuchsfeldes fur Workzeugmaschinen an der Tech. Hochschule Berlin* Vol. 8.

Krekeler, K. A. (1957). Dissertation T. H. Aachen.

Leverenz, R. V. (1973). *Tech Paper* MR73-924. Society of Manufacturing Engineers, Dearborn, Mi.

Rao, S. B., Kumar, K. V., and Shaw, M. C. (1978). *Int. J. Wear* **49**, 353.
Shaw, M. C. (1968). In *Cutting tool material selection* p. 1. Society of Manufacturing Engineers, Dearborn, Mi.
Shaw, M. C. (1980*a*). *Bull. Jap. Soc. mech. Engrs* **23**, 324.
Shaw, M. C. (1980*b*). In *Proc. of Third Int. Prod. Eng. Conf.* (Tokyo), p. 492.
Sorin, P. (1955). *Microtechnic* **9**, 125.
Takeyama, H. and Usui, E. (1958). *Trans. Am. Soc. mech. Engrs* **80**, 1089.
Taylor, F. W. and White, M. (1901). U.S. Patent 668 270.
Usui, E., Kikuchi, K., and Hoshi, K. (1964). *J. Engng Ind.* **86**, 95.
Usui, E. and Shaw, M. C. (1962). *Trans. Am. Soc. mech. Engrs* **84**, 87.
Wilson, F. G., Alworth, R., and Ramalingam, S. (1980). *Proc. Int. Conf. Manu. Eng.* (Melbourne), p. 3.

15

WORK MATERIAL CONSIDERATIONS

Machinability is a general term used to rate ease of machining a material relative to tool life, surface finish produced or specific power consumed. Due to the wide range of meaning involved it is best to state the basis of comparison when two materials are compared relative to their machinability.

The basic machinability of a material is a function of its

1. Chemistry.
2. Structure.
3. Compatibility with tool material.

The basic machining characteristics of iron, aluminum, titanium, nickel, cobalt, and copper, and their alloys are quite different due to the difference in chemical and physical properties of the base metal. Such a wide variety of materials will also have different machinability ratings depending upon the tool material used. Steels of the same chemical composition but different metallographic structure may have widely different machining characteristics. Some work material—tool-material combinations are well suited to each other while others are relatively incompatible.

All materials machined may be classified as:

1. Easy-machining materials (aluminum and copper alloys).
2. Ordinary wrought steels and cast irons.
3. Difficult-to-machine materials.

Since the wrought steels and cast irons are of major importance these will be considered first.

Steels

Steel is basically an alloy of iron and up to about 1.5 $^w/_o$ carbon. Alloy steels contain additions of one or more elements introduced to provide a variety of special properties. Low-carbon steels contain up to about 0.3 $^w/_o$, high-carbon steels contain 0.9 $^w/_o$ carbon or higher, and the medium-carbon steels are in between.

The hardness, flow, and fracture characteristics of steel can be altered appreciably by heat treatment provided there is a sufficient carbon content (greater than about 0.3 $^w/_o$ C). These characteristics may also be altered by cold working (strain hardening) and by recrystallization to alter grain size. All of these changes are due generally to a change in the structure of the material.

Pure iron is too soft for good machining due to the large strain associated with chip formation (low shear angle and low cutting ratio) and relatively high tool-face friction. Its machinability can be improved by cold working (large strain plastic deformation at a temperature below an homologous temperature of 0.5 = 1200 °F or 650 °C for iron) or by refinement of crystal size prior to machining. All practical metals are polycrystalline with a mean crystal diameter in the range of 0.001 to 0.010 in (0.025 to 0.250 mm). A decrease in crystal size may be accomplished by generating a large number of dislocations by plastically working the material followed by the removal of dislocations at a temperature above $T_H = 0.5$. This may be done by hot working the material or by a recrystallization heat treatment following cold work.

The properties of a steel capable of being heat treated ($C > 0.3$ w/o) may be appreciably altered by changing the shape and spacing of insoluble carbides distributed in an iron matrix. Iron undergoes an allotropic transformation from a body-centered cubic (BCC) atomic arrangement to a face-centered cubic (FCC) arrangement at a temperature of 1670°F (910°C). Carbon is essentially insoluble in BCC-iron but is quite soluble in FCC-iron. All of the carbon may be put into solid solution by heating and holding above the transformation temperature. Then, by altering the cooling path and rate to room temperature, a wide variety of carbide shapes and spacings may be obtained. Carbide particles (Fe_3C for a plain carbon steel) grow on nuclei, the number of nuclei increasing with an increase in cooling rate. Thus, the spacing of carbides generally decreases with an increase in cooling rate and the lower the carbide spacing, the greater the resistance to dislocation formation and motion and hence the greater the hardness of the material.

A phase diagram is a chart showing the number and nature of phases (physically identifiable constituents) present in a system at any temperature and composition under equilibrium conditions. The iron–carbon diagram (Fig. 15.1) is such a diagram for steel. This chart merely indicates which of the following phases are present at a given temperature and carbon content:

L = liquid
γ = austenite = a solid solution of up to 2 w/o C in FCC-iron
α = ferrite = essentially pure (< 0.02 w/o C) BCC-iron
$\bar{C}$ = iron carbide (Fe_3C), also called cementite.

This diagram shows that transformation from α to γ generally occurs on heating over a range of temperatures the magnitude of which depends on the carbon content. At the particular composition of about 0.8 w/o transformation begins and ends at the same temperature (727 °C). This is called the eutectoid composition.

An equilibrium phase diagram does not indicate the shape and sizes of the

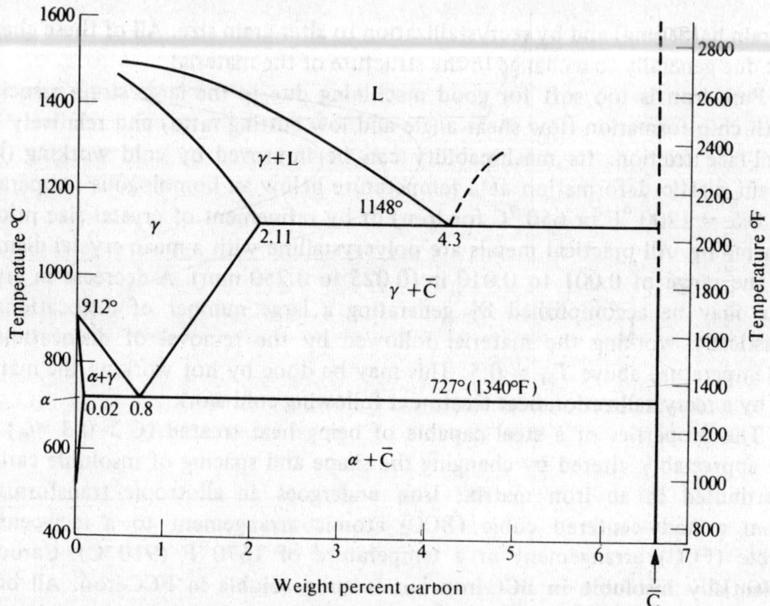

FIG. 15.1. Simplified iron–carbon equilibrium diagram. L = liquid, γ = austenite, α = ferrite, M = martensite, C = Fe_3C.

phases present. This depends on the rate of cooling. Another diagram that shows temperature plotted against time (the time-temperature-transformation, TTT diagram) for a given composition provides information on the shape and size of the precipitated carbides present. Fig. 15.2 is the TTT diagram for a eutectoid steel (AISI 1080 steel). The three dotted curves represent three cooling paths. Path A represents a quench in brine to produce a phase called martensite (M). Path B (quenched into a high-temperature bath and held until transformation is complete) leads to a soft structure called pearlite (P), while path C leads to a hard structure called bainite (B). Paths B and C are called isothermal transformations.

The various structures encountered in a heat-treated steel may be briefly summarized as follows:

Martensite (M) is a metastable body-centered tetragonal iron atomic arrangement with carbon in solid solution in the elongated tetrahedron. The M_s and M_f lines in Fig. 15.2 correspond to the temperatures at which the martensitic transformation starts and finishes respectively. Cooling path A (Fig. 15.2) will yield 100 percent M. Martensite is an extremely hard structure that is too brittle for use. The practical structure is tempered martensite (T) which is produced by heating M to an elevated temperature below the critical temperature where transformation of γ begins and holding for a given time such as

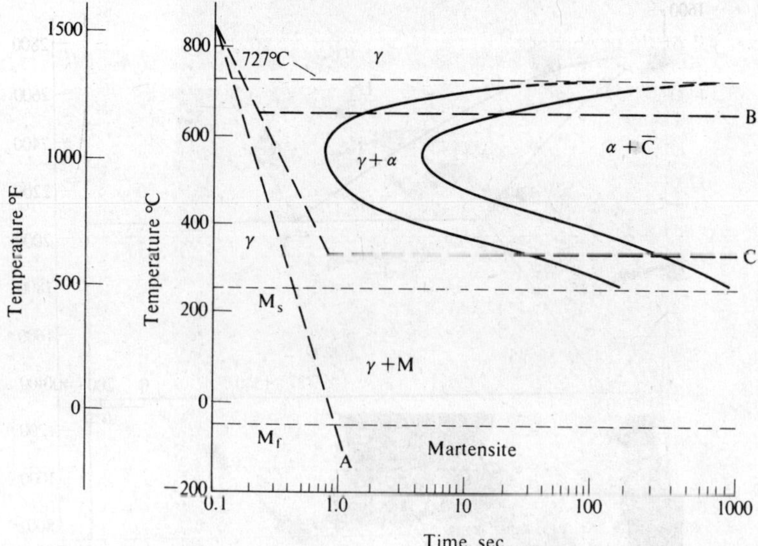

FIG. 15.2. Time–temperature–transformation diagram for plain carbon steel of eutectoid composition (0.8 w/o C). γ = austenite, α = ferrite, $\bar{C} = Fe_3C$, M = martensite.

one hour. This treatment is called tempering. During tempering, the body-centered tetragonal M collapses to a BCC structure as the carbon comes out of solution in the form of spherical iron carbide particles ($Fe_3C = \bar{C}$). The size and spacing of carbide particles in a T-structure can be adjusted over a wide range depending on the tempering temperature and time to produce a wide spectrum of hardness and ductility.

Pearlite (P) is a structure consisting of alternate plates of $\bar{C}$ and α. It is produced by slow cooling or by an isothermal transformation such as curve B in Fig. 15.2. It is a relatively soft structure—the coarser the spacing of platelets, the softer the material. Fine P is obtained by transforming isothermally at a low temperature near the knee of the TTT curve and it has a higher hardness and strength than coarse P. The entire structure will consist of P for a eutectoid steel (0.8 w/o C), but it will consist of islands of P in a matrix of α for a hypoeutectoid steel (C < 0.8 w/o) and of islands of P in a matrix of $\bar{C}$ for a hypereutectoid steel (C > 0.8 w/o).

Bainite (B) is a relatively hard structure consisting of fine jagged needles of $\bar{C}$ in an α matrix. It can be produced by an isothermal transformation at a temperature below the knee of the TTT curve (as for cooling path C in Fig. 15.2).

Spheroidite (S) is a structure consisting of large, widely spaced, spherical $\bar{C}$ particles in an α matrix. An S structure may be produced by heating P for a long time just below the initial transformation temperature or by tempering M for a long time at high temperature.

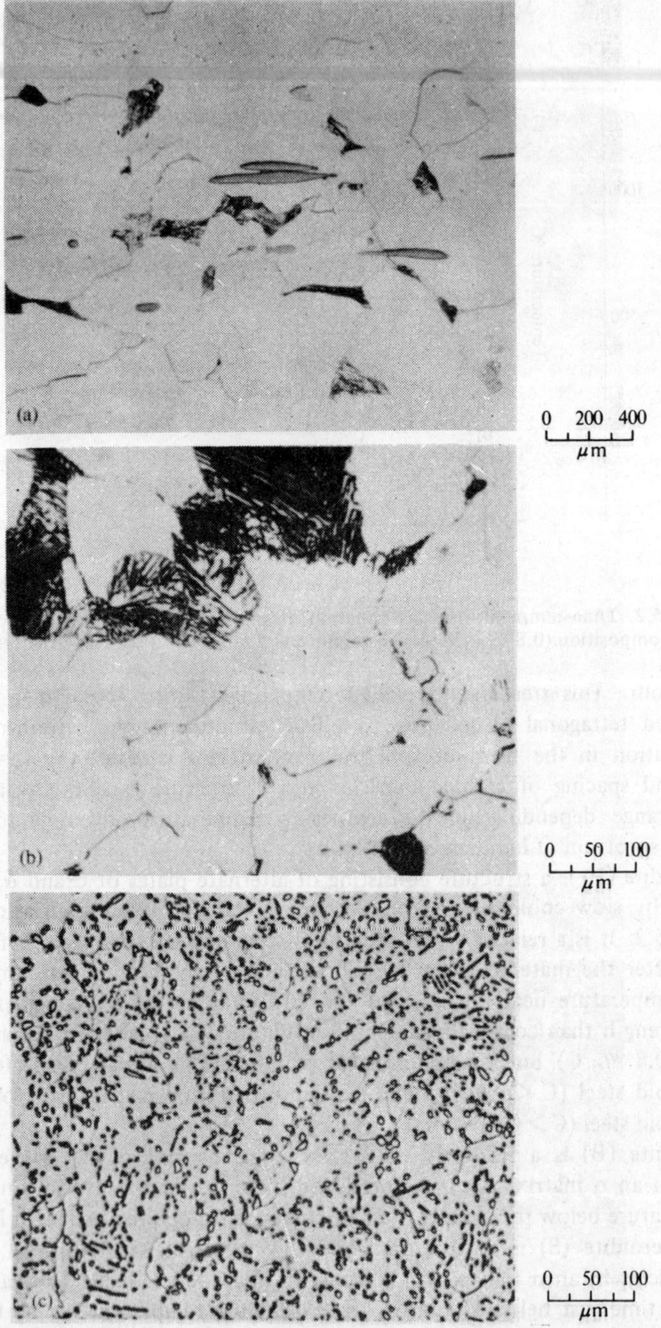

FIG. 15.3. Representative structures of wrought steels. (After Field 1963.) (a) AISI 1112 resulfurized steel $H_B = 135$. (b) AISI 8620 (30% P, 70% α) $H_B = 135$. (c) AISI 8640 (100% S) $H_B = 180$.

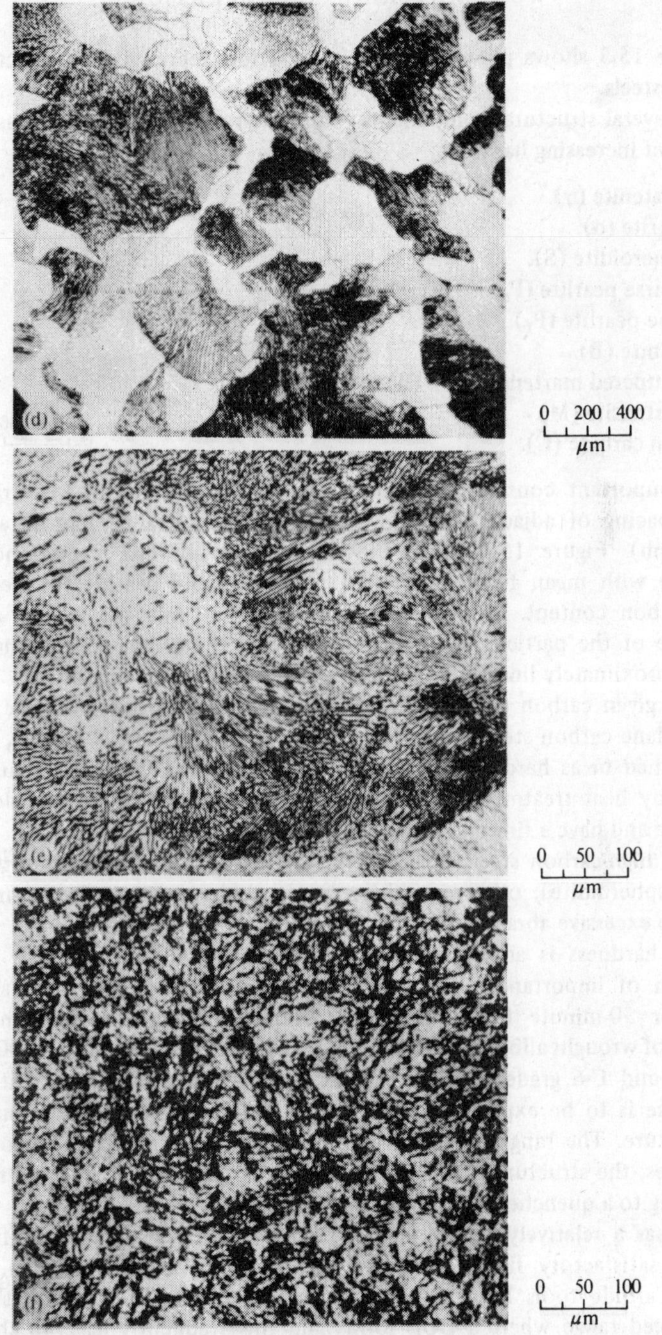

(d) AISI 8640 (50% P, 50% α) $H_B = 170$. (e) AISI 1340 (100% P) $H_B = 221$. (f) AISI 8640 (100% T) $H_B = 300$.

Figure 15.3 shows photomicrographs of a few representative structures for wrought steels.

The several structural components found in steel may be arranged as follows in order of increasing hardness:

1. Austenite (γ).
2. Ferrite (α).
3. Spheroidite (S).
4. Coarse pearlite (P_c).
5. Fine pearlite (P_f).
6. Bainite (B).
7. Tempered martensite (T). (Wide hardness range).
8. Martensite (M).
9. Iron carbide ($\bar{C}$).

The important consideration relative to hardness is the mean ferrite path (mean spacing of adjacent carbide particles along a straight line on a photomicrograph). Figure 15.4 shows the variation in ultimate stress (and hence hardness) with mean ferrite path for spheroidite and pearlite for steels of a given carbon content. From this, it is evident that only the spacing and not the shape of the particles is important and that the ultimate stress (hardness) varies approximately linearly with the log of the mean ferrite path.

For a given carbon content, there is an optimum hardness relative to tool life for plane carbon steels. In general, a low-carbon steel ($C < 0.3$ w/o) should be machined in as hard a state as possible. Since little increase in hardness is possible by heat treatment, this means that the material should be cold-drawn if possible and have a fine grain size.

For a high-carbon steel ($C = 0.8$ w/o or higher), the best structure is the softest (spheroidite); otherwise, the many carbides are too firmly anchored leading to excessive abrasive tool wear.

While hardness is an important variable relative to tool life, it is not the only item of importance. Figure 15.5 shows values of V_{30} (cutting speed in f.p.m. for 30-minute tool life) plotted against Brinell hardness number for a variety of wrought alloy steels (AISI 1020, 1112, 8630, 3140, 4140, 4340, 8640) for HSS and C-6 grade tungsten carbide turning tools. It is seen that a very large range is to be expected for hardnesses below 300 Brinell, depending on the structure. The range of values above 300 Brinell is relatively narrow since in all cases, the structure is the same (spherical $\bar{C}$ particles in an α matrix, corresponding to a quenched and tempered condition).

Steel has a relatively strong tendency to form a BUE and it is difficult to obtain a satisfactory finish at cutting speeds below about 100 f.p.m. (31 m min^{-1}). Carbide tools have a tendency to form strong bonds with the chip in the speed range where a BUE forms and this frequently leads to chipping. According to Trent (1963), this is particularly the case for steel-cutting grades

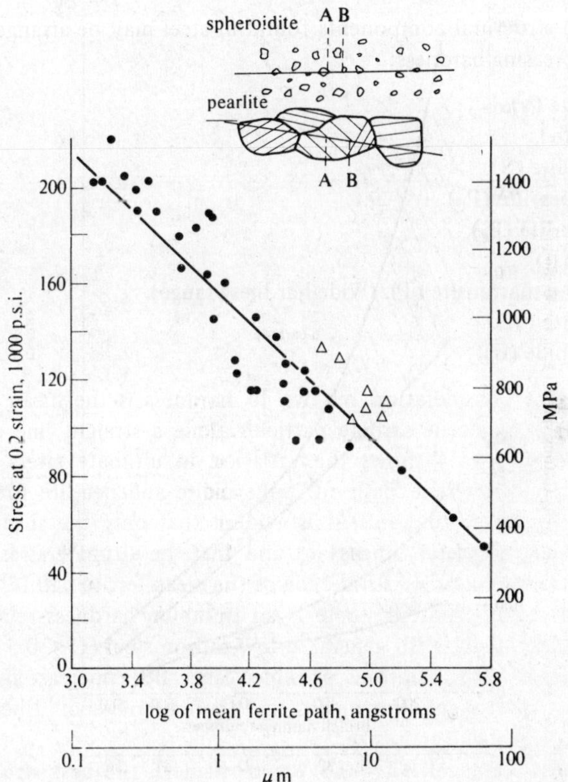

FIG. 15.4. Variation in tensile stress at strain (ϵ) of 0.2 (~ ultimate stress) with log of mean ferrite path (mean of AB in insert) for steels of pearlitic (●) and spheroidal (△) structure. (After Gensamer, Pearsall, Pellini, and Low 1942.)

of carbide. In general, low-carbon steels which tend to form a strongly adherent BUE are best machined with HSS tools in the speed range where a large BUE tends to form.

At high cutting speeds, the surface of a steel chip will be austenitic and this tends to reduce WC particles by removing carbon. The carbon absorbed thus strengthens the surface of the chip and the combined effect is the rapid formation of a crater if a straight WC tool is used to machine a steel. As discussed in Chapter 14, use of mixed carbides (WC + TiC + TaC + NbC) and coatings of TiC, TiN, or Al_2O_3 offer a useful solution to the cratering problem when machining steel.

If a hardened steel (quenched and tempered to above 350 Brinell) must be machined, a ceramic tool used with a light feed rate is often a good solution.

While it is difficult to predict accurately the best machining conditions for

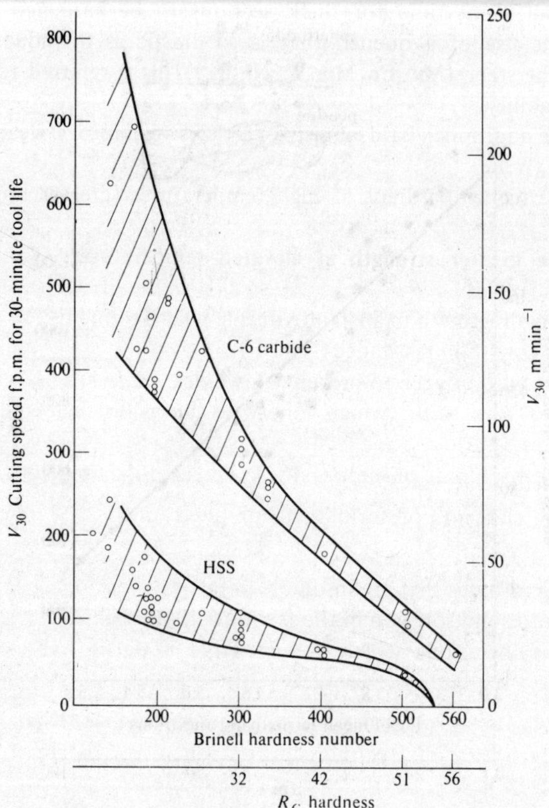

FIG. 15.5. Variation of thirty-minute tool life (V_{30}) with hardness depending on structure of wrought alloy steels. (After Field 1963.) Tool materials C-6 tungsten carbide, T-1 HSS; side rake 6° (WC), 15° (HSS); back rake, 0°; feed, 0.010 in; depth of cut, 0.100 (WC 0.062 (HSS); fluid, dry (WC), 1:20 sol. oil (HSS); life w = 0.015 in (WC), 0.030 in (HSS).

a given workpiece material in a given machining operation, handbook data usually represents a good starting point from which to proceed to the optimum by progressive change. The Machining Data Handbook (1980) is a very comprehensive source of such data.

Alloy Steels

Elements other than carbon are frequently alloyed with iron for one or more of the following reasons, and the elements used for each of these purposes are given in parenthesis.

1. To delay the rate at which austenite transforms to pearlite upon quenching

to allow sufficient time for thick sections to be hardened throughout without the use of a quench that is so drastic as to induce damaging cracks in the steel. (Mn, Cr, Mo, W, Ni, Si.) This is referred to as increasing hardenability.

2. To provide additional hard abrasive particles to improve wear resistance. (V, Mo, Cr, W.)
3. To provide higher hardness at high temperatures (greater hot-hardness). (Mo, Cr, V, W.)
4. To provide greater strength at elevated temperatures. (Mn, Ni, Si, C, Cr, Mo.)
5. To inhibit grain growth in austenite during heat treatment. (V, Al.)
6. To provide corrosion resistance (Cr, Ni.)
7. To combine with oxygen to prevent blowholes. (Si, Al, Ti.)
8. To combine with sulfur which otherwise causes brittleness at hot-rolling temperatures. (Mn)
9. To improve machining properties. (S, P, Pb, Te, Bi, graphite.)

These alloying elements perform their intended function in one or more of three basic ways:

1. By forming carbides that are insoluble in ferrite. (W, Mo, V, Cr.)
2. By going into solid solution in the ferrite. (Si, Mn, Ni, Co, Cr.)
3. By forming inclusions which are insoluble in ferrite. (Si, Al, Ti, Mn, S, P, Pb.)

Of the materials that go into solid solution in the ferrite, manganese and silicon tend to make the matrix more brittle (decrease the strain to rupture) while nickel makes it less brittle. All of the materials going into solution in the ferrite increase the tendency to strain-harden. Of the materials which form inclusions, silicon, aluminum, and titanium will be in the form of oxides MnS or MnP. Lead (insoluble in iron) will be present as small globules in elementary form.

The elements manganese, nickel, and cobalt have a pronounced tendency to lower the initial transformation temperature and if present in sufficient quantity may make it possible to have an austenitic structure at room temperature. Examples of alloy steels that can have an austenitic matrix at room temperature are Hadfield's steel, which contains 12 percent manganese, and 18-8 stainless steel which contains 18 percent chromium and 8 percent nickel.

The stainless steels contain chromium in amounts from 11 to 27 w/o to provide corrosion resistance. All stainless steels contain some carbon and the greater the carbon content, the greater the amount of chromium required for corrosion resistance, inasmuch as carbon combines with about 17 times its weight of chromium to form carbides. Chromium in the form of carbides is not available for corrosion resistance. Nickel is another important element in stainless steels, and if sufficient nickel is present the steel is austenitic at room temperature.

All stainless steels may be classified into three basic groups:

1. Low-chromium stainless steels
2. High-chromium stainless steels
3. Chromium–nickel stainless steels.

A low-chromium type stainless steel contains from 11 to 14 w/o chromium, less than 2.5 w/o nickel, and less than 0.4 w/o carbon. This material is magnetic and can be heat treated to produce a martensitic structure as with ordinary steels to provide a hardness range from about 140 to 600 Brinell.

High-chromium type stainless steels contain from 14 to 27 w/o chromium and less than 0.35 w/o C. They are also magnetic but cannot be heat treated. The matrix is ferritic and the hardness of materials in this class is usually from 140 to 187 Brinell.

The chromium–nickel stainless steels contain 16 to 26 w/o chromium, 6 to 22 w/o nickel, and up to 0.25 w/o C. They are nonmagnetic, have an austenitic structure at room temperature, and can only be hardened by cold work. The hardness of this type of stainless steel usually falls in the range from 140 to 187 Brinell. Representative alloys from each of the three basic categories are as follows:

Type	AISI No.	C	Cr	Ni	Cb	Fe
1	410	0.15	12			Balance
2	430	0.12	16			Balance
3	347	0.08	18	10	.8	Balance

From the foregoing considerations it is evident that just as the presence of carbon complicates pure elementary metals, the presence of additional alloying elements further complicates plain carbon steels in the following ways:

1. Additional hard carbides appear which have an even stronger tendency than cementite to precipitate in small size and hence tend to provide a smaller mean ferrite path and greater hardness and strength.
2. The harder inclusions such as SiO_2 and Al_2O_3 will provide additional hard particles to reduce the mean ferrite path.
3. Elements which go into solution in ferrite will cause a hardening of the ferrite and an increased tendency to strain harden. Nickel is particularly pronounced in this respect.
4. Certain elements such as manganese and silicon tend to make the ferrite more brittle which, as we will see later, is beneficial in machining while nickel tends to make the matrix less brittle.
5. The weakening effect of inclusions (such as manganese sulfide and lead) upon the ferrite is apt to be negligible unless the quantity present is excessive or the inclusions are poorly distributed.

6. The presence of manganese, nickel, and cobalt which increase the tendency to obtain an austenitic rather than ferritic matrix will tend to produce a material with greater tendency to strain harden as the percentage of retained austenite is increased.

It is thus evident that the general effect of alloying elements is to produce a harder and stronger matrix having a greater tendency to strain harden. The optimum hardness for an alloy steel will generally be somewhat lower than that for a plain carbon steel of the same carbon content.

Alloy steels have been classified and assigned four-digit numbers by the American Iron and Steel Institute (AISI). The first digit designates the principal alloying element and the last two digits the number of hundredths of one weight-percent of carbon. A few examples of common AISI steel numbers follow:

AISI 1020 plain carbon steel with 0.20 $^w/_o$ C.
AISI 1045 plain carbon steel with 0.45 $^w/_o$ C.
AISI 1112 resulfurized steel with 0.1 $^w/_o$ S and 0.12 $^w/_o$ C.
AISI 8620 Ni-Cr-Mo steel with about 0.5 $^w/_o$ Ni, 0.8 $^w/_o$ Mn, and 0.20 $^w/_o$ C.
AISI 3140 Ni-Cr steel containing about 1.25 $^w/_o$ Ni, 0.6 $^w/_o$ Cr, 0.8 $^w/_o$ Mn, and 0.40 $^w/_o$ C.
AISI 4140 molybdenum steel containing about 0.80 $^w/_o$ Cr, 0.20 $^w/_o$ Mo, and 0.40 $^w/_o$ C.
AISI 4340 Ni-Cr-Mo steel containing about 1.8 $^w/_o$ Ni, 0.65 $^w/_o$ Cr, 0.25 $^w/_o$ Mo, and 0.40 $^w/_o$ C.
AISI 8640 Ni-Cr-Mo steel containing about 0.50 $^w/_o$ Ni, 0.60 $^w/_o$ Cr, 0.15 $^w/_o$ Mo, and 0.40 $^w/_o$ C.
AISI 52100 ball-bearing steel containing about 1.75 $^w/_o$ Ni, 0.80 $^w/_o$ Cr, and 1.00 $^w/_o$ C.

The AISI numbers and complete compositions of all alloy steels in common use are to be found in the Metals Handbook (1964) published by the American Society for Metals (ASM).

Cast Irons

Iron alloys which contain from about 2 to 4 $^w/_o$ carbon are referred to as cast irons. Just as in the case of steel there are many characteristic structures that are found in cast iron (Fig. 15.6). If cast iron is cooled rapidly, a product known as white cast iron is obtained which consists of cementite particles in a matrix of pearlite. If the same iron is cooled at a slow rate, a product known as gray cast iron results, which consists primarily of graphite flakes in ferrite or pearlite. If the percentage of silicon or carbon is decreased (or if carbide forming alloys such as Cr, Mo, or V are present to tie up carbon) a slower cooling rate is required

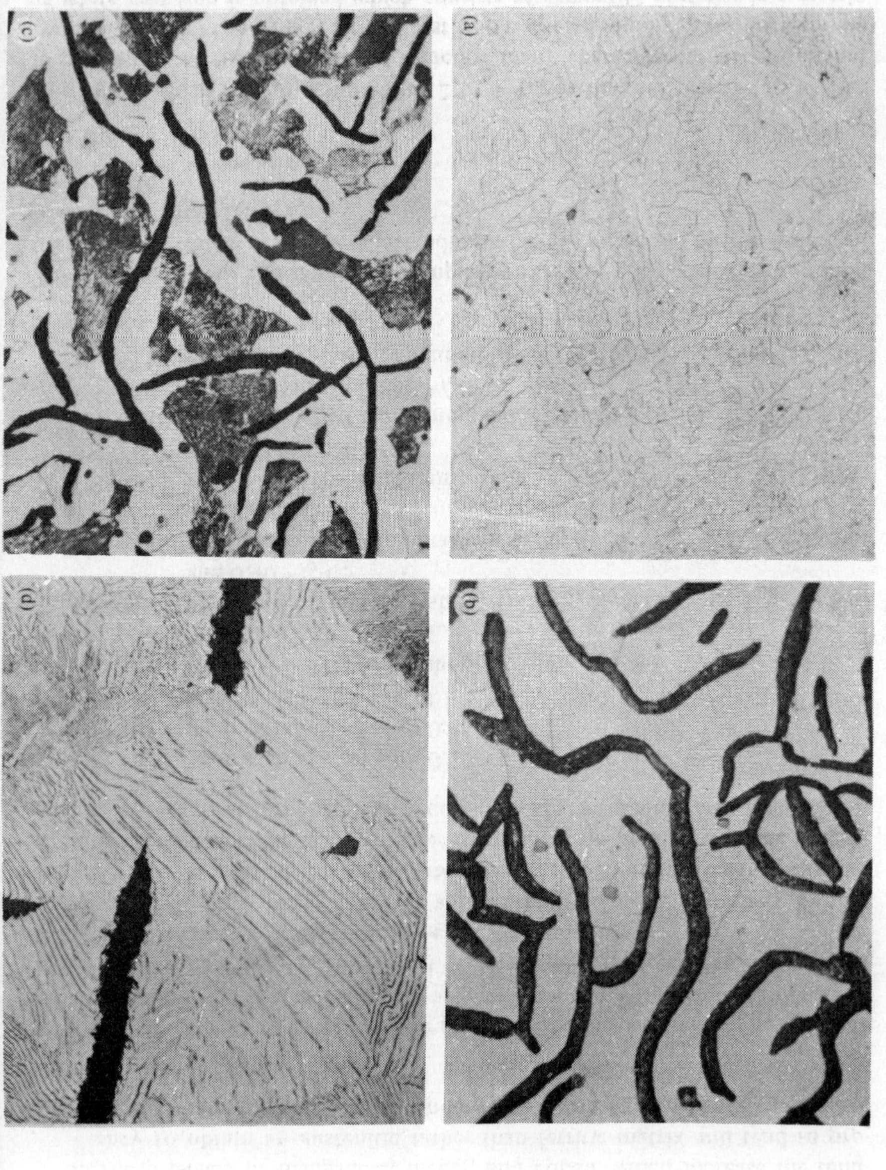
(a)
(b)
(c)
(d)

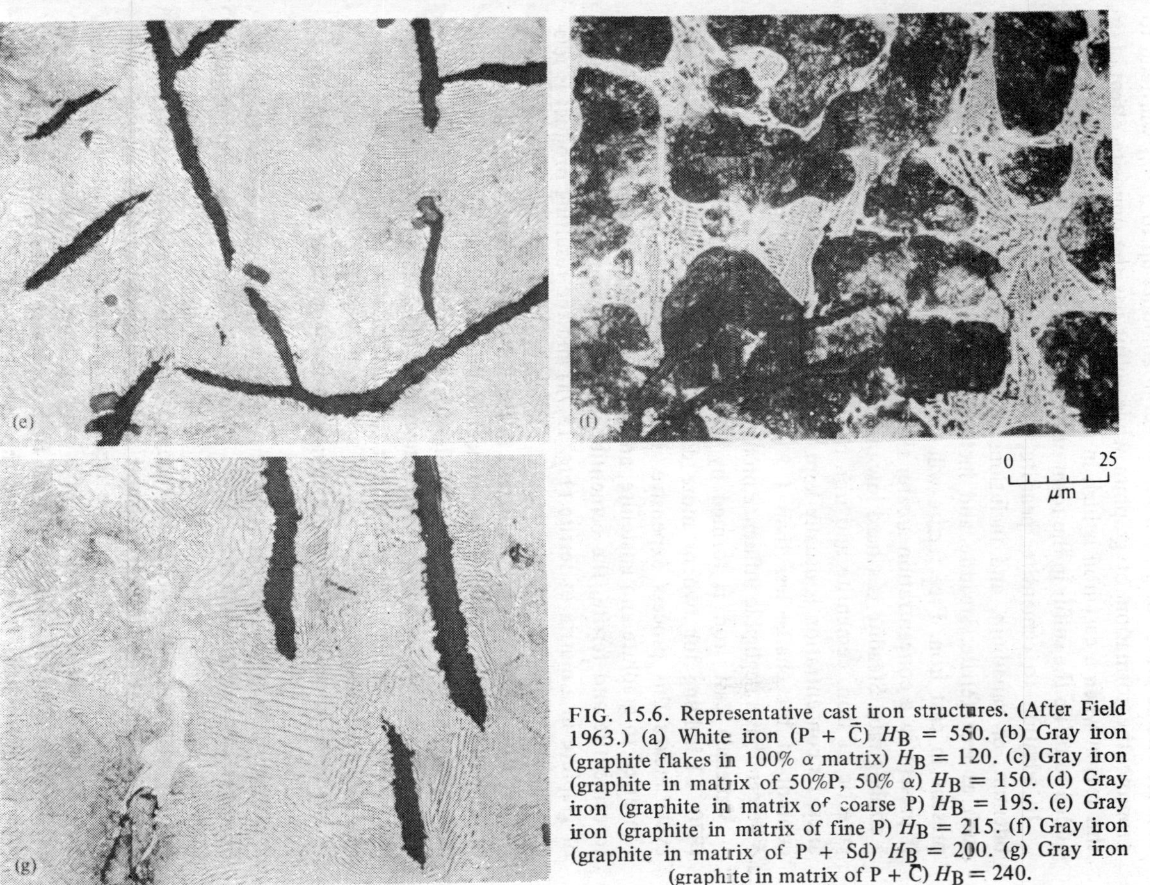

FIG. 15.6. Representative cast iron structures. (After Field 1963.) (a) White iron (P + $\bar{C}$) H_B = 550. (b) Gray iron (graphite flakes in 100% α matrix) H_B = 120. (c) Gray iron (graphite in matrix of 50%P, 50% α) H_B = 150. (d) Gray iron (graphite in matrix of coarse P) H_B = 195. (e) Gray iron (graphite in matrix of fine P) H_B = 215. (f) Gray iron (graphite in matrix of P + Sd) H_B = 200. (g) Gray iron (graphite in matrix of P + $\bar{C}$) H_B = 240.

to prevent the formation of cementite ($\bar{C}$) and to produce the graphite flakes that are characteristic of gray iron. Nickel in amounts up to 5 w/o will tend to promote the formation of graphite while sulfur has the opposite effect. For this reason when a cast iron is high in sulfur, an excess of manganese is usually added to tie up the sulfur in the form of harmless MnS inclusions.

In addition to cementite, pearlite, graphite, carbides of chromium, molybdenum, or vanadium, and inclusions of manganese sulfide, other structures such as free ferrite, steadite and inclusions of sand and slag are sometimes present in cast iron. Free ferrite will cause a marked weakening of cast iron if present in a concentration greater than about 10 w/o and indicates too slow a cooling rate. Steadite is a hard low-melting point (1800 °F or 982 °C) ternary eutectic of iron, cementite and iron phosphide that forms at the grain boundaries. Its concentration is usually kept below 10 w/o by keeping the phosphorus content of the charge less than 1 w/o. Steadite in amounts less than about 5 percent has a negligible influence on tool life.

Malleable cast iron is formed by slowly heating white iron to 1600 °F, (870 °C) holding for two or more days and slowly cooling. The long time cycle makes this process expensive. During the soaking period cementite is converted to graphite and austenite; and upon cooling, the austenite transforms to cementite and ferrite, the cementite in turn decomposing to small nodules of graphite in a matrix of ferrite (Fig. 15.7a) and the resulting product is relatively tough.

Cast iron is sometimes innoculated in the ladle to provide a finer dispersion of spheroidal graphite and a less critical cooling rate (Fig. 15.7b). Meehanite is the trade name for cast iron that is innoculated with calcium silicide. Powdered magnesium is another widely used innoculent. Innoculated irons are generally referred to as ductile irons and appear to provide better properties than malleable iron at decreased cost.

Hardness values of the major phases present in cast irons are given in Table 15.1.

Figure 15.8a shows the variation of tool life with cutting speed when turning different cast irons while Fig. 15.8b gives similar data for several malleable irons.

Gray cast iron does not tend to cause a crater to form on the tool face and consequently the more wear resistant straight tungsten carbides may be used. Wright (1976) has attributed the lack of crater formation when machining gray cast iron to the formation of a built-up layer of carbonaceous material on the tool face that unlike the usual BUE decreases in thickness as the cutting edge is approached. It is suggested this stationary carbide layer prevents diffusion wear from occuring when gray cast iron is machined. However, it should be noted that Wright's observations were made on a HSS tool operated at unusually high temperature (speed) and it has not been shown that similar results would be obtained under more normal operating conditions (a carbide

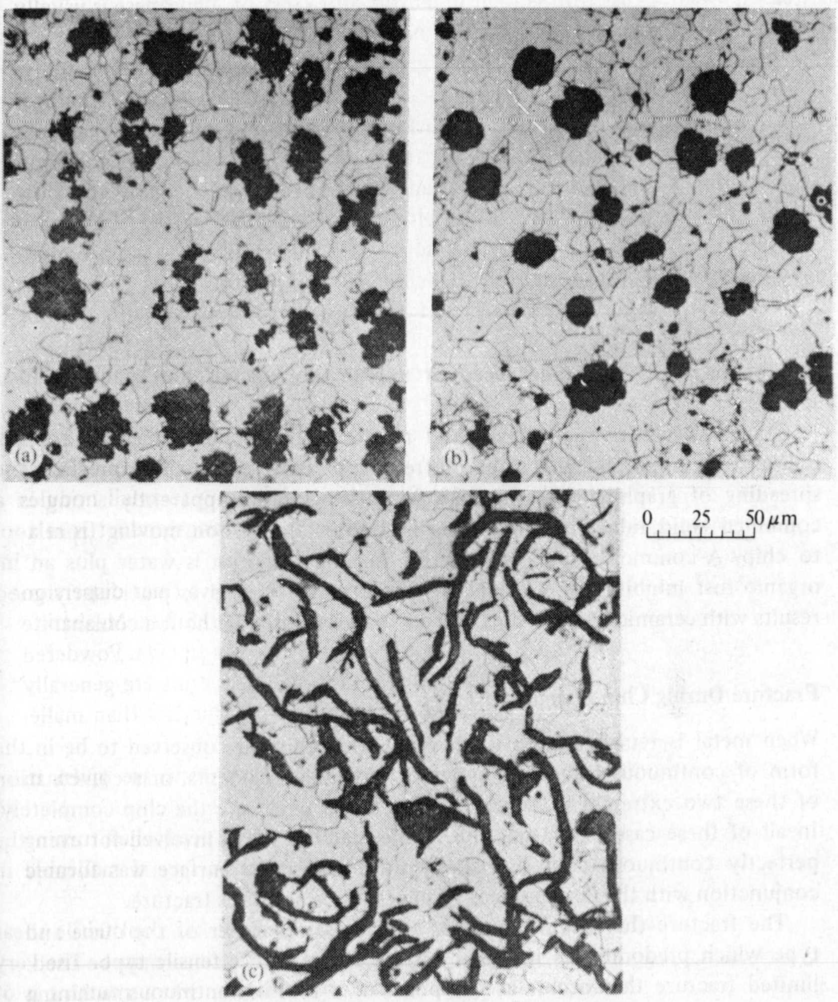

FIG. 15.7. Typical structures of three cast irons. (After Cook, Finnie, and Shaw 1954.) (a) Malleable cast iron (3.1 $^{w}/_{o}$ C, 2.39 $^{w}/_{o}$ Si, 0.37 $^{w}/_{o}$ Mn) $\sigma_u = 50\,000$ p.s.i. (345 MPa), $\sigma_{0.2} = 37\,000$ p.s.i. (255 MPa); $E = 21.7 \times 10^6$ p.s.i. (150 000 MPa); $H_B = 150$. (b) Innoculated cast iron (3.1 $^{w}/_{o}$ C, 2.41 $^{w}/_{o}$ Si, 0.43 $^{w}/_{o}$ Mn) $\sigma_u = 65\,000$ p.s.i. (448 MPa), $\sigma_{0.2} = 47\,500$ p.s.i. (328 MPa), $E = 23.5 \times 10^6$ p.s.i. (162 000 MPa) $H_B = 150$. (c) Gray cast iron (3.2 $^{w}/_{o}$ C, 2.43 $^{w}/_{o}$ Si, 0.32 $^{w}/_{o}$ Mn.) $\sigma_u = 20\,000$ p.s.i. (138 MPa), $\sigma_{0.2} = 17\,500$ p.s.i. (121 MPa) $E = 11.6 \times 10^6$ p.s.i. (79 980 MPa) $H_B = 105$.

TABLE 15.1
Range of hardness values of major constituents of cast iron. (After Field 1963)

Microconstituent	Knoop hardness (100 g load) kg mm^{-2}
Graphite	15-40
Ferrite	215-270
Pearlite	300-390
Steadite	600-1200
Carbide	1000-2200

tool operating at the same speed or a HSS tool operating at a much lower speed).

Gray cast iron normally gives best results when machined dry. If a coolant is used it is important that it be oil-free. Apparently an oily film prevents the spreading of graphite over the rake face, the graphite apparently acting as a combined solid lubricant and diffusion barrier for carbon moving from tool to chip. A commonly used coolant for turning cast iron is water plus an inorganic rust inhibitor (0.2 $^w/_o$ $NaNO_2$). Gray cast iron gives particularly good results with ceramic (Al_2O_3) cutting-tool materials used without a coolant.

Fracture During Chip Formation

When metal is removed by a single-point tool, chips are observed to be in the form of continuous ribbons, a series of individual segments, or a combination of these two extremes in which cracks do not penetrate the chip completely. In all of these cases some fracture of the material cut is involved for even the perfectly continuous chip that shows no cracks in its surface was formed in conjunction with the development of new surface and thus fracture.

The fracture that occurs in all of these cases is either of the ductile shear type which predominates in metal cutting or the brittle tensile type. The very limited fracture that occurs at the point of a tool in continuous cutting is of the tensile type and a possible mechanism for the generation of new surface at the tool tip is discussed below.

No matter how carefully a tool may be ground, there always will be a finite radius of curvature at its point as shown exaggerated in Fig. 15.9. If the metal cut consisted of a number of separate plates (**1**, **2**, **3**, etc.), then the shear surface might be expected to be as shown in Fig. 15.9. The rake angle associated with plate **1** is very small and hence the corresponding shear angle also should be

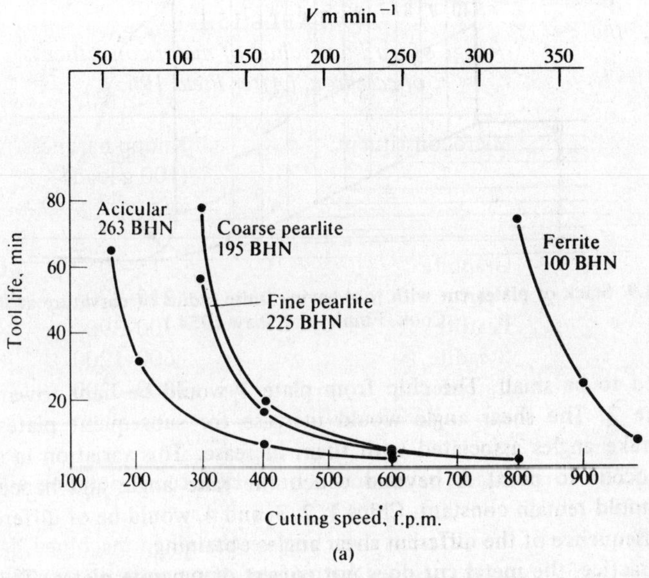

(a)

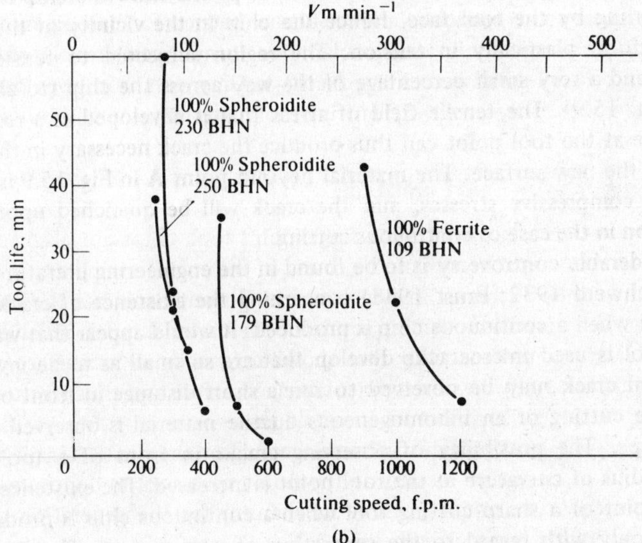

(b)

FIG. 15.8. Variation of tool life (0.030 in (0.75 mm) wear land) with cutting speed for different cast and malleable irons using a C-2 carbide tool dry. Side rake, 6°; back rake, 0°; feed, 0.010 in (0.25 mm/rev); depth of cut, 0.100 in (2.5 mm). (a) Cast irons. (b) Malleable irons. (After Field 1963.)

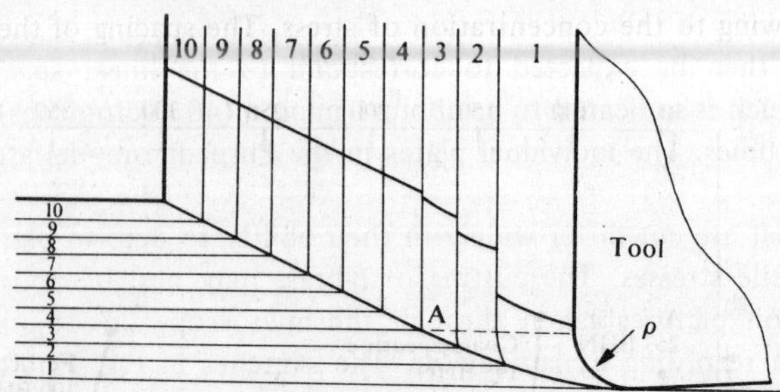

FIG. 15.9. Stack of plates cut with tool having finite radius of curvature at its tip. (After Cook, Finnie, and Shaw 1954.)

expected to be small. The chip from plate **1** would be bent toward the tool by plate **2**. The shear angle would increase for subsequent plates inasmuch as the rake angles associated with them increase. This variation in shear angle would occur to point **A** beyond which the rake angle and hence the shear angle should remain constant. Chips **1, 2, 3,** and **4**, would be of different lengths as a consequence of the different shear angles obtaining.

In practice, the metal cut does not consist of separate plates. The chip near the tool face cannot be shorter than at other points since the chip is restrained from curling by the tool face. Hence the chip in the vicinity of the tool face must deform plastically in tension. The region subjected to tension usually will extend a very small percentage of the way across the chip (to about point **A** in Fig. 15.9). The tensile field of stress that is developed as a result of the curvature at the tool point can thus produce the crack necessary in the development of the new surface. The material beyond point **A** in Fig. 15.9 is subjected to large compressive stresses, and the crack will be quenched upon reaching this region in the case of continuous cutting.

Considerable controversy is to be found in the engineering literature (Sturney 1925; Schwerd 1932; Ernst 1938) concerning the existence of cracks in front of a tool when a continuous chip is produced. It would appear that when a very sharp tool is used microcracks develop that are so small as to be invisible. An occasional crack may be observed to run a short distance in front of the tool when the cutting of an inhomogeneous ductile material is observed under the microscope. The possibility of observing cracks in front of a tool increases as the radius of curvature at the tool point is increased. The existence of cracks at the point of a sharp cutting tool when a continuous chip is produced is of interest only with regard to the generation of new surface. Their presence is of negligible importance with regard to the analysis of the shear process and the mechanics of cutting.

When a ductile metal that is free of gross inhomogeneities is cut, the microcracks responsible for the new surface will form at the microvoids present in all materials owing to the concentration of stress. The spacing of the individual cracks should then be expected to correspond to the linear spacing of the microvoids, which is indicated to be about 1 micron (40 microinches) by several independent studies. The individual plates in the Piispanen model are thus seen to be very thin.

Materials that are cut differ widely in their ability to deform plastically and to sustain tensile stresses. The cutting of β-brass may best be appreciated by studying motion pictures taken through the microscope in conjunction with records from a recording dynamometer. The sketches in Fig. 15.10 are reproduced from pictures taken at 16 frames per second when a 0.007 in (0.18 mm) depth of cut was taken at a cutting speed of 0.5 i.p.m. (12.7 mm min^{-1}), using a 15-deg-rake-angle tool in a two-dimensional planning operation. The width of cut was 0.161 in (4.09 mm) and no fluid was used. Frame numbers are indicated beneath the sketches, and the distance traversed by the tool per frame was 0.0005 in (0.01 mm). The chips were completely discontinuous and the horizontal and vertical components of force varied periodically as shown in Fig. 15.10, with a period of 39 frames.

In frame 1 rupture along the dotted line is impending and actually occurs between frames 1 and 2. When failure occurs the load on the tool is released suddenly and it is driven rapidly forward by the elastic energy stored in the system as a result of tool deflection. The depth of cut is, in effect, never zero and hence the forces do not drop to zero. Frame 2 shows the tool after its elastic motion has occurred, and metal is seen piled up at the tool point. As the tool moves forward metal is not removed by the normal shearing process represented by the Piispanen model, but is rather extruded between the tool and free surface into a parabolic shape in the manner of Fig. 15.11. During

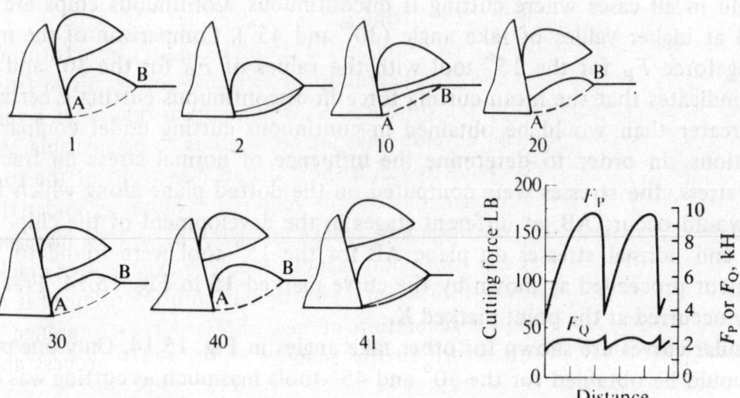

FIG. 15.10. Sketches of discontinuous cutting action of beta-brass with corresponding cutting force data. (After Cook, Finnie, and Shaw 1954.)

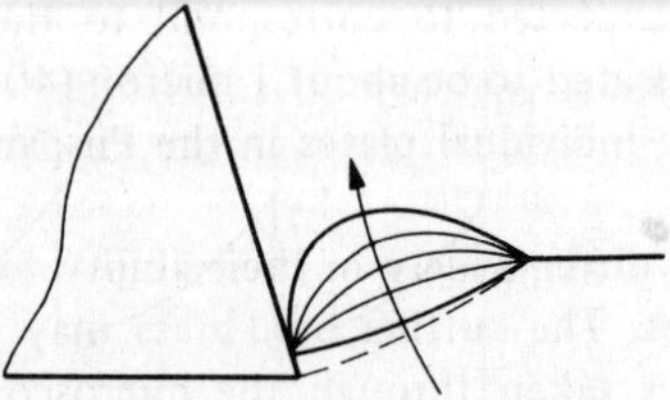

FIG. 15.11. Extrusion of chip into parabolic shape in discontinuous cutting operation. (After Cook, Finnie, and Shaw 1954.)

this extrusion the chip does not slide along the tool face but, rather, rolls down upon it. The shear stress on the dotted line, which represents the region of plastic flow, increases as extrusion proceeds, until a value sufficient to cause rupture is obtained. The entire process is then repeated.

The force data in Fig. 15.10 was obtained from a strain-gage-type dynamometer and, when used to compute the coefficient of friction on the tool face, the results shown in Fig. 15.12 were obtained. Here it is evident that the coefficient of friction actually decreases as the cut proceeds. The values of coefficient of friction given in Fig. 15.12 are unusually low for ordinary metal cutting with a 30-deg-rake-angle tool. This should not be disturbing, however, when it is realized that the significance of the static friction on the tool face in the extrusion process considered here is entirely different from the dynamic coefficient of friction in continuous cutting.

Sketches and force data are given in Fig. 15.13 for cuts made at different values of rake angle. The views shown are those corresponding to the frame just preceding rupture. The previously described extrusion process is seen to hold in all cases where cutting is discontinuous. Continuous chips are obtained at higher values of rake angle (30° and 45°). Comparison of the mean cutting force F_P for the 15° tool with the values of F_P for the 30° and 45° tools indicates that the mean cutting force in discontinuous cutting is certainly not greater than would be obtained in continuous cutting under comparable conditions. In order to determine the influence of normal stress on fracture shear stress, the stresses were computed on the dotted plane along which fracture would occur, AB, at different stages in the development of the chip. The shear and normal stresses on plane AB for the 15° tool were found to vary as the cut proceeded as shown by the curve marked **15** in Fig. 15.14. Fracture finally occurred at the point marked X.

Similar curves are shown for other rake angles in Fig. 15.14. Only one point each could be obtained for the 30° and 45° tools inasmuch as cutting was continuous in these cases. The fracture point for each rake angle has an X over it, and it is evident that a line D–E can be drawn which separates the conditions where no fracture occurred from those where a crack ran clear across the chip.

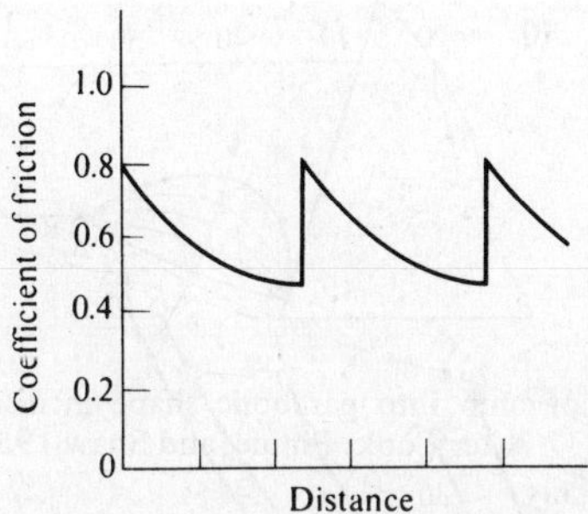

FIG. 15.12. Periodic variation of coefficient of friction in discontinuous cutting tests of Fig. 15.10. (After Cook, Finnie, and Shaw 1954).

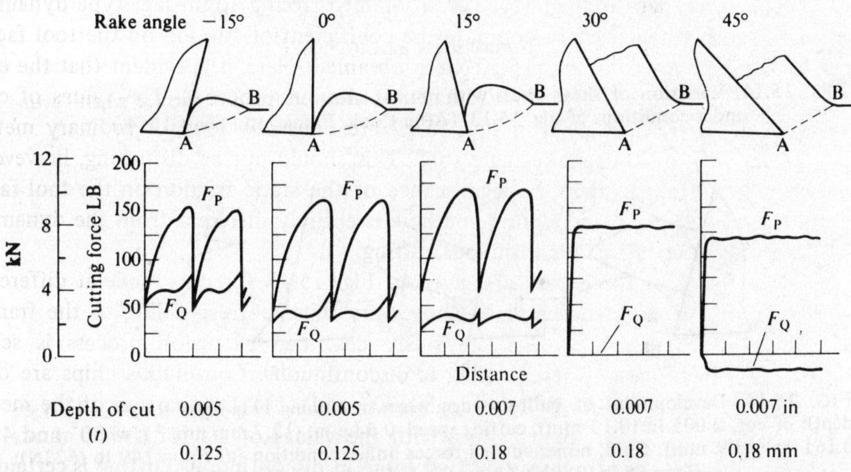

FIG. 15.13. Effect of rake angle upon chip discontinuity. Material, beta brass; cutting speed, 0.5 i.p.m. (12.7 mm min^{-1}); width of cut, 0.161 in (4.09 mm); fluid, none. (After Cook, Finnie, and Shaw 1954.)

Line D–E is seen to be inclined upward, which means that as the compressive stress on the plane where rupture will occur is increased, there is a corresponding higher value of critical fracture stress. Inasmuch as the normal stress on the shear plane is inherently relatively less for small values of rake angle, the tendency to cut discontinuously increases with decreased rake angle.

When the depth of cut was reduced from 0.007 to 0.005 in (0.18 to 0.13 mm) in cutting with the 15-deg-rake-angle tool, the chip formation was initially as shown in Fig. 15.15a which is seen to be similar to the action in Fig. 15.10.

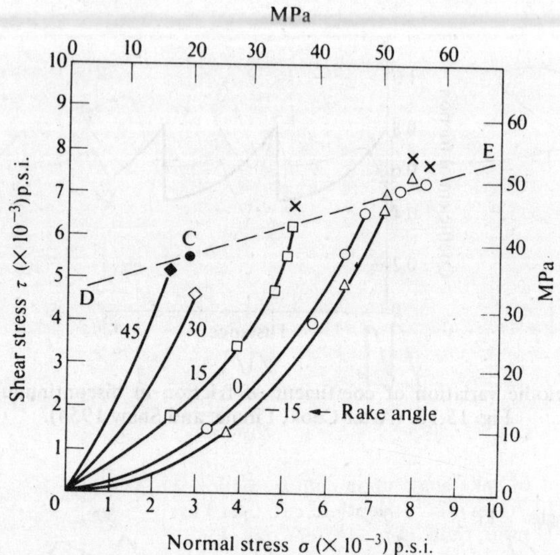

FIG. 15.14. Variation of shear stress with normal stress on fracture surface when cutting under conditions of Fig. 15.13. (After Cook, Finnie, and Shaw 1954.)

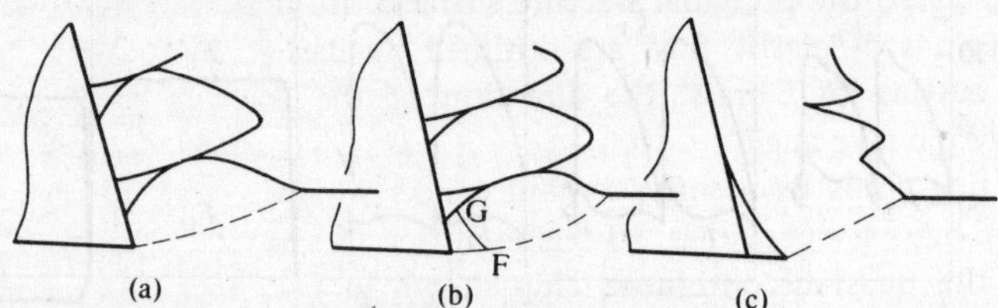

FIG. 15.15. Development of built-up edge when machining beta-brass Rake angle, 15°; depth of cut, 0.005 in (0.13 mm); cutting speed, 0.5 i.p.m. (12.7 mm min⁻¹); width of cut, 0.161 in (4.09 mm); fluid, none; cutting forces under condition (c) $F_P = 140$ lb (623N); $F_Q = 65$ lb (289N). (After Cook, Finnie, and Shaw 1954.)

However, occasionally a crack was found to occur in a direction roughly perpendicular to the potential fracture surface, i.e. along FG in Fig. 15.15b, to suddenly produce a built-up edge on the tool. Since the surface of this built-up edge corresponded to an increased rake angle, the chip formation tended to become more continuous as shown in Fig. 15.15c. Only the back surface of the chip was then irregular. This example illustrates how a built-up edge may quickly develop as a result of plastic fracture within the chip.

Data for magnesium similar to those previously given for β-brass are shown in Fig. 15.16. The action in this case is similar in many respects to that for β-brass. The fracture surface is, however, much higher and consequently the

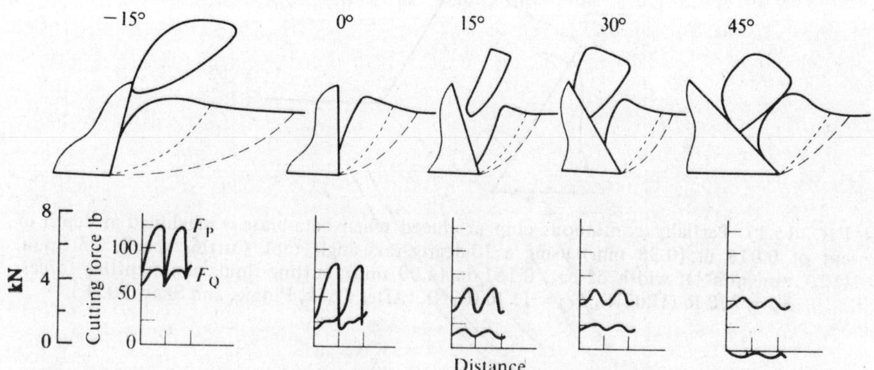

FIG. 15.16. Effect of rake angle upon chip formation when cutting magnesium. (Cutting speed, 0.5 i.p.m. (12.7 mm min^{-1}); depth of cut, 0.010 in (0.25 mm); width of cut, 0.159 in (4.04 mm), fluid, none. (After Cook, Finnie, and Shaw 1954.)

extrusion action is even more pronounced. In the case of magnesium, fracture does not occur at the boundary of plastic flow but within the region of plastic flow. In Fig. 15.16 the potential fracture surfaces are indicated by dotted lines while the regions of plastic flow are outlined by dashed lines. As in the discontinuous cutting of β-brass, the direction of the fracture surface is little influenced by the rake angle.

Metal-cutting chips frequently contain cracks which go part way across the chip. When β-brass is cut with a 30-deg-rake tool at a depth of cut of 0.015 in (0.38 mm) the quasi-discontinuous chip shown in Fig. 15.17 is obtained. In this case a crack is found to develop periodically, but the crack goes only part way across the chip. A ripple in the force trace is observed, but the large fluctuations characteristic of the completely discontinuous cut are absent. When the normal and shear stresses for this case are plotted in Fig. 15.14, the point **C** is found to lie on line **DE**. This is as it should be, for in this case there is a shear failure, even though the crack does not extend all the way across the chip. In order for a crack to propagate spontaneously across a chip as in the case of completely discontinuous cutting, there must be sufficient elastic energy stored in the tool to supply the energy associated with the propagation of the crack. If the system is very stiff and there is not sufficient stored elastic energy, the tool may lose contact with the chip as the crack grows and hence there may be no source for the energy required for crack propagation.

Several operating variables are found to influence the nature and extent of the fracture involved during chip formation:

1. Tool sharpness (fracture tendency decreases with sharpness).

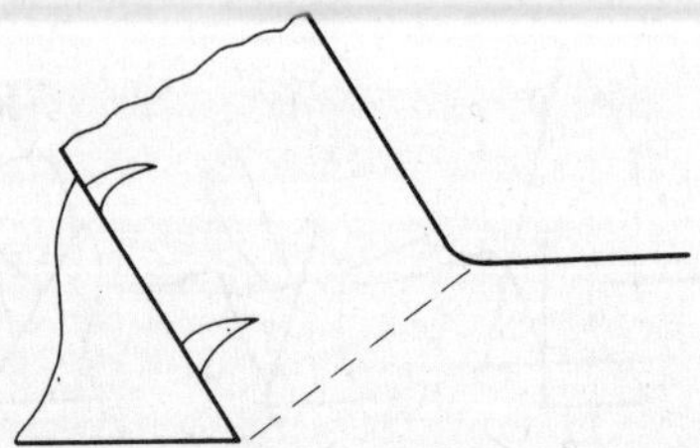

FIG. 15.17. Partially continuous chip produced when beta-brass is machined at depth of cut of 0.015 in (0.38 mm) using a 30-degree-rake-angle tool. Cutting speed, 0.5 i.p.m. (12.7 mm min^{-1}); width of cut, 0.161 in (4.09 mm); cutting fluid, none; cutting forces; F_P = 270 lb (1202N), F_Q = 15 lb (67N). (After Cook, Finnie, and Shaw 1954.)

2. Rigidity and capacity of tool work system to store elastic strain energy (fracture tendency increases with energy stored).
3. Rake angle (fracture tendency increases with decrease in rake angle).
4. Tool face friction (fracture tendency increases with increase in tool face friction).
5. Cutting speed (fracture tendency decreases with increase in cutting speed).
6. Depth of cut (fracture tendency increases with depth of cut).
7. Material characteristics (number, size, shape and relative hardness of inhomogeneities, hardness produced by heat treatment or by strain hardening).

The importance of inhomogeneities in the metal cut is immediately evident when the cutting characteristics of gray cast iron, nodular iron, and malleable iron are compared. In Fig. 15.7 the structures of specimens of these materials are shown. The carbon content of each of these materials is essentially the same. In each case we observe a ferrite matrix with inclusions of graphite. The graphite in the gray iron is in the form of plates, that in the innoculated iron is in the form of spheres, while that in the malleable iron is in the form of rougher spherical clumps of very fine graphite particles. The tensile strength of the gray iron is relatively low inasmuch as it is not difficult to find plates oriented across the direction of maximum normal stress. A much larger stress concentration is thus associated with the plates than with either of the spherical structures. Similar behavior is observed when hardness readings are taken on these materials or when they are cut.

The gray iron produces a brittle chip as a result of the stress concentration associated with the graphite plates. While most cracks are in the direction of maximum shear stress to the very point of the tool and hence indicate that fracture was of the plastic type, cast iron is sometimes observed to fail in tension, a crack extending downward from the tool point and then swinging into the direction of maximum shear stress as shown in Fig. 15.18.

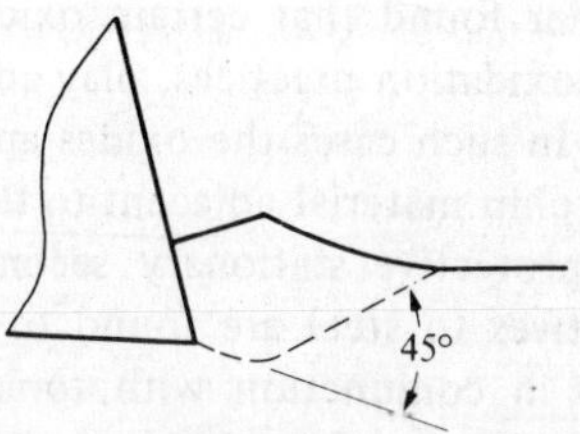

FIG. 15.18. Direction of tensile crack (dotted curve) sometimes observed in brittle materials. (After Cook, Finnie, and Shaw 1954.)

It is only with very brittle materials that a tensile crack is observed initially to extend downward from the tool point. When this occurs, it is difficult to maintain dimensional accuracy because of the variable depth of cut. By far the majority of chip fracture observed in cutting is of the shear type. While this is true even with such materials as cast iron, tensile-type cracks may be observed when (a) the inhomogeneities are unusually large; (b) the cutting temperature is low; and (c) large depths of cut are taken.

In addition to graphite, inclusions of other soft materials at elevated temperatures such as lead or manganese sulfide frequently lead to discontinuous cutting action.

Secondary Shear Zone

The secondary shear zone which forms at high cutting speeds along the tool face has been extensively studied by Trent (1963) and Opitz (1963). The secondary layer is a portion of the chip that has been rendered unusually plastic as a result of high temperatures. The temperatures in this layer will definitely be above the strain recrystallization point and even lie above the $\alpha \rightarrow \gamma$ phase-transformation temperature, in the case of steels. Temperatures in the layer will be sufficiently high to enable carbon and tungsten from the tool material to diffuse into the chip. (Carbon will diffuse particularly rapidly if the zone has transformed to γ-iron.) If the secondary shear zone is moving relative to the tool face, wear is apt to be rapid since the carbon will be rapidly convected away by the chip. A crater will then rapidly form which may weaken the cutting edge as it grows and ultimately cause tool point breakage. It is highly advisable that the secondary shear zone become affixed to the tool face in order to avoid the rapid removal of diffused material. Titanium carbide appears to help form a thin stationary secondary shear zone on the tool face and is used as an additive to tungsten carbide tools to decrease the crater-forming tendency of such tools. At the same time TiC lowers the thermal conductivity of the tool and hence the speed at which the protective (stationary) secondary shear zone will form.

Opitz (1963) has further found that certain oxides, when present in steel as a result of particular deoxidation practices, play an important role in secondary shear zone formation. In such cases the oxides apparently lower the softening (melting) point of the chip material adjacent to the tool face and thus make it possible to provide a protective stationary secondary shear zone at lower cutting speeds. Such additives to steel are found to be effective in increasing tool life only when used in conjunction with tools containing TiC. This is apparently due to the fact that early secondary shear zone formation is important only if the zone can be rendered stationary at the tool face. Since titanium has a greater affinity for oxygen or oxides than for iron or itself, it forms a strong adhesive bond with oxide-bearing steels.

Manganese sulfide appears to influence secondary shear zone formation in a way similar to that for the special oxides (i.e. by lowering the softening point of the metal immediately adjacent to the tool face). Manganese sulfide also appears to have a strong affinity for both WC and TiC, and hence is effective in promoting the formation of a protective stationary shear zone with both types of tools.

From the point of view of crater formation the properties of the work material immediately adjacent to the tool face are more important than the bulk properties of the work. The material in the secondary shear zone is subjected to enormous strains (20–50 in in^{-1} or more), and if oxides or sulfides are present as inclusions they will be drawn out into extremely thin plates. Thus, the manganese sulfide particles in the secondary shear zone will be transformed into a layered structure of very thin plates. Such a structure may have very different strength, thermal, and chemical properties from those of the usual undeformed material or even the deformed material in the main body of the chip. In addition, the temperature of the secondary shear zone will be very high. It is therefore important that the properties of highly strained materials containing layered structures of sulfides and oxides be studied at a wide variety of temperatures. It is to be expected that the properties of materials in this special structural condition may prove to be more useful in explaining the behavior of work materials containing different additives than the ordinary bulk physical properties of these materials.

It would appear that useful properties of a secondary shear zone include:

1. High affinity for the tool material.
2. Poor thermal conductivity.
3. Low melting point.
4. Low solubility for tungsten or carbon.

Some of these are the reverse of those to be expected from the more conventional view that the chip should glide freely and with low friction over the tool face. The easy sliding view is correct for low-speed cutting (broaching) but not for high-speed cutting (turning) where crater formation is a major problem. To avoid crater formation a stationary or low-speed protective layer is essential to prevent the major tool-face shear energy from reaching the tool face. The secondary shear zone should be capable of acting not only as a thermal barrier but also as a diffusion barrier. A highly oriented, relatively low-melting point, layered structure of an inorganic inclusion (oxide, sulfide, phosphide, etc.), with the layers running parallel to the tool face, should be useful from all points of view.

Properties of the tool must also be considered in relation to the layered structure of the secondary shear zone and not relative to the work material in bulk. A good tool material should have constituents possessing a high affinity for the layered structure material, low conductivity (to promote secondary shear zone formation), and a low tendency for its constituents to diffuse into the layered structure. It is conceivable that the best material for low-speed cutting (no secondary shear zone present) may not be the best for high-speed cutting.

At very high cutting speeds a secondary flow zone will also form on the clearance face, and attention should also be directed to this region of the cutting tool.

Behavior of Sulfur and Lead

The role of manganese sulfide has been touched upon in the previous section in connection with crater formation, and the results in Fig. 15.19 illustrate this action. Here five steels from the same heat, but having different sulfur contents, were machined using a tungsten carbide tool of 6° rake angle and a feed of 0.0046 in/rev (0.12 mm/rev). The cutting performance as measured by either V_{60} (speed corresponding to a 60-minute tool life) or V_{240} is seen to improve with sulfur content (Table 15.2). In these tests the cutting speeds are high, no built-up edge is present, and crater wear plays an important role.

The overall role of manganese sulfide is far more complex than this influence on secondary shear zone formation would suggest. This is illustrated by some similar tests on the same steels using high-speed steel (HSS) tools (Fig. 15.20). Here it is evident that sulfur does not always lead to improved tool life. Table 15.3 shows that although V_{60} increases with sulfur content, V_{240} does not. The reason for this lies in the fact that manganese sulfide actually promotes BUE formation and when running at speeds where the BUE is apt to be large (speeds in the vicinity of V_{240} for HSS) the presence of sulfur is apt to reduce tool life.

This result, interesting as it is, is not in agreement with general experience, where it is found that the presence of manganese sulfide in steel caused an improvement in the life of automatic screw machine tools. The cause of the discrepancy is to be found in the feed used in the tests of Fig. 15.20 (0.0046 in/rev or 0.12 mm/rev). Screw machines actually operate at much smaller feeds. When

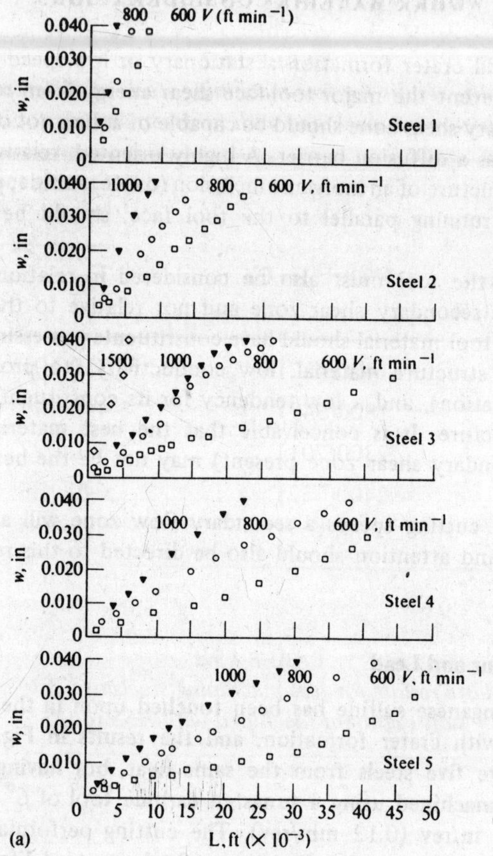
w, in
800
600 V (ft min⁻¹)
Steel 1
1000
800
600 V, ft min⁻¹
Steel 2
1500
1000
800
600 V, ft min⁻¹
Steel 3
1000
800
600 V, ft min⁻¹
Steel 4
1000
800
600 V, ft min⁻¹
Steel 5
0 5 10 15 20 25 30 35 40 45 50
(a)
L, ft (× 10⁻³)

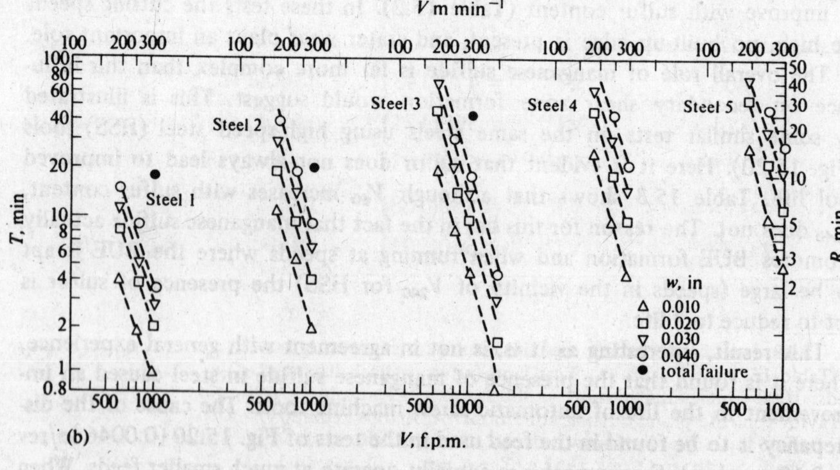
V m min⁻¹
100 200 300
T, min
Steel 1
Steel 2
Steel 3
Steel 4
Steel 5
R, min
w, in
0.010
0.020
0.030
0.040
total failure
500 1000
(b)
V, f.p.m.

TABLE 15.2
Tool-life results for free-machining steels of different sulfur content machined with tungsten carbide tools

Steel	Sulphur content, %	$VT^{0.33}$ (= C)	Cutting speed, ft min^{-1} (m min^{-1}) V_{60}	V_{240}
1	0.033	1400(427)	330(101)	230(70)
2	0.11	1800(549)	490(149)	295(90)
3	0.18	2300(701)	630(192)	377(115)
4	0.26	2400(732)	640(195)	394(120)
5	0.37	2700(823)	680(207)	442(135)

TABLE 15.3
Tool-life results for free-machining steels of different sulfur content machined with HSS tools

Steel	Sulphur content, %	Equation	Cutting speed, ft min^{-1} (m min^{-1}) V_{60}	V_{240}
1	0.033	$VT^{0.05} = 470(143)$	380(116)	360(110)
2	0.11	$VT^{0.05} = 560(171)$	440(134)	350(107)
3	0.18	$VT^{0.07} = 600(183)$	430(131)	320(100)
4	0.26	$VT^{0.10} = 550(168)$	450(137)	320(100)
5	0.37	$VT^{0.10} = 660(201)$	460(140)	220(67)

FIG. 15.19. (*opposite*) Carbide tool life results for steels with five different sulfur contents (see Table 15.2 for w/o S. (a) Wear land (*w*) vs. helical distance cut (*L*) for three cutting speeds: 1000 f.p.m. (304 m min^{-1}). 800 f.p.m. (247 m min^{-1}), 600 f.p.m. (185 m min^{-1}). (b) Taylor tool life plots. Cutting conditions: ASA tool geometry, −7, −7, 7, 7, 15, 0.03; feed, 0.0046 i.p.r. (0.12 mm/rev); depth of cut, 0.100 in (2.54 mm); fluid, none. (After Shaw, Smith, and Cook 1961.) (The *R* value will be explained in Chapter 19.)

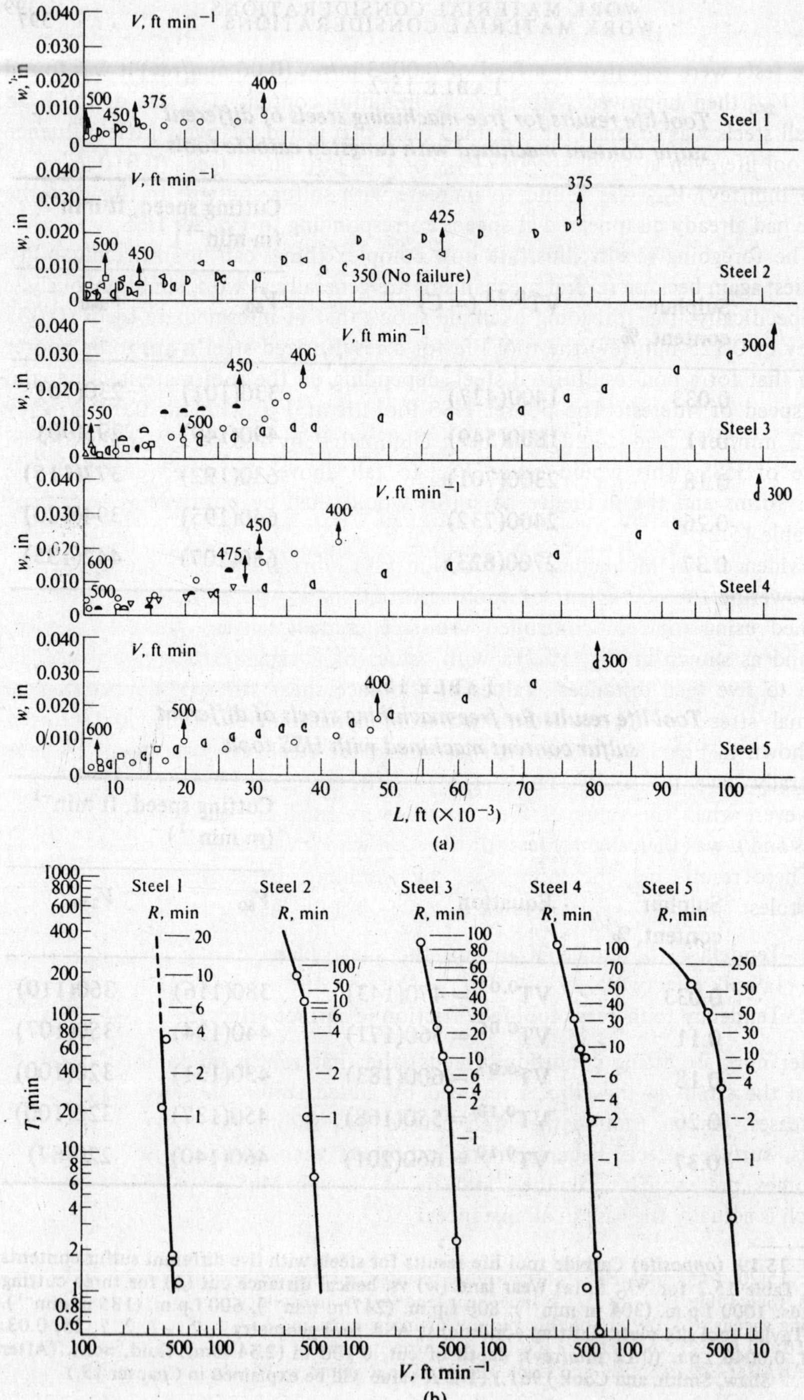

V, ft min⁻¹
w, in
Steel 1
Steel 2
350 (No failure)
Steel 3
Steel 4
Steel 5
L, ft (×10⁻³)
(a)
Steel 1
Steel 2
Steel 3
Steel 4
Steel 5
R, min
T, min
V, ft min⁻¹
(b)

these tests were repeated at a feed of 0.0023 in/rev (0.06 mm/rev) it was found that V_{240} then improved with increase in sulfur content. The size of the BUE for all steels was so small, at the small feed, that it had relatively little influence on tool life even at small speeds (V_{240} or V_{480}). At larger feeds (0.010 in/rev or 0.25 mm/rev) V_{240} was found to increase with sulfur content. In this case the BUE had already disappeared at speeds corresponding to V_{240} for HSS.

The foregoing results illustrate how complex things can be in machinability studies; again because several mechanisms are generally at work simultaneously.

Specifically, the foregoing example shows that at intermediate feeds (0.005 in/rev or 0.125 mm/rev) the tool life for a resulfurized steel is apt to be poorer than that for a non-resulfurized steel, depending on the tool material used, and the speed of interest. The poorer HSS tool life at V_{240} with the 0.0046 in/rev (0.12 mm/rev) feed could have been eliminiated by use of a carbide tool in place of HSS. This would cause V_{240} to fall above the speed at which the BUE forms and the influence of sulfur would then be positive (as is evident in Table 15.2).

Evidence that MnS does not function as a lubricant in the usual sense (i.e. by lowering the coefficient of tool-face friction) is offered by some tests performed using tools of controlled tool-face contact length. These tools were ground as shown in Fig. 15.21a with values of C ranging from one feed distance to five feed distances. Values of tool-face shear stress (τ_c) and tool face normal stress (σ_c) were determined from dynamometer results and plotted as shown in Fig. 15.21b. Here it is seen that the steel of lowest sulfur content generally gave the lowest ratio $\tau_c/\sigma_c = \mu$ (coefficient of tool-face friction). However, when the value of feed (t) was very small the curves were found to cross and μ was then slightly less for the steel with lowest sulfur content.

These results may be interpreted by assuming manganese sulfide to have two roles:

1. Tendency to reduce strain on the shear plane during chip formation (a bulk effective).
2. Tendency to increase tool-face friction (a surface effect).

Under normal cutting conditions (feeds) the first role is predominant. Hence, when the strain in the chip is reduced by added sulfur, the ratio of τ_c to σ_c decreases, just as though the MnS caused low sliding friction. At very low feeds, surface effects become more important than bulk effects and item 2 becomes predominant. It then appears as though MnS is a poor lubricant, which is actually the case in all instances.

FIG. 15.20. (*opposite*) HSS (M2) tool-life results for steels with five different sulfur contents. (See Table 15.3 for $^w/_o$ S.) (a) Wear land (w) vs. helical distance cut (L) for different cutting speeds. (b) Taylor tool-life plots. Cutting conditions: ASA tool geometry 0, 15, 10, 10, 10, 20 0.032; feed, 0.046 i.p.r. (0.12 mm/rev); depth of cut, 0.100 in (2.54 mm); fluid, none. (After Shaw, Smith, and Cook 1961.) (The R values will be explained in Chapter 19.)

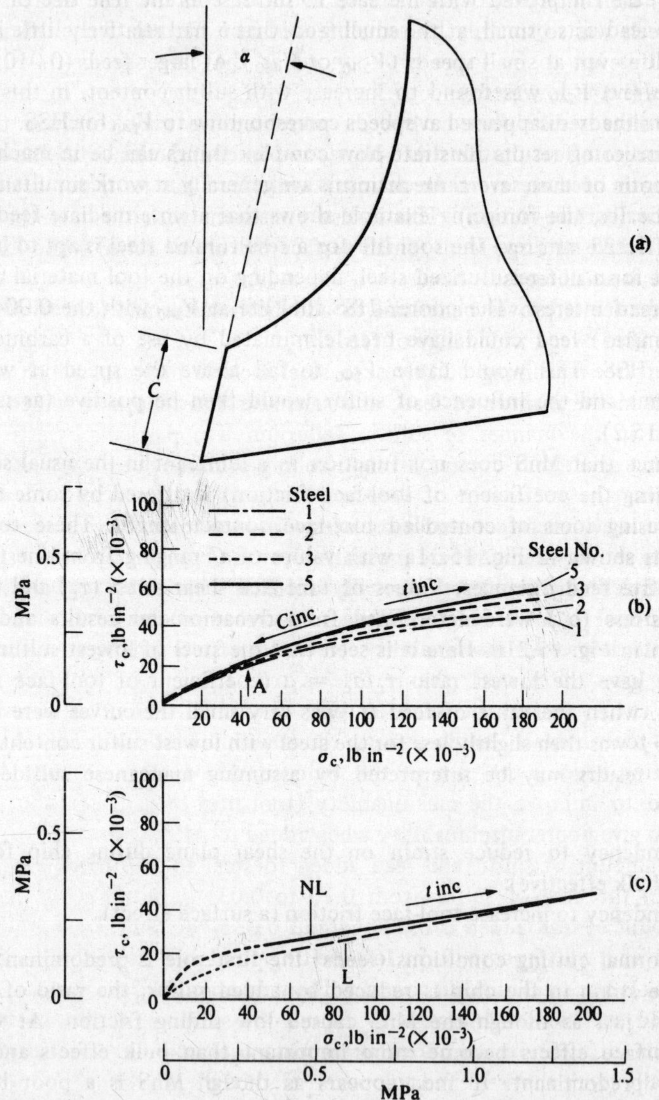

FIG. 15.21. Variation of mean shear stress on tool face τ_c with mean normal stress on tool face σ_c for various artifical contact lengths C and feeds t. (After Shaw, Usui, and Smith 1961.) (a) Tool with controlled contact length (b) τ_c vs. σ_c for resulfurized steels. (See Table 15.2 for S-contents.) (c) τ_c vs. σ_c for leaded steel (L) and nonleaded steel (NL)

Similar results are shown in Fig. 15.21c for leaded and non-leaded steels. In this case the action of lead is seen to increase as the feed decreases. This is because lead acts primarily by reducing tool-face friction. Since this is a surface effect it is apt to be more pronounced as the chip surface-to-volume ratio increases (i.e. at low feeds).

From the results that have been presented we may conclude that sulfur has its major effect on the shear plane, and in conjunction with the size of the BUE (both of these are bulk actions) while lead has its major action on the tool face (surface action). This is also consistent with the fact that sulfur is most effective at large feeds (small surface to volume ratio) while lead is most effective at small feeds.

In practice it proves beneficial to add both sulfur and lead to steels which must be used over a wide range of feeds.

Other free-machining additives to steel include bismuth, which appears to function in the manner of lead, and selenium and tellurium, which appear to function in the manner of sulfur. Tellurium is a particularly effective free-machining agent, but creates a major rolling problem even when used in small quantities.

There is evidence that lead associates with MnS making the latter more effective in its action when both are present.

Free-machining additives normally increase chip-curl and hence decrease the chip-tool contact length. While a decrease of chip-tool contact length represents improved machinability (tool life) for a relatively weak material such as low-carbon steel, the reverse is true for a material involving higher specific cutting energy (u) such as AISI 4340 steel. This is because a decrease in chip-tool contact length brings the maximum temperature too close to the cutting edge in the case of the stronger AISI 4340 steel. Thus, while lead is found to improve the machinability (tool life) of a low-carbon steel, it is found to give poorer machinability when added to AISI 4340 steel.

The amount of sulfur and lead added to steel to improve machinability is small. In the case of sulfur from 0.10 to 0.03 w/o is normally added while the amount of lead added is usually about 0.035 w/o. Sulfur has a high affinity for iron and will tend to form low-melting-point FeS in the grain boundaries. This makes the steel difficult to roll without intergranular fracture. The solution is to add sufficient manganese to tie up the sulfur as MnS and prevent the formation of FeS. As an added precaution the amount of manganese added is generally twice that required to convert all of the sulfur added to MnS.

Special Deoxidized Steels

An important development that was pioneered in Germany (Opitz and Koenig 1965) and later extended in Japan (Japanese Working Group 1968) and the USA (Tipnis and Joseph 1971) is the use of special deoxidation methods to produce steel that has a lesser tendency to cause carbide tools to crater in high-speed machining. This technique involves the use of calcium and ferro silicon as deoxidizing materials resulting in a relatively low-melting-point ternary (SiO_2-Al_2O_3-CaO) inclusion which spreads over the tool face in high-speed machining and acts as a diffusion barrier.

Calcium-deoxidized steels appear to be most effective in turning, face milling, and gun drilling operations but show little improvement in drilling with HSS tools or grinding using white aluminum oxide wheels (Japanese Working Group 1968, 1969). The protective layer is very thin and the presence of TiC appears to be necessary in the tool for the layer to become firmly attached. Aluminum oxide (ceramic) tools also give good results with calcium-deoxidized steels but HSS tools do not. A similar MnS protective layer is reported to form at lower speeds than required for calcium-deoxidized steels (Opitz and Koenig 1965).

It has been suggested (DeSalvo and Shaw 1969) that hydrodynamic action is possible on the tool face. However, this is not a result of penetration of a liquid film of cutting fluid between chip and tool, but rather due to the formation of a wedge-shaped liquid or semi-liquid layer between chip and tool due to thermal softening and spreading of an ingredient in the work material. The shape of such a 'fluid' film will be as shown in Fig. 15.22. At first glance it may be concluded that the inclination present is such as to give a negative rather than a positive hydrodynamic pressure. However, the resultant motion of the solid portion of the chip will be parallel to the tool face (AE in Fig. 15.23a) and not to the inclined surface of the molten layer (AF). This gives rise to a squeeze-film action (DeSalvo and Shaw 1969) that more than offsets the sliding action associated with the negative inclination present to provide the generation of a positive hydrodynamic pressure between chip and tool. An alternative way of looking at this problem is to draw the kinematically equivalent diagram in Fig. 15.23b where it is immediately clear that a positive pressure will be generated.

Figure 15.23c shows a developed view of the rake and clearance portions of the tool which suggests that the role of the wear land is to provide a composite bearing configuration and a shroud action. Composite bearings are analyzed by Shaw and Macks (1949).

While the role of the inclusion material present in specially deoxidized steels is generally believed to be that of a diffusion barrier for carbon, it could also tend to act as a thermal diffusion barrier just as the glass layer in the Sejournet (1954, 1962) process used in hot extrusion of steel. In this case a hot billet is coated with glass powder which melts to provide a continuous glass coating on the surface of the material to be extruded. An important function of the glass is to decrease heat transfer between the hot billet and the cold die. Similarly, the molten layer of oxide coming from the specially deoxidized steel provides a poor thermal conducting layer that tends to decrease heat transfer from chip to tool.

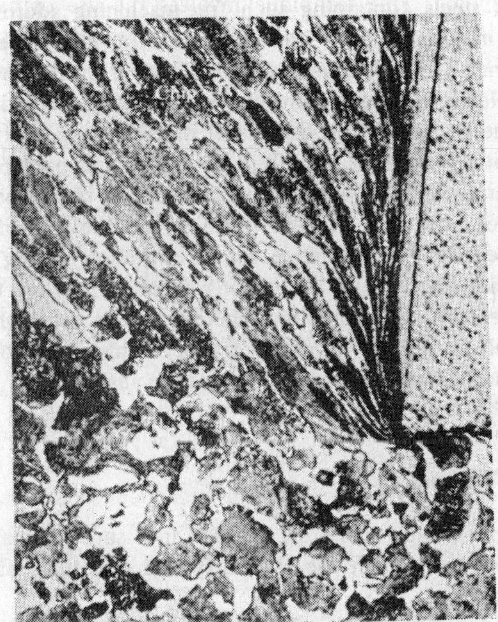

FIG. 15.22. Formation of 'fluid' layer immediately adjacent to tool face in high-speed machining. (After Schaller 1962.)

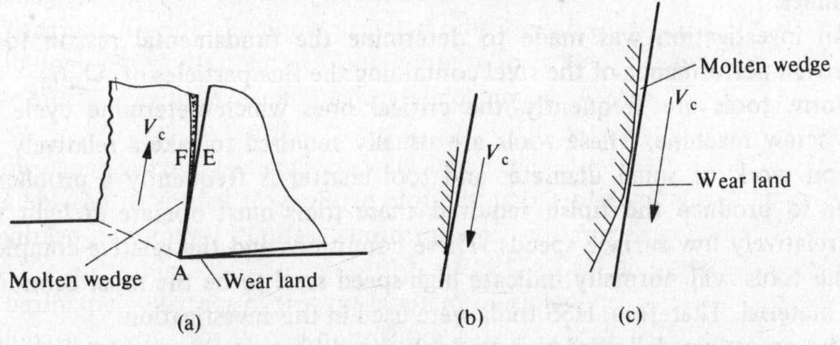

FIG. 15.23. Mechanism of positive-pressure generation on tool face.

It has been reported that specially deoxidized steels tend not only to reduce crater formation on the tool face but also to reduce the rate of wear-land formation. However, this is not believed to be due to a diffusion barrier action, since the temperature on the clearance face is very much lower than that on the tool face. A possible alternative explanation for the beneficial effect of oxide on the clearance surface is a hydrodynamic one associated with the liquid oxide that would leak past the cutting edge (Fig. 15.23c).

It is possible that a hydrodynamic action is also responsible for the improved machinability of steels containing such free-machining additives as lead and bismuth. These materials which are insoluble in steel will be squeezed from the chip to provide a thin wedge-shaped liquid layer on the tool face as in the case of the complex low-melting-point oxides in specially deoxidized steels.

Analysis of the hydrodynamic aspects of metal cutting is extremely difficult since the inclination angle of the fluid wedge is not predetermined by the geometry of bearing surfaces but by a melting action that depends on the rate of heat generation, the rate of heat transfer, and the rate of melting. From this one might expect the existence of an optimum cutting speed for each tool–work–inclusion system. If the cutting speed is too low the energy level will be insufficient to cause melting; whereas if it is too high there will not be time for melting during chip–tool contact. This is obviously a problem that is best approached experimentally rather than analytically.

Hot-Hardness of Workpiece

It frequently happens that one steel behaves in a superior manner to another in the workshop even though the chemistry and structures of the two steels are essentially the same.

This was found to be the case for two stainless steels which are designated 303 and 303A. The latter steel contained a small amount of very-finely divided (unresolvable microscopically) Al_2O_3 which improved tool life in screw machines.

An investigation was made to determine the fundamental reason for the improved performance of the steel containing the fine particles of Al_2O_3.

Form tools are frequently the critical ones which determine cycle time in a screw machine. These tools are usually required to take a relatively wide cut on work of small diameter and tool chatter is frequently a problem. In order to produce the finish required these tools must operate at light feeds and relatively low surface speeds. These conditions and the relative complexity of the tools will normally indicate high-speed steel to be the most economical tool material. Therefore, HSS tools were used in this investigation.

The procedure followed was to look for differences in machining performance of the two steels. The items investigated included chemistry, physical properties, surface finish, cutting forces, chip formation, tool temperatures, and wear rates.

The chemistry and physical properties of the two stainless steels studied are given in Tables 15.4 and 15.5.

The analyses show no significant differences except for sulfur and aluminium. Only small differences in mechanical properties are evident in Table 15.5.

The work material was in the form of cold-drawn bars, $1\frac{1}{16}$ in dia (27 mm).

TABLE 15.4
Chemical analyses of two stainless steels

	C	Mn	Si	S	Cr	Ni	Cu	Mo	Co	Al
303A	0.085	1.63	0.65	0.15	18.14	9.24	0.22	0.56	0.21	0.82
303	0.094	1.67	0.62	0.31	17.36	8.80	0.21	0.32	0.13	0.023

TABLE 15.5
Mechanical properties of two stainless steels

	303A	303		303A	303
Ultimate tensile strength (lb in^{-2})	100 300	105 000	(MPa	692	724)
Yield strength (lb in^{-2})	71 500	68 000	(MPa	493	469)
Elongation (%)	49.5	55.3			
Reduction of area (%)	67	66			
Charpy impact (ft lb)	79.0	58.5	(Nm	107	79.3)
Hardness (Rockwell B)					
Outside	102.8	103.6			
Mid-radius	98.6	97.0			
Centre	95.1	89.1			

The tools investigated were M-2, HSS (5% Mo, 4% Cr, 2% V, 6% W) having a rake angle of 10° and a clearance angle of 8°.

The nature of the cut taken is shown in Fig. 15.24. After each plunge cut was completed the remaining material was removed and a new groove produced by an auxiliary tool.

No significant differences were observed between the two materials for surface finish, cutting forces, specific energy, cutting ratio, shear angle, chip-tool interface temperature, or chip root details obtained from the quick-stop arrangement shown in Fig. 15.25. Only the difference in tool wear shown in Fig. 15.26 was observed. This was found to correlate with a difference in hot-hardness for the two materials (Fig. 15.27). These results suggest that the 303A-material is more refractory than the 303-material and that the tool material at the wear land resists softening longer for the 303A-material than the 303-material. When thermal softening occurs, the ratio of A_R/A increases abruptly as does the wear rate.

This work suggests that the high wear rates at the beginning and end of a tool life test are due to the fact that the mean pressure on the wear land then exceeds the effective flow stress of the work which is approximately one-third of its indentation hardness. It is thus not the tool that softens when the wear rate increases abruptly at the end of a test but the work.

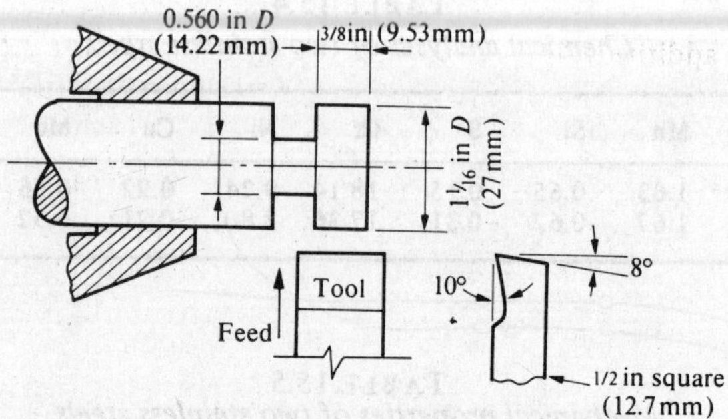

FIG. 15.24. Test arrangement for stainless steel machinability tests. (After Vilenski and Shaw 1970.)

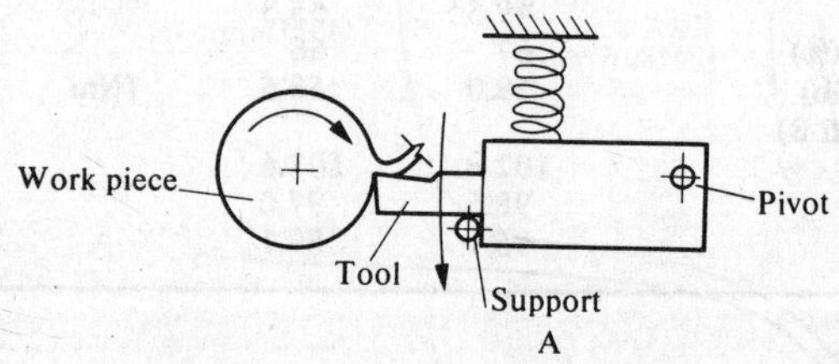

FIG. 15.25. Schematic arrangement for quick-stop tests.

Titanium Alloys

Titanium is a member of the tin group of the periodic system. In many ways it is similar to iron. Although sixth in abundance in the earth's crust, titanium is unusually difficult to convert from the naturally found oxides (TiO_2 and $FeTiO_2$) to the metallic state. Titanium ore is first converted to $TiCl_4$ by treatment with chlorine in conjunction with carbon at high temperature. This in turn is reduced to Ti by treatment with molten magnesium or sodium. The sodium or magnesium chloride is leached from the titanium which is then milled to a fine powder, melted in an arc furnace (m.p. = 1760 °C compared with 1430°C for structural steel), and cast into ingots, Titanium is a very expensive structural material compared with steel because of the high cost of winning the metal from its ore.

Titanium alloys have a high strength to weight ratio and are unusually corrosion resistant. The specific weight of titanium is only $\frac{2}{3}$ that of steel and only 60 percent greater than aluminum. Titanium has a strong tendency to react with oxygen, nitrogen, carbon, and halogens particularly at high temperatures. Titanium normally has a hexagonal close packed structure and unusually low

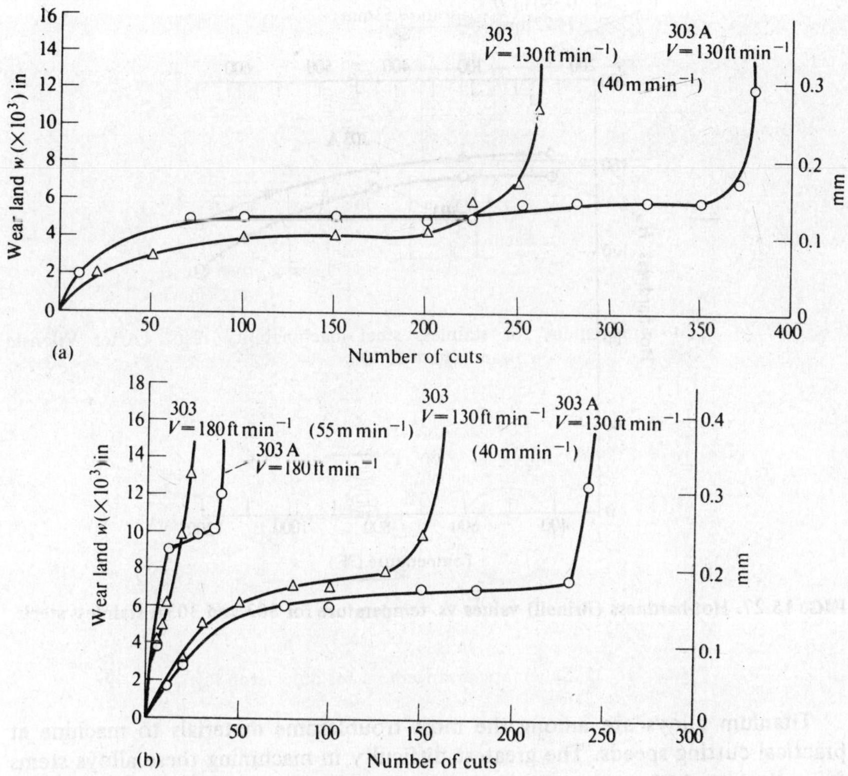

FIG. 15.26. Variation of wear land (w) with number of cuts for cutting speed of 130 f.p.m. (39.6 m min^{-1}). (a) For 0.002 in/rev (0.05 mm/rev) feed. (b) For 0.004 in/rev (0.10 mm/rev) feed.

thermal conductivity and volume specific heat. Titanium alloys have about the same strain-hardening tendency as ordinary structural steels.

While the coefficient of friction of two identical metals in dry sliding contact is normally about one, dissimilar metals will have a coefficient of friction of about 0.2. When steel slides over a titanium surface the initial coefficient of friction is about 0.2 but this soon rises to a value of about 0.45. This is due to the formation of a built-up layer of titanium oxide on the steel surface resulting in titanium oxide sliding on titanium oxide instead of the dissimilar pair—steel on titanium. Boundary and extreme boundary lubricants that perform well on steel are relatively ineffective on titanium surfaces. This is due to the strong tendency of titanium to form an oxide and the poor tendency for lubricants to adsorb or react with the oxidized surface. For this reason, titanium surfaces are very difficult to lubricate.

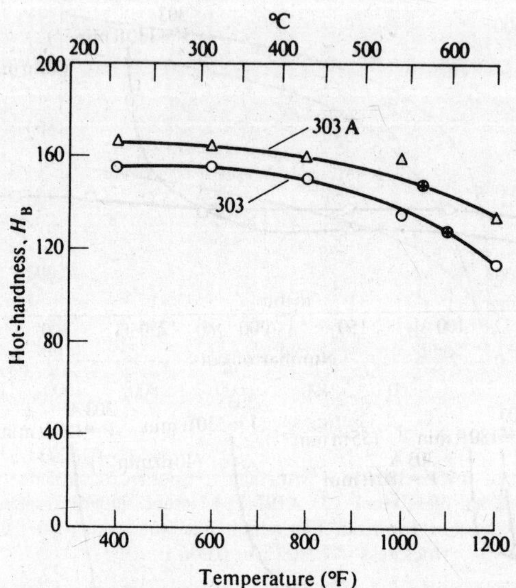

FIG. 15.27. Hot-hardness (Brinell) values vs. temperature for 303 and 303A stainless steels.

Titanium alloys are among the most troublesome materials to machine at practical cutting speeds. The greatest difficulty in machining these alloys stems from the very high cutting temperatures experienced under conditions that are ordinary for most other materials. The extent to which cutting temperatures for titanium alloys are found to exceed those for other metals is shown in Fig. 15.28. The fact that the stainless steel temperatures are greater than those for the AISI 1113 steel may be attributed to the greater specific energy (u) obtained when stainless steel is cut. However, the still-greater titanium temperatures cannot be explained on the basis of a greater specific energy alone.

The very high tool temperatures experienced when machining titanium have been shown to be due to the very low value of ($k\rho C$) for titanium (Shaw 1958). When different materials are machined with the same tool, cutting fluid, and feed rate, eqn 12.42 suggests that

$$\bar{\theta}_t \sim u\left(\frac{Vt}{k\rho C}\right)^{\frac{1}{2}}. \tag{15.1}$$

The speed required to give the same tool temperature will be

$$V_\theta \sim \frac{k\rho C}{u^2}. \tag{15.2}$$

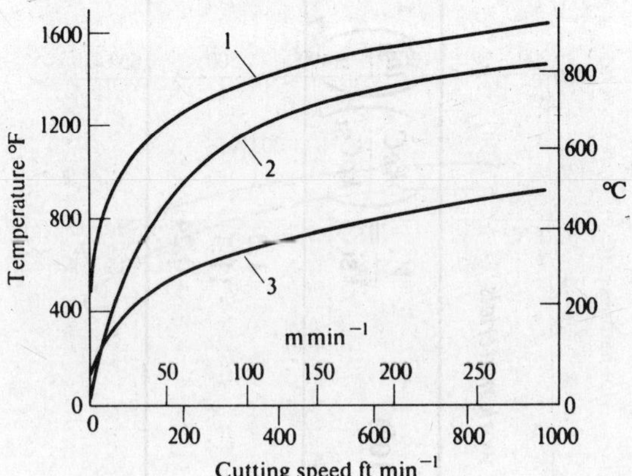

FIG. 15.28. Variation of mean chip–tool interface temperature with cutting speed for (1) titanium alloy, (2) 18–8 stainless steel, (3) AISI 1113 steel. Turning with K2S carbide tool with rake angle of 0° except for item (3) which has 20° rake angle. Undeformed chip thickness = 0.0023 in (0.06 mm).

Table 15.6 gives approximate values of u and $k\rho C$ for steel, aluminum, and a titanium alloy as well as values for the ratio of

$$\frac{V}{V_{St}} = \frac{k\rho C}{k\rho C_{St}} \left(\frac{u_{St}}{u}\right)^2 \tag{15.3}$$

From this comparison it is obvious why aluminum may be machined at very high speeds relative to steel and why titanium must be machined at very much lower speeds than steels.

When a titanium alloy is machined at a speed of about one-fourth the satisfactory speed for steel of about the same hardness, comparable tool life results are obtained. However, the rate of removal (Vbt) will then be only one-fourth that for steel. The basic reason for this is the relatively low $k\rho C$ for titanium. As long as this is the case it is fruitless to try to machine titanium at speeds that are comparable to those used for steel. At the same time the comparison given in Table 15.6 suggests that high-speed machining efforts, where economical should concentrate on aluminum alloys and other materials having high values of $(k\rho C/u^2)$.

One of the characteristics of titanium is to give very inhomogeneous chips relative to strain. This is due to two reasons as illustrated in Fig. 3.21. Fig. 3.21c shows a titanium chip produced at very low speed (1 i.p.m. = 0.03 m min^{-1}).

TABLE 15.6

Relative cutting speeds for equal tool temperature for different work materials

Work material	u in lb in^{-3} (10^{-6})	$k\rho C$(B.T.U in^{-2} °F^{-1})^2 × s^{-1} × 10^6)	$\frac{V}{V_{St}} = \left(\frac{k\rho C}{k\rho C_{St}}\right)\left(\frac{u_{St}}{u}\right)^2$
1020 steel	0.3	27	1.0
75ST aluminum	0.1	34	11.3
140A titanium	0.3	6.4	0.24

This chip fractures periodically and block-wise sliding occurs followed by rewelding as a new fracture occurs. The result is a very inhomogeneous chip with a saw-toothed back.

Figure 3.21a shows a titanium chip produced at relatively high cutting speed of 175 f.p.m. (53 m min^{-1}). At first glance, this chip resembles that of Fig. 3.21c. However, on closer inspection, it is seen that this chip does not involve gross block-wise sliding and rewelding but instead a band of very concentrated shear develops as the higher-magnification photomicrograph in Fig. 3.21b reveals. This is called adiabatic shear since the heat associated with initial slip causes thermal softening and hence more shear strain and more deformation, etc., until the shear band jumps to the next point of weakness. The concentrated shear does not extend clear along the chip but is concentrated in a rather narrow band. The result is again a very inhomogeneous saw-toothed chip but for a different reason than at (a).

Figure 15.29 shows a titanium chip produced at the intermediate speed of 50 f.p.m. (15 m min^{-1}). This chip is a continuous ribbon that is extremely homogeneous relative to shear strain and relatively smooth on the back surface.

It is thus evident that a wide variety of chip types relative to the homogeneity of strain may be obtained with titanium. The adiabatic shear type of chip is more likely to be observed when machining titanium than with other metals due to the unusually low value of $k\rho C$ for titanium.

Nickel

Nickel is a member of the platinum group of the periodic system. Nickel-base alloys are refractory materials used in the gas-turbine industry. Nickel is also an important ingredient of stainless steels where it stabilizes austenite, decreases the rate of strain hardening, provides a negative thermal coefficient of friction, and decreases the tendency for E to decrease with rise in temperature. Nickel is about as abundant in the earth's crust as copper (Ni is 26th in abundance). Nickel is not particularly difficult to reduce from sulfide and silicate ores and is moderate in cost.

Nickel has a somewhat higher specific weight than iron (0.322 lb in^{-3} compared with 0.284 for steel (0.087 N cm^{-3} compared with 0.076 N cm^{-3} for steel). Nickel has a face centered cubic structure and a melting point of 2650 °F (1454 °C) compared with 2793 °F (1534 °C) for iron.

Nickel and nickel-base alloys are relatively difficult to machine because of high tool tip temperatures. This is not apparently due to unusual strength or hardness (pure nickel has about the same hardness as ingot iron $\cong$ 85 kg mm^{-2} Brinell hardness).

Smart and Trent (1975) have found experimentally that when machining nickel, the highest tool face temperature is at the cutting tool tip. Figure 15.30 shows isotherms obtained when machining steel and nickel. These were obtained

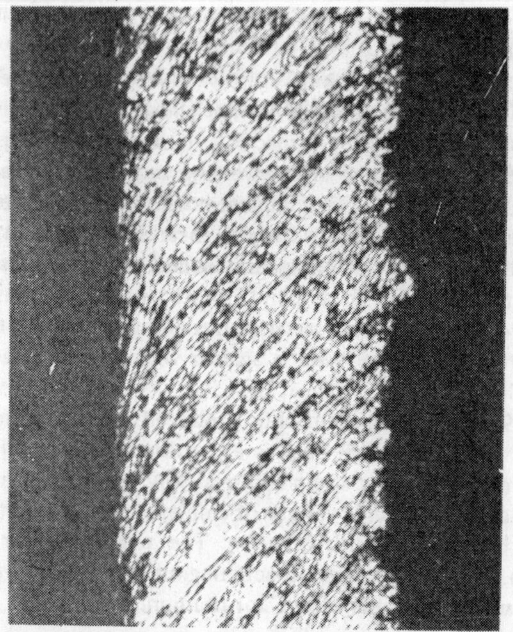

FIG. 15.29. Continuous titanium alloy chip showing homogeneous shear. Cutting speed, 50 f.p.m. (15 m min^{-1}).

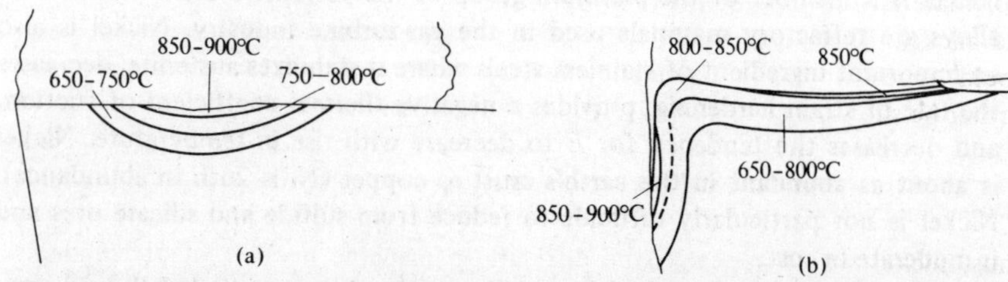

FIG. 15.30. Sections through HSS tools used at ultra-high-speed for a short time (30 s) to reveal isotherms by thermal softening. (a) Low-carbon iron cut at 500 r.p.m. (152 m min^{-1}). (b) CP nickel cut at 175 f.p.m. (53 m min^{-1}). Feed, 0.010 i.p.r. (0.25 mm/rev) and cutting fluid, none in both cases. (After Smart and Trent 1975.)

by the thermal-softening method of Wright and Trent (1973) described in Chapter 12 and clearly show that whereas the maximum temperature occurs above half-way along the contact length for steel, the peak temperature occurs at the tool point or even on the clearance face when similarly machining nickel. This is apparently due to the fact that instead of the chip heating up as it passes along the tool face, it is cooled. Childs (1978) has pointed out that when this occurs the partition coefficient (R_2) between chip and tool will be negative and has suggested an inequality that determines when this will be so. The following derivation produces the equivalent of the Childs inequality to produce a negative R_2 (condition for cooling of chip as it passes over tool face) in terms of the notation used here.

From eqn 12.34, for R_2 to be negative

$$\frac{u_F V t \bar{A}}{J k_3} + \theta_o' < \bar{\theta}_S. \tag{15.4}$$

Substituting from eqn (12.22) for $\bar{\theta}_S$:

$$\frac{u_F V t \bar{A}}{J k_3} + \theta_o' < \frac{R_1 u_S}{J (C\rho)_1} + \theta_o. \tag{15.5}$$

Ignoring the two ambient temperatures θ_o' (tool) and θ_o (work) and rearranging the inequality we have for negative R_2

$$\frac{u_F}{u_S} \frac{V t \bar{A}}{R_1} \frac{(C\rho)_1}{k_3} < 1. \tag{15.6}$$

For ingot iron cutting with a chip of the same proportion m/l or $\bar{A}$ as one for nickel, (Vt) will be somewhat larger for iron than for nickel but so will the value of R_1 (see Fig. 12.16). Therefore, the quantity ($Vt\bar{A}/R_1$) will be about the same for iron and nickel. The value of ($C\rho$) for nickel will be only about 15 percent greater than for steel and hence when cutting with the same tool material (same k_3), the quantity $(C\rho)_1/k_3$ will be about the same for iron and nickel. Therefore, the unusual tendency for nickel to give a negative R_2 would appear to lie in a tendency for u_S to be large relative to u_F. It may further be shown that to a relatively good approximation

$$\frac{u_F}{u_S} \cong r. \tag{15.7}$$

Thus, nickel would appear to be difficult to machine because of a tendency for the maximum tool face temperature to be at the very tip of the tool which in turn is due to the combination of unusually low value of coefficient of tool-face friction or cutting ratio (r) or both.

If eqn (12.45) is used for $\bar{\theta}_S$ instead of eqn (12.22), it still follows that R_2 will tend to become neagtive as u_F/u_S becomes smaller.

It is also evident from eqn (15.6) that a low value of tool conductivity (k_2) will tend to offset a low value of u_F/u_S. Table 15.7 gives values of k_2 for a number of tool materials. This explains in part why HSS tools (low k_3) often give unexpectedly good results when machining nickel-base alloys. A more refractory tool material such as polycrystalline SiC should be even more effective than HSS provided it may be prevented from chipping. A high-cobalt-containing

TABLE 15.7

Thermal conductivities of cutting-tool materials

Tool material	k(BTU in^{-2} sec^{-1} (°F in^{-1})$^{-1}$ (× 10^4))	Relative values
T-1 HSS (18-4-1)	3.7	2.3
Steel grade carbides†	3.6 to 7.8	2.3 to 4.9
C 1 grade carbides‡	7.3 to 12	4.6 to 7.5
Silicon carbide	1.6	1

† The higher the TiC and TaC content, the lower the k.
‡ The higher the Co content, the lower the k.

tungsten carbide containing relatively large amounts of TiC and TaC should be the most effective cemented carbide type for machining refractory nickel-base alloys.

Machining High-Strength Materials

The trend in the aircraft industry during the past twenty-five years has been toward materials of construction that are:

1. Of greater strength.
2. Of lighter weight.
3. Resistant to oxidation, particularly at high temperatures.
4. Exhibit small deformations at high temperatures.
5. Are not brittle at low temperatures.

The materials that have been introduced to meet these requirements have generally been more difficult to machine.

While an ever-increasing quantity of ordinary structural materials will be machined in mass-production industries in the future, an increasing percentage of the productive effort will involve the processing of less conventional materials. This section is concerned with some of the basic principles associated with the machining of materials that might be classified as difficult to machine for one reason or another. Such materials are sometimes also referred to as space-age materials, or high-strength temperature-resistant (HSTR) materials, or corrosion and oxidation resistant materials.

In approaching this problem it is difficult to know which materials to include. It has been decided to consider only those materials that have been used in actual engineering structures and whose usefulness has been demonstrated. The materials to be considered may be classified as follows:

1. Low-alloy and hot-work die steels.
2. Stainless steels.

3. Light-weight metals.
4. High-temperature alloys.
5. Refractory metals.

There are many materials and alloys that could be discussed under each of these headings. However, at the risk of over-simplification only a few representative examples from each category will be considered. Table 15.8 gives the chemical compositions of a number of representative materials.

Next in importance to chemical composition is workpiece structure and hardness. The alloy steels may be hardened to strengths in the vicinity of 300 000 lb in^{-2} (210 kg mm^{-2}) and hardness levels of 550 Brinell. To machine such materials is obviously a very challenging problem which has not been completely solved. Some of the difficult-to-machine materials have low thermal conductivity or low specific heat, both of which give rise to higher cutting temperatures. Other materials are brittle, tend to strain harden badly, have a low Young's modulus of elasticity, or tend to weld to ordinary tool materials. A high coefficient of thermal expansion makes it difficult to maintain dimensional accuracy. A few of the physical properties of the materials of Table 15.8 are given in Table 15.9.

The principal machining characteristics of these difficult-to-machine materials are also encountered when machining ordinary metals, but usually to a lesser degree. Each class of materials has its particular combination of difficulties.

The alloy steels must often be machined after heat treatment; this imposes very large forces on cutting tools, and may result in tool fracture and high cutting temperatures. Feeds and speeds must normally be reduced and tools that are very refractory are called for. Ceramic tools have been particularly successful for this class of work. Hard materials call for rigid machine tools that are stable when operating at high energy levels.

The austenitic stainless steels have somewhat lower conductivities and this causes temperatures to be higher than for ordinary steels. These materials also tend to strain harden appreciably and may even transform to martensite while being cut. This results in relatively high values of cutting force unless the strain in the chip can be reduced. Tools of relatively large rake angle and low tool-chip friction are called for. Going hand-in-hand with large strain-hardening is built-up-edge (BUE) formation and poor finish. The machinability of austenitic stainless steels decreases markedly with carbon content.

Another problem associated with the austenitic stainless steels is the tendency to leave a disturbed subsurface layer and residual stresses in the finished surface. If a second cut must be taken it is then necessary to deform previously strain-hardened material. In order that the weakest portion of the tool (the tip) will not have to engage strain-hardened materials, it is important that the undeformed chip thickness exceed the depth of the disturbed layer. The chips produced when machining austenitic stainless steel are ductile and chip disposal

TABLE 15.8
Chemical compositions of a variety of materials
(After Shaw and Nakayama 1967)

Designation	Composition w/o C	Mn	Si	Cr	Ni	Co	Mo	W	Cb	Ti	Al	Fe	Other
Alloy steels													
AISI 4340	0.4	0.7	0.3	0.8	1.8	–	0.2	–	–	–	–	Bal	–
AISI 610 (Mod H-11)	0.4	0.4	1.0	5	–	–	1.4	–	–	–	–	Bal	0.5V
Stainless steels													
Austenitic													
AISI 302 (18–9)	0.15	2	1	18	9	–	–	–	–	–	–	Bal	–
Martensitic													
AISI 420	0.15	1	1	13	–	–	–	–	–	–	–	Bal	–
Ferritic													
AISI 405	0.08	1	1	13	–	–	–	–	–	–	0.2	Bal	–
Precipitation-hardening													
AISI 631 (17–7 PH)	0.07	0.5	0.3	17	7	–	–	–	–	–	1	Bal	–

Light-weight metals													
Ti–8A1–1Mo–1V							1			Bal	3		IV
Zirconium													100Zr
Beryllium													100Be
High-temperature alloys													
Iron-base													
AISI 662 (Discaloy)	0.04	0.9	0.8	13.5	26		2.75			1.75	0.07	Bal	–
Nickel-base													
AISI 685 (Waspaloy)	0.07	0.1	0.1	20	Bal	13.5	4.45	–	–	3	1.4	0.75	–
Cobalt-base													
AISI 671 (S 816)	0.42	1	0.45	20	20	43.5	4	4	4	–	–	–	–
Refractory metals													
Columbium									100				
Molybdenum							100						
Tatalum													100Ta
Tungsten								100					

TABLE 15.9

Mechanical and thermal properties of a variety of materials
(After Shaw and Nakayama 1967)

Material	σ_u kg cm^{-2} ($\times 10^4$)	σ_y kg cm^{-2} ($\times 10^4$)	$E \times 10^{-6}$ kg cm^{-2}	k cal cm^{-2} cm^{-1} °C^{-1} s^{-1}	ρ g cm^{-3}	ρC cal cm^{-3} °C^{-1}	$k\rho C$ cal^2 cm^{-4}(°C)$^{-2}$ s^{-1}	α °C^{-1} $\times 10^6$	m.p. °C
Alloy steels									
AISI 4340	2.02	1.80	2.1	0.089	7.8	0.83	0.074	11.3	1500
AISI 610	2.18	1.69	2.15	0.069	7.8	0.90	0.062	11.9	
Stainless steels									
Austenitic									
AISI 302	0.63	0.25	1.95	0.039	7.8	0.94	0.037	17.3	1400
Martensitic									
AISI 420	1.75	1.41	2.0	0.060	7.8	0.86	0.052	10.3	
Ferritic									
AISI 405	0.70	0.62	2.0	0.064	7.8	0.86	0.055	10.8	1480
Precipitation-hardening									
AISI 631	1.86	1.83	2.1	0.040	7.8	0.94	0.038	11.2	1400
Light-weight alloys									
Ti–8A1–1Mo–IV	1.12	1.05	1.1	0.017	4.5	0.59	0.010	8.5	1760
Zr	0.56	0.49	0.96	0.039	6.5	0.44	0.017	5.85	1850
Be	0.35	0.53	3.1	0.35	1.8	0.81	0.28	15.0	1280

High-temperature alloys									
Iron-base									
AISI 662	1.02	0.75	1.99	0.038	8.1	0.97	0.037	15.6	
Nickel-base									
AISI 685	1.34	0.91	2.15	0.022	8.23	1.02	0.023	12.2	
Cobalt-base									
AISI 671	0.98	0.46	2.48	0.029	8.68	0.77	0.022	13.4	
Refractory metals									
Columbium	0.28	0.21	1.05	0.12	8.6	0.60	0.072	7.4	2468
Molybdenum	0.67	0.53	3.30	0.32	10.2	0.67	0.21	5.9	2610
Tantalum	0.37	0.28	1.90	0.13	16.6	0.56	0.073	6.5	2996
Tungsten	0.84	–	3.66	0.28	19.3	0.64	0.18	4.6	3410

σ_u = ultimate tensile strength
σ_y = yield stress
E = Young's modulus of elasticity
k = thermal conductivity
ρ = specific weight

ρC = volume specific heat
$k\rho C$ = cal^2 cm^{-4} °C^{-1} s^{-1}
α = coefficient of linear expansion
m.p. = melting point

or chip breaking is apt to be a problem. The large thermal coefficient of expansion and the dimensional change with phase-transformation makes it difficult to machine to close tolerances. An increased rake angle helps both of these problems.

Best results are therefore obtained with austenitic steels when they are machined at moderate values of feed using tools of high rake angle and having a low tendency to weld to stainless steel.

Ceramic tools are not well suited to these materials. Cast-iron grades of carbide (straight tungsten carbide) give best results and the tool hardness should be as high as possible, consistent with edge chipping, in order to decrease the tendency for weld formation.

The martensitic and ferritic stainless steels do not have as great a tendency to strain harden and their machining characteristics are similar to those of AISI 4340 or a hot-work die steel (AISI 610) at the same hardness level. The precipitation-hardened stainless steels can be austenitic or martensitic. The AISI 631 variety considered here is normally partly austenitic when used. However, by heat treatment it may be machined in either the austenitic or martensitic states. If machined in the austenitic state it behaves like an austenitic stainless steel and strain-hardens greatly. If machined in the martensitic state it behaves like a martensitic stainless steel and the problems are associated with high cutting forces as with a hardened alloy steel. AISI 631 is best machined in the semi-austenitic state and under such a condition is relatively easy to machine.

The titanium alloys are difficult to machine for several reasons:

1. Low thermal conductivity and specific heat gives high cutting temperatures.
2. Strong tendency for chips to weld to tool and cause attritious wear.
3. Low strain in the chip which yields an unusually high shear angle and hence a high chip velocity and small contact length between chip and tool.
4. Low value of Young's modulus.

The strength and strain hardening characteristics of titanium alloys are about the same as those for steel of the same hardness. Hence, high cutting forces are not a major problem. The main problem is associated with an unusually high cutting temperature due to items 1 and 3 above. The best solution is to use a good coolant and to lower the cutting speed. A water base fluid is preferable due to the tendency for titanium chips to burn.

Despite the strong tendency for welding, a large unstable BUE (such as is formed with steel at low speed) is not produced. Surface finish at low speed is thus not a problem. Clearance angles should be 10° instead of 5° due to the welding tendency when titanium is machined.

Ceramic tools form particularly strong bonds to titanium and hence have a poor life. Cast-iron-cutting grades of carbide generally perform better than grades containing TiC, since they tend to weld less to the work material.

Titanium alloys are relatively expensive and hence trial-and-error experimental work is costly.

The less refractory titanium alloys tend to form chips with inhomogeneous strain. Due to the poor thermal properties the heat developed on a given shear plane is not dissipated and softens the metal causing further shear deformation at the same point. With the less refractory alloys (such as commercially pure titanium) this procedure can continue to the point where the chip consists of alternate bands of high and low strain. This, of course, gives rise to a periodic fluctuation in cutting force and can lead to a forced vibration.

Pure titanium has a hexagonal close packed (HCP) structure at room temperature. At 890 °C an allotropic transformation occurs where the low-temperature HCP structure (α) changes to a body centred cubic structure (β) similar to that found in iron at room temperature. Titanium alloys may be all α, all β, or a mixture of the two. The all-β-alloys are generally more difficult to machine than the mixed structure, and all alloys are more difficult to machine than the commercially pure material.

Zirconium is next to titanium in the periodic table and has essentially the same machining characteristics as titanium.

Beryllium has an unusually low density (0.065 lb/in^3 or 1.80 g cm^{-3}) for its Young's modulus (44×10^6 lb in^{-2} or 31×10^3 kg mm^{-2}) and tensile strength and therefore is sometimes used where a high strength-to-weight ratio is desired. Its main characteristics of importance to machining are its toxicity and brittleness due to its hexagonal close packed structure. Machining dust of beryllium is extremely harmful if ingested into the lungs and special ventilation and clothing procedures must be followed. In addition, the metal is expensive.

The brittleness of beryllium gives rise to discontinuous cutting and hence a tendency for chatter. There is also a tendency for sub-surface cracks to develop in machined beryllium surfaces which may be highly undesirable for certain space applications. Otherwise beryllium is not particularly difficult to machine and can be turned at 200–300 f.p.m. (60–90 m min^{-1}) using ordinary carbide tools.

Austenitic stainless steels are oxidation-resistant only to about 1200 °F (650 °C). For higher operating temperatures the so-called high-temperature alloys must be used. The nickel-base alloys, which are presently the most stable against oxidation, may be used to temperatures of 1850 °F (1010 °C). In addition to stability against oxidation at high temperatures the high-temperature alloy must have high structural strength at elevated temperatures. The cobalt alloys exhibit the greatest strength at operating temperatures in the vicinity of 1800 °F (980 °C). The iron-base high-temperature alloys are not as effective as the nickel-base and cobalt-base alloys with respect to oxidation stability and high-temperature strength, but are more easily machined and less expensive.

The iron-base high-temperature alloys are austenitic and hence behave like austenitic stainless steel in machining but to an even greater degree. Therefore, essentially all of the discussion pertaining to austenitic stainless steels applies to these materials as well. The best structure for machining is one produced by annealing, cold-working, and finally stress-relieving. This structure machines better than either an annealed or solution-treated material, since it strain hardens to a far smaller degree than the former and is less hard than the latter.

Work-hardening, welding and BUE formation, and low thermal properties are the chief difficulties when machining the high-temperature alloys.

High rake angles should be used due to the strong tendency for strain hardening. Since larger rake angles may generally be used with HSS tools than with carbide, HSS tools frequently give the best results when machining high-temperature alloys. In addition, low values of speed and feed should be used and machine tools should be as rigid as possible. Cutting fluids containing active chlorine and sulfur generally aid in the machining of high-temperature alloys. As the maximum operating temperature of a high-temperature alloy rises, the difficulty of machining increases. The nickel-base alloys presently have the highest operating temperatures and they are among the worst materials to machine. They tend to strain harden very badly and give rise to very large values of shear stress at machining strain rates. Although all materials tend to strain harden in the hot working region when strained at high rates, this is particularly true of the nickel and cobalt high-temperature alloys. This strong tendency to strain harden gives rise to deep grooves at both edges of a chip. The nickel-base alloys machine best in the solution-treated state, since they then strain harden less, but this best structure is far from satisfactory. Also, wrought materials tend to machine better than cast structures. The main difficulties in machining the nickel- and cobalt-base materials are due to:

1. Tendency for maximum tool-face temperature to be close to tool tip (particularly for Ni).
2. High work-hardening rates at machining strain rates leading to high machining forces.
3. Strong tendency to weld to the tool and to form a BUE.
4. Low thermal properties leading to high temperatures.

The cobalt high-temperature alloys have machining characteristics that parallel the nickel-base alloys but are not quite so difficult.

The high-temperature alloys and high-strength steels should not be machined with tools having flank wear values that are as high as those normally employed in cutting steel. Tools should be taken from service when flank wear reaches 0.25–0.38 mm instead of the usual 0.75 mm.

The refractory metals (Cb, Ta, Mo, W) generally have melting points in excess of 2200 °C. All four metals have low values of volume specific heat. In addition, columbium and tantalum have low values of thermal conductivity. This results in high machining temperatures for all four materials, and particularly for columbium and tantalum. Columbium and tantalum are quite ductile and give rise to a large and troublesome BUE. They are usually not machined with carbide tools since the speeds employed are usually not high enough to avoid BUE formation.

Molybdenum and tungsten on the other hand are brittle and give rise to discontinuous chips. Like beryllium it is difficult to machine these materials without leaving small cracks in the finished surface. HSS tools are usually not used for machining molybdenum or tungsten.

Tungsten which is the most refractory of the group is used for rocket nozzles. Additions of 10 percent silver infiltrated into a powder compact of tungsten greatly improves its machinability and at the same time cools the surface of the nozzle in use by ablation of the silver. Tungsten has been successfully machined at low temperatures (−40 °F, −40 °C) which tends to offset its tendency to give high tool-face temperatures due to its high strength and poor thermal properties. It has been discovered that dispersion of 2 percent thorium oxide in a tungsten matrix triples the machinability of tungsten without changing its refractory character.

Specific recommendations concerning feeds and speeds for use in machining high-strength materials are to be found in many publications and hence no attempt will be made here to give more than a general impression concerning the relative machinability of the various materials. The results obtained in any particular case depend not only on the chemistry of the workpiece but also on its structure (hardness), the tool materials and geometry used, and the rigidity and condition of the machine tool. Tables of recommended operating conditions should always be looked on as being approximate and as points of departure for further study.

Table 15.10 gives optimistic values of cutting speed for turning and drilling. The turning values are based on a depth of cut of 2.5 mm and a feed of 0.25 mm/rev with the exception of the values for Cb, Ta, and Mo which are for a depth of cut of 1.25 mm and a feed of 0.125 mm/rev. The drilling speeds are for high-speed steel drills (M-2 or M-10) having a feed consistent with the drill diameter from the point of view of chip disposal. For this class of materials the feed should be about 1 percent of the drill diameter (i.e. a 0.2 in drill should have a feed of about 0.002 i.p.r. or a 5 mm drill should have a feed of 0.05 mm/rev). Drills made of T-15 will yield results that are about 30 percent greater than those for the more usual high-speed steels (M-2 or M-10).

The expected tool life in Table 15.10 is about $\frac{1}{2}$ hr in turning, or 50 holes of two-diameters depth in drilling.

The amount of nickel in a nickel-base high-temperature alloy is very important. For example, the alloy of Table 15.10 has a nickel content of about 60 percent and a recommended carbide tool speed of 13 m min^{-1}. This speed would be increased to about 20 m min^{-1} for work containing about 50 percent Ni and to 26 m min^{-1} for work of 45 percent nickel. This shows the extremely important role of workpiece chemistry, which is the most important variable relative to machinability.

TABLE 15.10

Recommended machining speeds for a variety of materials

Material	Brinell hardness	Turning speed HSS (f.p.m.)	Turning speed HSS ($m\ min^{-1}$)	Turning speed Carbide (f.p.m.)	Turning speed Carbide ($m\ min^{-1}$)	Drilling speed (f.p.m.)	Drilling speed ($m\ min^{-1}$)
Alloy steels							
AISI 4340	300	80	25	500	165	70	23
AISI 610							
Stainless steels							
Austenitic							
AISI 302	160	95	30	300	100	30	10
Martensitic and precipitation-hardening							
AISI 420							
AISI 631	200	100	35	450	150	30	25

Light-weight metals							
Ti–8A1–1Mo–1V	320	50	15	150	50	40	13
Zirconium		200	65	450	150	80	25
Beryllium		90	30	300	100	40	13
High-temperature alloys							
Iron-base							
AISI 662	200	40	13	125	40	20	6
Nickel-base							
AISI 685	250	15	5	40	13	15	5
Cobalt-base							
AISI 671	200	25	8	75	25	30	10
Refractory metals							
Columbium	150	50	15	–	–	75	25
Molybdenum	200	–	–	300	100	80	25
Tantalum	150	50	15	–	–	40	13
Tungsten	250	–	–	200	65	150†	50

† Tungsten carbide

The next most important variable is hardness. High-speed steel tools should not be used to machine martensitic low-alloy steels or hot-work die steels at hardness levels above 450 Brinell. For very hard steels (550 Brinell) the feed and depth of cut should be reduced to 0.125 mm and 0.0125 mm/rev respectively and the carbide cutting speed dropped to about 50 m min^{-1}. In general as workpiece hardness increases the feed and depth of cut should be reduced to avoid tool chipping or breakage. Ceramic tools are well suited for turning high-hardness alloy steels at light feeds.

In general, carbide tools are not recommended for use in machining columbium or tantalum due to the strong tendency for these materials to form welds at the low cutting speeds that must be used with these materials. On the other hand high-speed steel tools are not recommended for use in machining molybdenum or tungsten except in drilling. Even in drilling it is highly preferable to use carbide in connection with tungsten.

Tapping of all of these materials is normally done at 1.5–8 m min^{-1}.

References

Childs, T. H. C. (1978). *Int. J. Wear* **50**, 321.

Cook, N. H., Finnie, I., and Shaw, M. C. (1954). *Trans. Am. Soc. mech. Engrs* **76**, 153.

DeSalvo, G. J. and Shaw, M. C. (1969). In *Advances in machine tool design and research*, p. 961. Pergamon Press, Oxford.

Ernst, H. (1938). In *Machining of metals.* Am. Soc. for Metals, Metals Park, Cleveland, Ohio.

Field, M. (1963). In *International Research in Production Engineering,* p. 188 Am. Soc. Mech. Engrs, New York.

Gensamer, M., Pearsall, E. S., Pellini, W. S., and Low, J. R. (1942). *Trans. Am. Soc. Metals* **30**, 1003.

Japanese Working Group on Machinability (1968). *J. Jap. Soc. Precs Engrs* **34**, 680 (1st report); **35**, 169 (second report—1969); **35**, 227 (third report—1969).

Machining Data Handbook (1980). 3rd edn Vols I and II. Compiled by U.S. Machinability Data Center. Available from Metcut Research Associates, Inc., 980 Rosslyn Drive, Cincinnati, OH 45209.

Metals Handbook (1964). 8th edn, Vol. II. American Society for Metals, Metals Park, Ohio. (Vol I 1961.)

Opitz, H. (1963). In *International research in production engineering*, p. 107. American Society of Mechanical Engineers, New York.

Opitz, H. and Koenig, W. (1965). *Ind. Anz.* **87**, 26 (part I); **87**, 43 (part II); and **87**, 51 (part III).

Schaller, E. (1962). Dissertation T. H. Aachen.

Schwerd, F. (1932). *Z. Ver. deutscher Ing.* **76**, 1257.

Sejournet, J. (1954). *Engineering* **177**, 463.

Sejournet, J. (1962). *Lubr. Engng* **18**, 324.

Shaw, M. C. (1958). *Technische Mitteilungen.* Essen.

Shaw, M. C. (1967). *Machinability.* Report 94 p. 1. The Iron and Steel Institute, London.

Shaw, M. C. and Macks, E. F. (1949). *Analysis and lubrication of bearings.* McGraw-Hill, New York.

Shaw, M. C. and Nakayama, K. (1967). *Annals of CIRP* **15**, 45. International Institution for Production Engineering Research, Paris.

Shaw, M. C., Smith, P. A., and Cook, N. H. (1961). *Trans. Am. Soc. mech. Engrs* **83**, 163.

Shaw, M. C., Usui, E., and Smith, P. A. (1961). *Trans. Am. Soc. mech. Engrs* **83**, 181.

Smart, E. F. and Trent, E. M. (1975). *Int. J. Prod. Res.* **13**, 265.

Sturney, A. C. (1925). *Proc. Instn Mech. Engrs* **1**, 141.

Tipnis, V. A. and Joseph, R. A. (1971). *J. Eng. Ind. Trans. ASME.* **93**, 571.

Trent, E. M. (1963). In *International Research in Production Engineering,* p. 161. ASME, New York.

Vilenski, D. and Shaw, M. C. (1970). *Annals of CIRP* (International Institution for Production Engineering Research, Paris) **18**, 623.

Wright, P. K. 1976). *J. Aust. Inst. Metals* **21**, 34.

Wright, P. K. and Trent, E. M. (1973). *J. Iron Steel Inst.* **211**, 364.

16

COMPLEX TOOLS

Thus far only two-dimensional cutting operations have been considered in which the cutting edge is perpendicular to the velocity vector. Such cutting operations are referred to as orthogonal. While most commercial tools are three-dimensional, there are some practical orthogonal cases. These include some finishing planer cuts, surface broaching, the lathe cut-off operation, and some plain-milling operations. Representative three-dimensional cutting tools are a conventional lathe tool, a face-milling cutter, and a twist drill. The geometrical aspects of these more complex tools are considered in this chapter.

The simplest three-dimensional cutting tool is a straight cutting edge that is inclined to the velocity vector (see Fig. 16.1b). Inclination i is the distinguishing feature of all three-dimensional cutting operations and represents the point of departure from orthogonal cutting. Inclination significantly alters the chip flow direction and hence the performance of a tool.

Two distinct rake angles are frequently distinguished:

1. The normal rake angle α_n is the angle measured from a normal to the finished surface in a plane perpendicular to the cutting edge (plane 0A in Fig. 16.2). This is the angle most frequently specified and most easily measured, but it is not the rake angle of greatest significance. With the clearance angle, α_n determines the amount of material available at the tool point to support the load and absorb the heat generated during cutting. A greater included angle at the tool point (smaller α_n) is generally required for cutting edges subjected to shock (i.e. milling or other interrupted cuts).

2. The velocity rake angle α_v unfortunately sometimes referred to as the true rake angle, is the angle measured from a normal to the finished surface

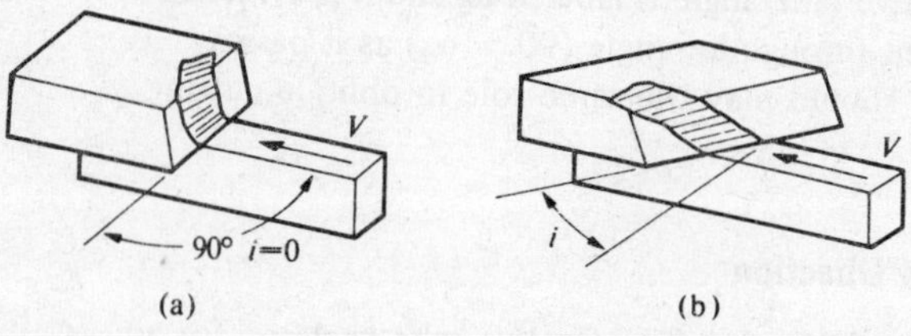

FIG. 16.1. Comparison of orthogonal and oblique cutting operations. (a) Orthogonal cut. (b) Oblique cut with inclination angle i.

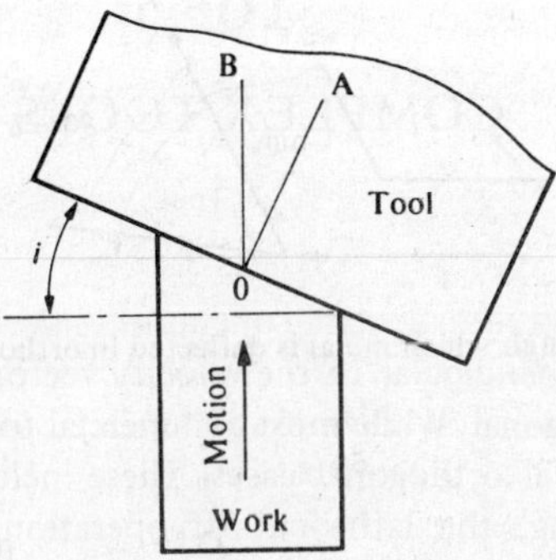

Fig. 16.2 Plan view of cutting tool with inclination i.

in a plane containing the cutting-velocity vector (plane 0B in Fig. 16.2). This particular rake angle is insignificant and is mentioned here only because it has received so much attention from previous investigators.

Angle α_n may be expressed in terms of α_v as follows:

$$\tan \alpha_v = \frac{\tan \alpha_n}{\cos i} . \tag{16.1}$$

While the normal and velocity rake angles are easily measured, they are not of fundamental significance in the cutting process.

In Fig. 16.3 the complement to the rake angle in orthogonal cutting $(90 - \alpha)$ is seen to measure the angle through which the metal is deflected by the tool. In general, the greater this angle, the greater will be the cutting work required. It is unfortunate that the importance of this more natural angle was not originally recognized and the rake angle so defined. In comparing three-dimensional cutting data with orthogonal data, an effective rake angle comparable to that used in the two-dimensional case should be adopted. The effective rake angle for the oblique tool of Fig. 16.4 should be related to the cutting-velocity vector D0 and the chip-flow direction 0C, and measured in the plane of D0 and 0C. The effective rake angle is labeled α_e and it is evident in Fig. 16.4 that the metal is deflected through an angle $(90 - \alpha_e)$ as it passes across the oblique tool face. Hence α_e should play the same role in oblique cutting as α does in orthogonal cutting.

Chip-Flow Direction

In order to determine the effective rake angle α_e for any tool with inclination, it is necessary to know the direction the chip takes as it crosses the tool face. This is most effectively specified by the angle η_C between the chip-flow direction and the normal to the cutting edge, in the plane of the tool face (Fig. 16.5).

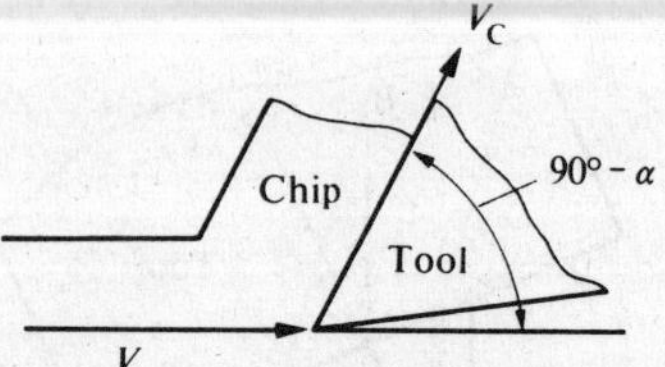

FIG. 16.3. Angle through which metal is deflected in orthogonal cutting (90 – α).

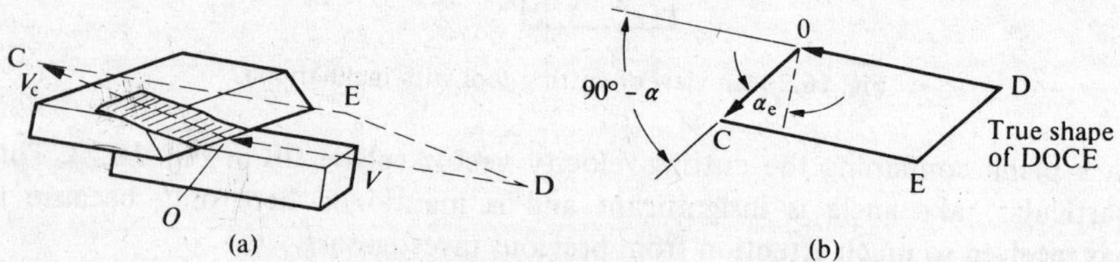

FIG. 16.4. Effective rake angle for oblique cutting tool.

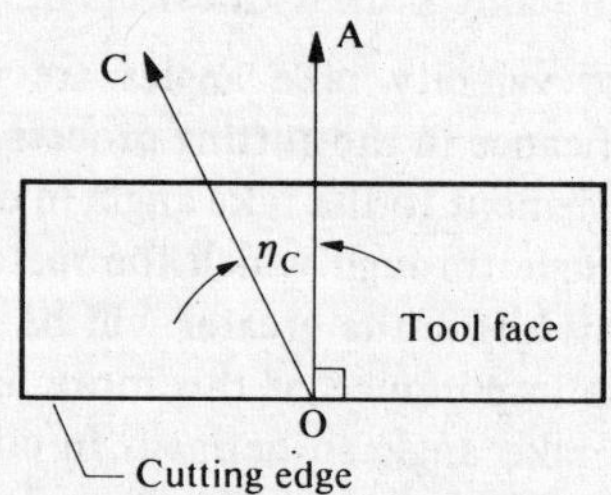

FIG. 16.5. View of tool normal to tool face showing manner of specifying chip flow direction 0C.

Stabler (1951) has shown that when angle η_C is known, the effective rake angle α_e may be determined as follows:

$$\sin \alpha_e = \sin \eta_C \sin i + \cos \eta_C \cos i \sin \alpha_n. \tag{16.2}$$

Angle η_C may be directly measured in a number of ways, including:

1. Directly with a protractor as the chip flows across the tool face.
2. From chip-width measurements.

The second method has been found the most precise, particularly at higher cutting speeds and hence will be described in detail. In Fig. 16.6 it is evident that for any arbitrary chip-flow direction η_C

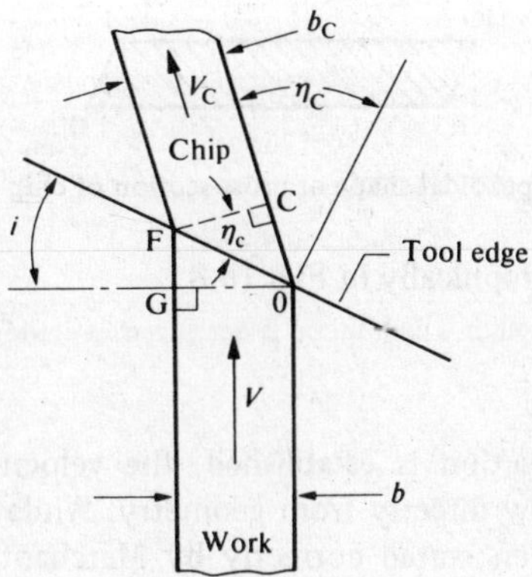

FIG. 16.6. Plan view of oblique cutting tool.

$$\cos \eta_C = \frac{b_C}{b/\cos i} \tag{16.3}$$

where b_C is the width of the chip (FC in Fig. 16.6); b is the width of the work (0G in Fig. 16.6). Thus chip flow direction η_C may be determined simply from measurements of chip and work width. In making the chip-width measurement (b_C), it is preferable to use a measuring microscope to determine distance 0F on the face of the chip, (Fig. 16.7), rather than a micrometer, which will measure horizontal distance 0K.

Stabler (1951) reported that angle η_C was approximately equal to the inclination angle i for a variety of tool and work materials, rake angles, and speeds. In the following discussion this relationship will be referred to as Stabler's rule of chip flow. Although Stabler's rule is valid to a first approximation, it has been found that η_C increases relative to inclination angle i when:

1. Rake angle α_n is decreased.
2. A more efficient cutting fluid is used.
3. The friction characteristics of the metal-cut improve (cold-drawn steel as opposed to soft aluminum for example).

The physical significance of Stabler's rule may be appreciated by reference to eqn 16.3. Here it is evident that when η_C is equal to i, this is the same as taking chip width b_C the same as work width b. Thus Stabler's rule merely means that the chip will take a direction relative to the cutting edge such that there is no change in width as the metal crosses the cutting edge. Negligible side flow is not only a good approximation in orthogonal cutting, it also represents a first approximation for a tool having inclination.

When eqn (16.2) is simplified by use of Stabler's rule, eqn (16.4) is obtained

$$\sin \alpha_e = \sin^2 i + \cos^2 i \sin \alpha_n. \tag{16.4}$$

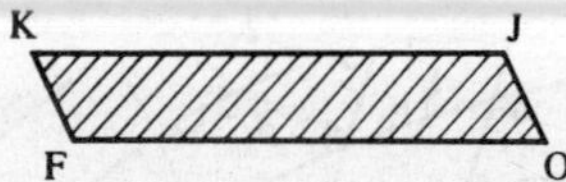

FIG. 16.7. Trapezoidal shape of cross-section of chip in Fig. 16.6.

This equation is shown graphically in Fig. 16.8.

Velocity Relations

Once the chip-flow direction is established, the velocity and strain relations for oblique cutting follow directly from geometry. While most of the equations of this section have been stated correctly by Merchant (1944), they will be briefly reviewed as a matter of convenience. As in orthogonal cutting, there is a shear plane when cutting with an oblique tool. This shear plane will contain the cutting edge and will rise from the finished surface when proceeding into the workpiece in front of the cutting edge. The direction of the shear plane is most conveniently defined in terms of a normal shear angle (ϕ_n) measured in a plane normal to the cutting edge, i.e. in a vertical plane along 0A in Fig. 16.2. In Fig. 16.9 a section view of a partially formed chip is shown in a plane normal to the cutting edge. The angles, ϕ_n and α_n, are as shown and a relationship between these two angles and the chip and workpiece thicknesses may be derived just as in orthogonal cutting (Chapter 3).

$$\tan \phi_n = \frac{(t/t_C)\cos \alpha_n}{1-(t/t_C)\sin \alpha_n} \tag{16.5}$$

where t and t_C are the undeformed chip thickness and chip thickness, respectively. If the back of the chip is very rough, it is usually more precise to use a cutting ratio rather than a chip-thickness ratio. This may be done by applying the volume-continuity relationship of plasticity.

$$lbt = l_C b_C t_C \tag{16.6}$$

where l and l_C are corresponding work and chip lengths. From eqn (16.6)

$$\frac{t}{t_C} = \frac{l_C b_C}{lb}. \tag{16.7}$$

The direction of shear flow is given by a line in the shear plane that makes an angle η_S with a line that lies in the shear plane and is normal to the cutting edge (see Fig. 16.10). This shear-flow angle η_S is

$$\tan \eta_S = \frac{\tan i \cos(\phi_n - \alpha_n) - \tan \eta_C \sin \phi_n}{\cos \alpha_n}. \tag{16.8}$$

In orthogonal cutting the shear strain γ is

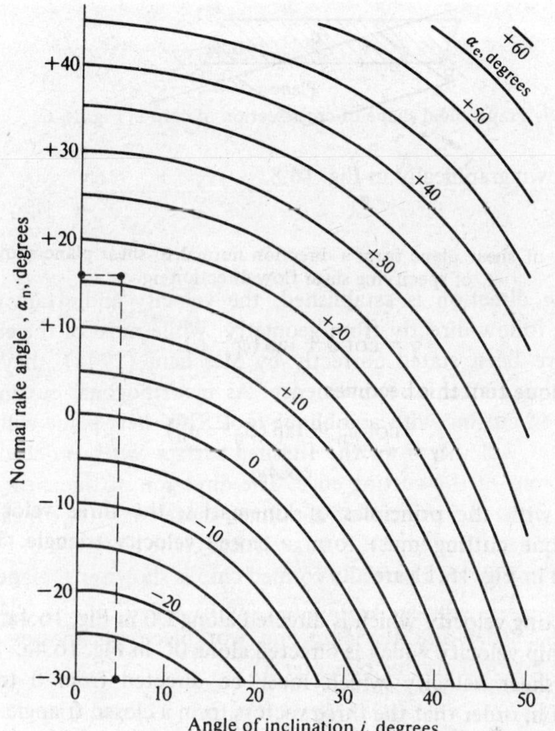

FIG. 16.8. Chart for obtaining the effective rake angle α_e for any combination of inclination i and normal rake α_n angles.

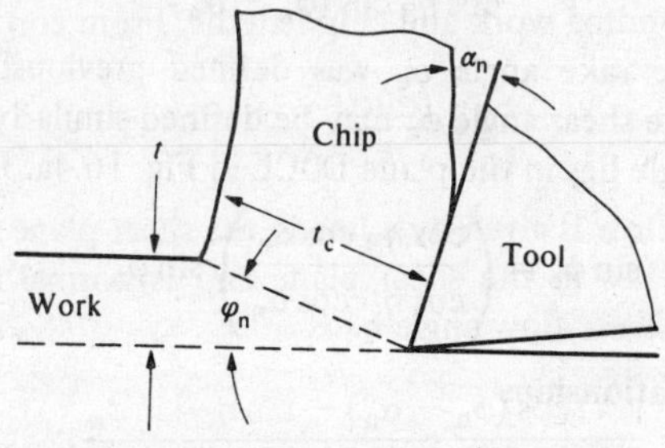

FIG. 16.9. Section of partially formed chip in plane normal to cutting edge.

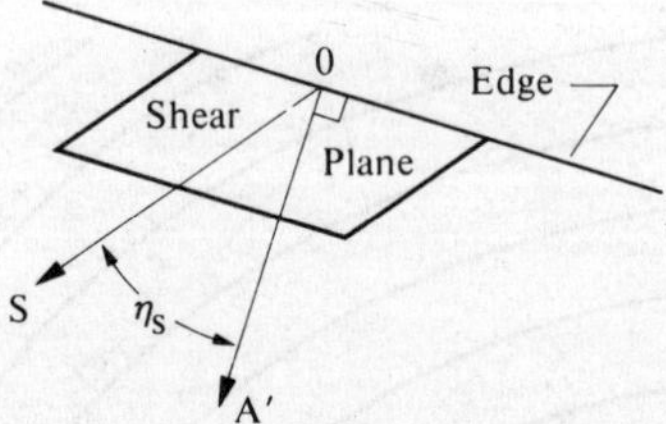

FIG. 16.10. View of shear plane from a direction normal to shear plane showing manner of specifying shear flow direction η_S.

$$\gamma = \cot\phi + \tan(\phi - \alpha) \tag{16.9}$$

while for an oblique tool this becomes

$$\gamma = \frac{\cot\phi_n + \tan(\phi_n - \alpha_n)}{\cos\eta_S}. \tag{16.10}$$

In accordance with the principles of kinematics, the three velocity vectors involed in oblique cutting must form a closed velocity triangle. These three velocities shown in Fig. 16.11 are

1. V, the cutting velocity which is directed along D0 in Fig. 16.4a.
2. V_C, the chip velocity which is directed along 0C in Fig. 16.4a.
3. V_S, the shear velocity which must be directed from 0 toward E in Fig. 16.4a in order that the three vectors form a closed triangle.

The relationship between these velocities may be found readily by use of a three-dimensional model of the velocity triangle to give

$$\frac{V_C}{V} = \frac{\cos i \sin\phi_n}{\cos\eta_C \cos(\phi_n - \alpha_n)} \tag{16.11}$$

$$\frac{V_S}{V} = \frac{\cos i \cos\alpha_n}{\cos\eta_S \cos(\phi_n - \alpha_n)}. \tag{16.12}$$

Just as an effective rake angle α_e was defined previously for an oblique cutting tool, an effective shear angle ϕ_e may be defined similarly (see Fig. 16.11). This effective shear angle lies in the plane D0CE in Fig. 16.4a. From Fig. 16.11

$$\sin\phi_e = \left(\frac{\cos\eta_S \cos\alpha_e}{\cos\eta_C \cos\alpha_n}\right)\sin\phi_n \tag{16.13}$$

Shear and Friction Relationships

As in orthogonal cutting the shear and friction processes occur simultaneously and in close proximity. In order that static equilibrium of forces be satisfied it is necessary that

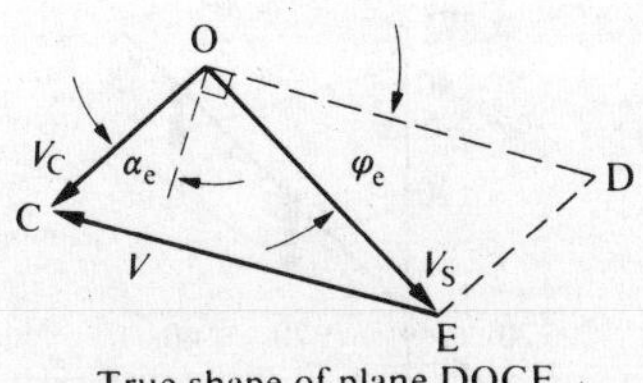

FIG. 16.11. Velocity components shown in plane DOCE in Fig. 16.4.

$$R' = R \tag{16.14}$$

where R' is the resultant force on the shear plane and R is the resultant force on the tool face. In accordance with continuity, eqn (16.8) must also be satisfied. In order to combine eqns (16.8) and (16.14) the relationships between force and velocity for both the shear and friction processes must be known. For this, either of the following may be assumed:

1. That force and velocity vectors are collinear for both the friction and shear cases, or
2. That in either or both cases the vectors are not collinear.

If the first assumption (which is intuitively appealing) is adopted and, at the same time Stabler's rule is employed, then eqns (16.8) and (16.14) lead directly to the following shear angle relationship.

$$\phi_n = \Psi(\beta, i, \alpha_n) \tag{16.15}$$

where Ψ is some function of the friction angle (β), inclination angle (i), and normal rake angle (α_n). However, eqn (16.15) is not found to be in agreement with experiment. This can only mean that assumption **1** and/or Stabler's rule is invalid.

Since in many cases Stabler's empirical rule is found to represent a good approximation (for example see Fig. 16.12), the validity of assumption **1** is questioned despite its intuitive appeal.

Force Relations

In order to evaluate a three-dimensional cutting operation, it is necessary to determine three mutually perpendicular components of force. Consider the measured force components to be those along the x-, y-, and z-axes of Fig. 16.13. Here y extends along the work surface parallel to cutting velocity V; x is along the work surface normal to V, while z is in the vertical direction. In order to find the component of the resultant force in the plane of the tool face, it is convenient to change to the x', y', z' right-hand orthogonal set in Fig. 16.13 where x' is along the cutting edge, y' is normal to the tool face, and z' is normal

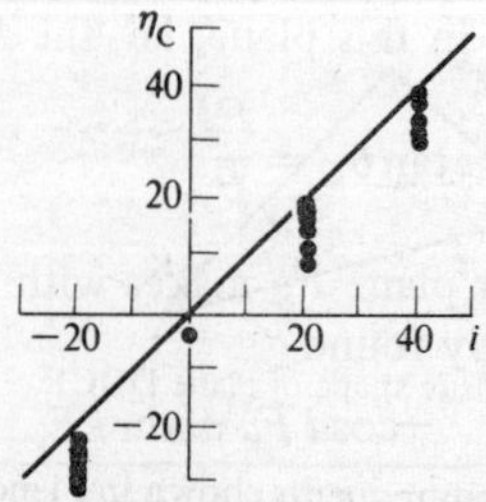

FIG. 16.12. Data illustrating degree of approximation associated with Stabler's rule ($\eta_C = i$). Material, leaded steel; tool 18–4–1 HSS; cutting speed, 100 f.p.m. (30.5 m min^{-1}); depth of cut, 0.100 in (2.54 mm); back rake, –20, 0, 20, 40; side rake –20, 0, 20, 40; side-cutting edge, 0; feed, 0.0026, 0.0052, 0.0104 i.p.r. (0.07, 0.18, 0.26 mm/rev).

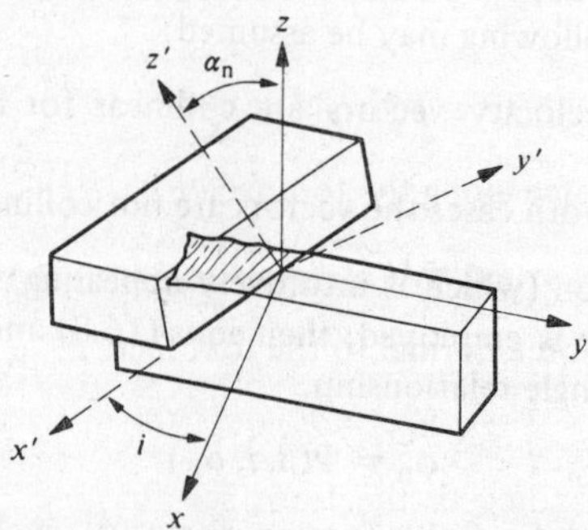

FIG.16.13. Sketch of inclined cutting tool showing axes of reference.

to the cutting edge, but in the plane of the tool face. If a_1, b_1, and c_1 are the direction cosines of the x'-axis relative to the x-, y-, z-coordinate system, and a_2, b_2, c_2, and a_3, b_3, c_3, the direction cosines of y' and z', respectively, then the transformation may be readily made as follows:

$$\begin{aligned} F'_x &= a_1F_x + b_1F_y + c_1F_z \\ F'_y &= a_2F_x + b_2F_y + c_2F_z \\ F'_z &= a_3F_x + b_3F_y + c_3F_z \end{aligned} \tag{16.16}$$

where

$$\begin{aligned} a_1 &= \cos i & b_1 &= -\sin i & c_1 &= 0 \\ a_2 &= \cos\alpha_n \sin i & b_2 &= \cos\alpha_n \cos i & c_2 &= \sin\alpha_n \\ a_3 &= \sin\alpha_n \sin i & b_3 &= -\sin\alpha_n \cos i & c_3 &= \cos\alpha_n \end{aligned}$$

If δ_c is the angle that the force in the plane of the tool face F_C makes with the normal to the cutting edge in this plane, i.e. the z'-axis in Fig. 16.13 then

$$\tan \delta_C = \frac{F_{x'}}{F_{z'}} \tag{16.17}$$

The angle the force in the shear plane F_S makes with the normal to the cutting edge in this plane may be similarly found

$$\tan \delta_S = \frac{-\cos i\, F_x + \sin i\, F_y}{\cos \phi_n \sin i\, F_x + \cos \phi_n \cos i\, F_y + \sin \phi_n\, F_z} \tag{16.18}$$

where in this case the positive z'-axis is in the direction opposite that used in the derivation of eqn (16.17).

There are several forces of interest in an oblique cutting process. These include the following:

1. The resultant force acting on the tool face R, or on the shear plane R', $(R = R')$

$$R^2 = F_x^2 + F_y^2 + F_z^2. \tag{16.19}$$

2. The component of force along the tool face F_C

$$F_C = (F_{x'}^2 + F_{z'}^2)^{\frac{1}{2}}. \tag{16.20}$$

3. The components of force normal to the tool face, F_{Cn}

$$F_{Cn} = F_{y'}. \tag{16.21}$$

4. The components of force in the tool face that are parallel to the chip flow direction ($F_C \cos \zeta_C$), and normal to the chip-flow direction ($F_C \sin \zeta_S$), respectively, where $\zeta_C = \eta_C - \delta_C$ and $\zeta_S = \eta_S - \delta_S$.
5. The component of force along the shear plane, (F_S) is

$$F_S = [(-F_x \cos i + F_y \sin i)^2 + (F_y \cos \phi_n \cos i - F_z \sin \alpha_n)^2]^{\frac{1}{2}}. \tag{16.22}$$

6. The component of force normal to the shear plane (F_{Sn}) is

$$F_{Sn} = -F_x \sin \phi_n \sin i - F_y \sin \phi_n \cos i + F_z \cos \phi_n. \tag{16.23}$$

7. The components of force in the shear plane that are parallel to the shear flow direction ($F_S \cos \zeta_S$), and normal to the shear flow direction ($F_S \sin \zeta_S$), respectively.

The coefficient of friction μ is defined as the tangential force divided by the normal force. We must generalize this definition here and refer to the tangential-force component in the direction of relative motion to take care of the fact that the friction force and the velocity vectors are not colinear; thus

$$\mu = \frac{F_C \cos \zeta_C}{F_{Cn}}. \tag{16.24}$$

There are two stress components associated with the shear plane that are of interest—a shear component τ and a normal component σ. These may be computed as follows:

$$\tau = \frac{F_S}{bt} \sin \phi_n \cos i \tag{16.25}$$

$$\sigma = \frac{F_{Sn}}{bt} \sin \phi_n \cos i. \tag{16.26}$$

Finally, there are three specific energies involved in the oblique cutting process. These are as follows:

1. The friction energy per unit volume

$$u_F = \frac{F_C \cos \zeta_C V_C}{btV} = \frac{F_C}{A_C} \cos \zeta_C \tag{16.27}$$

where A_C is the cross-sectional area of the chip.

2. The shear energy per unit volume

$$u_S = \frac{F_S \cos \zeta_S V_S}{btV} = \tau\gamma \cos \zeta_S \tag{16.28}$$

3. The total energy per unit volume

$$u = u_F + u_S = \frac{F_y}{by} \tag{16.29}$$

Experimental Results

Representative cutting data for oblique cutting tools are given in Table 16.1, and the following observations may be made.

1. The quantity ($\eta_C - i$) which is a measure of the deviation from Stabler's rule is seen to decrease with increased coefficient of friction, or to decrease with decreased shear strain.

2. As would be expected from geometrical consideration, η_S increases whenever η_C increases, but at a lower rate.

3. The inclincation angle i is seen to have a very significant influence upon the effective rake angle, particularly at high values of the inclination angle. As might be expected, the effective shear angle ϕ_e is seen to increase with α_e.

4. The deviation between the chip-flow vector and the resultant-force vector in the plane of the tool face (as measured by angle ζ_S) is seen to increase with both an increase in the effective rake angle α_e and an increase in the inclination angle i.

5. The deviation between the shear-velocity vector and the resultant-force vector in the shear plane (as measured by angle ζ_S) is seen to be insignificant in all of the tests.

6. The fact that ζ_S is insignificant while ζ_C is substantial would indicate that the shear process predominates over the friction process. That this is actually so may be seen by a comparison of the ratio u_F/u.

7. The coefficient of friction μ on the tool face is seen to decrease with increased inclination angle i. This is particularly significant inasmuch as the coefficient of friction will increase with increased rake angle in orthogonal cutting. However, as already mentioned, when inclination angle is increased, the effective rake angle is seen to increase. Thus the special geometry associated with an inclination angle has a greater effect upon the coefficient of friction than does the effective rake angle.

8. The shear stress on the shear plane is seen to increase with inclination angle while the shear stain decreases. An explanation for this paradox will be considered later.

The total energy per unit volume, u, is seen to decrease with increased inclination. This decrease is due to a decrease in the shear strain which in turn is due to an increase in the effective rake angle with increased inclination. Obviously, the decrease in u is not due to a decrease in the friction energy per unit volume, u_F, with increased inclination.

Views of representative chips obtained using the three angles of inclination of the tests of Table 16.1 are shown in Fig. 16.14. All chips are for a length of cut of approximately 4 in (102 mm). Here the change from the flat spiral chip of orthogonal cutting to the helical chip that is characteristic of a tool with inclination is evident.

One of the functions of inclination is to alter the chip-flow direction, and the manner in which side flow increases with the inclination of the cutting edge is evident in Fig. 16.14.

While there are some production tools that employ simple inclined cutting edges, such as those just discussed, most tools in use are geometrically more complex. However, when carefully studied, these tools may be reduced to equivalent inclined cutting tools. Were this not so, it is unlikely that we should be as interested in the inclined tool to the extent the foregoing discussion would indicate.

The various angles that are used in practice to describe a lathe tool, milling cutter, or drill will now be considered and the relationships between these angles and

1. Normal rake angle (α_n).
2. Inclination angle (i).
3. Effective rake angle (α_e).

will be presented.

TABLE 16.1.

Representative cutting data for oblique cutting tools

Work material: AISI 1015 cold-rolled steel; tool material, T-1 HSS; cutting fluid, CCl_4; undeformed chip thickness, 0.005 in (0.125 mm)

α_n degrees	0	10	30	10	10	10
i degrees	30	30	30	0	15	30
b in (mm)	0.25 (6.35)	0.25 (6.35)	0.25 (6.35)	0.25 (6.25)	0.25 (6.35)	0.25 (6.35)
b_C in (mm)	0.19 (4.83)	0.22 (5.59(	0.24 (6.10)	0.249 (6.32)	0.24 (6.10)	0.22 (5.59)
l in (mm)	3.95 (100.3)	3.95 (100.3)	3.95 (100.3)	3.95 (100.3)	3.95 (100.3)	3.95 (100.3)
l_C in (mm)	1.71 (43.3)	2.30 (58.4)	2.84 (72.1)	1.32 (33.5)	1.96 (49.8)	2.30 (58.4)
F_x lb (N)	−130 (−578.2)	−100 (−444.8)	−90 (−400.3)	0 (0)	−50 (−222.4)	−100 (−444.8)
F_y lb (N)	−330 (−1468)	−270 (−1201)	−200 (−890)	−300 (−1334)	−290 (−1290)	−270 (−1201)
F_z lb (N)	80 (356)	55 (244.6)	20 (89.0)	120 (534)	95 (422.6)	55 (244.6)
V i.p.m. (m min^{-1})	20 (0.51)	20 (0.51)	20 (0.51)	20 (0.51)	20 (0.51)	20 (0.51)
V_C i.p.m. (m min^{-1})	9.56 (0.24)	11.4 (0.29)	13.6 (0.35)	6.74 (0.17)	9.84 (0.25)	11.4 (0.29)
V_S i.p.m. (m min^{-1})	18.6 (0.97)	18.1 (0.46)	15.5 (0.39)	20.0 (0.51)	20.0 (0.51)	18.1 (0.46)
η_C degress	48.6	40.0	33.3	0	21.2	40.0
δ_C degrees	33.3	25.6	9.9	0	10.9	25.6
ζ_C degrees	15.3	14.4	23.4	0	10.3	14.4

η_S degrees	11.0	8.6	8.2	0	4.7	8.6
δ_S degrees	9.8	12.1	8.6	0	7.1	12.1
ζ_S degrees	1.2	−3.5	−0.4	0	−2.4	−3.5
ϕ_n degrees	18.1	28.5	42.0	19.4	27.0	28.5
ϕ_e degrees	25.3	34.2	44.2	19.4	28.4	34.2
α_e degrees	22.1	25.9	39.6	10	14.5	25.9
F_C lb (N)	95.6 (425)	114 (507)	124 (552)	170 (756)	147 (654)	114 (507)
F_{Cn} lb (N)	−351 (−1561)	−270 (−1201)	−179 (−796)	−274 (−1219)	−271 (−1205)	−270 (−1201)
$F_C \sin \zeta_C$ lb (N)	25.3 (112.5)	28.5 (126.8)	49.4 (219.7)	0 (0)	26.3 (117.0)	28.5 (126.8)
$F_C \cos \zeta_C$ lb (N)	92.2 (410.1)	111 (493.7)	114 (507.1)	170 (756.2)	144 (640.5)	111 (493.7)
F_S lb (N)	313 (1392)	236 (1050)	151 (671.7)	244 (1085)	219 (974.1)	236 (1050)
F_{Sn} lb (N)	185 (822.9)	184 (818.4)	180 (800.6)	213 (947.4)	218 (970.0)	184 (818.4)
μ	0.26	0.41	0.64	0.62	0.53	0.41
σ p.s.i. ($\times 10^{-3}$) (MPa)	39.8 (274)	62.7 (432)	83.5 (576)	56.7 (391)	82.5 (569)	62.7 (432)
τ p.s.i. ($\times 10^{-3}$) (MPa)	67.2 (463)	80.7 (556)	70.0 (483)	65 (448)	77 (531)	80.7 (556)
γ	3.44	2.20	1.33	2.98	2.28	2.20
u_F p.s.i. ($\times 10^{-3}$) (MPa)	35.8 (247)	50.8 (350)	62.2 (429)	46.2 (319)	57.0 (393)	50.8 (371)
u_S p.s.i. ($\times 10^{-3}$) (MPa)	233 (1607)	171 (1179)	94 (579)	196 (1351)	176 (1214)	171 (1179)
u p.s.i. ($\times 10^{-3}$) (MPa)	268 (1848)	222 (1531)	156 (1076)	242 (1669)	233 (1607)	222 (1531)
u_F/u	0.13	0.23	0.40	0.19	0.25	0.23

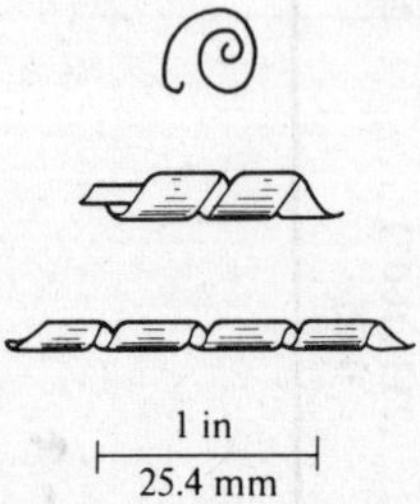

Chip	Inclination, i, degrees	Chip length in	(mm)
	0	1.32	(33.53)
	15	1.96	(49.78)
	30	2.30	(58.42)

1 in
25.4 mm

FIG. 16.14. Representative chips. Rake angle, α_n = 10°; length of cut, 4 in (102 mm); depth of cut, 0.005 in (0.13 mm); cutting speed, V = 20 i.p.m. (50.8 cm min^{-1}); cutting fluid, carbon tetrachloride.

Turning

Since each three-dimensional cutting tool evolved during a different period of technological development, the nomenclature that has come to be accepted is far from uniform among the different tools. The same angle on one tool is frequently found to have an entirely different name for a second type of tool. Furthermore the quantities that have come to be recognized in two-dimensional cutting as the most important ones are frequently obscurely given by two or more other quantities that are more conveniently measured but are far less significant from the fundamental point of view.

The lathe tool offers a good example. The nomenclature adopted by the American Standards Association is defined in Fig. 16.15 for a representative lathe-tool and the standard shorthand method of specifying the several dimensions of a tool is illustrated by an example. It will be noted that two quantities are necessary in this case to define the inclination of the tool face to the tool base which latter plane is perpendicular to the cutting velocity vector. These two quantities are the side rake (α_s) and the back rake (α_b) angles. These two angles together with the side cutting edge angle (C_s) determine the effective rake and inclination angles of the lathe tool, the other angles being of secondary importance.

It is of interest to note that British and German engineers thought it preferable to define the position of the tool face relative to the tool base in terms of other pairs of quantities than the back- and side-rake angles adopted in the USA. The German and British systems together with the American system are compared in Fig. 16.16. Here the plane of the paper is taken parallel to the tool base and the lines on the figure show the directions in which the various rake angles are measured. In the German system the back- and side-rake angles are measured parallel and perpendicular to the cutting edge respectively rather than along and perpendicular to the axis of the tool shank as in the American

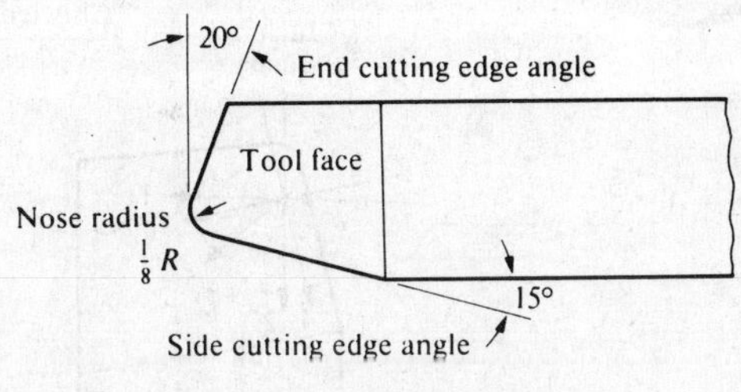

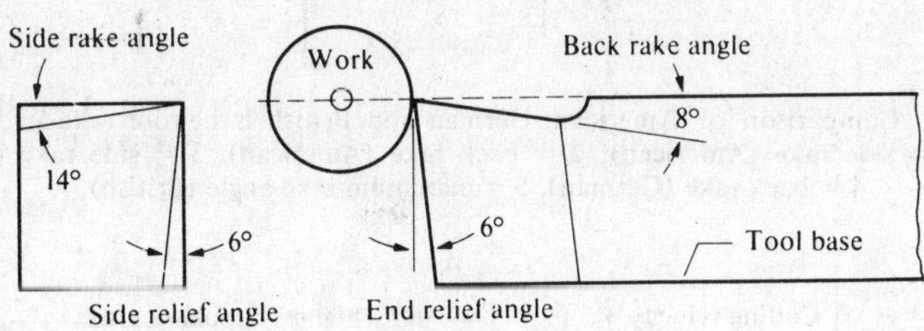

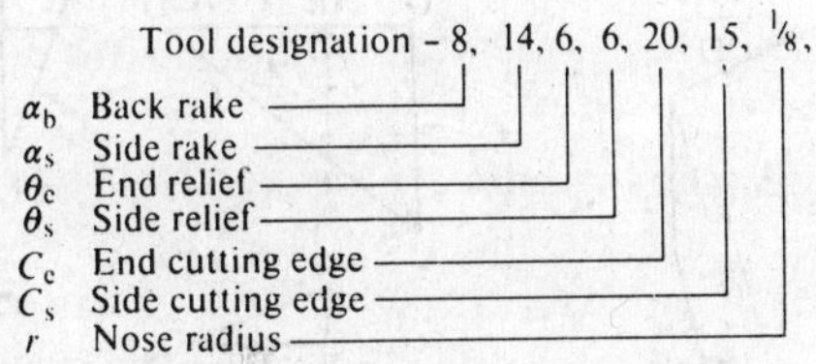

FIG. 16.15. Representative lathe-tool showing American Standards Association nomenclature.

system. The tool face is located according to the British standard by specifying the maximum rake angle and the maximum rake plan angle (see Fig. 16.16). All this illustrates how complex things can be without standardization.

Tool Geometry

Most practical three-dimensional tools are found to have two edges that cut simultaneously. Probably the most common tool of this type is the ordinary lathe tool. Here the primary cutting edge is the side cutting edge while the secondary edge is the end cutting edge. Since the depth of cut is usually much greater than the feed per revolution, what happens along the secondary cutting edge is relatively negligible compared with what happens along the primary cutting edge for cuts of normal aspect ratio. If the nose radius is small compared with the depth of cut the side cutting edge may be considered to be a simple oblique cutting tool.

In Fig. 16.17a a lathe tool is shown without clearance or end cutting edge angles for simplicity. In this figure plane OBDC is drawn perpendicular to the

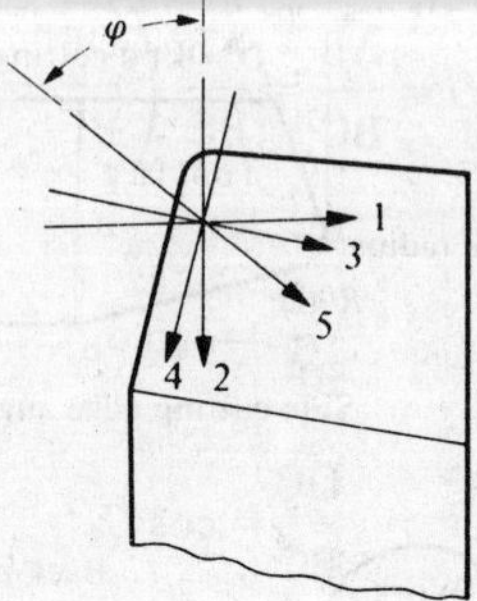

FIG. 16.16. Comparison of American, German and British lathe-tool rake-angle nomenclature. 1 = side rake (American), 2 = back rake (American), 3 = side rake (German), 4 = back rake (German), 5 = maximum rake-angle (British).

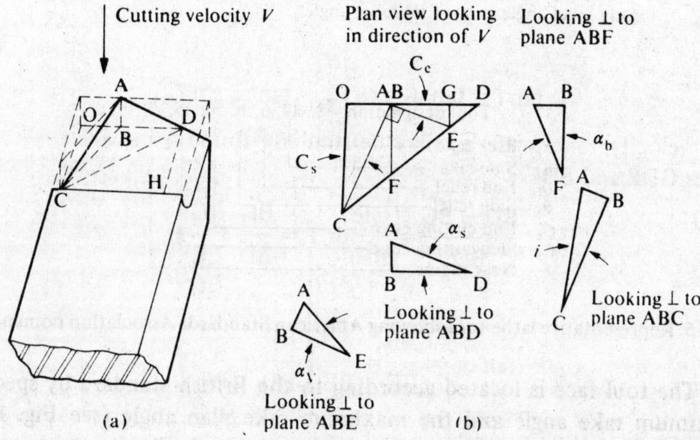

FIG. 16.17. Relationship between lathe tool angles.

velocity vector V, the curvature of which may be ignored if the depth of cut is small compared with the work diameter. Line AB is drawn parallel to V. In Fig. (16.17b) a plan view of the construction shown in Fig. (16.17a) is given, and the known angles α_s, α_b, and C_s are indicated on auxiliary diagrams as are the unknown quantities i and α_v.

The inclination angle i may be found in terms of the known angles as follows. Since triangles ODC and BDF are similar:

$$\frac{OB + BD}{BD} = \frac{OC}{BF} \text{ or } 1 = \frac{OC}{BF} - \frac{OB}{BD} \tag{16.30}$$

and

$$\tan i = \frac{AB}{BC} = \frac{AB}{BC}\left(\frac{OC}{BF} - \frac{OB}{BD}\right) \tag{16.31}$$

but

$$\frac{AB}{BF} = \tan \alpha_b$$

$$\frac{OC}{BC} = \cos C_s$$

$$\frac{AB}{BD} = \tan \alpha_s$$

$$\frac{OB}{BC} = \sin C_s$$

hence

$$\tan i = \tan \alpha_b \cos C_s - \tan \alpha_s \sin C_s. \tag{16.32}$$

The velocity rake angle α_v may be similarly found by working with similar triangles GDE and BDF:

$$\frac{BD - BG}{BD} = \frac{GE}{BF} \text{ or } 1 = \frac{BG}{BD} + \frac{GE}{BF} \tag{16.33}$$

and

$$\tan \alpha_v = \frac{AB}{BE} = \frac{AB}{BE}\left(\frac{BG}{BD} + \frac{GE}{BF}\right) \tag{16.34}$$

but

$$\frac{AB}{BD} = \tan \alpha_s$$

$$\frac{BG}{BE} = \cos C_s$$

$$\frac{AB}{BF} = \tan \alpha_b$$

$$\frac{GE}{BE} = \sin C_s$$

hence

$$\tan \alpha_v = \tan \alpha_s \cos C_s + \tan \alpha_b \sin C_s. \qquad (16.35)$$

Figure (16.18) is a nomograph for graphically solving eqn (16.32) and (16.35) for inclination angle (i) and velocity rake angle (α_v). For the tool of Fig. 16.15 the dotted lines in the nomograph give

$$i = 4.1°$$

$$\alpha_v = 15.6°$$

From eqn (16.1), α_n is found to be 15.5°, and if Stabler's rule is assumed to hold, the dotted lines on Fig. (16.18) give $\alpha_e = 15.7°$. Thus, to a first approximation the relatively complex tool of Fig. (16.15) is seen to be equivalent to a simple oblique tool having a 4° inclination angle and a 16° effective rake angle.

Representative Tool Dimensions

From the foregoing discussion of the lathe tool, it is evident that there are a number of geometrical quantities that may be varied as well as several operating variables. Many of the more important considerations pertaining to the choice of these quantities will now be considered.

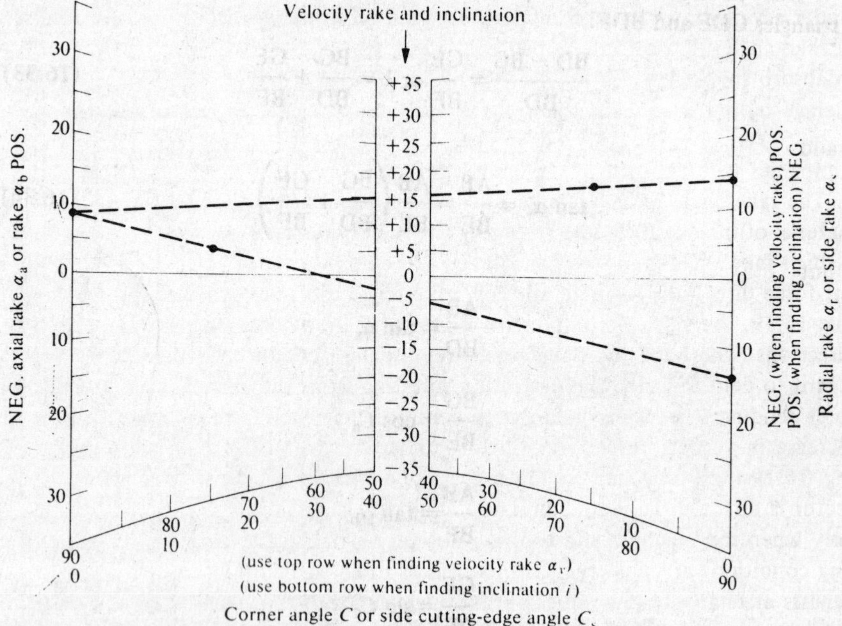

FIG. 16.18. Nomograph for finding inclination angle i and velocity rake angle α_v for three-dimensional cutting tools. (Based on Kronenberg 1966.)

Due to the interdependence of the many variables influencing the performance of a lathe tool, it is not possible to state exact rules pertaining to the choice of operating conditions. One must be content with a presentation of good average practice and the consequence of changing each variable from the recommended value. An additional reason why it is not possible to precisely specify optimum operating variables is that the best choice of variables depends upon whether tool life or finish is of major interest. The method of arriving at optimum operating conditions in practice is to use general recommendations such as those presented here as a starting point and then proceed to improve performance by trial. Of course, a general appreciation of the relative importance of each of the many operating variables is invaluable in reducing the effort involved in approaching the optimum.

Representative values of rake and relief angles for turning tools are given in Table 16.2. Values of α_n, i, and α_e are also given in this table for a range of side cutting edge angles. The recommended values of rake angle are seen to be such that α_n generally decreases as the strength of the metal cut increases. This decrease in α_n provides a greater wedge angle at the tool point to support a greater load. When very hard materials are machined or when intermittent cuts must be taken the values of α_s and α_b should be reduced to values below those given in Table 16.2 to provide a greater normal rake angle (α_n) to absorb the shock. Under severe shock conditions it is frequently desirable to use negative values of rake angle with carbide-tipped tools. Due to the greater ability of HSS tools to absorb shock it is preferable to provide them with larger rake angles (on the average $\alpha_e = 10$ to 15°) than is desirable with carbide tools (on the average $\alpha_e = 5^\circ$) in order to take advantage of the decreased cutting energy that accompanies a shift to a higher effective rake angle.

The rake angle recommended for brass in Table 16.2 are unusually low for a material of this hardness and strength. The reason for this lies in the fact that tools of high rake angle tend to dig into brass and exhibit a gouging action. The resultant force directions for high and low rake angle tools are generally as shown in Fig. 16.19, the vertical components of force in the two cases being in opposite directions. The large rake angle case represents a condition of instability with regard to depth of cut, the deflection resulting from the vertical component of force tending to cause an increase in depth of cut which in turn causes a further increase in vertical force, etc. Cutting under the high-rake-angle conditions of Fig. 16.19 may thus lead to gouging. Rake angles for which the resultant force vector is directed below the direction of cut are generally to be avoided, particularly when the machine and tool are not very rigid. In the case of brass the cutting conditions are such that a force vector directed below the direction of cut persists at relatively low values of rake angle and thus the values of α_e to be used for brass are unusually small. Tools that are used to cut aluminum and copper, which have similar hardness and strength to brass, do not exhibit a tendency to dig in; and hence aluminum and copper are cut with larger values of α_e.

TABLE 16.2

Representative turning tool angles (degrees)

Work material	α_s Side rake	α_b Back rake	Side and end relief	For $C_s = 0°$			For $C_s = 15°$			For $C_s = 30°$		
				α_n	i	α_e	α_n	i	α_e	α_n	i	α_e
High-speed steel tools												
Aluminum and magnesium alloys	20	20	10	18.9	20	23.7	23.4	14.5	25.8	26.3	7.6	26.9
Copper	15	15	10	14.5	15	17.5	17.9	10.8	19.3	20.0	5.6	20.4
Brass and bronze	0	0	8	0	0	0	0	0	0	0	0	0
Cast iron	10	5	8	9.9	5	10.3	10.9	2.3	11.0	11.1	−0.7	11.1
Steel (C < 0.3)	15	15	8	14.5	15	17.5	17.9	10.8	19.3	20.0	5.6	20.4
Steel (C > 0.3)	12	12	8	11.8	12	13.8	14.5	8.6	15.5	16.2	4.5	16.4
Stainless steels	10	15	8	9.7	15	13.0	13.2	12.0	15.1	15.8	8.2	16.6
Carbide tools												
Aluminum and magnesium alloys	10	0	8	10	0	10	9.7	−2.6	9.8	8.6	−5.0	9.1
Copper	15	0	7	15	0	15	14.5	−4.0	14.7	13.0	−7.7	13.8
Brass and bronze	4	0	7	4	0	4	3.9	−1.0	3.9	3.5	−2.0	3.6
Cast iron	8	0	6	8	0	8	7.8	−2.1	7.9	7.0	−4.1	7.2
All steels	6	0	6	6	0	6	5.8	−1.6	5.9	5.2	−3.0	5.4

All positive value of inclination angle (i) will cause the chip to be directed away from the workpiece, while a negative value of i will direct the chip toward the workpiece.

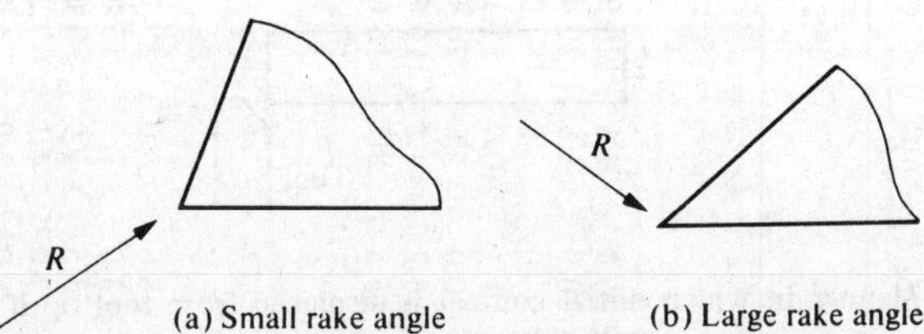

FIG. 16.19. Comparison of directions of resultant force for tools with (a) small and (b) large rake angles.

The relief angles given in Table 16.2 are seen to decrease slightly as the hardness of the metal cut increases. In cutting very hard materials the relief angle should be as low as possible; and when the feed is also reduced, relief angles as low as 5° may be successfully used. In cutting very soft metals or materials with low elastic moduli at very high rates of feed it is sometimes necessary to increase relief angles to as high as 12°. It is preferable to make the end- and side-relief angles the same in the interest of simplicity and to make it easier to blend the nose radius into the flanks of the tool. On carbide tools a secondary relief angle of from 10 to 15 degrees is frequently provided.

The role of the side cutting edge angle (C_s) in influencing the inclination angle (i) and the velocity rake angle (α_v) may be seen in the nomograph of Fig. (16.18). If C_s is zero, the back rake angle becomes the inclination angle (i.e. $\alpha_b = i$) and the side rake angle equals the velocity rake angle (i.e. $\alpha_s = \alpha_v$) which in turn is essentially the effective rake angle (α_e). As the side cutting edge angle (C_s) is increased, α_s plays an increasing role in determining the value of i just as α_b plays an increasing role with α_s in determining α_e. In Table 16.2 it is seen that C_s plays a small part in influencing α_n, i, and α_e for carbide tools in the range of C_s extending from 0 to 30°. Table 16.2 on the other hand shows C_s to have a significant influence on these values for HSS tools over the same range of C_s.

Examination of the values of Table 16.2 reveals that a large value of C_s makes it possible to have a large value of α_e while at the same time the corresponding value of α_n is unusually small. This will enable a cut to be made at low power with a tool of large included angle and hence provide improved strength. It would thus appear that tools should be ground with large values of side cutting edge angle (C_s), and this would be the case were it not for the fact that the tendency for a tool to chatter increases with C_s. In order to avoid chatter it is customary to limit C_s to about 15°, although values as high as 45° may be used to advantage if the system is unusually rigid.

Unless it is necessary to turn to a radial shoulder, it is advisable to have a side cutting edge angle other than zero. The shock is thus taken at a point away from the tool tip as the tool engages the work (Fig. 16.20). When machining

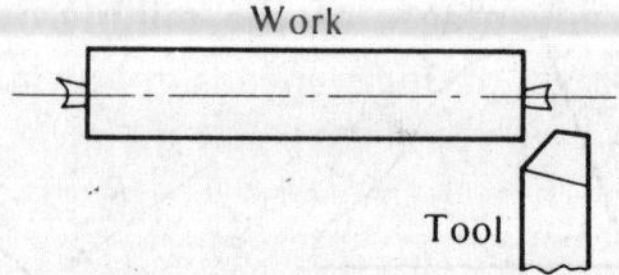

FIG. 16.20. Manner in which initial contact is displaced from tool tip if tool is provided with side cutting edge angle.

very rough forgings having a ragged surface, it is particularly important to provide a sufficient side cutting edge angle to protect the tool point. Carbide tools are very susceptible to damage under such conditions. A side cutting edge angle is also useful in releasing the stored elastic energy less abruptly, thus reducing the danger of edge chipping as the tool completes a cut.

Like a relief angle, the end cutting edge angle (C_e) should be as small as possible in the interest of increased tool strength. The normal range of values for C_e is 5 to 15 degrees. The lower values can only be used with very rigid systems since any increase in radial force (due to there not being sufficient end-clearance) will tend to cause tool chatter. Values in the vicinity of 15° may be required at high feeds, particularly when the depth of cut is increased while the feed is engaged.

Milling

The milling process is distinguished by a tool with one or more teeth which rotate about a fixed axis while the workpiece is fed into the tool. The process is further distinguished from other metal cutting operations by the chips which are produced. These are generally short, discontinuous segments. It should be emphasized that this type of discontinuous chip formation is a result of geometry alone whereas in the turning operation, discontinuous chips are usually due to fracture of the chip. The undeformed chip thickness in most milling operations varies from one end of the chip to the other. It will be shown later, that depending upon the particularly process, the maximum chip thickness may occur at either end or near the middle of the chip. The regular chip discontinuity associated with milling leads to non-steady-state cyclic conditions of force and temperature. As a tooth engages the work, it receives a strong shock followed by a varying force. At the same time the tool chip is relatively cool as it enters the work, is heated as the chip is formed, and cools while awaiting the next engagement. The entering shock is detrimental to tool life, while the cooling period is usually beneficial, unless the tool material is sensitive to thermal shock.

The cyclic variation of force can provide the necessary energy to excite a natural mode of vibration in any part of the machine. Such vibrations can, and frequently do, lead to poor surface finish and decreased tool life.

One of the greatest advantages of the milling operation, is that small, light tools can easily be rotated at high speeds to produce flat or curved surfaces on workpieces of a wide variety of sizes and shapes.

Types of Milling

Important types of milling are plain-, end- or face-milling. The basic plain-milling cutter used to produce flat surfaces has cutting edges on the periphery which are parallel with the axis of rotation. The performance of this cutter is usually enhanced if the teeth are arranged in helical fashion rather than parallel to the axis of rotation (Fig. 16.21). Use of helical teeth not only provides smoother cutting (in the same way that helical gears run smoother than spur gears) but chip formation is also improved.

An end-milling cutter (Fig. 16.22) has teeth on one end as well as on the periphery. In use, its axis of rotation is perpendicular to the surface produced while, in plain-milling, the cutter axis is parallel to the finished surface. The end-milling cutter is used for producing flat surfaces and profiles—frequently simultaneously. The term end-milling cutter is misleading because in normal operation cutting is done on the periphery rather than by the teeth on the end of the cutter.

One could continue to amplify the number of types of milling cutters almost indefinitely. These would include side-milling cutters, milling saws as well as angle, fly, T-slot, and formed milling cutters. However, all of these are adaptations of the basic types considered above.

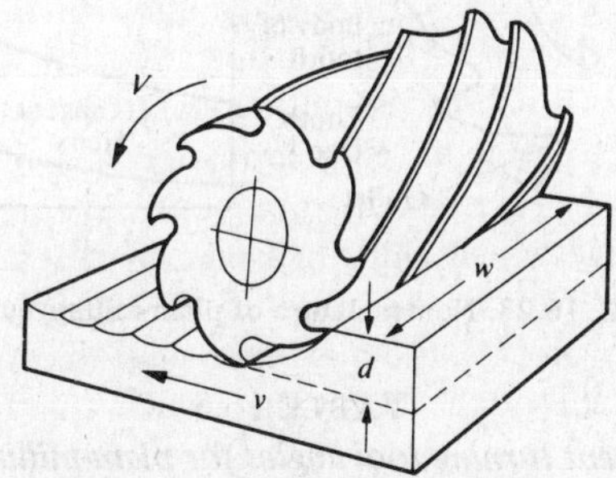

FIG. 16.21. Plain-milling operation.

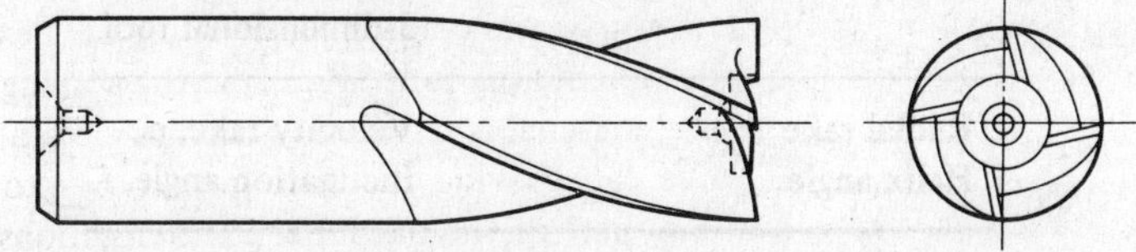

FIG. 16.22. Standard end-milling cutter with straight shank.

Milling cutters are generally made of high-speed steel or have cemented carbide teeth. It is common practice with large milling cutters to use inserted teeth which may be removed for grinding and replacement. Smaller cutters are frequently equipped with carbide tips that are brazed in place. Figure 16.23 shows the accepted nomenclature for a plain- or helical-milling cutter. The analysis of this chapter may be applied directly with the aid of Table 16.3.

The normal rake angle is readily computed from eqn (16.1), while the nomograph of Fig. 16.8 will give the effective rake angle α_e. The helix or inclination angle may be determined by rolling the cutter on a sheet of carbon paper that rests on white paper.

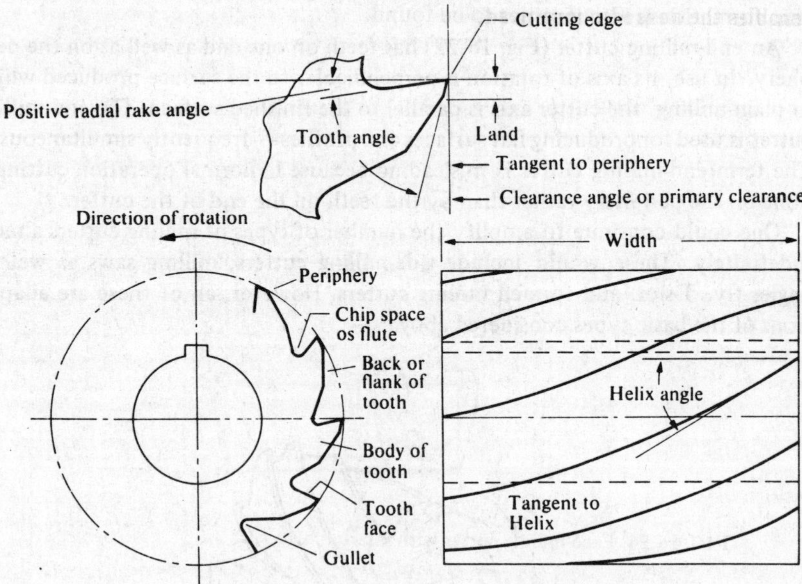

FIG. 16.23. Nomenclature of plain-milling cutter.

TABLE 16.3

Equivalent turning tool angles for plain-milling cutter

Plain milling cutter	Equivalent 3-dimensional tool
Radial rake	Velocity rake, α_v
Helix angle	Inclination angle, i

As an example, consider a plain milling cutter with the following angles:

Radial rake, $\alpha_r = 10°$

Helix angle, $i = 25°$

From eqn (16.1) the normal rake angle (α_n) is found to be 9.1° while the effective rake angle (α_e) is found to be 18° by use of Fig. 16.8.

Nomenclature for an end or face-milling cutter is shown in Fig. 16.24. The inclination angle (i) and effective rake angle (α_e) may be found if the face-milling cutter angles are replaced by their equivalent turning tool angles given in Table 16.4. With these substitutions, the previous analysis of a turning tool enables the desired quantities to be found.

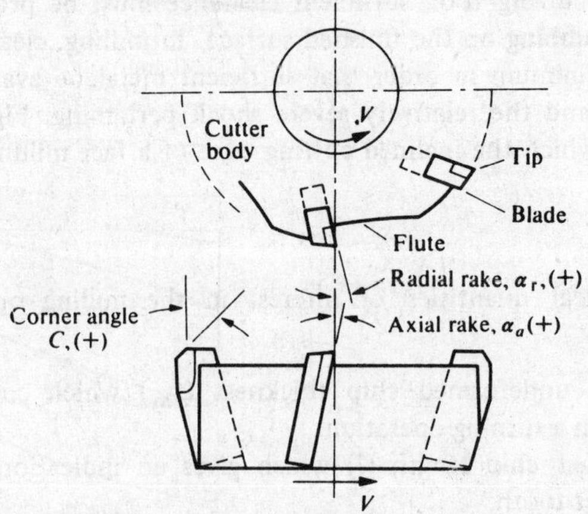

FIG. 16.24. Face-milling cutter with inserted blades, showing principal angles.

TABLE 16.4
Equivalent turning tool angles for face-milling cutter

Face-milling cutter	Turning Tool
Radial rake, (α_r)	Side rake, (α_s)
Axial rake, (α_a)	Back rake, (α_b)
Corner angle, (C)	Side cutting edge angle (C_s)

As an example consider the following case:

Axial rake angle, $(\alpha_a) = 10° = \alpha_b$

Radial rake angle, $(\alpha_r) = -10° = \alpha_s$

Corner angle, $(C) = 30° = C_s$

By following steps similar to those employed for a turning tool, the pertinent values are found to be:

Normal rake angle $(\alpha_n) = -3.6°$

Inclination angle $(i) = 13.5°$

Effective rake angle $(\alpha_e) = 0.8°$

As for any cutting tool, sufficient clearance must be provided to prevent the tool from rubbing on the finished surface. In milling, clearance angles must be held to a minimum in order that sufficient metal be available at the tool point to withstand the relatively severe shock pertaining. Figure 16.25 shows the manner in which the inclined cutting edge of a face milling cutter produces a chip.

Chip Geometry

Three geometrical quantities of interest in the milling operation are (see Fig. 16.26):

1. Maximum undeformed chip thickness (t_m) which corresponds to the feed rate in a turning operation.
2. Undeformed chip length (l) which gives an indication of the time of contact per tooth.
3. Peak-to-valley distance (h) of the scallops which are left on the finished surface when plain milling.

Unlike turning where the depth of cut and feed can be directly set on the machine in milling, these important quantities must be computed from the operating variables. Before determining the desired quantities, one must specify the direction of motion of tool and workpiece. When these two velocities have the same direction, the process is called 'climb' or 'down' milling; however when the two velocities are in opposite directions, a 'conventional' or 'up' milling process is involved. Figure 16.27 clearly shows the difference between these two types of milling. It is readily seen that with otherwise similar conditions, up-milling gives a slightiy smaller t_m, a larger l, and smaller h than does down-milling, because of the smaller average radius of curvature of the chip when down-milling.

The actual shpae of an undeformed plain-milling chip is rather complex as the cutting edge traverses a trochoidal path. Equations for milling chip geometry

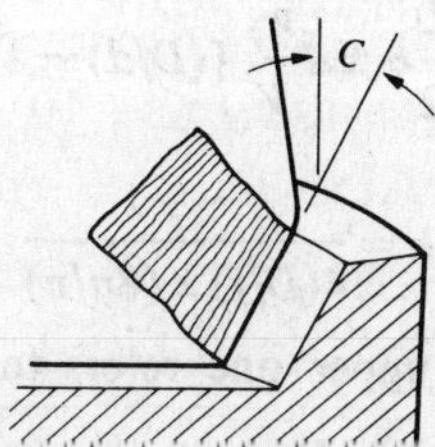

FIG. 16.25. Action of face-milling cutter showing manner in which inclined peripheral cutting edge produces chip. (i = inclination angle; n = number of teeth in cutter.)

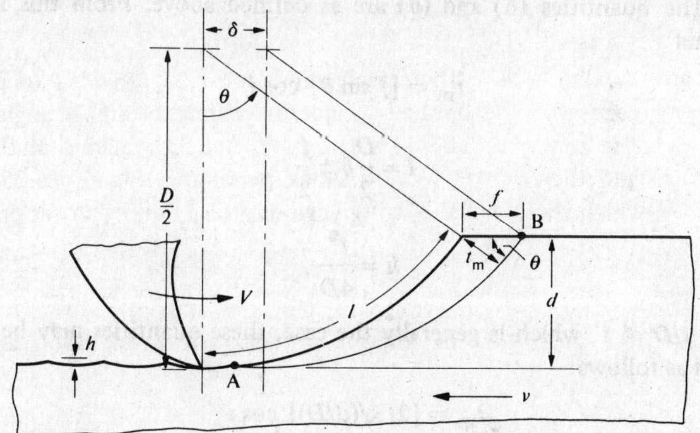

FIG. 16.26. Approximate shape of plain-milling chip.

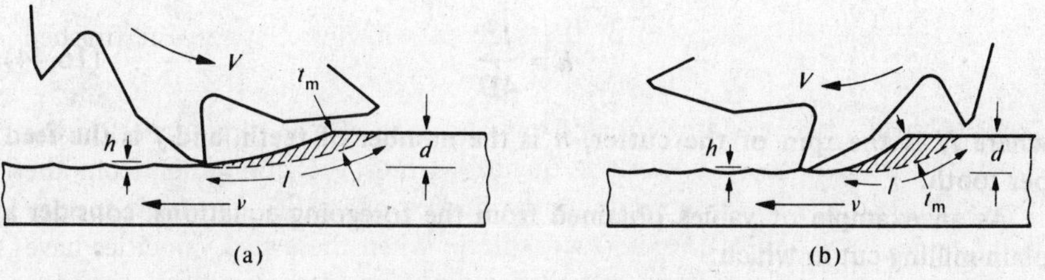

FIG. 16.27. Difference between (a) up-, and (b) down-milling.

have been developed by Martellotti (1941) but since these derivations are quite tedious, only the final results are given here (see Fig. 16.26 for meaning of symbols):

$$t_m = \left[f \left\{ \frac{(D/d) - 1}{(D/2d)^2 (1 \pm v/V)^2 \mp vD/Vd} \right\}^{\frac{1}{2}} \right] \cos i. \qquad (16.36)$$

$$l = \frac{D}{2}\theta \pm d\frac{v}{V}\{(D/d)-1\}^{\frac{1}{2}}. \tag{16.37}$$

$$h = \frac{f}{4(D/f) \pm (8n/\pi)} \tag{16.38}$$

When two signs are given, the upper one refers to up milling while the lower one refers to down milling.

If the cutting speed (V) is much greater than (v) (i.e. $v/V \ll 1$), as it always is in practice, a simplified solution can be obtained. Figure 16.27 shows the trochoidal chip shape replaced by a circular arc which is permissible when $V \gg v$. The quantities (δ) and (θ) are as defined above. From this figure it follows that:

$$t_m = [f \sin\theta] \cos i \tag{16.39}$$

$$l = \frac{D}{2}\theta \pm \frac{f}{2} \tag{16.40}$$

$$h = \frac{f^2}{4D}. \tag{16.41}$$

When $d/D \ll 1$, which is generally the case, these quantities may be further simplified as follows:

$$t_m = [2f\sqrt{(d/D)}] \cos i \tag{16.42}$$

$$l = (Dd)^{\frac{1}{2}} \pm \frac{v}{2nN} \tag{16.43}$$

$$h = \frac{f^2}{4D} \tag{16.44}$$

where N is the rpm of the cutter, n is the number of teeth, and f is the feed per tooth.

As an example of values obtained from the foregoing equations, consider a plain-milling cut in which:

$N = 200$ r.p.m.
$v = 12$ i.p.m. (305 mm min^{-1})
$n = 10$ teeth
$D = 4$ in (102 mm)
$d = \frac{1}{4}$ in (6.35 mm)
$i = 0°$.

For these values we obtain:

$$f = \frac{12}{(200)(10)} = 0.006 \text{ in/tooth } (0.15 \text{ mm/tooth})$$

$$\theta = \cos^{-1}\left((1 - \frac{0.5}{4}\right) = 29^\circ = 0.506 \text{ radians.}$$

The values of t_m, l, and h obtained from three foregoing approximations are given in Table 16.5 where it is apparent that the approximate methods are good in the practical range.

For face- or end-milling the chip geometry is easily approximated. From Fig. 16.28 it is seen that if the axis of the cutter intersects the workpiece the maximum chip thickness (t_m) is just equal to the feed per tooth.

$$t_m = \frac{v}{Nn} = f. \tag{16.45}$$

The chip length is then approximated by

$$l = \frac{D}{2}\Psi \tag{16.46}$$

where

$$\Psi = \sin^{-1}\left(\frac{2m}{D}\right) + \sin^{-1}\left\{\frac{2(w-m)}{D}\right\}. \tag{16.47}$$

If a corner angle (C) exists, the value of t_m should be modified as follows

$$t_m = \frac{v}{Nn}\cos C \tag{16.48}$$

Practical Considerations

The effect of ***rake angle*** upon tool life, surface finish, and power consumed is the same with a milling cutter as it is with a simple two-dimensional tool. However, the shock loadings to which milling cutters are subjected limits the maximum useful rake angle. This is particularly true with carbide-tipped milling cutters. Negative rake angles are frequently used in carbide cutters to provide a greater included angle for increased strength and shock resistance. However, use of negative rake angles is accomplished by an increase in the energy consumed per unit volume of metal cut and the temperature at the tool point.

A composite rake angle tooth has been found to combine some of the better qualities of both positive and negative rake angle tools. The composite rake angle tool shown in Fig. 16.29 was tested by Armitage and Schmidt (1957) with the results shown in Table 16.6. These data show that the addition of a positive secondary rake angle reduces the cutting force (F_P) although it is still higher

TABLE 16.5

Comparison of exact and approximate equations

Method	t_m *inches*†		l inches		h ($\times 10^6$) inches	
	up	down	up	down	up	down
Martellotti's analysis (eqns 16.36 to 16.38)	0.00290	0.00291	1.017	1.007	2.225	2.270
If $v/V \ll 1$ (eqns 16.39 to 16.41)	0.00291	0.00291	1.015	1.009	2.250	2.250
If $v/V \ll 1$ and $d/D \ll 1$ (eqns 16.42 to 16.44	0.0030	0.0030	1.003	1.003	2.250	2.250

† Metric values omitted since the purpose of this table is only to compare values obtained from the approximate and exact equations

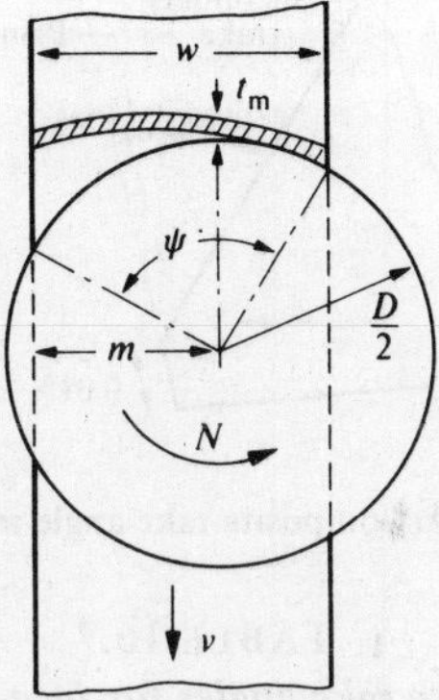

FIG. 16.28. Approximate shape of face-milling chip.

TABLE 16.6

Cutting force values (F_P) for composite rake angle tools. (After Armitage and Schmidt 1957)

Rake angles (α) degrees	Cutting force (F_P) lb	N
+30	64	285
+15	82	365
−12	115	512
−12 primary, +15 secondary	96	427
−12 primary, +30 secondary	76	328

than when a positive rake angle is used alone. The greatest advantage of this type of tool occurs after a small amount of tool wear. The cutting force (F_P) for the composite rake angle tool is lower than that for either the purely positive rake angle tool or the purely negative rake angle tool.

Naturally, no hard and fast rules can be given for milling cutter rake angles but Table 16.7 gives representative values for several materials. When three-dimensional theory is applied to Table 16.7 the values given in Table 16.8 are obtained.

Clearance angles should be as small as possible consistent with good surface

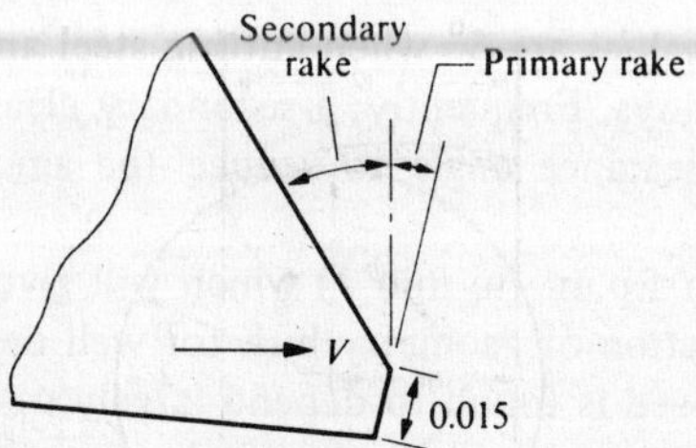

FIG. 16.29. Composite rake angle milling tooth.

TABLE 16.7
Representative rake angles for face-milling cutters

Work material	Tool materials: H.S.S. α_r	H.S.S. α_s	Cast alloys α_r	Cast alloys α_s	Cemented carbides α_r	Cemented carbides α_s
Steel and cast iron	12	12	6	6	0	0
Steel and cast iron†	−5	25	−5	25	−10	10
Brass and bronze	10	10	6	6	4	4
Aluminum alloys	25	25	12	12	15	15

† Cutting with a 45° corner angle; in all other cases corner angle is 0°.

TABLE 16.8
Values of angles i, α_n, and α_e for the tools of Table 16.7

Work material	Tool materials: H.S.S. i	H.S.S. α_n	H.S.S. α_e	Cast alloys i	Cast alloys α_n	Cast alloys α_e	Cemented carbides i	Cemented carbides α_n	Cemented carbides α_e
Steel and cast iron	12.0	11.8	13.8	6.0	6.0	6.5	0	0	0
Steel and cast iron†	21.4	14.0	19.4	21.4	14.0	19.4	14.1	0	3.4
Brass and bronze	10	9.9	11.4	6.0	6.0	6.5	4.0	4.0	4.3
Aluminum alloys	25	22.9	29.9	12.0	11.8	13.8	15.0	14.5	17.6

† Cutting with a 45° corner angle; in all other cases corner angle = 0°.

finish. Representative values are 5° when cutting steel and 10° when machining aluminum or similar alloys. Frequently, a secondary clearance angle is provided behind the primary clearance angle to reduce the amount of grinding when sharpening.

The cutter speed in f.p.m. (m min^{-1}) which will give reasonable tool life is of course a strong function of geometry but, for well designed cutters properly used, the maximum speed is found to depend largely upon the hardness of the metal cut and the tool material. Figure 16.30 gives representative cutting speeds as a function of workpiece hardness and material.

Normally the feed per tooth (δ) is considered an important quantity and is specified in milling operations. However, as previously mentioned, a more meaningful quantity, but one which must be computed, is the maximum chip thickness (t_m). Representative values of δ are given in Table 16.9 to illustrate the order of magnitude of this variable in practice.

In face milling, initial contact should not occur at the cutting edge of the tooth in order to avoid chipping of the edge. In order to ensure this the cutter should be larger than the work. A good working rule is to make the cutter diameter at least $1\frac{1}{2}$ times the work width.

In order to cause chips to flow away from the cutter body in face milling it is desirable that the teeth be designed to have a substantial positive inclination. A large corner angle (*C*) combined with a large positive inclination angle will usually provide long tool life. The large corner angle distributes the wear over a long cutting edge and hence a given volume of metal may be removed with a minimum wear per unit of edge length.

It is particularly important in milling operations that the work be rigidly supported and that the machine be in good mechanical condition. Cutters should be mounted on large arbors and the arbor supported close to the cutter to further ensure a rigid system. Flywheels should be used whenever possible and should be mounted as close to the cutter as is feasible.

When milling with carbide cutters, it is best to cut steel and cast iron dry using compressed air to blow the chips away. Under such conditions the use of a water-base cutting fluid will actually give significantly shorter tool life than when cutting dry. Aluminum on the other hand should be cut with a copious supply of a water-base cutting fluid.

Drilling

The drill is a tool designed to produce holes in metal parts quickly and easily. While very precise work can be done with a drill, it is a roughing operation and the primary items of interest are usually long life and high penetration rate. Very frequently drilling is a preliminary operation to reaming, boring or grinding where final finishing and sizing takes place. In addition to the common two flue twist drill there are several other types in use including flat drills,

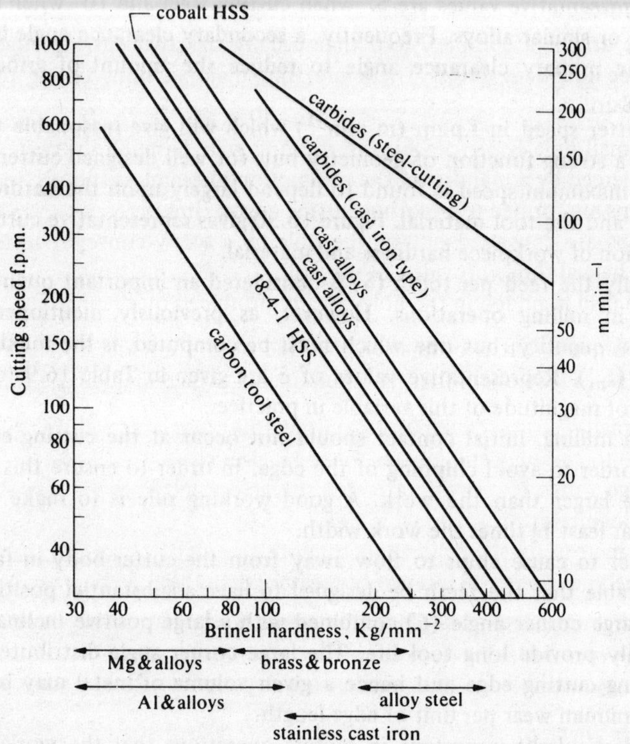

FIG. 16.30. Representative values of cutting speed as a function of work hardness and tool material.

TABLE 16.9

Representative values of feed per tooth (δ) for milling cutters, i.p.r. (mm/rev)

Material	H.S.S.	Carbide and cast alloy
Stainless steel	0.005 (0.13)	0.008 (0.20)
Steels	0.005–0.010 (0.13 to 0.25)	0.010–0.015 (0.25 to 0.38)
Cast iron	0.010–0.015 (0.25 to 0.38)	0.0.10–0.20 (0.25 to 0.50)
Brass	0.008–0.020 (0.13 to 0.50)	0.010–0.020 (0.25 to 0.50)
A1 and Mg alloys	0.02 (0.50)	0.02 (0.50)

drills with one, three or four flutes, core, shell, and spade drills. The discussion here will be limited to the two-flute twist drill since this type adequately illustrates the nature of the problems associated with drills and their analysis.

Nomenclature

The twist drill is a complex tool that usually has two cutting edges designed to produce identical chips. There are four major actions taking place at the point of a drill:

1. A small hole is pierced by the rotating web, in much the same way a tube is pierced by a mandrel in the production of seamless tubing.
2. Chips are formed by rotating cutting edges. (lips)
3. Chips are conveyed out of the hole by a screw conveyor in the form of the helical flutes provided in the drill.
4. The drill is guided in the hole already produced by the margins.

The geometry that is ordinarily provided on a drill (Fig. 16.31) represents a compromise of several conflicting requirements, which include the following:

1. A small web to reduce thrust on the drill but a large web for greater resistance to chipping and greater torsional rigidity.
2. Large flutes to provide a larger space for chip transport but small flutes in the interest of torsional rigidity.
3. An increase in the helix angle to more quickly remove chips but a decrease in helix angle in the interest of greater strength of cutting edges.

Less obvious compromises are associated with the choice of the geometric parameters that influence the effective rake and relief angles.

The more important drill quantities from the analytical point of view are:

1. Helix angle, δ.
2. Point angle, $2p$.
3. Web thickness, w.
4. Clearance angle, θ
5. Drill diameter, d.

The helix angle varies with the radius (r) to any particular point on the cutting edge. A helix angle without specification refers to the helix angle at the circumference of the drill. The pitch length of the helix (L) is constant for all points along the cutting edge, and the helix angle at any point may therefore be determined by use of the following equation.

$$\delta = \tan^{-1}\left(\frac{2\pi r}{L}\right) \tag{16.49}$$

For example, a 0.750 inch (19 mm) drill having a helix angle of 32° is found to have a pitch length (L) of 3.77 inches (95.8 mm). This pitch length may then be used to find the helix angle at the point on the drill where r = 0.250 inch (6.35 mm). When this is done the helix angle is found to be 22.6°. The helix angle may be most easily determined by rolling a drill across a piece of carbon paper that rests on a sheet of white paper.

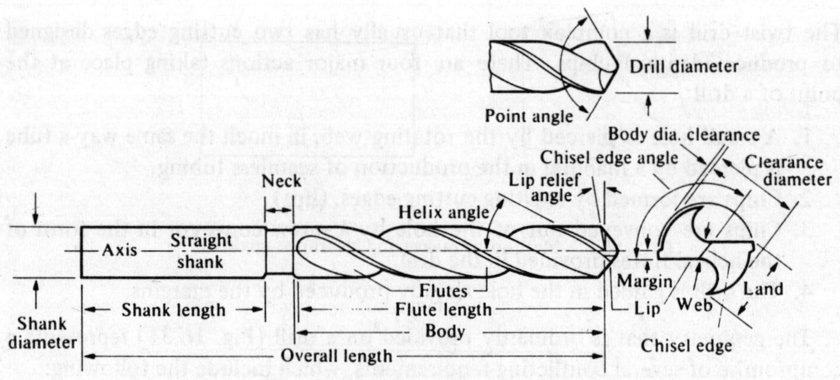

FIG. 16.31. Drill nomenclature.

The point angle, web thickness, and diameter of a drill may be readily determined by direct measurement.

The clearance angle (θ) at the periphery of a drill may be measured by wrapping a sheet of paper around the drill and marking it as shown in Fig. 16.32. The clearance angle at the center of the drill will be considerably greater than this value. Depending upon the shape of the point that is ground on the drill the increase in clearance angle from periphery to center may be as much as 100 percent. The matter of clearance angles for drills presents a special problem in that a drill is fed downward upon the cut surface. The feed angle (ϵ) that is generated by any point on the cutting edge at radius r may be obtained from the following expression in terms of the feed per revolution of the drill (f)

$$\epsilon = \tan^{-1}\left(\frac{f}{2\pi r}\right) \tag{16.50}$$

The clearance angle (θ) at any radius must provide this much clearance before there is anything left to take care of elastic recovery. The drill clearance that corresponds to that of a conventional tool is ($\theta - \epsilon$). The quantity ϵ is seen to increase as the point of the drill is approached.

In Fig. 16.33, values of helix angle (δ), clearance angle (θ), and feed angle (ϵ) are given for different points on the cutting edge of a representative $\frac{9}{16}$ inch (14.33 mm) diameter drill operating under conventional conditions. The effective clearance angle ($\theta - \epsilon$) is seen to be particularly ample at the center of the drill, and the feed angle (ϵ) is seen to play a rather insignificant role. The feed angle becomes important only on larger drills and drills that are fed at abnormally high rates.

The quantity (f), the feed rate in inches per revolution (mm/rev) is the feed for two cutting edges and therefore $f/2$ (for a two-flute drill) corresponds to

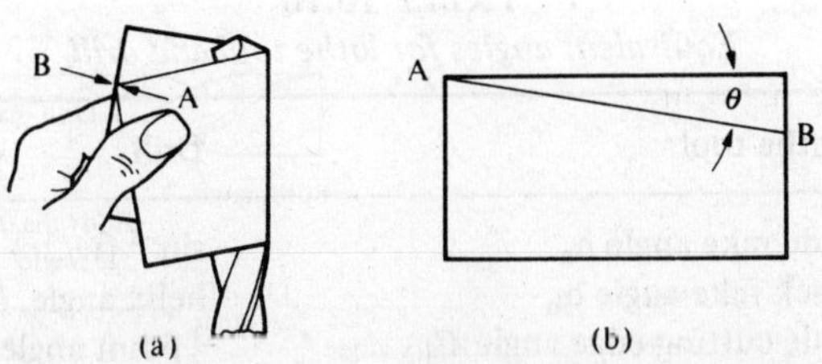

FIG. 16.32. Method of measuring clearance angle θ.

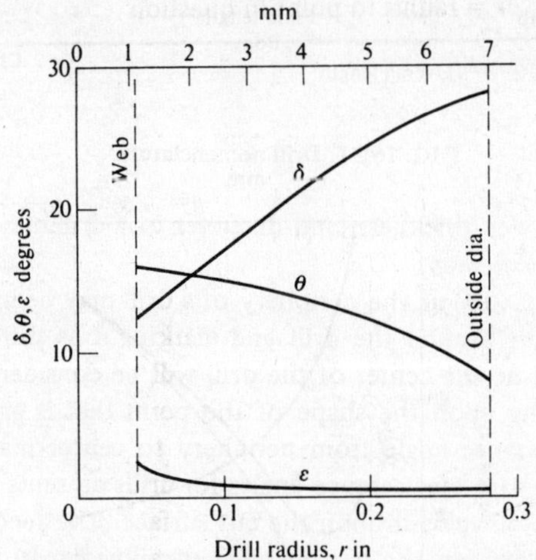

FIG. 16.33. Variation of helix angle δ clearance angle θ and feed angle ϵ across the cutting edge of a representative $\frac{9}{16}$ (14.3 mm) diameter drill. Helix angle, 28°; clearance angle, 8°; point angle, 118°; chisel edge angle, 120°; web thickness, 0.08 in (2.03 mm); feed rate, 0.009 i.p.r. (0.23 mm/rev).

the feed of a lathe tool. The relief provided at the outside of most drills will be from 8 to 12° (larger values for soft metals).

Analysis

The analysis of a drill point is much more involved than that for other tools inasmuch as the quantities of interest vary with radial position across the cutting edge. Table 16.10 lists angles that are approximately equivalent for a lathe tool and a drill. The first two drill items in Table 16.10 vary with radius (r) across the cutting edge. By use of the equivalence noted in Table 16.10, the variation of normal rake angle (α_n), effective rake angle (α_e), and inclination angle (i) with radius r may be obtained. For example, the values of Fig. 16.34 were obtained (Shaw 1954) for a standard drill of 0.750 in (19.05 mm) diameter.

TABLE 16.10

Equivalent angles for lathe tool and drill

Lathe tool	Drill
Side rake angle α_s	$\sin^{-1}(w/2r)$
Back rake angle α_b	helix angle, δ
Side cutting edge angle C_s	$\frac{1}{2}$ point angle, p

where w = web thickness
r = radius to point in question

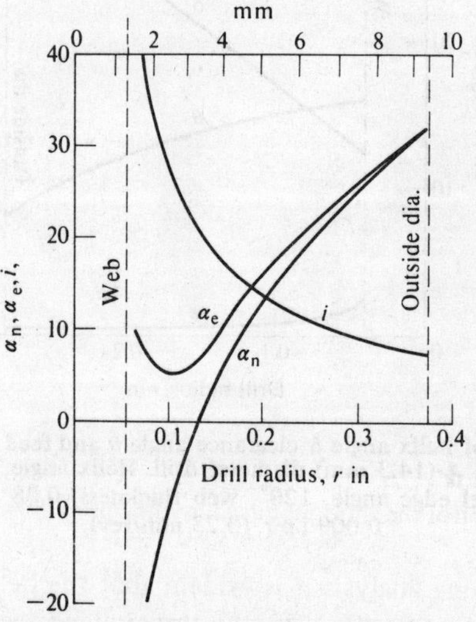

FIG. 16.34. Variation of inclination angle i, normal rake angle α_n and effective rake angle α_e across cutting edge of representative 0.750 in (19 mm) drill. Helix angle, 32°; point angle, 118°; web thickness, 0.110 in (2.79 mm)

The inclination angle (i) is seen to increase as the center of the drill is approached, while the normal rake angle (α_n) decreases and even assumes large negative values. However, the effective rake angle (α_e) does not become negative as a consequence of the large inclination angles near the drill point. The action of the drill near the web might be described as a slicing action due to the large inclinations obtaining. The support at the cutting edge, which varies inversely with the magnitude of α_n is seen to be greatest near the point of the drill. This is fortunate since the action of a drill at its very center does not involve cutting but might better be described as extrusion.

The manner in which the metal directly under the chisel edge of the drill is extruded is shown in Fig. 16.35. In producing this specimen a pilot hole the size of the web thickness of a larger drill was made in a brass block and filled with lead. When the larger drill was suddenly stopped the situation shown in Fig. 16.35 resulted, where two distinct types of chips are evident—the one produced by cutting and the other by a combination of cutting and extrusion.

FIG. 16.35. Comparison of the chips produced by the lips of a drill point and the chisel edge.

From the foregoing analysis it is evident that the helix angle (δ), web thickness (w), and point angle ($2p$) are the three quantities of greatest significance in determining drill performance. The first of these is controlled by the drill manufacturer while the latter two quantities can be altered to a certain extent by the user. The effect upon α_n and α_e of a variation of each of the quantities δ, p, and w may be different at the periphery and at the point of a drill. In Table 16.11 the influence of changes in δ, p, and w upon the significant drill angles are given for positions near the periphery and also near the chisel edge. In this table a plus sign signifies an increase, a minus sign a decrease, while zero indicates a insignificant change in the angle in question.

The orders of magnitude of the changes in Table 16.11 are illustrated by the numerical values of Tables 16.12 to 16.14. These quantities were computed for the numerical example involving a 0.750 inch (19 mm) drill given

TABLE 16.11

Influence of a change of drill dimensions upon drill angles

For increase of	For points near the drill periphery						For points near the chisel edge					
	i	C_s	α_b	α_s	α_n	α_e	i	C_s	α_b	α_s	α_n	α_e
Web thickness (w)	+	0	0	−	0	0	+	−	0	−	−	0
Point angle ($2p$)	0	+	0	−	0	0	+	+	0	0	+	+
Helix angle (δ)	0	0	+	+	+	+	0	0	+	0	+	+

TABLE 16.12

Representative changes in drill point angles due to changes in web thickness (w) only. Other conditions are the same as those for Fig. 16.34

Web thickness w, in (mm)	Radius r, in (mm)	degrees					
		i	C_s	α_b	α_s	α_n	α_e
0.05 (1.27)	0.375 (9.53)	3.2	58.9	32.0	17.3	33.0	33.1
0.110 (2.79)	0.375 (9.53)	7.2	58.7	32.0	13.2	31.8	32.3
0.150 (3.81)	0.375 (9.53)	9.9	58.5	32.0	10.2	30.6	31.5
0.050 (1.27)	0.100 (2.54)	12.4	58.1	9.5	−8.9	2.9	5.5
0.110 (2.79)	0.100 (2.54)	28.1	54.3	9.5	−28.2	−10.2	5.5
0.150 (3.81)	0.100 (2.54)	40.0	47.9	9.5	−44.3	−21.3	12.0

TABLE 16.13

Representative changes in drill point angles due to changes in point angle (2p) only. Other conditions are the same as those for Fig. 16.34.

Point angle $2p$ deg	Radius r in (mm)	degrees					
		i	C_s	α_b	α_s	α_n	α_e
100	0.375 (9.53)	6.3	49.5	32.0	21.1	34.9	35.5
118	0.375 (9.53)	7.2	58.7	32.0	13.2	31.8	32.3
140	0.375 (9.53)	7.9	69.7	32.0	4.8	29.8	30.5
100	0.100 (2.54)	24.9	44.9	9.5	−26.1	−11.5	1.0
118	0.100 (2.54)	28.1	54.3	9.5	−28.2	−10.2	5.5
140	0.100 (2.54)	31.1	66.5	9.5	−29.9	−4.3	12.5

TABLE 16.14

Representative changes in drill point angles due to changes in helix angle (δ only). Other conditions are the same as those for Fig. 16.34.

Helix angle δ deg	Radius r, in (mm)	degrees i	C_s	α_b	α_s	α_n	α_e
20	0.375 (9.53)	7.2	58.7	20	4.2	17.4	18.0
32	0.375 (9.53)	7.2	58.7	32	13.2	31.8	32.3
40	0.375 (9.53)	7.2	58.7	40	19.9	43.7	44.5
20	0.100 (2.54)	28.1	54.3	5.6	–30.4	–13.3	2.5
32	0.100 (2.54)	28.1	54.3	9.5	–28.2	–10.2	5.5
40	0.100 (2.54)	28.1	54.3	12.7	–26.4	–5.3	9.0

previously. From this analysis the following observations may be made regarding the influence of changes of w, p, and δ.

1. An increase of web thickness (w) has an insignificant influence at the periphery of the drill. The main effect of an increase in web thickness is a strengthening of the cutting edge near the point (as indicated by a decrease in α_n) and a smaller increase in effective rake angle. Larger values of web thickness would thus be indicated when strong materials must be drilled, for under such conditions failure at the drill point is most common. It would appear advisable to increase w for any drill that tends to fail at the center.

2. An increase of point angle also has a much smaller influence at the periphery than at the center of a drill. When p is increased the effective rake angle is increased near the drill center while the normal rake angle is increased a corresponding amount. This provides freer cutting at the point of the drill but less support for the cutting edge. The smaller change at the periphery of the drill leads to a slightly stronger edge but slightly higher forces. A small point angle would be indicated for the drilling of relatively weak metals such as aluminum alloys where support of the cutting edge can be sacrificed in the interest of decreased cutting forces at the periphery. While a drill usually tends to fail at the periphery when operating in a soft metal, failures at the center of the drill are most common when drilling hard metals and the results of Table 16.13 would indicate that a large point angle would want to be used for a strong metal.

3. Unlike the two previous quantities the helix angle (δ) has its greatest effect near the periphery of the drill. An increase in helix angle causes a large increase in both α_n and α_e at the circumference of the drill and a smaller increase in both quantities near the center. Thus, increased helix angle results in freer cutting but correspondingly less support all across the cutting edge but particularly at the drill periphery. A large helix angle with small point angle and thin web would be indicated for weak materials and the reverse for strong metals such as the high-temperature alloys and stainless steels.

Drill Dimensions

The dimensions of the standard drill point are as follows:

helix angle = 28° to 32°
point angle = 118°
clearance angle = 8° to 12°

The shape of the milled flutes are such that the cutting edge is straight when the proper point angle is ground on the drill. When very hard and strong metals are ground it is customary to use a high point angle (approaching 140°) and a smaller clearance angle (as low as 5°). In such a case it would probably be advisable to use a lower helix angle to obtain greater support at the cutting edge, but drills of other than standard helix angle are not easily obtained.

For drilling hard alloy steels the following angles are frequently used:

helix angle = 28° to 32°
point angle = 135°
clearance angle = 6° to 9°

For drilling very soft metals point angles as low as 100° and clearance angles as high as 12° are not uncommon.

The chisel edge angle is a rough measure of the clearance angle of the drill. With the standard 118° point an 8° clearance angle should result in a chisel edge angle of 120° while a 12° clearance angle corresponds to a 135° chisel edge angle. Similarly for the 135° point angle, the chisel edge angle would be 115° for a 6° clearance and 125° for a 9° clearance angle.

The web thickness of the average drill represents a decreasing percentage of the drill diameter as the size of the drill increases. This is illustrated in Table 16.15 where recommended web thicknesses are given. The web of a drill, which is the metal that connects the flutes, is generally made to increase as the shank of the drill is increased in the interest of greater rigidity. This increase in web thickness may be as much as 50 percent and therefore when a drill is sharpened a number of times it may be found to have too large a web thickness. Web thinning is therefore practiced, but a better solution is for manufacturers to provide the useful portion of the drill with a constant web thickness. If a drill has a web that is thicker than necessary the thrust force will be needlessly high and if such a drill is used at excessive feed rates a crack may develop extending up the web of the drill. In such instances a decrease in feed or web thickness is indicated. The clearance at the web should also be checked in cases where the web tends to split since insufficient clearance can also cause this difficulty.

It is customary to provide a drill with decreasing diameter toward the shank (0.0005 to 0.001 in in^{-1}) in order to provide clearance of the drill body in the finished hole. When a drill is sharpened a number of times it must therefore be expected that the drill will cut undersize. The drill is normally supplied with margins for purposes of guiding. These margins are without circumferential clearance and should be as small as possible in the interest of decreased friction. The margin of a one-inch drill is normally about 0.1 inch (2.54 mm).

In the interest of good drill performance and life it is extremely important that both lips of a drill do equal work. The length of the two cutting edges should be checked for equality as well as the height of the two lips. An asymmetrical drill point operating at normal feed rate will produce inaccurate holes and will last a relatively short time since one lip will be overloaded.

TABLE 16.15
Recommended web thickness values

Drill dia., in (mm)	Web thickness % of dia.
$\frac{1}{8}$ (3.18)	20
$\frac{1}{4}$ (6.3)	17
$\frac{1}{2}$ (12.5)	14
1 (25.4)	12
>1 (>25.4)	11

Operating Conditions

It is customary to feed drills at a rate dependent on the size of the drill but independent of material, the speed of the drill being adjusted to the particular metal cut. The reason for this is that the flutes of a drill can handle only a certain chip volume and this volume is about proportional to the diameter of the drill. It is therefore common practice to feed drills at a rate that is proportional to the drill diameter in accordance with the relation:

$$f = \frac{d}{65} \tag{16.51}$$

where f is the feed rate and d is the drill diamter. A one-inch diameter drill would thus be fed at about 0.015 i.p.r. (0.38 mm/rev). When holes that are long than $3d$ are drilled the feed should be reduced.

Recommended speeds for HSS drills are given in Table 16.16. Carbon steel drills should be operated at speeds of 50 percent those given in Table 16.16.

TABLE 16.16
Recommended speeds for HSS drills

Material	f.p.m.	(m min^{-1})
Soft cast iron	150	45.7
Medium hard cast iron	80	24.4
Mild steel	90	27.4
Alloy steel	60	18.3
Tool steel	50	15.2
Brass and bronze	100–300	30.5–91.5
Copper	150	45.9
Aluminum and magnesium alloys	300–400	91.5–122

While the best combination of speed and feed may differ somewhat from the values recommended above due to differences in relative chip volume, material structure, cutting fluid effectiveness, depth of hole, and condition of drill and machine, the foregoing values are of use as a general guide. If the outer corner of the drill is found to wear prematurely too high a speed is indicated, while if the cutting edges are found to chip, too great a feed or too much clearance on the drill point may be the cause.

It is customary to drill all of the metals of Table 16.16 with a copious supply of an oil or soluble cutting fluid with the exception of cast iron and magnesium alloys which are normally cut in air. If deep holes are drilled it may be necessary to supply the fluid under high pressure (10 to 20 atm) in order to lubricate the cutting point and help convey the chips along the flutes of the drill. In such applications an oil-hole drill that is provided with a hole extending down the drill is useful.

Drilling Forces

Although an ordinary twist drill operates in essentially the same way as a single point tool, it is extremely complex geometrically. Analyses of the mechanics of a twist drill are to be found in Oxford (1955). The following quantities influence the torque M and thrust T to which a drill is subjected in producing a hole:

Work material and structure
Drill diameter (d), in
Helix angle (θ), deg

$$\text{Length of chisel edge } (c) = \frac{\text{web thickness}}{\text{cosine (chisel-edge angle)}}$$

$$= \frac{w}{\cos(120)} = 1.15\, w, \text{ approx.}$$

Point angle ($2p$), deg
Number of cutting edges (n)
Feed (f), i.p.r. (mm/rev)
Cutting fluid
Drill sharpness

From the treatment of cutting-tool forces in Chapter 3, we should not expect the drilling speed (V), the drill clearance, or small changes in flute shape to be important. The flute geometry behind the cutting edge mainly influences the conveyance of chips and assumes an important role only when chips tend to jam in the flutes. Thus flute-shape details for a normally functioning drill may be ignored.

Useful empirical equations relating torque and thrust with feed and drill diameter have been developed by Boston and Oxford (1930) and Kronenberg (1935). A more analytical approach has been presented by Shaw and Oxford, Jr. (1957) which will be summarized here:

Three basic actions are involved in the performance of a two-flute twist drill:

1. Cutting by the two lips.
2. Cutting action at the web.
3. Extrusion by the web.

Only the first two of these contribute significantly to the torque (M) while all three are important to the thrust (T).

Variables of primary important to drill torque are listed in Table 16.17. Before dimensional analysis:

$$M = \Psi_1(u, c, f, d) \tag{16.52}$$

After dimensional analysis

$$\frac{M}{d^3u} = \Psi_2\left(\frac{f}{d}, \frac{c}{d}\right) \tag{16.53}$$

In the absence of a size effect (increase in u with volume deformed at one time) the torque should be expected to vary as the projected area of the drill or as d^2. Thus, in the absence of a size effect

$$\frac{M}{d^3u} = \frac{f}{d}\,\Psi_3\left(\frac{c}{d}\right) \tag{16.54}$$

Shaw and Oxford (1957) have shown that just as the specific cutting energy (u) varies inversely with $t^{0.2}$ in ordinary two-dimensional orthogonal cutting, in drilling, u varies inversely with $(fd)^{0.2}$. Therefore, when drilling a given work material having a Brinell hardness H_B

$$u \sim \frac{H_B}{(fd)^{0.2}} \tag{16.55}$$

Combining equations 16.54 and 16.55

$$\frac{M}{d^3H_B} = \frac{f^{0.8}}{d^{1.2}}\,\Psi_3\left(\frac{c}{d}\right) \tag{16.56}$$

A large number of experiments (about five-hundred holes) were run on heat-treated chrome–nickel steel (AISI 3245) with sharp drills cutting dry with

Workpiece hardness (H_B) = 200 kg mm^{-2}
Drilling speed 20 f.p.m. (6.1 m min^{-1})
$c/d \cong 0.18$
Drills of different diameter ($\frac{1}{4}$ to 1 inch = 6.35 to 25.4 mm)

TABLE 16.17
Variables of primary importance to torque

Quantity	Symbol	Dimensions
Drill torque	M	FL
Feed per revolution	f	L
Drill diameter	d	L
Length of chisel edge	c	L
Specific cutting energy of work	u	FL^{-2}

Figure 16.36a shows measured torque (M) plotted against feed (f) while Fig. 16.36b shows measured torque (M) plotted against drill diameter (d). The curves have slopes of 0.8 and 1.8 respectively in agreement with eqn (16.56). The relation between drill torque (M) and specific drilling energy (u) may be readily shown to be

$$u = \frac{8M}{fd^2} \tag{16.57}$$

Figure 16.37 shows u obtained from eqn (16.57) for measured values of M plotted against the product (fd) on log-log coordinates. The experimental points are seen to lie on a straight line in excellent agreement with eqn (16.55). For these tests eqn (16.56) may be written

$$\frac{M}{d^3 H_B} = 0.087 \frac{f^{0.8}}{d^{1.2}} \tag{16.58}$$

if H_B is in p.s.i. (1420 times H_B in kg mm^{-2}) and all other quantities are in inch and inch-pound units. While this equation is for the conventional value of c/d of 0.18, further studies have shown that drill torque is relatively insensitive to changes in c/d. Therefore eqn 16.58 may be used to estimate drilling torque (M) for any material having a Brinell hardness H_B. A similar equation in SI units is:

$$\frac{M}{d^3 H_B} = 0.0031 \frac{f^{0.8}}{d^{1.2}}. \tag{16.58a}$$

The non-dimensional quantity for thrust (T) corresponding to ($M/d^3 H_B$) for torque will be ($T/d^2 H_B$). Since in orthogonal cutting the thrust component of force (F_Q) is usually proportional to the cutting component ($F_P \cong 2F_Q$), it is reasonable to consider the cutting part of the thrust force to be proportional to the torque component. This leads to the following relation for drill thrust where the second term on the right (Ψ_4) is the contribution to thrust due to 'extrusion' at the web of the drill.

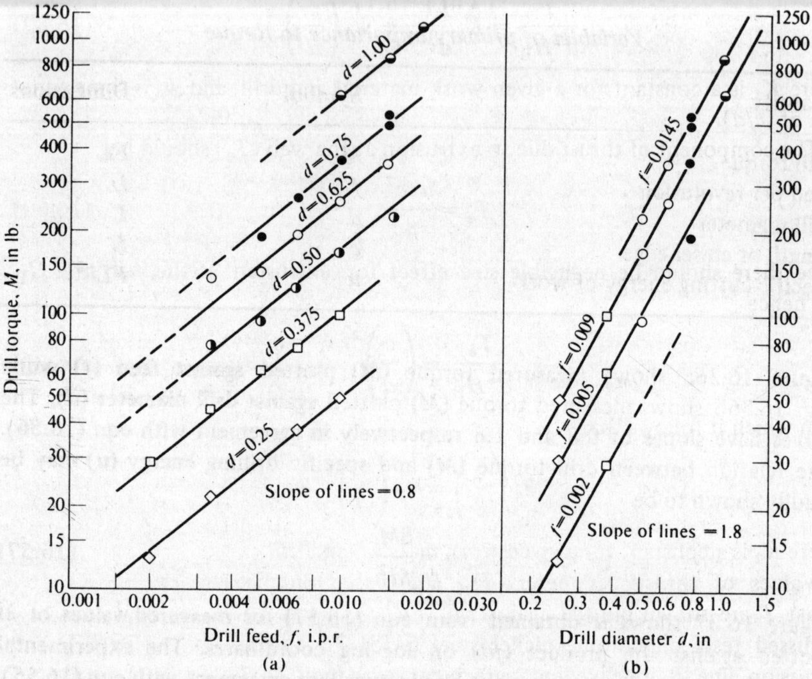

FIG. 16.36. Variation of drill torque with (a) feed and (b) diameter. Work material AISI 3245 of Brinell hardness 196–207; drill speed, 20 f.p.m. (6.1 m min^{-1}); cutting fluid, none.

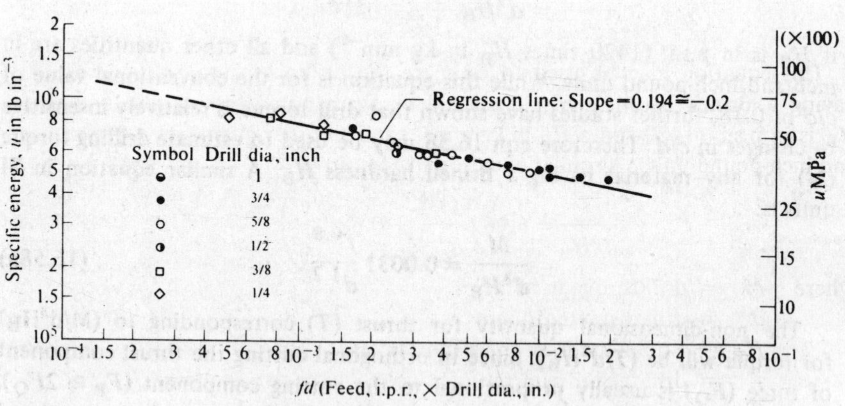

FIG. 16.37. Plot of u against fd for data of Fig. 16.36.

$$\frac{T}{d^2 H_B} = K_1 \frac{f^{0.8}}{d^{1.2}} + \Psi_4\left(\frac{c}{d}\right) \tag{16.59}$$

where K_1 is a constant for a given work material and drill, and Ψ_4 is some function of (c/d).

The component of thrust due to extrusion at the web (T_e) should be

$$T_e = \frac{u\pi c^2}{4} \tag{16.60}$$

Since there should be negligible size effect for 'extrusion' at the web $u \sim H_B$ and hence

$$\frac{T_e}{d^2 H_B} \sim \left(\frac{c}{d}\right)^2 \tag{16.61}$$

from which it follows that

$$\Psi_4\left(\frac{c}{d}\right) = K_2\left(\frac{c}{d}\right)^2 \tag{16.62}$$

where K_2 is a constant for a given work material and drill.

Values of thrust parameter (T/d^2H_B) based on measured values of drill thrust T are shown plotted against $(f^{0.8}/d^{1.2})$ in Fig. 16.38 for the previously discussed tests on chrome-nickel steel (AISI 3245). Here it is seen that the regression line is in excellent agreement with eqn 16.59. Since the value of c/d for all these drills was approximately 0.18, it follows from the regression equation that

$$\frac{T}{d^2 H_B} = 0.195 \frac{f^{0.8}}{d^{1.2}} + 0.68\left(\frac{c}{d}\right)^2 \tag{16.63}$$

This equation may be used to estimate the drilling thrust for any material having a Brinell hardness H_B. It should be again noted that in this equation H_B is in p.s.i. (1420 times H_B in kg mm^{-2}) and all other quantities are in inches and inch-pound units. A similar equation in SI units is

$$\frac{T}{d^3 H_B} = 6.962 \frac{f^{0.8}}{d^{1.2}} + 0.68\left(\frac{c}{d}\right)^2 \tag{16.63a}$$

where M = drill torque in N m
d = drill diameter in mm
H_B = Brinell hardness in kg mm^{-2}
f = drill feed in mm/rev
T = drill thrust in N

When the helix angle was varied over the wide range from 18° to 30°, the results shown in Fig. 16.39 were obtained. This shows that an increase in helix

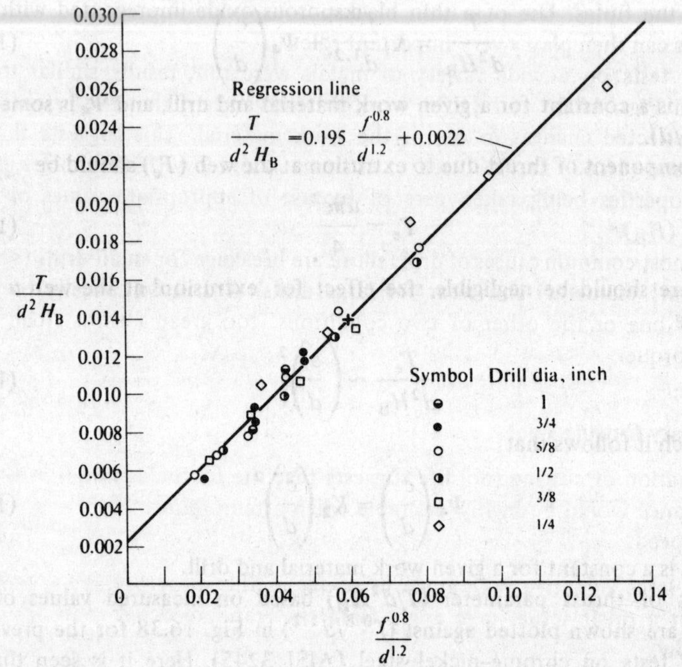

FIG. 16.38. Nondimensional plot of T/d^2J_B against $f^{0.8}/d^{1.2}$ when drilling AISI 3245 steel.

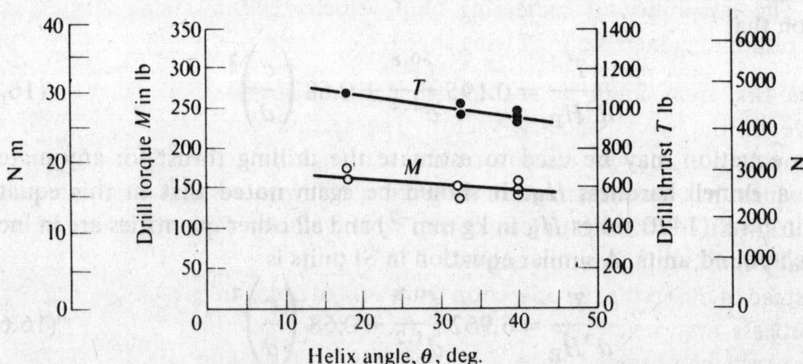

FIG. 16.39. Variation of drill thrust and torque for drills of different helix angle. Work material AISI 3245 of Brinell hardness 196–207 kg mm^{-2}; drill speed, 16 f.p.m. (4.9 m min^{-1}) feed, 0.008 i.p.r. (0.20 mm/rev); drill diameter, $\frac{1}{2}$ in (13 mm); cutting fluid, none; $c/d = 0.245$.

angle causes a relatively small decrease in both T and M. However, it should be noted that this is the case where no chip jamming in the flutes occurred. If chip jamming is encountered both thrust and torque will rise greatly and under such conditions helix angle can have a profound influence on promoting chip flow up the flutes. Use of a thin black porous oxide impregnated with oil on the flutes can then play a very important role.

When tests on a wide variety of metals were run, results similar to those shown in Figs. 16.36 and 16.38 were obtained except that the heights of the curves reflected changes in H_B of the work material. This suggests that eqns (16.58) and (16.63) may be used to estimate M and T with differences in workpiece properties being taken care of by use of appropriate values of Brinell hardness (H_B).

The most common causes of drill failure are breakage for small drills ($< \frac{1}{16}$ inch [1.59 mm] diameter) and wear for large drills. A drill will usually break as a result of one or the other of two conditions: too great a thrust load or too great a torque.

The Rotary Cutting Tool

Consideration of cutting tool life suggests that the following conditions of tool performance would be highly desirable when attempting to cut metal at a high rate of speed.

1. A large effective rake angle in so far as the shear process is concerned but a small actual rake angle from the standpoint of strength and heat flow.
2. Positive means for carrying the fluid to the tool point at high cutting speeds as in the case of a journal bearing.
3. The possibility of increasing chip velocity without necessitating a corresponding increase in cutting speed.

The first item could be accomplished by means of inclination of the tool relative to the work velocity vector. The second two items can be achieved by providing a sidewise motion of the tool relative to the work. Indeed, if a tool which has zero inclination is given a sidewise component of motion V_r (Fig. 16.40a) as well as the customary work velocity V_W, there will be a kinematically induced inclination angle i.

Instead of observing the operation from a fixed point in space as in Fig. 16.40a, the process may be observed from a fixed point on the cutting tool. Such a picture may be obtained by simply translating everything in Fig. 16.40a to the left with a velocity V_r. This will bring the tool to rest, will cause the workpiece to move in an inclined direction with a velocity V, and will cause the chip to appear to flow off to the left rather than to the right, as shown in Fig. 16.40b.

The velocity of the chip across the tool face will be equal to V_{Cr} which is the vector sum of velocities V_C and V_r. The solid lines in Fig. 16.40b represent the cutting process that is kinematically equivalent to that in Fig. 16.40a. This equivalent process is seen to be that of cutting with a simple inclined cutting tool. Thus one of the significant features of a cutting tool with sidewise motion

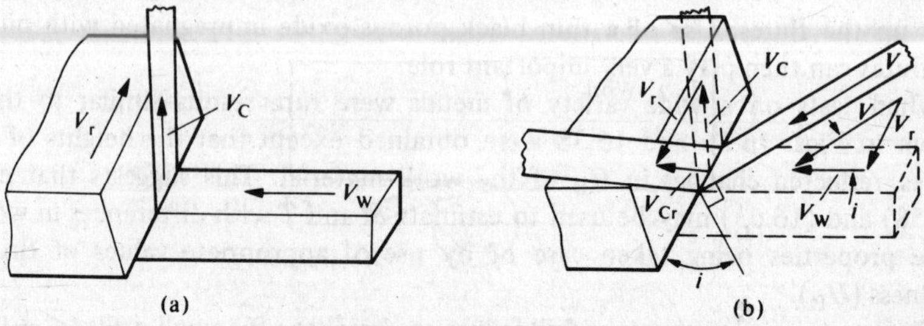

FIG. 16.40. Rotary cutting process. (a) As observed from fixed point in space. (b) From fixed point on tool.

is to provide what is equivalent to an inclined cutting edge, the effective inclination angle i being given by

$$\tan i = \frac{V_r}{V_W} \tag{16.64}$$

The practical way of providing the sidewise motion of the tool is to use a continuous cutting edge in the form of the rotary tool shown in Fig. 16.41. Here, a particular portion of the cutting edge is in operation for a very brief period, which is followed by a much longer rest period during which the thermal energy associated with cutting has ample opportunity to be dissipated to the bulk of the cutter. In this respect the rotary tool resembles a multitoothed tool such as a milling cutter.

A detailed analysis of the power-driven rotary tool has been given by Shaw, Smith, and Cook (1952). Experimental tests were performed on a rotary tool as shown in Fig. 16.42. Representative data are shown in Fig. 16.43 where the principal cutting variables are shown plotted vs. the kinematically induced inclination angle. It is evident that the angle of inclination (i) is the predominant variable that controls performance. The following general observations may be made regarding this motor driven rotary tool.

1. The effective rake angle α_e is seen to increase as the angle of inclination is increased by increasing the tool velocity V_r. It is evident that effective rake angles of over 60° may be obtained with a rotary tool having an actual normal rake angle of but 10°.

2. The effective shear angle ϕ_e is seen to increase as the effective rake angle increases, from a value of about 10° for the stationary tool to a value of over 40° when the angle of inclination was about 70°.

3. The shear stress on the shear plane is seen to increase with i, while the shear strain decreases. This same paradoxical behavior was also observed for conventional oblique cutting tools.

4. The coefficient of friction on the tool face is seen at first to decrease with increased inclination and then to increase. The minimum value of μ appears

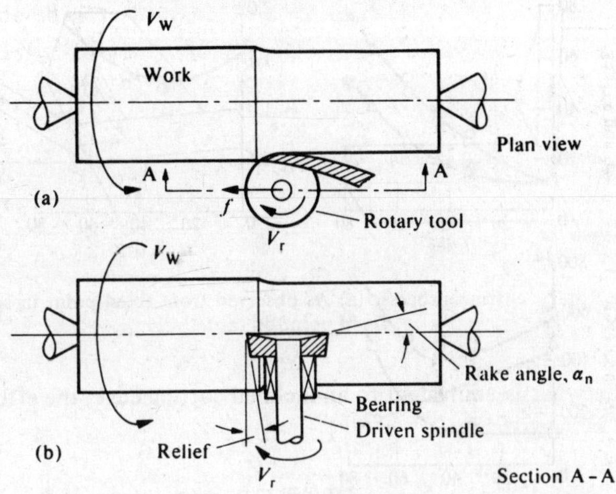

FIG. 16.41. Rotary tool shown in use in a turning operation.

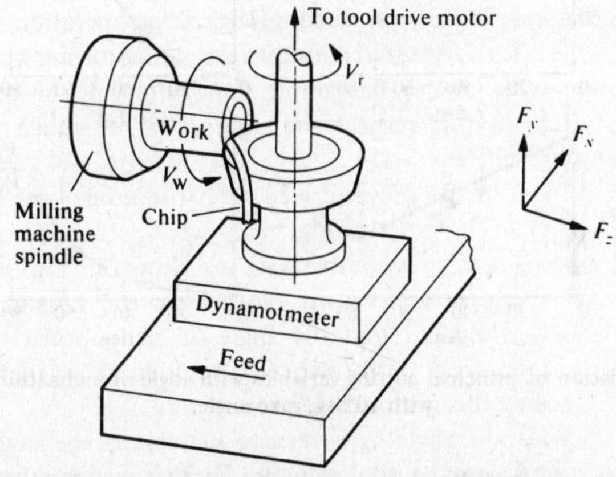

FIG. 16.42. Schematic diagram showing relative arrangement of workpiece and motor-driven rotary tool.

to occur at a value of i of about 30°. This result is also in agreement with the friction characteristic of the previous investigation of a conventional oblique tool, where μ was observed to decrease for values of i up to 30°.

5. The total energy per unit volume (u) has a minimum value at an inclination angle of about 45°. This minimum value is seen to be due to a tendency for the shear energy per unit volume u_S to decrease with increased i (due predominantly to the decrease in γ with increased α_e), while the friction energy

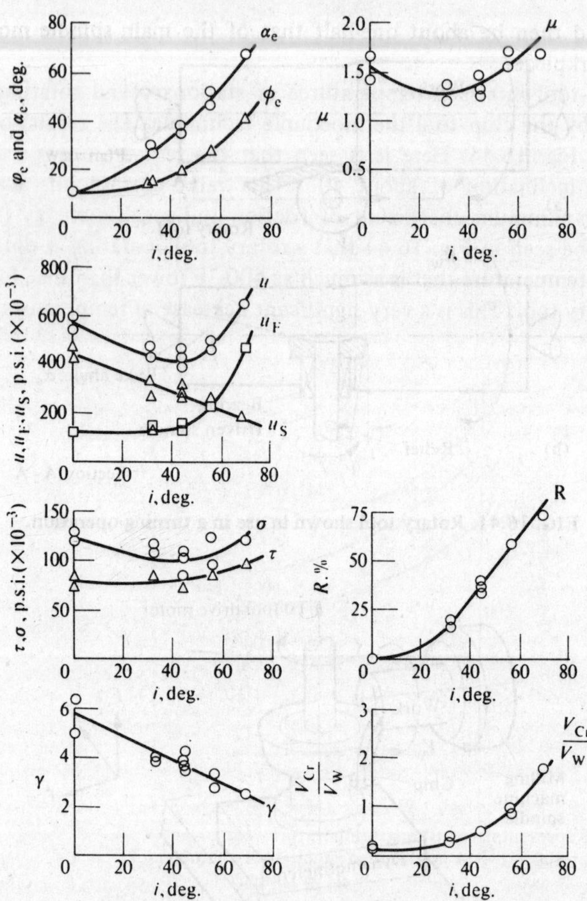

FIG. 16.43. Variation of principal cutting variables with angle of inclination for rotary tool with 10 deg. rake angle.

per unit volume u_F exhibits the opposite trend (due to the tendency for the coefficient of friction to increase with increased α_e). The friction energy begins to increase at a rapid rate with increased inclination at a value of *i* of about 50°.

6. The quantity *R* is the ratio of the power required to drive the tool to the total power in the rotary-tool process. For the tool investigated here (α_n = 10°) about one-third of the total power goes to drive the tool, while the remaining two-thirds is consumed in driving the workpiece, when the angle of inclination is about 45°. From the aforementioned minimum points in the curves of *u* and μ, it would appear that the optimum angle of inclination is somewhat under 45°. Therefore a rotary tool should be operated with a surface speed V_r just under that of work speed, V_W, and the capacity of the tool-drive

motor should then be about one-half that of the main spindle motor used to drive the workpiece.

The chip-tool interface temperatures of stationary and rotating tools were determined by the chip-tool thermocouple technique. The results of this study are given in Fig. 16.44. Here it is seen that the temperature is a minimum at an angle of inclination of about 40°. This value corresponds approximately to those for minimum coefficient of friction and specific energy (Fig. 16.43). It may also be seen in Fig. 16.44 that a rotary tool operating at optimum speed may have a temperature that is as much as 400 °F lower than that for an equivalent stationary tool. This is a very significant decrease in temperature.

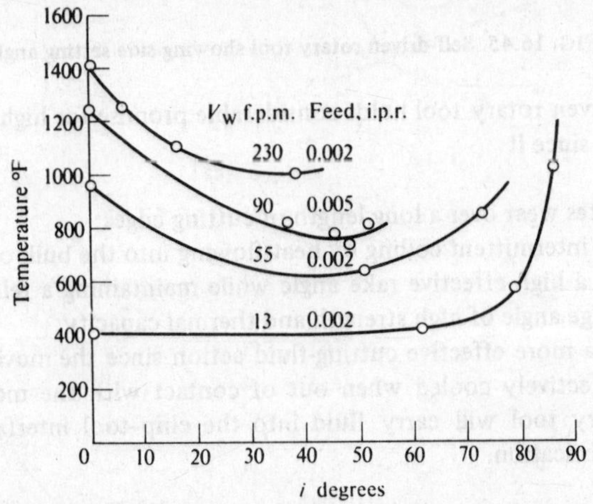

FIG. 16.44. Representative cutting temperatures for rotary tool. Work material SAE 1015 steel; tool material, 18–4–1 high-speed steel; tool diameter, 2.5 in (63.5 mm); rake angle $\alpha_n = 10$ deg.

If a rotary tool is mounted so that its axis is inclined to the vertical (with respect to Fig. 16.41b), no drive motor is required since the tool will then be driven by the chip (Shaw, Smith, Cook, and Loewen, 1963). This corresponds to a negative side rake angle that is equal to the side setting angle if the normal rake of the tool is zero (i.e. if the tool face is perpendicular to the tool axis). As illustrated in Fig. 16.45, as the side setting angle of the rotary tool is increased the rotational speed of the tool increases and cutting forces and temperatures drop. Optimum performance is obtained at an amazingly high setting angle of about minus 70°. This corresponds to a tool axis inclined only 20 degrees above the horizontal in Fig. 16.41b (in effect a side rake angle of −70°). At this steep side-setting angle, the speed of the chip just equals the surface rotational speed of the tool. There is then no slip between chip and tool and this is believed responsible for optimum performance occurring at such a high setting angle.

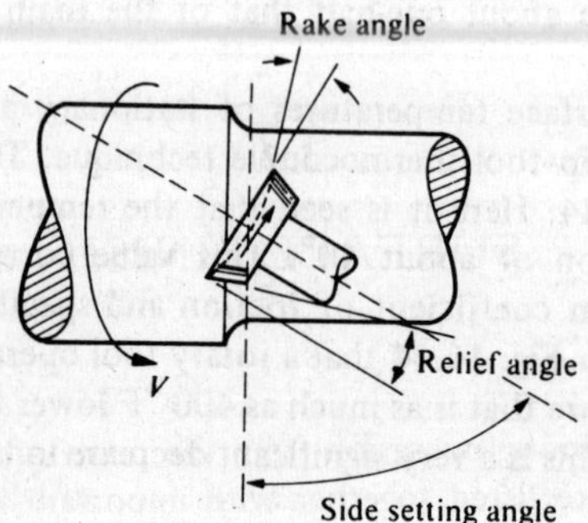

FIG. 16.45. Self-driven rotary tool showing side setting angle.

The self-driven rotary tool holds considerable promise for high-speed machining operations since it

1. Distributes wear over a long length of cutting edges;
2. Provides intermittent colling by heat flowing into the bulk of the tool;
3. Provides a high effective rake angle while maintaining a relatively massive tool wedge angle of high strength and thermal capacity;
4. Enables a more effective cutting-fluid action since the moving tool can be more effectively cooled when out of contact with the moving tool and the rotary tool will carry fluid into the chip-tool interface to provide better lubrication.

References

Armitage, J. B. and Schmidt, A. O. (1957). *Tool Engr* **27**, 36.
Boston, O. W. and Oxford, C. J. Sr. (1930). *Trans. Am. Soc. mech. Engrs* **52**, 5.
Kronenberg, M. (1935). *Machinery* **45**, 661.
Kronenberg, M. (1966). *Machining science and application.* Pergamon Press, Oxford.
Martelloti, M. E. (1941). *Trans. Am. Soc. mech. Engrs* **63**, 677.
Merchant, M. E. (1944). *Trans. Am. Soc. mech. Engrs* **66**, A-168.
Oxford, C. J., Jr. (1955). *Trans. Am. Soc. mech. Engrs* **77**, 103.
Shaw, M. C. (1954). *Metal cutting principles* (3rd edn). The Technology Press, MIT, Cambridge.
Shaw, M. C., Cook, N. H., and Smith, P. A. (1952). *Trans. Am. Soc. mech. Engrs* **74**, 1055.
Shaw, M. C. and Oxford, C. J., Jr. (1957). *Trans. Am. Soc. mech. Engrs* **79**, 139.
Shaw, M. C., Smith, P. A., and Cook, N. H. (1952). *Trans. Am. Soc. mech. Engrs* **74**, 1065.
Shaw, M. C., Smith, P. A., Cook, N. H., and Loewen, E. G. (1963). *Canadian Patent No. 665 459.*
Stabler, G. V. (1951). *Proc. Instn mech. Engrs* **165**, 14.

17

SURFACE FINISH

Table 14.1 shows a turning operation from the point of view of a system. Here a number of inputs are listed together with important internal items and outputs. The outputs are most important since they represent the end-goal of the operation. The integrity of the machined surface is frequently one of the important outputs.

Surface integrity is a term that involves several considerations: surface finish and freedom from cracks, chemical change, thermal damage (burn, transformation, and overtempering), and adverse (tensile) residual stress. The first of these (surface finish) is by far the most important for machining operations. The others are mainly a concern relative to ground surfaces. Where surfaces finish is relatively important, a finishing operation is usually involved.

Measurement and Specification

In engineering design and production, it is important to be able to specify the degree of surface roughness desired. Before 1930 this was done by use of tactual standards. This involved the use of a series of specimens that had different finishes. The man in the shop used these specimens by running his fingernail first across a standard tactual surface and then across the surface he was producing. When the two surfaces were felt to have the same roughness, the workpiece was considered to be smooth enough. In those applications where surface finish is important to the functioning of a device such as in face seals, ball bearings, gears, or cam surfaces or journals, it is found that performance varies linearly as the surface roughness varies logarithmically. This means that for there to be a twofold improvement in performance there must be a tenfold reduction in the mean peak-to-valley roughness. Because of this, a set of tactual standards will have peak-to-valley roughness values that stand in the following ratios: 1:2:4:8:16:32:64:128:256:512:1024 instead of being linearly spaced.

Surface finish is specified on a drawing as shown in Fig. 17.1. If a smoother surface is required than one of 64, a finish of 32 would be specified instead of one between 32 and 64.

In the early 1930s surface roughness was made quantitative by the appearance of stylus instruments that traversed a surface to be measured with a diamond stylus, using transducers to convert the vertical and horizontal motions of the diamond into recorded traces. In the USA, the vertical motion of the

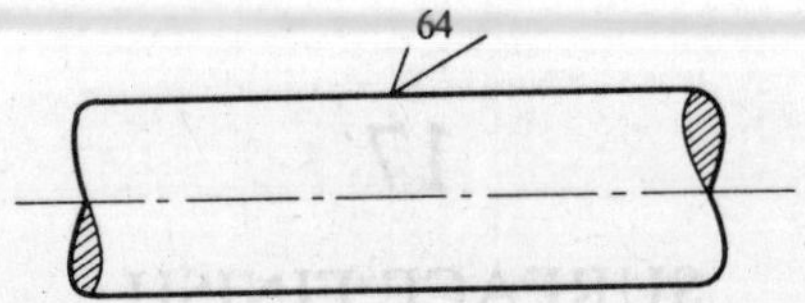

FIG. 17.1. Method of indicating the surface finish required on a machine part.

stylus was analyzed automatically and a meter read the square root of the mean of the vertical motions squared. This so-called root mean square (RMS) value in micro-inches (μ in) was specified on drawings as shown in Fig. 17.1 in terms of the progression 1, 2, 4, 8, 16, 32, 64, 128 μ in.

On the continent of Europe and in Japan, a similar development took place but the characteristic maximum peak-to-valley roughness (R_t) is used in place of the RMS value. In Great Britain, the centerline average (CLA) value of the surface is used in place of the RMS value. This is now also the standard in the USA and is called arithmetic average (AA) roughness. The CLA or AA roughness (R_a) is obtained by measuring the mean deviation of the peaks from the centerline of a trace, the centerline being established as the line above and below which there is an equal area between the center line and the surface trace. There is little difference between the CLA and RMS values for a given surface. (RMS = 0.90 AA for a sine wave.) The only virtue of the RMS value is that it is easily measured using an a.c. voltmeter. However, the CLA value is more readily interpreted physically. Modern tracer instruments have the capability of giving the AA value as a meter reading and the shape of the surface as a magnified tracing. The vertical magnification of the tracing is usually arranged to be five to ten times as great as in the horizontal direction since practical surfaces have asperities that are very widely spaced relative to their differences in height. Figure 17.2 is a typical stylus trace of a relatively smooth surface (AA = 10 μ in = 0.25 μ m).

The peak-to-valley roughness for a ground surface is usually about five times the AA value. However, other ratios pertain for other finishing methods. For example, this ratio is closer to 10:1 for lapped surfaces.

The radius of curvature of the standard diamond stylus is 0.0005 in (12.5 μ m) although in research, instruments with stylii having a radius as small as 0.0001 inch (2.5 μ m) are used.

In order to obtain the true profile of a surface without the influence of the curvature of the tracer, a taper section of the surface may be used. After plating the surface with nickel to protect it, a new flat surface is ground on the nickel which makes a small angle with the original surface as shown in Fig. 17.3. The interface between the steel and nickel may then be studied.

A taper section magnifies the peak-to-valley distance and makes it possible to observe the profile of a surface under study at different magnifications in

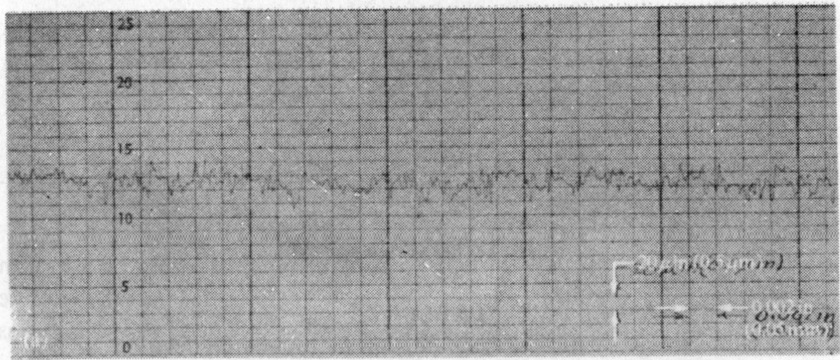

FIG. 17.2. (a) Stylus trace of a relatively smooth ground surface, AA $= 10\ \mu$ in (0.25 μ m) (b) Scanning electron photograph of surface.

the horizontal and vertical directions. By making the inclination of the two surface differ by $a = 2^\circ\ 17'$, a magnification of cse$(2^\circ\ 17') = 25$ times is obtained. If the polished new surface is viewed with a microscope at 100X, then the horizontal magnification of the nickel surface will be 100X while the vertical magnification is 2500X.

Table 17.1 gives values of surface finish R_t and R_a for a variety of surfaces of different roughness.

In addition to surface roughness, surfaces may exhibit waviness. The waviness of a surface is the vertical distance between peaks and valleys of relatively long-wavelength irregularities while the corresponding peak-to-valley height for short-wavelength irregularities is surface roughness. The maximum wavelength for surface roughness is usually taken to be 0.03 in (0.75 mm). By filtering the

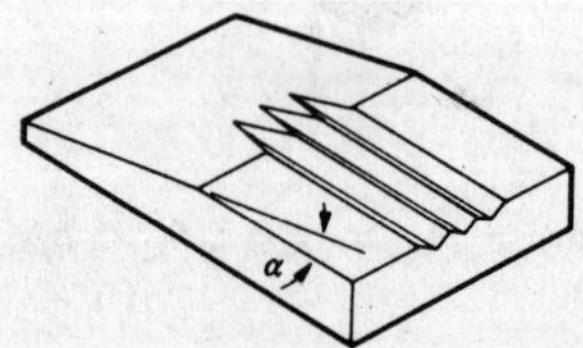

FIG. 17.3. Tapered section.

TABLE 17.1
Range of surface roughness values obtainable with different finishing operations

Finish	R_t μ in (μm)	R_a μ in (μm)
Very rough surfaces (brick, castings, etc.)	>40 000 (> 10^3)	~ 0.5 × 10^6 (~ 12500)
Rough machining	10^4 to 10^5 (250 to 2500)	>10^3 (>25)
Medium machining (furrows visible–rough grind)	2000–10 000 (50 to 250)	>400 (>10)
Fine machining (furrows visible–fine grind)	200–400 (5 to 10)	30–100 (0.8 to 2.5)
Very fine grind (turned)	20–200 (0.5 to 5.0)	5–30 (0.13 to 0.8)
Honed	40–80 (1 to 2)	2–10 (0.05 to 0.3)
Lapped, polished	2–40 (0.05 to 1)	>4 (>0.1)
Fine metallographic polish	>10 (>0.3)	~ 1 (~ 0.03)
Cleavage plane (mica)	>1 (>0.03)	<1 (<0.03)

output from a stylus-type tracer instrument, irregularities having a wavelength greater than 0.03 inch may be removed and it is possible to measure surface roughness without interference due to waviness. Figure 17.4 illustrates the difference between surface roughness and waviness. Figure 17.4a is a flat surface with roughness while Fig. 17.4b is a wavy surface with roughness.

Finish-turning is characterized by relatively small depths of cut and light feeds. The depth of cut will normally be less than 0.06 in (1.5 mm) and the feed less than 0.006 i.p.r. (0.15 mm/rev). The items of major interest in finish-machining are:

1. Dimensional accuracy.
2. Surface finish.
3. Tool life.

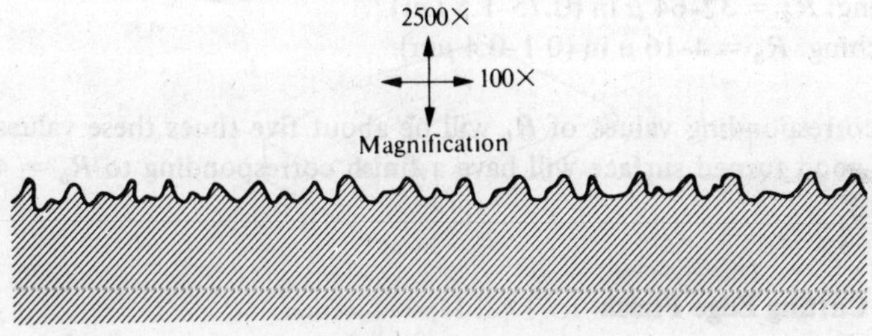

(a) Flat surface with roughness

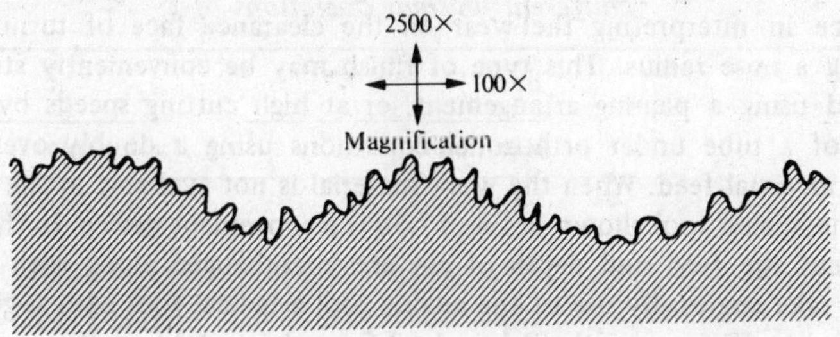

(b) Wavy surface with roughness

FIG. 17.4. Flat and wavy surfaces with surface roughness.

Dimensional accuracy is mainly a matter of avoiding errors in longitudinal and circumferential form. Errors of longitudinal form result from static deflection of the spindle and workpiece under cutting forces and thermally induced stresses in the machine. Errors in circumferential form result from run out of the spindle and from vibration of the tool or workpiece. The accuracy achieved in any particular case mainly depends upon the stiffness and stability of the machine tool. Representative studies of errors in form have been presented by Schuler (1957) and the dynamic aspects of machine-tool performance are to be found in many places, for example: Hahn (1953), Doi and Kato (1956), Tobias and Fishwick (1958), Danek, Polacek, Sparek, and Tlusty (1962), Tobias (1965), and Peters (1963).

There are two distinct types of finish encountered in a turning operation:

1. The finish produced by the primary cutting edge;
2. The finish produced by the secondary cutting edge.

The first of these normally pertains to surface broaching and form-turning, while the second is characteristic of conventional turning using a tool having a nose radius.

The range of roughness values to be expected in finish-machining is as follows:

Turning: $R_a = 32\text{-}64\ \mu$ in (0.75-1.5 μm)
Broaching: $R_a = 4\text{-}16\ \mu$ in (0.1-0.4 μm).

The corresponding values of R_t will be about five times these values of R_a. (A very good turned surface will have a finish corresponding to $R_a = 4\ \mu$m or 160 μ in.)

Primary Cutting Edge Finish

Knowledge of the finish produced by the primary cutting edge is not only of direct importance in those processes where the resultant surface is generated by the major cutting edge (i.e. form-cutting with radial feed) but it is also of importance in interpreting the wear on the clearance face of turning tools employing a nose radius. This type of finish may be conveniently studied at low speed using a planing arrangement, or at high cutting speeds by cutting the end of a tube under orthogonal conditions using a doubly overhanging tool and an axial feed. When the work material is not available in the form of a tube, the analog tool shown in Fig. 17.5b is a convenient device. In the analog arrangement the feed and depth of cut are interchanged along with the side and back rake angles. By use of the analog tool a helical tool path is generated which has a sufficient width (0.1 inch, 2.5 mm) to enable surface roughness to be measured with a tracer instrument without interference from the feed marks. Cutting-force values obtained from an analog tool are found to be identical with those from the prototype tool.

Surface profiles produced by the major cutting edge of a tool are shown in Fig. 17.6 and comparable values produced by a secondary cutting edge machining the same material are shown in Fig. 17.7.

The profiles of Fig. 17.6 were obtained by tracing in the circumferential direction while those of Fig. 17.7 were obtained by tracing in the axial direction.

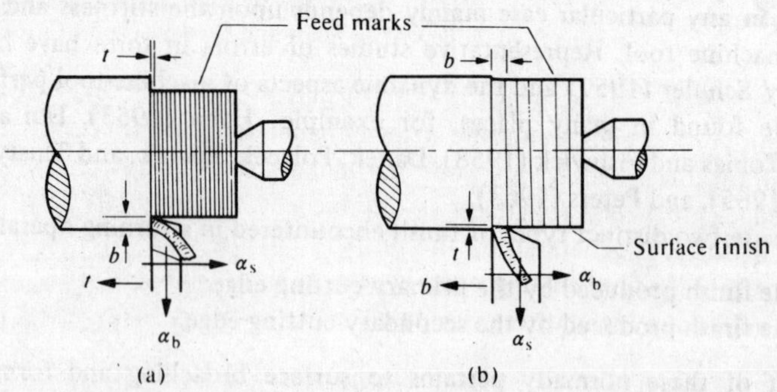

FIG. 17.5. Schematic diagram of (a) conventional and (b) analog cutting processes with an interchange of feed and depth of cut and back (α_b) and side (α_s) rake angles. (After Shaw and Crowell 1965.)

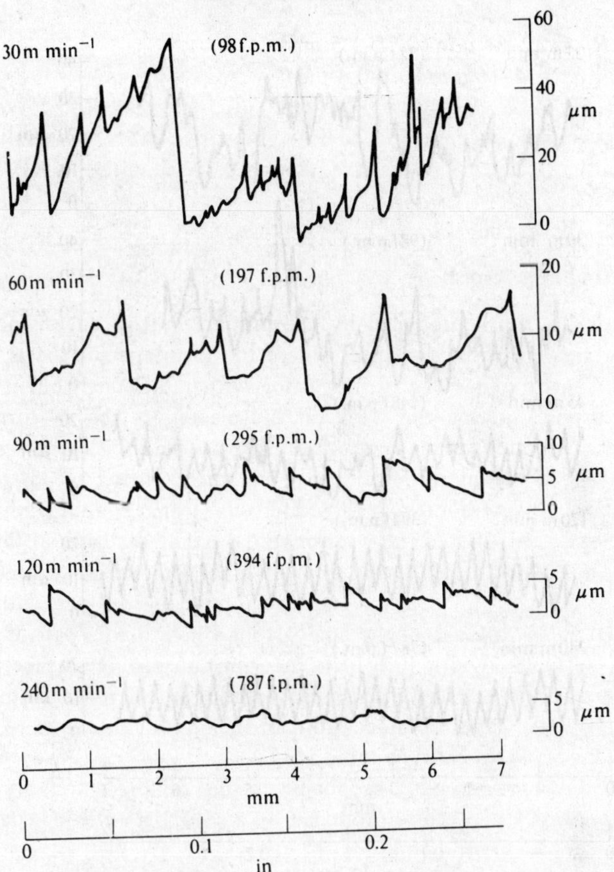

FIG. 17.6. Surface finish produced by a primary cutting edge at different cutting speeds. Work material: normalized AISI 1045 steel; tool material, tungsten carbide; tool geometry: rake and clearance angles = 6°; depth of cut = 2.5 mm (0.1 in); feed = 0.2 mm/rev (0.008 i.p.r.); cutting fluid, none. (After Shaw and Crowell 1965.)

The finish produced is seen to be strongly dependent on the cutting speed (V). The finish for low speeds is very poor, particularly that produced by the primary cutting edge. Much less roughness is obtained at higher speeds, particularly in the case of surfaces generated by the primary cutting edge. The finish obtained by use of a secondary cutting edge is seen to approach that corresponding to the feed marks at a high cutting speed (V).

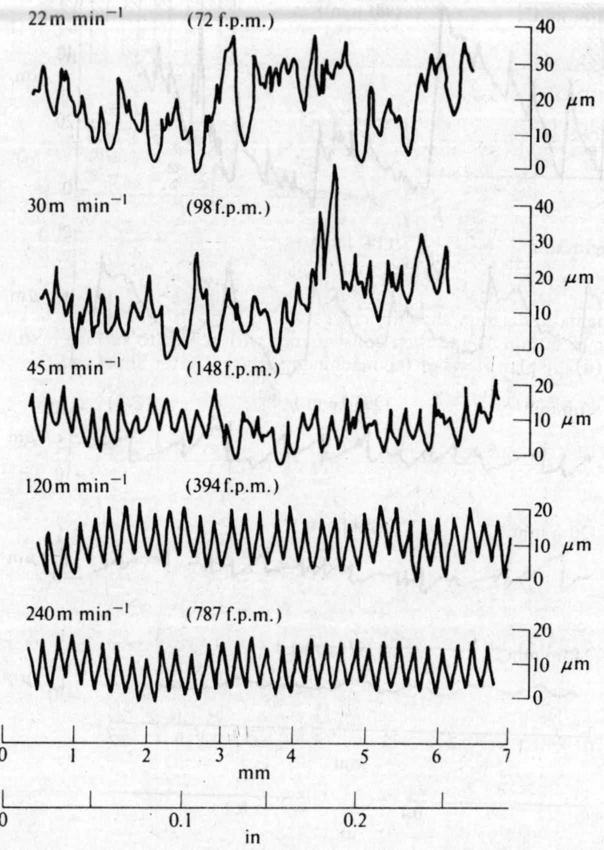

FIG. 17.7. Surface profiles for conventionally turned surfaces. Cutting conditions same as Fig. 17.6 except tool had 15° side- and end-cutting edge angles and the nose radius was 0.8 mm (0.032 in). (After Shaw and Crowell 1965.)

Fracture Roughness

When cutting steel at very low speeds a special type of roughness is frequently observed due to subsurface fracture. Figure 17.8 shows the origin of this type of roughness. When the metal cut is in the vicinity of room temperature, the chip is frequently found to fracture periodically. When it fractures, the crack is frequently found to follow approximately a path of maximum shear stress. Figure 17.9 shows the elastic stress pattern for a concentrated load applied to a plane surface. In this diagram the circles (such as E′) represent points of constant shear stress magnitude while the curved lines (such as OG) give the direction of maximum shear stress. Curved line OG in Fig. 17.9 has a direction similar to line AB in Fig. 17.8.

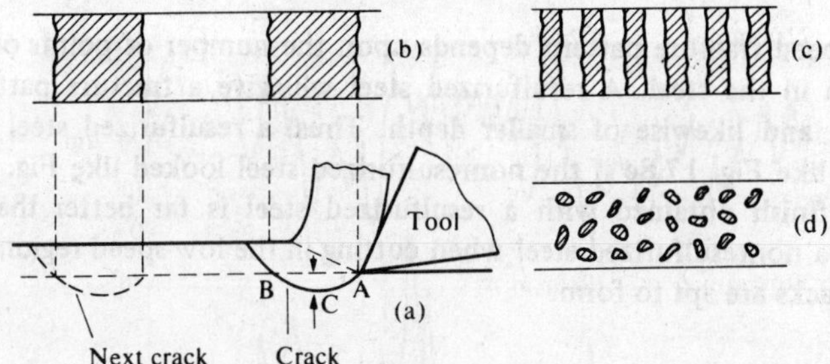

FIG. 17.8. Schematic diagram showing cracks which form at tool tip at low cutting speeds (< 50 f.p.m. or 15 m min^{-1}) and their consequence with regard to surface texture. Diagrams (b), (c), and (d) are plan views of the machined surface. (After Shaw and Crowell 1965.)

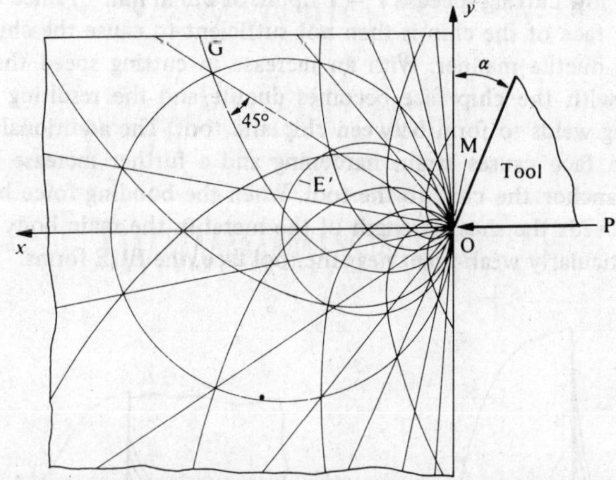

FIG. 17.9. Elastic stress pattern existing at the beginning of a cut. Circles represent lines of constant shear stress magnitude and curved lines are lines of constant shear stress direction. (After Martelloti 1941.)

When the cracks run completely across the chip, the surface has a Moire appearance (Fig. 17.8c) consisting of alternate dull and shiny areas. The shiny areas are those which are actually cut while the dull (grey) areas are the uncut regions of the finished surface. As the speed (temperature) is increased the cracks may not go all the way across the chip. Such chips will appear to be continuous, but careful examination of the finished surface reveals many small gray craters as shown in Fig. 17.8d. These craters are due to the same subsurface fracture. Above a certain cutting speed a subsurface pattern will no longer be found and this source of surface roughness disappears.

The low-speed fracture pattern depends upon the number of points of stress concentration in the steel. A resulfurized steel will give a fracture pattern of shorter pitch and likewise of smaller depth. Thus, a resulfurized steel would tend to look like Fig. 17.8c if the nonresulfurized steel looked like Fig. 17.8b. The surface finish obtained with a resulfurized steel is far better than that obtained for a nonresulfurized steel when cutting in the low-speed region where subsurface cracks are apt to form.

The Built-up Edge

As the cutting speed is increased, the friction between chip and tool will increase and when this becomes large enough to cause a shear fracture in the vicinity of the tool face, a built-up edge (BUE) will form (Fig. 17.10). There is no BUE at very low cutting speeds ($V = 1$ f.p.m. or 0.3 m min^{-1}) since the temperature on the face of the chip is then not sufficient to cause the chip surface to behave in a ductile manner. With an increase in cutting speed the chip metal in contact with the chip face becomes ductile and the resulting plastic flow causes strong welds to form between chip and tool. The additional plastic flow on the chip face causes strain hardening and a further increase in the force tending to anchor the chip to the tool. When the bonding force between chip and tool exceeds the shear strength of the metal in the main body of the chip, at some particularly weak point near the tool face, the BUE forms.

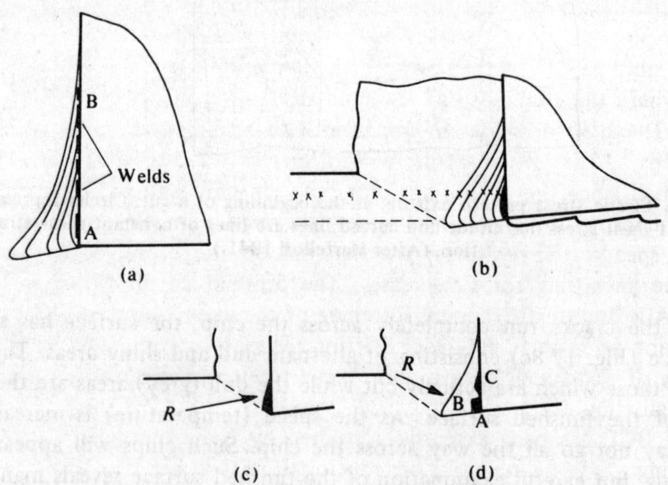

FIG. 17.10. The built-up edge. (a) Welds required to anchor BUE to tool face. (b) Origin of the layered structure and growth and decay of BUE. (c) Magnitude and direction of resultant cutting force (R) for small BUE. (d) Mangitude and direction of resultant cutting force (R) for large BUE. (After Shaw and Sanghani 1963.)

The BUE is a dynamic body that has a layered structure (Shaw and Sanghani 1963). When a point of particularly high stress concentration approaches the outer face of the BUE a new shear surface forms and the BUE grows in size. The origin of the layered structure and the gradual growth of the BUE due to a statistical array of imperfections in the work material is shown diagramatically in Fig. 17.10b. As the BUE grows forward it will usually also grow downward, causing the finished surface to be undercut. The BUE causes an increase in the rake angle which in turn causes a decrease in the magnitude of the resultant force on the tool and a clockwise rotation of the resultant force vector. For a small BUE the direction of the resultant force is such as to put the BUE into compression and to make it stable. When the BUE becomes large the resultant force loads the BUE as a cantilever and eventually the moment at the base of the BUE becomes sufficient to pry it loose. The BUE then passes off partly with the chip and partly on the finished surface. The gradual growth and rapid decay of the size of the BUE causes a sawtoothed finished surface (Fig. 17.10b), which is characteristic of the BUE component of surface roughness.

Influence of Sulfur, Lead, and Cold Work

The BUE has a major influence on the cutting forces encountered at intermediate cutting speeds. Figure 17.11 shows representative cutting force results for analog tools using five steels from the same heat but containing different amounts of manganese sulfide. The solid curve is for a non-resulfurized steel. The tangential or power component of cutting force is designated F_P and F_Q is the force component corresponding to the feed force, but which in this case is directed radially since an analog tool is used. Force component F_P is large at very low speeds since no BUE is then present to alter the rake angle. It falls to a minimum at **B** where the BUE has its maximum size and influence on the effective rake angle. The BUE decreases in size from **B** to **A** and there is essentially no BUE at speeds to the left of point **A**. With increase in cutting speed the BUE is at first small since the anchoring force between chip and tool is then small. As this anchoring force becomes larger, the maximum size of BUE grows. At higher speeds (temperatures) the material on the tool face begins to soften and the anchoring force decreases. The drop in cutting force to the left of point **A** is due to the thermal softening of the material on the tool face and the attendant decrease in tool face friction. The asymptotic level of force in the vicinity of 500 f.p.m. (175 m min^{-1}) is a result of the surface of the chip being essentially liquid.

The BUE has been extensively studied by Nakayama (1957) who cites the following ways of eliminating or reducing the size of the BUE:

1. Increase the cutting speed.
2. Make metal that is less ductile (i.e. cold-work or heat-treatment). Brittle metals do not form BUE.

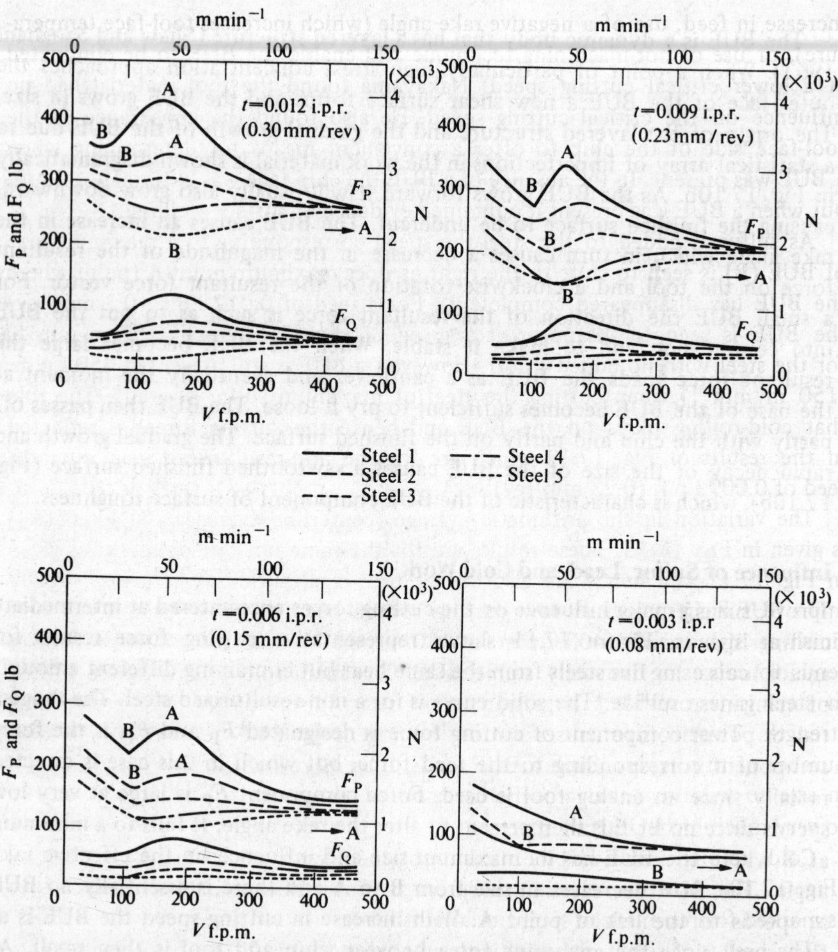

FIG. 17.11. Variation of cutting forces F_P (tangential) and F_Q (radial) for an analog-tool cutting hot-rolled steels of different sulfur content. All materials are plain-carbon steels and 0.08% carbon content, 1% manganese and the following sulfur contents: steel 1, 0.033%; steel 2, 0.11%; steel 3, 0.18%; steel 4, 0.26%; steel 5, 0.37%. Rake angle, 10°; clearance angle, 5°; nose radius, 0.005 in (0.125 mm); width of cut, 0.10 in (2.5 mm); undeformed chip thickness = t (variable); cutting fluid, none; tool material, M-2 HSS. (After Shaw, Usui, and Smith 1961.)

3. Increase the rake angle.
4. A fluid will eliminate the BUE but only at low speeds.

Item (1) may be generalized when it is realized that an increase in speed is effective because it causes the chip temperature to reach a critical value. An increase in feed, use of a negative rake angle (which increases tool-face temperature), or use of hot-machining technique will enable the BUE to be eliminated at a lower critical cutting speed. Nakayama found the tool material had no influence on the critical cutting speed. He also found the appearance of the tool-face side of the chip to offer a convenient means for determining when a BUE was present. In the absence of a BUE the chip face was extremely smooth but when a BUE was present it contained evidence of surface welding.

As sulfur is added to the steel the speed corresponding to a maximum size of BUE (**B**) is seen to shift to the right as does maximum point **A** (point where the BUE has disappeared completely). For a feed of 0.012 i.p.r. (0.3 mm/rev) the BUE is seen to disappear at a speed of about 150 f.p.m. (50 m min^{-1}) for the steel with no added sulfur. However, a BUE is still present at 450 f.p.m. (150 m min^{-1}) when cutting the steel of maximum sulfur content. The role that cold-rolling plays on the BUE and hence the cutting forces is reflected in the results of Fig. 17.12 which are for the non-resulfurized steel cut at a feed of 0.009 i.p.r. (0.23 mm/rev).

The variation in the arithmetic average (centerline average) roughness (R_a) is given in Fig. 17.13. These results are for the same hot-rolled steel tests given in Fig. 17.11 and hence are also analog tool results. Sulfur is seen to greatly improve surface finish at low cutting speeds but to cause a slightly poorer finish at high speeds (400 f.p.m. or 130 m min^{-1}), due to the fact that sulfur tends to cause an unusually stable BUE. The main role of manganese sulfide in steel is to provide a large number of stress-raisers of approximately equal strength. This in turn causes a more homogeneous BUE consisting of a large number of thin layers. Such a BUE tends to be more stable and not to fluctuate in size so wildly in the large BUE region. It also tends to remain on the tool to very much higher cutting speeds.

Cold rolling improves the finish obtained with a non-resulfurized steel (Fig. 17.12), however, the improvement in finish due to cold working is not as marked for resulfurized steel cut at light feed.

The best tool finish obtained with an analog tool (12–14 μin) is remarkably independent of undeformed chip thickness (feed) and is obtained at high speeds (> 300 f.p.m. or 100 m min^{-1}) and with a cold-drawn non-resulfurized steel. Sulfur has its greatest influence in improving finish at low and intermediate cutting speeds. For example, at a cutting speed of 200 f.p.m. (61 m min^{-1}) and a feed of 0.003 i.p.r. (0.075 mm/rev) the analog tool finish for the cold-drawn resulfurized steel (5C) was 20 μin (0.5 μm) compared with 35 μ in (0.9 μm) for the non-resulfurized cold-drawn steel. At a speed of 50 f.p.m. (15 m min^{-1}) and the same feed, the corresponding values of finish were 15 μ in (0.4 μm) for the 5C steel and 350 μ in (8 μm) for the 1C steel. Inasmuch as a steel must usually operate over a wide speed range the best overall performance will be obtained with a cold-drawn resulfurized steel. An analog tool finish R_a of from 15 to 20 μ in (0.4 to 0.8 μm) may be obtained with such a steel

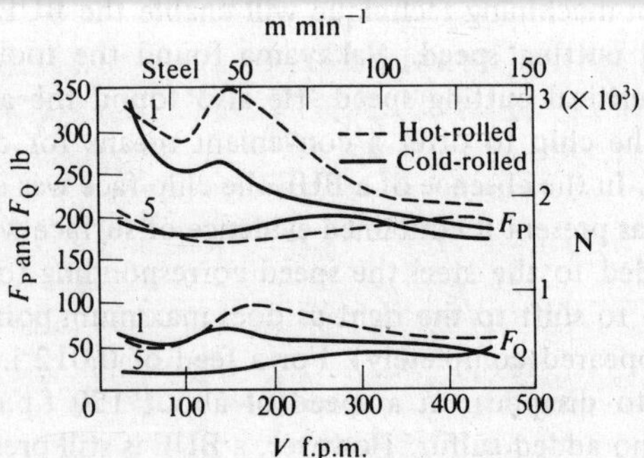

FIG. 17.12. Comparison of cutting-force values for hot-rolled and cold-drawn [two drafts of $\frac{1}{8}$ in (3.18 mm) each from a diameter of $4\frac{1}{2}$ in (114.3 mm)] steels. Cutting conditions are the same as in Fig. 17.11 except for the undeformed chip thickness (t) which was 0.009 i.p.r. (0.23 mm/rev). (After Shaw, Usui, and Smith 1961.)

over the entire practical range of speeds and feeds (20–800 f.p.m. or 6–244 m min^{-1}; 0.001–0.012 i.p.r. or 0.025–0.3 mm/rev).

Additions of lead to steel are found to improve surface finish and lead is relatively more effective at light feeds (t) which is the reverse of the trend with sulfur additions. Also, lead does not increase the stability of the BUE at high speeds, and unlike sulfur, lead gives better finish at all speeds. Cold-drawing improves the finish obtained with a leaded steel, as in the case of a resulfurized steel.

One of the results of adding sulfur or lead to steel is to cause the chip to curl more and to decrease the chip-tool contact length. This in turn lowers cutting forces and the tendency to form a large BUE at low speeds. The chip-tool contact length may be reduced artificially, by grinding away the tool-face beyond the cutting edge. When this is done a large part of the benefit of sulfur relative to surface finish is lost and a non-resulfurized steel machined using a cut-away tool will give a fairly good finish even at low cutting speeds. However, the poorer finish obtained with sulfur at high cutting speeds becomes even poorer when a cut-away tool is used. A cut-away tool also gives poorer finish at high speeds when machining a leaded steel (Shaw and Usui, 1962).

At low cutting speeds best finish is obtained when a resulfurized steel is cut using a cut-away tool having a chip-tool contact length of $3t$ (where t is the undeformed chip thickness). For a non-resulfurized steel of low-carbon content the best chip-tool contact length appears to be about $2t$. From the point of view of finish a cut-away tool should be used at low cutting speeds but not at high speeds.

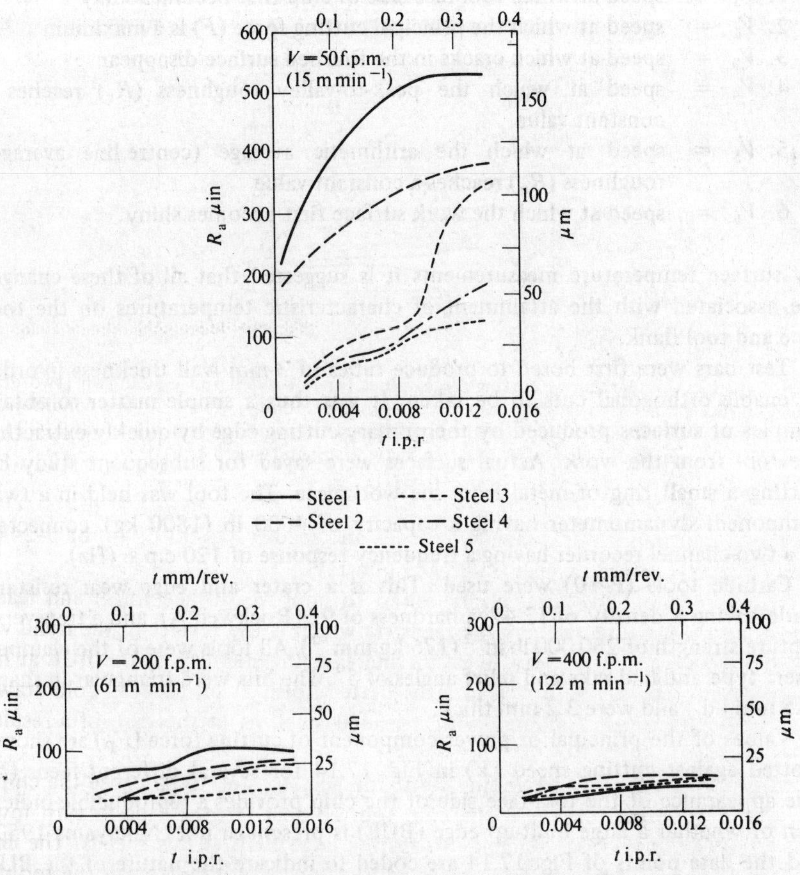

FIG. 17.13. Variation of arithmetic average (centerline average) and cutting roughness (R_a) with undeformed chip thickness (t) and cutting speed (V) for steels of different sulfur content. Cutting conditions same as Fig. 17.11. (After Shaw, Usui, and Smith 1961.)

Experimental Results—Normalized AISI 1045 Steel

A better understanding of primary cutting edge roughness may be obtained by simultaneously recording surface roughness, cutting forces, tool–chip interface temperature (by tool–chip thermocouple measurements), and chip formation for a wide range of cutting speeds (V). This is done in the following discussion for normalized AISI 1045 steel.

The following characteristic cutting speeds (V) may be identified when cutting forces, surface roughness values, and chip and workpiece surfaces are examined:

1. V_1 = speed at which tool-face side of chip first becomes shiny
2. V_2 = speed at which the principal cutting force (F) is a maximum
3. V_3 = speed at which cracks in the finished surface disappear
4. V_4 = speed at which the peak-to-valley roughness (R_t) reaches a constant value
5. V_5 = speed at which the arithmetic average (centre line average) roughness (R_a) reaches a constant value
6. V_6 = speed at which the work surface first becomes shiny.

By surface temperature measurements it is suggested that all of these changes are associated with the attainment of characteristic temperatures on the tool face and tool flank.

Test bars were first bored to produce tubes of 3-mm wall thickness in order to enable orthogonal cuts to be taken. It was thus a simple matter to obtain samples of surfaces produced by the primary cutting edge by quickly extracting the tool from the work. Actual surfaces were saved for subsequent study by parting a small ring of metal from the workpiece. The tool was held in a two-component dynamometer having a capacity of 4000 lb (1800 kg), connected to a two-channel recorder having a frequency response of 120 c.p.s. (Hz).

Carbide tools (P-10) were used. This is a crater and edge wear resistant grade having a density of 12.42, a hardness of 92 (Rockwell A), and a transverse rupture strength of 250 000 lb in^{-2} (176 kg mm^{-2}). All tools were of the clamped insert type and had rake and relief angles of 5°. The bits were triangular in shape (9.5 mm i.d.) and were 3.2 mm thick.

Values of the principal or power component of cutting force (F_P) are shown plotted against cutting speed (V) in Fig. 17.14 for several different feeds (t). The appearance of the tool-face side of the chip provides a reproducible indication of whether a large built-up edge (BUE) is present or not (Nakayama 1957) and the data points of Fig. 17.14 are coded to indicate the nature of the BUE that was present. The V_2 (maximum cutting force) and V_1 (lowest speed to produce a shiny chip) values were found to be identical as shown in Fig. 17.14.

Figures 17.15 and 17.16 show plots of peak-to-valley roughness (R_t) and arithmetic average roughness (R_a) vs. cutting speeds (V) for different values of feed (t). These curves are seen to fall rapidly with increase in speed, reaching a constant asymptotic value of speed V_4 (for R_t) and V_5 (for R_a). Values of V_4 and V_5, which may only be approximately estimated are shown in Fig. 17.14. The V_5 curve is not extended to a feed of 0.4 mm/rev since chatter occurred with this feed in the vicinity of speed V_5.

The line marked V_6 in Fig. 17.14 shows the speed at which burnishing of the finished surface is first observed at each feed. A smooth burnished chip surface was not produced until a speed of 32 m min^{-1} was reached. For speeds below this value, a BUE present and a continuous layer of flowed metal was not present on the side of the chip adjacent to the tool face. As the speed was

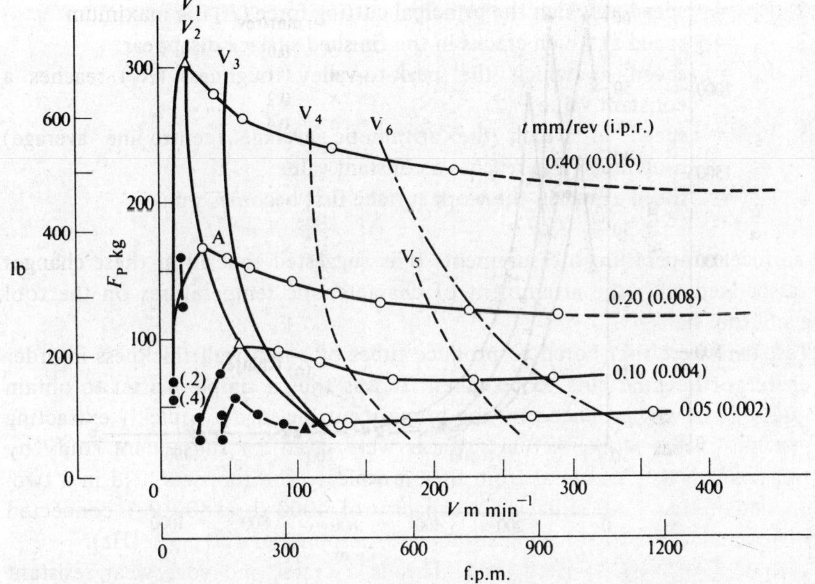

FIG. 17.14. Variation of principal cutting force (F_P) with cutting speed (V) for several feeds (t) when cutting in air. Work material, AISI 1045 steel. (After Nakayama, Shaw, and Brewer 1966.)

increased above V_1, the thickness of the surface layer decreased as the temperature gradient became steeper. Finally at a speed of 289 m min^{-1}, the burnished layer on the face of the chip became very thin, probably due to the time of contact between chip and tool being very short.

Measured values of chip-tool interface temperature are given in Fig. 17.17. Values are not shown for the 0.4 mm/rev feed since the chips caused a short circuit. Lines corresponding to the V_1 to V_6 curves of Fig. 17.14 are labelled θ_1 to θ_6.

The values of F_P in Fig. 17.14 are seen to reach an approximately constant value at high values of speed. This constant value of F_P may be used to compute the energy per unit volume (u). Values of u obtained in this way from Fig. 17.14 are shown plotted in Fig. 17.18. The slope of this curve is approximately minus 0.2 in accordance with the relationship that usually holds in turning.

Finally, values of $\bar{\theta}$ from Fig. 17.17 are shown plotted against the quantity $u\,(Vt)^{\frac{1}{2}}$ for each experimental point in Fig. 17.19. A line corresponding to θ versus $u\ (Vt)^{\frac{1}{2}}$ is found to pass close to all experimental points except those obtained when a BUE was present. In Fig. 17.17, the curves are seen to break and fall off very rapidly below the θ_1 line corresponding to the appearance of the BUE. The presence of a BUE decreased the thermoelectric e.m.f. generated and thus

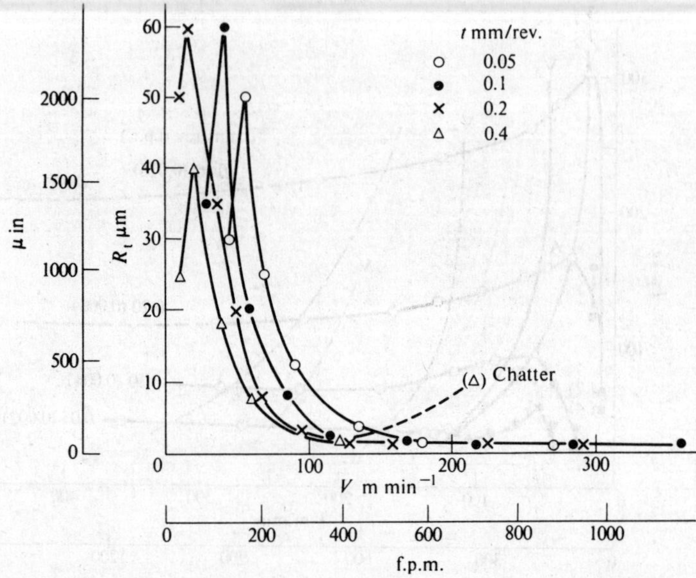

FIG. 17.15. Variation of peak-to-valley roughness (R_t) with cutting speed (V). (After Nakayama, Shaw, and Brewer 1966.)

suggests a temperature that is too low. Therefore, these points associated with a BUE in Fig. 17.19 should be ignored.

Discussion

The entire range of surface finishes produced by a primary cutting edge appear to depend on the temperatures pertaining on the face and flank of the cutting tool.

At very low tool-face temperatures, the junctions on the tool face come apart where they went together and no built-up edge is usually formed. At moderate speeds the temperature on the tool face becomes sufficiently high so that strong bonds are established, and the metal in the chip adjacent to the tool face then flows in a direction close to that corresponding to the direction of maximum shear stress. A BUE is thus established. Pieces of the BUE periodically leave with the chip and with the workpiece, and the face of the chip as well as the workpiece surface are very rough. Values of cutting force will usually be low when a BUE is present due to the large effective rake angle associated with the BUE.

As the speed is further increased a point is reached where the face of the chip becomes shiny. This is speed V_1. At the same speed (V_2) the cutting forces show

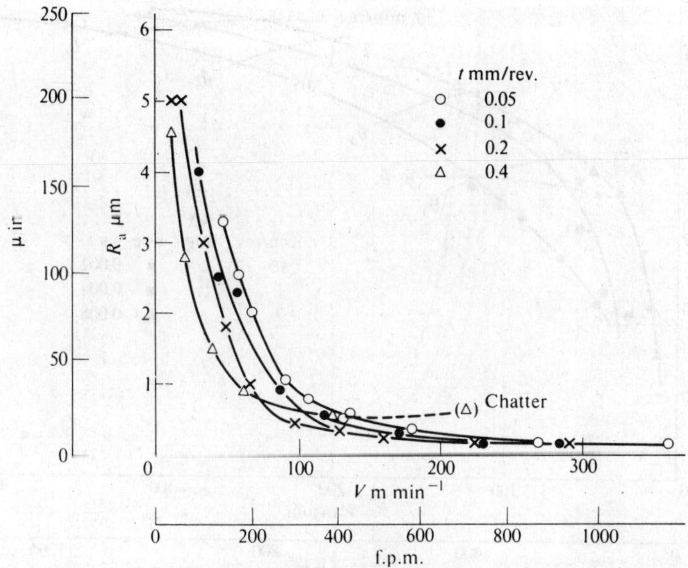

FIG. 17.16. Variation of centerline average roughness (R_a) with cutting speed. (After Nakayama, Shaw, and Brewer 1966.)

a maximum. The surface roughness is still quite high at this speed but has begun to decrease with speed. It is believed that V_1 and V_2 correspond to a mean tool-face temperature which equals the strain recrystallization temperature of the chip (T_H = 0.5). In Fig. 17.17 $\bar{\theta}_1$ and $\bar{\theta}_2$ are between 650 and 700 °C. These somewhat-approximate mean temperatures are just below the equilibrium temperature in the iron–carbon diagram (Fig. 15.1) where austenite is first formed (727 °C). As the speed rises slightly above V_1, the α–γ phase transformation will be reached on the tool face. The fact that the ($\bar{\theta}_1$) temperature rises with an increase in V_1 (i.e. for decreased feed) may be explained by the fact that recrystallization is a time–temperature reaction. At a higher cutting speed (V_1) and smaller feed (t) the chip is in contact with the tool for a shorter time and hence there is less time for recrystallization to occur. Since recrystallization and the α–γ transformation will occur more rapidly the higher the temperature, and hence the farther the material is removed from equilibirum, the critical temperatures should become higher as the feed decreases and as V_1 increases (i.e. as the tool–chip contact time decreases).

Ostermann (1960) and Opitz and Gappisch (1962) have clearly identified the presence of an austenitic layer on the surfaces of chips produced at high speeds, and have discussed this phenomenon in terms of tool wear.

At speeds above V_1 a zone of secondary plastic flow is evident on the tool-face

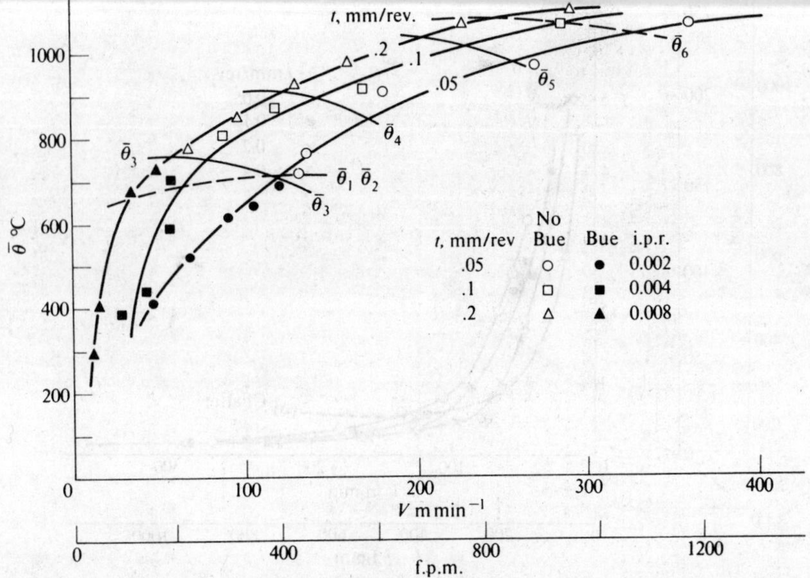

FIG. 17.17. Values of chip–tool interface temperature measured by the chip–tool thermocouple method, for different cutting speeds and feeds. (After Nakayama, Shaw, Brewer 1966.)

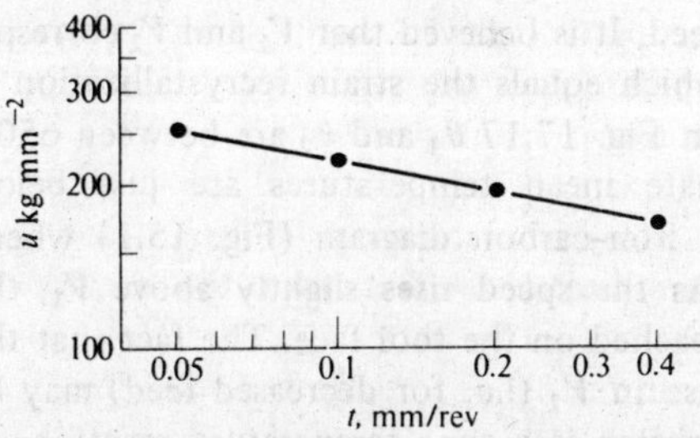

FIG. 17.18. Variation in energy per unit volume (u) with feed (t). (After Nakayama, Shaw, Brewer 1966.)

side of the chip. This zone is at first thick and then becomes thinner as the speed is increased. The thickness of the zone corresponds to the layer that is partially or completely transformed, and hence softened. The face of the chip is highly burnished and shiny whenever a secondary shear zone is present.

The cutting forces will usually drop as the secondary shear zone becomes thinner. The BUE is probably rapidly lost in going from speed V_2 to V_3. At speeds to the left of V_3 cracks are found in the finished surface which are

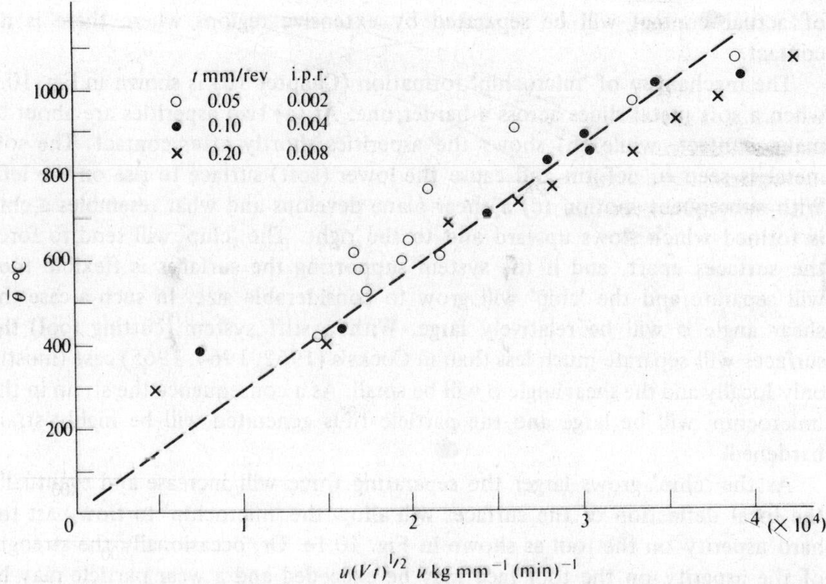

FIG. 17.19. Variation of measured chip–tool interface temperature (θ) with the quantity $u(VT)^{\frac{1}{2}}$. (After Nakayama, Shaw, and Brewer 1966.)

probably due to the presence of a small, stable, residual BUE. The BUE is probably missing completely to the right of V_3.

It is well known that austenite deforms by grain boundary rearrangement and has a viscous character. Creep is a result. Since grain boundary rearrangement is a more time-dependent mechanism than slip, the deformation of austenite is more strain rate dependent than the deformation of ferrite. The deformation of austenite may be better approximated by a liquid model, where the motion of holes is analogous to grain boundary rearrangement with regard to strain rate dependence.

Trent (1963) has suggested that the secondary shear zone often appears as though it were a liquid when sheared.

Schaller (1964) has assumed that the secondary shear zone on the surface of the chip behaves as though it were a viscous liquid that obeys Newton's law of viscous shear.

When the temperature on the tool face has reached the temperature ($\bar{\theta}_1, \bar{\theta}_2$) at which the BUE disappears and the face of the chip becomes burnished due to the appearance of a secondary shear zone, the temperature on the tool flank will be far below this temperature. Asperities on the newly cut surface will contact asperities on the tool flank and the real area of contact (A_R) on the clearance face will be far less than the apparent area (A). The high points of actual contact will be separated by extensive regions where there is no contact.

The mechanism of 'microchip' formation (Chapter 10) is shown in Fig. 10.1 when a soft metal slides across a harder one. At (a) two asperities are about to make contact, while (b) shows the asperities shortly after contact. The soft metal is seen to deform and cause the lower (soft) surface to rise on the left. With subsequent motion (c) a shear plane develops and what resembles a chip is formed which flows upward and to the right. The 'chip' will tend to force the surfaces apart, and if the system supporting the surfaces is flexible they will separate and the 'chip' will grow to considerable size. In such a case the shear angle ϕ will be relatively large. With a stiff system (cutting tool) the surfaces will separate much less than in Cocks's (1962, 1964, 1965) case (mostly only locally and the shear angle ϕ will be small. As a consequence the strain in the 'microchip' will be large and the particle thus generated will be highly strain hardened.

As the 'chip' grows larger the separating force will increase and eventually the local deflection of the surfaces will allow the 'microchip' to flow past the hard asperity on the tool as shown in Fig. 10.1e. Or, occasionally the strength of the asperity on the tool face may be exceeded and a wear particle may be removed from the tool surface as shown in Fig. 10.1d. In this manner a large number of hard particles will be formed on the finished surface which will constitute the main component of roughness between speeds V_1 and V_3 or V_4.

As the speed is increased, the metal in the finished surface will gradually soften and asperities in the surface will recede downward into the surface on the workpiece. Fewer smaller microchips will be developed. Finally, at speed V_6 the metal in the finished surface will reach the strain recrystallization temperature and then the transformation temperature (727 °C) and a burnished workpiece surface will be formed just as a burnished chip surface is formed at speeds above V_1. At speeds above V_6 the finished surface will possess a layer of secondary shear just as the chip face does above V_1.

The real area of contact on the tool flank abruptly changes from a small value relative to the apparent area to one that approaches the apparent area at speeds in the vicinity of V_4 and V_5. When this occurs it is to be expected there will be a sudden increase in the rate of tool flank wear.

Adhesion between the tool flank and the workpiece surface is unusually strong in the case of soft (grade 60 = 60 000 lb in^{-2} or 400 MPa) nodular iron. Trigger, Zylstra, and Chao (1952) have observed a ribbon-like flow from the clearance face of the tool in such cases and similar ribbon-like flows have been observed by Ham, Hitomi, and Thuering (1961, 1962 and 1964). It is believed that these ribbons are an extreme example of the 'microchip' formation described here in which the individual chips are so numerous they join up to form a small continuous auxiliary chip.

The number of 'microchips' that are formed in a given instance will depend upon many factors including the following.

1. The roughness of the tool flank surface and the extent of the wear land.
2. The inherent roughness of the freshly generated surface.
3. The hardness and flow characteristics of the workpiece material and the response of these quantities to temperature changes.
4. The compatibility as measured by the relative tendency for strong bonds to form when clean, dry tool and workpiece surfaces come into contact.
5. The atmosphere, lubricant, or contaminant that may be present to prevent weld formation between chip and tool asperities upon contact.

This investigation of the surface produced by a primary cutting edge has shown that a sequence of events which resemble those occurring on the tool face at low speeds may also occur on the tool flank at high speeds. A 'microchip' that plays the same role on the tool flank as the BUE does on the tool face is responsible for the best finish not being obtained immediately when the BUE is lost and the chip face becomes shiny.

Secondary Cutting Edge Finish

All of the results presented thus far are for the type of roughness obtained in a cutting operation in which there is no tool nose and hence no feed marks. In a conventional turning operation the finish left on the bar is produced by a secondary cutting edge that is separated from the primary cutting edge by a nose radius (Fig. 17.20). Use of a secondary cutting edge to generate the finished surface introduces several complications:

1. Ridges corresponding to the geometry of the tool at its nose and having a pitch equal to the axial feed rate are left behind on the finished surface.
2. The undeformed chip thickness goes gradually to zero at the secondary cutting edge and this causes uncertainty in the geometry of the cut at the trailing edge, since for a given edge sharpness there is minimum undeformed thickness that will be removed.
3. A concentration of wear occurs at both free surfaces of the cut. The groove thus formed on the end-cutting edge of the tool acts as a forming tool and leaves behind a highly cold-worked ridge on the surface that not only contributes to the roughness but further complicates matters by producing additional grooves on the end-cutting edge as it is recut on subsequent revolutions.
4. The metal at the trailing edge of the tool is subjected to unusually high normal stress and will flow to the side to relieve this stress. This in turn produces a furrow that contributes to the roughness, particularly in the case of a soft, ductile metal.

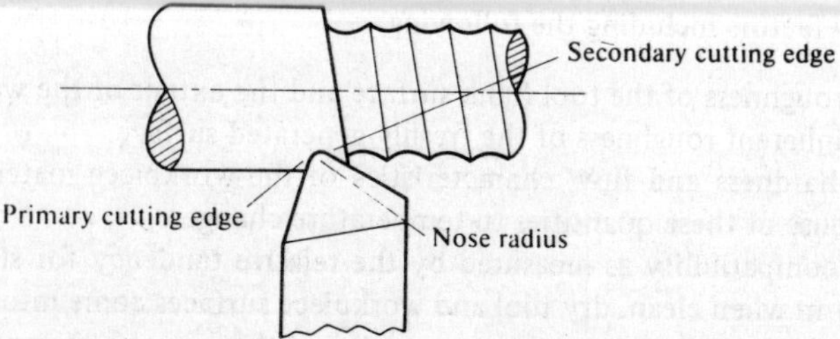

FIG. 17.20. Diagrammatic sketch of cutting operation showing primary and secondary cutting edges and characteristic wave-form left on the finished surface.

In addition to these special roughness components, BUE roughness, roughness due to an imperfect cutting edge, and roughness due to tool vibration will also be present.

Surfaces generated by a secondary cutting edge are thus far more complex than those produced by a primary cutting edge.

Surface roughness will generally have different values in different directions and the direction normally employed is that corresponding to the maximum roughness. For a surface produced by a primary cutting edge this will be in the direction of cut (parallel to the cutting velocity vector). Consequently all roughness values presented above are for the circumferential direction. For surfaces generated by a secondary edge the roughness will be greatest in a direction perpendicular to the cutting velocity vector (V) and roughness values should be obtained using a stylus traversed in an axial direction.

Geometrical Contribution to Roughness

The component of surface roughness due to tool nose geometry may be readily calculated. Figure 17.20 shows a plan view of a conventional turning operation with ridges left behind on the finished surface. Figure 17.21 shows an enlarged view of the tool tip which is defined in terms of three quantities:

1. Nose radius ($r = \mathrm{OT}$ in Fig. 17.21).
2. End-cutting edge angle (C_e).
3. Side-cutting edge angle (C_s).

When the nose radius is large and the feed is very small, the surface will be generated by the nose radius alone (Fig. 17.22). This will be referred to as case I and it is evident that

$$R_{th} = \mathrm{OT} - \mathrm{OU} = r - (r^2 - t^2/4)^{\frac{1}{2}} \cong \frac{t^2}{8r} \qquad (17.1)$$

or in nondimensional form

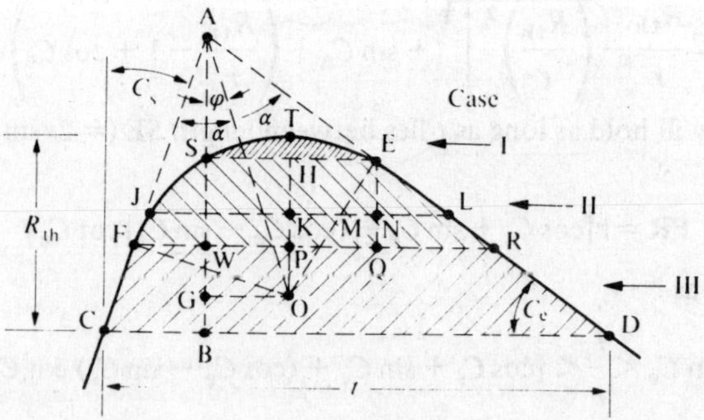

FIG. 17.21. Plan view of cutting tool showing the three cases to be considered in computing the theoretical peak-to-valley roughness (R_{th}) corresponding to any value of feed (t).

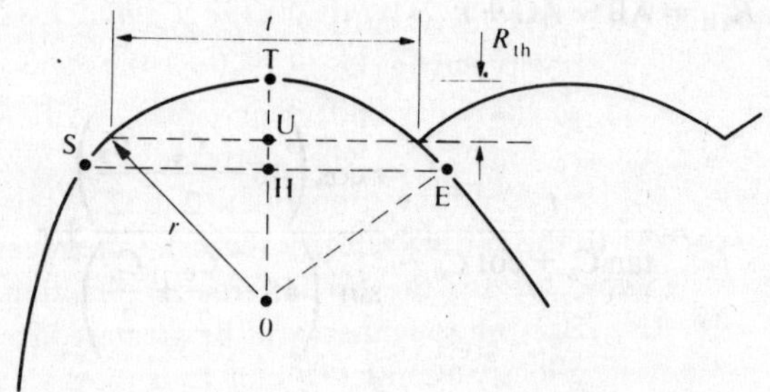

FIG. 17.22. Enlarged view for case I of Fig. 17.21.

$$\frac{R_{th}}{r} = \frac{1}{8}\left(\frac{t}{r}\right)^2. \tag{17.2}$$

From Fig. 17.22 it is evident that this equation will hold as long as $t/2$ is less than **EH** or

$$0 < (t/r) < 2 \sin C_e. \tag{17.3}$$

For larger values of feed both r and C_e must be considered in computing R_{th} and this will be called case II. If t is equal to **JL** (Fig. 17.21) this corresponds to case II and

$$t = \text{JK} + \text{HE} + \text{NL}$$

$$= [r^2 - (r - R_{th})^2]^{\frac{1}{2}} + r \sin C_e + r\left(\frac{R_{th}}{r} - 1 + \cos C_e\right) \cot C_e \tag{17.4}$$

$$\frac{t}{r} = \left[2\frac{R_{th}}{r} - \left(\frac{R_{th}}{r}\right)^2\right]^{\frac{1}{2}} + \sin C_e + \left(\frac{R_{th}}{r} - 1 + \cos C_e\right)\cot C_e.$$

This equation will hold as long as t lies between length **SE** ($= 2r \sin C_e$) and length **FR**,

$$\mathbf{FR} = r[\cos C_s + \sin C_e + (\cos C_e - \sin C_s)\cot C_e]$$

that is, as long as

$$2 \sin C_e < \frac{t}{r} < [\cos C_s + \sin C_e + (\cos C_e - \sin C_s) \cot C_e]. \qquad (17.5)$$

For still larger values of t, C_s in addition to r and C_e must be considered and this will be called case III. If t is equal to **CD** this corresponds to case III and from Fig. 17.21

$$R_{th} = \mathrm{AB} - \mathrm{AG} + r$$

$$= \frac{t}{\tan C_s + \cot C_e} - \frac{r\cos\left(45 - \frac{C_e}{2} - \frac{C_s}{2}\right)}{\sin\left(45 - \frac{C_e}{2} + \frac{C_s}{2}\right)} + r$$

or in nondimensional form

$$\frac{R_{th}}{r} = \frac{t/r}{\tan C_s + \cot C_e} - \frac{\cos\left(45 - \frac{C_e}{2} - \frac{C_s}{2}\right)}{\sin\left(45 - \frac{C_e}{2} + \frac{C_s}{2}\right)} + 1 \qquad (17.6)$$

which holds when

$$t/r > \cos C_s + \sin C_e + (\cos C_e - \sin C_s) \cot C_e. \qquad (17.7)$$

If the nose radius is very small compared with t then the last two terms of eqn (17.6) become negligible compared with the first and

$$\frac{R_{th}}{r} = \frac{t/r}{\tan C_s + \cot C_e}, \quad \left(\frac{t}{r} \gg 1\right) \qquad (17.8)$$

for all values of t.

Equations 17.2, 17.4, and 17.6 may be used to produce a nondimensional plot as shown schematically in Fig. 17.23, which enables R_{th} to be found for

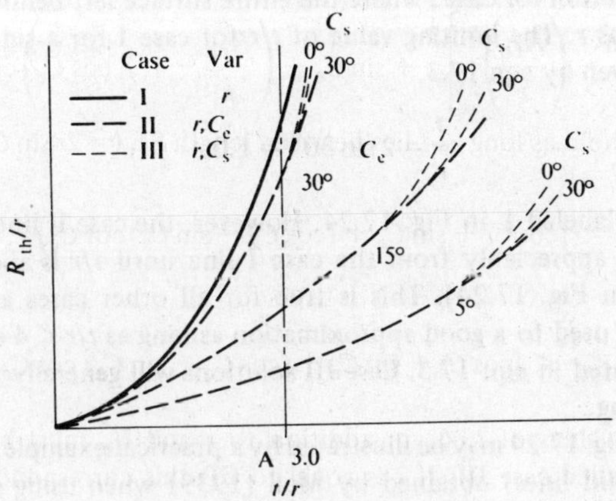

FIG. 17.23. Schematic diagram showing relation between R_{th}/r and t/r for all three cases. (After Shaw and Crowell 1965.)

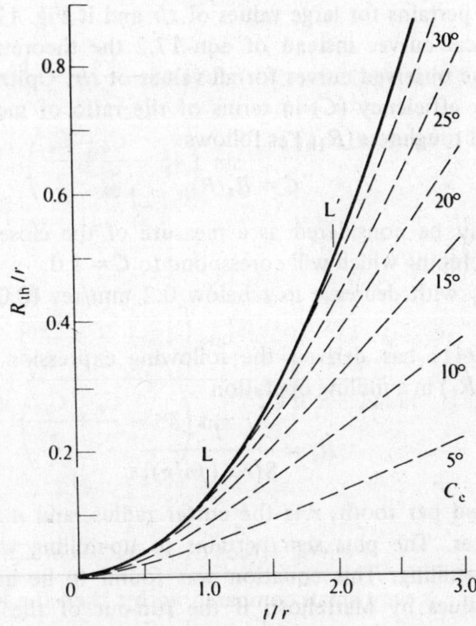

FIG. 17.24. Actual plot of practical region of Fig. 17.23 to the left of line A–A. (After Shaw and Crowell 1965.)

any combination of t, r, C_e, and C_s. However, only the part of Fig. 17.23 to the left of A–A is of interest in finish machining, and therefore only this part of the diagram is shown to scale in Fig. 17.24. The solid line in Fig. 17.24

represents the solution for Case I where the entire surface left behind is generated by the nose radius r. The limiting value of t/r for case I for a side-cutting edge angle of 30° is given by eqn 17.3.

$$\frac{t}{r} = 2 \sin 30^\circ = 1.0$$

and this point is labeled L in Fig. 17.24. However, the case II line for $C_e = 30°$ does not deviate appreciably from the case I line until t/r is about twice this value (point L′ in Fig. 17.24). This is true for all other cases and the case I equation may be used to a good approximation as long as $t/r < 4 \sin C_e$, instead of $2 \sin C_e$ as stated in eqn 17.3. Case III solutions will generally not be needed in finish-machining.

The need for Fig. 17.24 may be illustrated by a practical example. Figure 17.25 shows results (solid lines) obtained by Moll (1939) when using sharp tools of different nose radius r to machine a steel at different values of feed t. The theoretical curves (assuming case I to hold) are shown by the dashed lines. These are seen to lie above the observed curves for large values of feed (i.e. t/r). Actually, case II pertains for large values of t/r and if Fig. 17.24 were used to plot the theoretical curves instead of eqn 17.2 the theoretical curves would then lie below the observed curves for all values of t/r. Opitz and Moll (1941) defined a cutting efficiency (C) in terms of the ratio of measured roughness (R_t) to theoretical roughness (R_{th}) as follows:

$$C = R_t/R_{th} \tag{17.9}$$

This quantity may be considered as a measure of the closeness of approach to ideal finish-machining which will coresspond to $C = 1.0$.

The rise in R_t with decrease in t below 0.2 mm/rev (0.008 i.p.r.) will be discussed later.

Martellotti (1941) has derived the following expression for the peak-to-valley roughness (R_t) in a milling operation

$$R_t = \frac{f^2}{8[r \pm (fn/\pi)]} \tag{17.10}$$

where f is the feed per tooth, r is the cutter radius, and n is the number of teeth in the cutter. The plus sign pertains to up-milling while the negative sign is for down-milling. This equation was found to be in good agreement with measured values by Martellotti if the run-out of the spindle was held to a very low value.

Representative Finish Results

Representative values of peak-to-valley roughness (R_t) and arithmetic average roughness (R_a) are given in Figs. 17.26 for normalized AISI 1045 steel machined

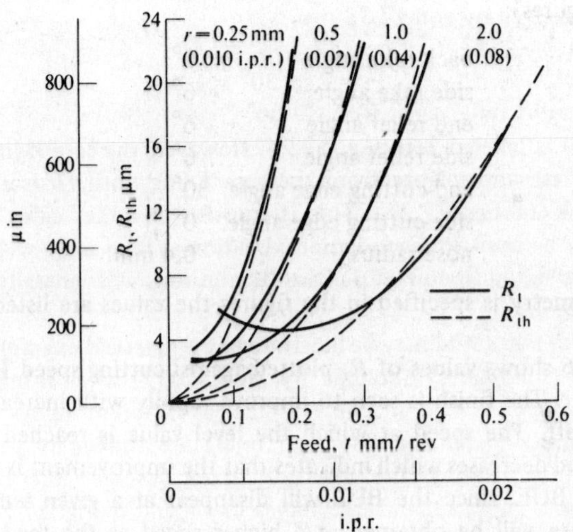

FIG. 17.25. Comparison of measured (R_t) and computed (R_{th}) values of peak-to-valley roughness for tools of different nose radius *r* used at different feeds. (After Moll 1939.)

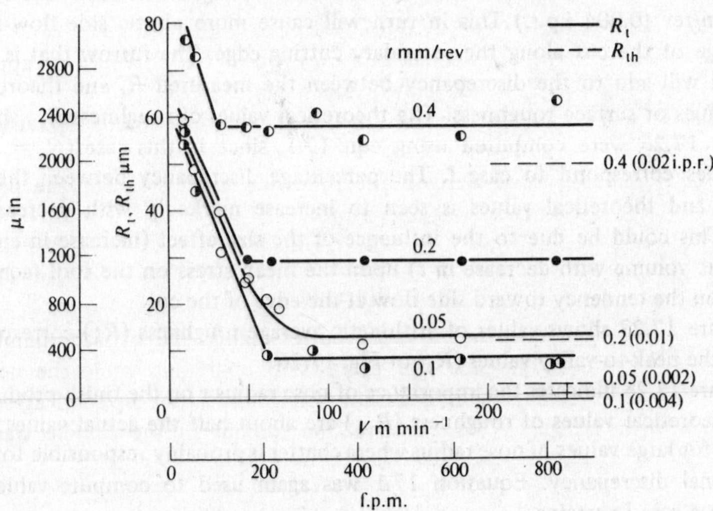

FIG. 17.26. Peak-to-valley roughness (R_t) versus speed and feed for a normalized AISI 1045 steel machined without fluid using sharp carbide tools of geometry: 0°, 6°, 6°, 6°, 30°, 0°, 0.4 mm. The theoretical values of R_t (R_{th}) computed from eqn (17.1) are shown by the dotted lines. (After Shaw and Crowell 1965.)

using sharp tungsten carbide tools having the following geometry (except as noted in the figures):

back rake angle	0°
side rake angle	6°
end relief angle	6°
side relief angle	6°
end-cutting edge angle	30°
side-cutting edge angle	0°
nose radius	0.4 mm.

When tool geometry is specified in the figures the values are listed in the above order.

Figure 17.26 shows values of R_t plotted against cutting speed V for different values of feed t. The finish is seen to improve rapidly with increased speed and then to level off. The speed at which the level value is reached is seen to increase as the feed decreases which indicates that the improvement is due primarily to loss of the BUE, since the BUE will disappear at a given temperature and this temperature will be obtained at a higher speed as the feed is decreased. It should be noted that the best finish does not correspond to the lowest feed (0.05 mm/rev) but instead corresponds to a feed of 0.1 mm/rev (0.004 i.p.r.). This is due in part to the fact that the cutting energy per unit volume (u) and hence the mean stress on the tool face increases rapidly as the feed falls below 0.1 mm/rev (0.004 i.p.r.) This in turn will cause more plastic side flow from the edge of the cut along the secondary cutting edge. The furrow that is thus formed will add to the discrepancy between the measured R_t and theoretical R_{th} values of surface roughness. The theoretical values of roughness R_{th} shown in Fig. 17.26 were computed using eqn 17.1, since in this case ($C_e = 30°$) all values correspond to case I. The percentage discrepancy between the observed and theoretical values is seen to increase markedly with decrease in feed. This could be due to the influence of the size effect (increase in energy per unit volume with decrease in t) upon the mean stress on the tool face and hence on the tendency toward side flow at the edge of the cut.

Figure 17.27 shows values of arithmetic average roughness (R_a) corresponding to the peak-to-valley values (R_t) of Fig. 17.26.

Figure 17.28 indicates the importance of nose radius r on the finish produced. The theoretical values of roughness (R_{th}) are about half the actual values (R_t) except for large values of nose radius where chatter is probably responsible for the additional discrepancy. Equation 17.1 was again used to compute values of R_{th} since case I pertains.

Figure 17.29 shows that the depth of cut has a negligible influence on surface roughness.

Figure 17.30 compares theoretical and actual values of surface roughness for different values of feed. The finish R_t is again seen to improve with decrease

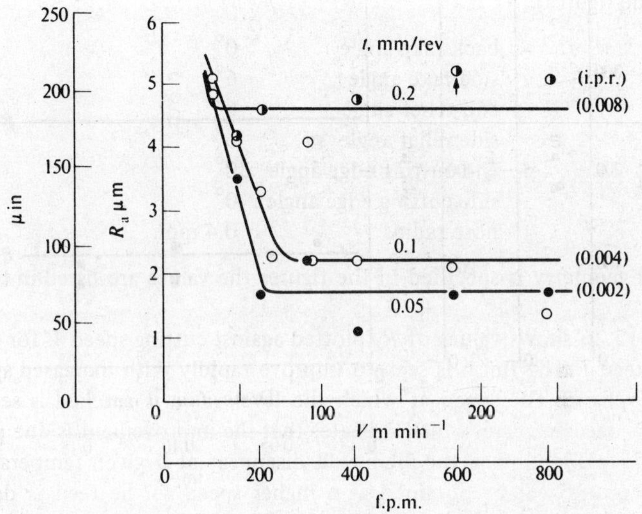

FIG. 17.27. Values of arithmetic average roughness (R_a) corresponding to the peak-to-valley values (R_t) in Fig. 17.26. (After Shaw and Crowell 1965.)

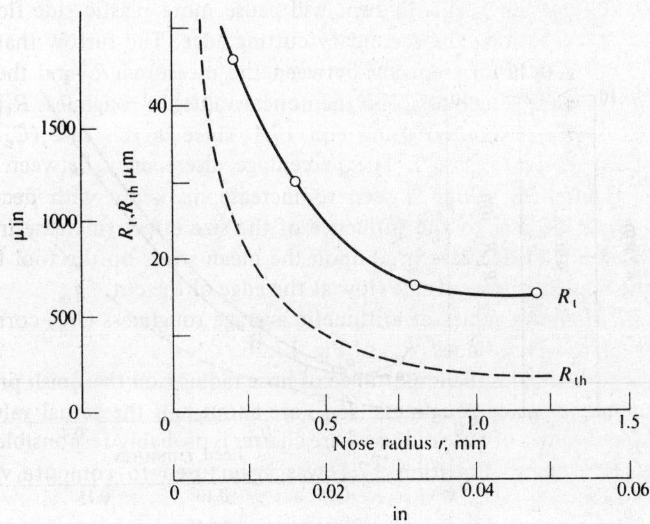

FIG. 17.28. Peak-to-valley roughness (R_t) for tools of different nose radius *r* when cutting the steel of Fig. 17.25 without a fluid. Tool geometry: 0°, 6°, 6°, 6°, 30°, 0°, var; tool material, carbide; depth of cut, 2.5 mm (0.10 in); feed, 0.2 mm/rev (0.008 i.p.r.); cutting speed, 240 m min^{-1} (800 f.p.m.). (After Shaw and Crowell 1965.)

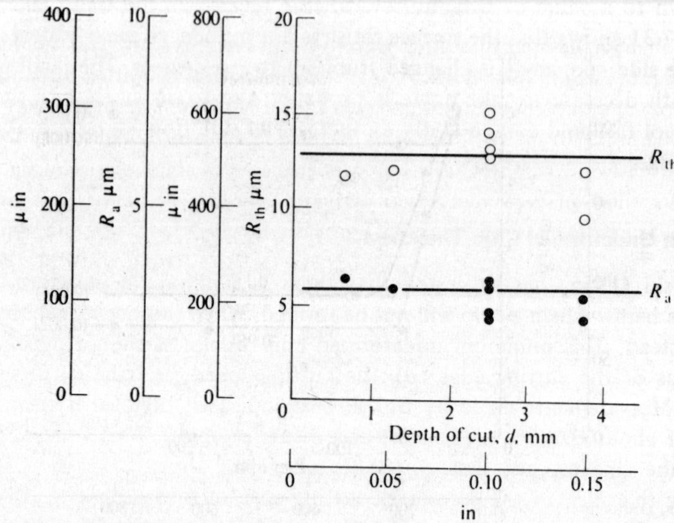

FIG. 17.29. Values of R_t and R_a for different depths of cut. Tool geometry: 0°, 6°, 6°, 6°, 15°, 15°, 0.5 mm (0.020 in); feed = 0.2 mm/rev (0.008 i.p.r.); speed = 120 m min^{-1} (394 f.p.m.); work material same as in Fig. 17.26; cutting fluid, none. (After Shaw and Crowell 1965.)

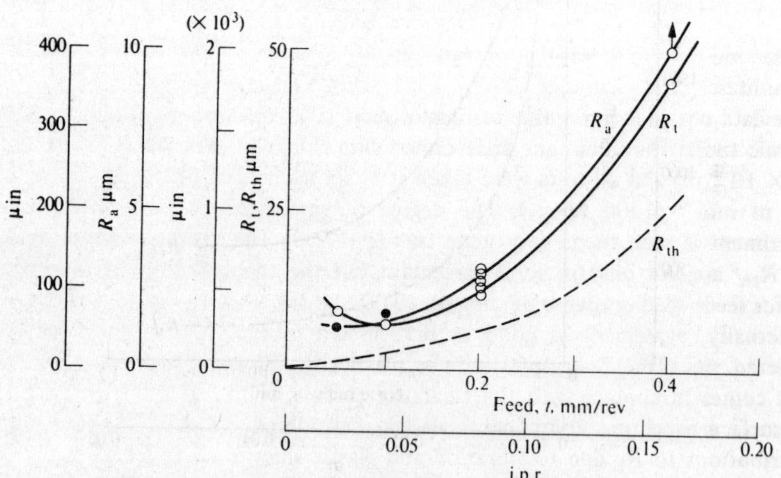

FIG. 17.30. Values of R_t and R_a for different feeds. Cutting conditions the same as for Fig. 17.28 except for the depth of cut, which = 2.5 mm (0.10 in). (After Shaw and Crowell 1965.)

in t down to a value of 0.1 mm/rev, when a reversal in the trend takes place. Figure 17.31 shows that the surface finish is not influenced in an important way when the side rake angle is changed from +6 to −6 degrees. The slight increase in R_t with decrease in rake angle is probably due to an increase in the stress on the tool face and consequently an increase in side flow along the secondary cutting edge.

Minimum Undeformed Chip Thickness

Sokolowski (1955) has suggested that there is a minimum undeformed chip thickness below which a chip will not be formed. When this occurs rubbing takes place instead. This minimum undeformed chip thickness should depend upon the radius of the cutting edge (ρ), the cutting speed, and the stiffness of the system. Measurements made by Sokolowski indicated that for a cutting edge radius of about 0.0005 in (12 μm) and a cutting speed of 650 f.p.m. (210 m min^{-1}) the smallest cut that could be taken was about 0.00016 in (4 μm). Applying this idea to the secondary cutting edge of a turning tool, it is found that a small triangular portion of the material that should be removed will be left behind (Fig. 17.32a). The portion left behind has been analyzed by Bramertz (1960, 1961) and called a Spanzipfel. Figure 1732b shows an enlarged view of the Spanzipfel used by Brammertz to derive the following expression for the theoretical surface roughness:

$$R'_{th} = \frac{t^2}{8r} + \frac{t_m}{2}\left(1 + \frac{rt_m}{2}\right) \tag{17.11}$$

The second term in this equation represents the contribution of the Spanzipfel.

Pahlitzsch and Semmler (1960, 1961, 1962) have applied this concept to some data obtained in the fine turning of AISI 1045 steel using newly sharpened ceramic tools. The minimum undeformed chip thickness was taken to be 1 μm (40×10^{-6} in) and all tests were taken after six seconds of cutting at a speed of 400 m min^{-1} (1200 f.p.m.). The degree of agreement between theory and experiment is seen to be very good in Fig. 17.33. The absolute values for R_t and R_{th}' are not only in good agreement but the theoretical curve is seen to rise for feeds smaller than 0.05 mm/rev (0.002 i.p.r.)

Actually, agreement as good as that shown in Fig. 17.33 should not be expected since the Spanzipfel will be plastically deformed and made smaller as it comes in contact with the clearance surface of the tool. In addition to the surface roughness components already considered, we should expect small contributions to R_t due to vibration and plastic side flow. The latter is apt to be most significant at very small values of feed and could be partly responsible for the rise in the curve for feeds less than 0.05 mm/rev (0.002 i.p.r.) as previously noted.

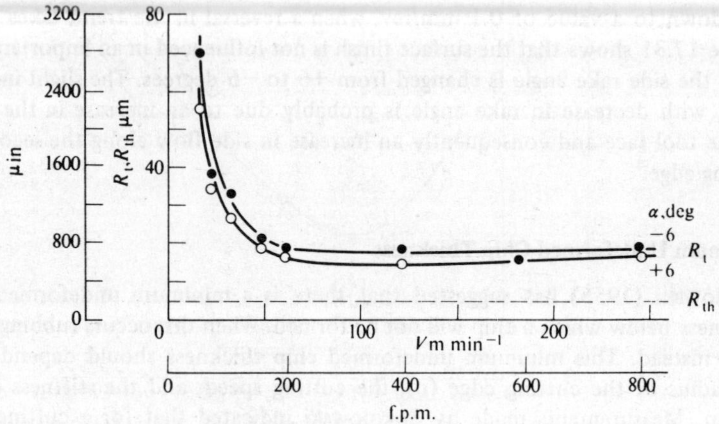

FIG. 17.31. Variation of roughness values (R_t) with rake angle of tool. Work material: normalized AISI 1045 steel; tool geometry: 0°, var. 6°, 6°, 15°, 15°, 0.8 mm (0.03 in); tool material, cemented carbide; cutting fluid, none; feed, 0.2 mm/rev (0.01 i.p.r.); depth of cut, 2.5 mm (0.10 in). (After Shaw and Crowell 1965.)

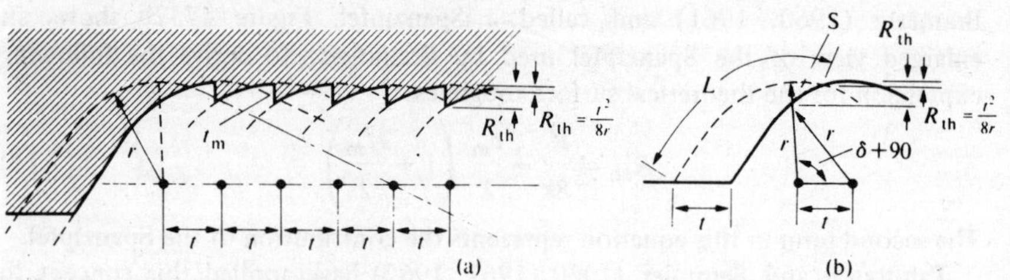

FIG. 17.32. Material left beind on finished surface due to the existence of a minimum undeformed chip thickness (t_m). The theoretical roughness, in the absence of a minimum undeformed chip thickness is $R_{th} = t^2/8r$ in this case (case I). (a) Shows relation between R_{th} and R'_{th}. (b) Shows Spanzipfel (S) shaded.

Groove Wear

Equation 17.11 will hold for a relatively short time, since it only applies to sharp tools. Grooves are very frequently formed at the edges of the cut (A and B in Fig. 17.34) on both the primary and secondary cutting edges. There will usually be a single groove at (A) but there may be several grooves at (B) spaced at a distance equal to the feed. The number of grooves increases with decrease in the end cutting edge angle and with an increase in time.

It is tempting to explain the presence of the grooves on the secondary cutting edge in terms of the Spanzipfel, since this will be a hard body with just the right shape and spacing to account for the grooves. However, there will be no

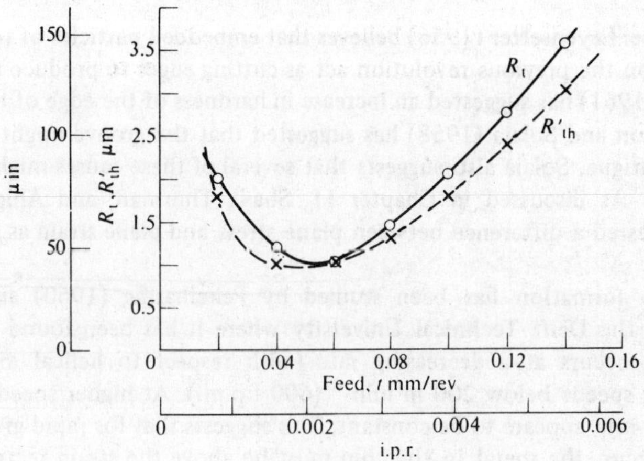

FIG. 17.33. Comparison between measured (R_t) and theoretical (R_{th}) values of surface roughness for sharp ceramic tools operating at 400 m min^{-1} (1200 f.p.m.) when cutting AISI 1045 steel dry at various values of feed (t). Nose radius = 1 mm (0.04 in); cutting time = 0.1 min. (After Pahlitzsch and Semmler 1960.)

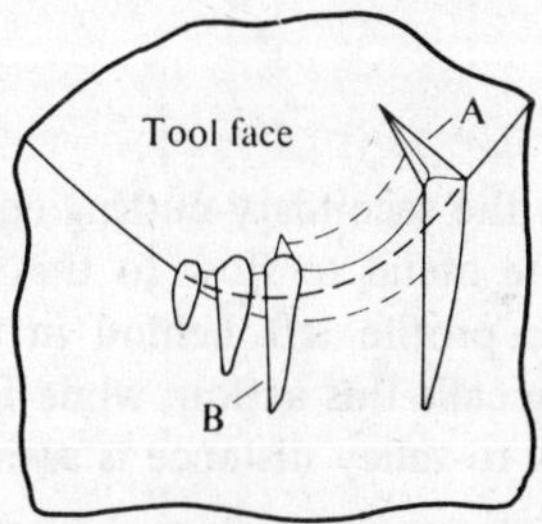

FIG. 17.34. Worn cutting tool showing large groove on primary cutting edge at (A) and smaller grooves on secondary cutting edge at (B).

Spanzipfel on the primary cutting edge and a groove is also found here (often larger than the ones on the secondary cutting edge). This does not mean to say that the Spanzipfel has nothing to do with the groove, but that it is probably not the sole cause for the groove.

There has been much speculation concerning the origin of the concentrated wear which gives rise to the grooves found at the edges of a chip. Hovenkamp and Van Emden (1953) have attributed the action to a work-hardened layer which gives rise to a concentration of stress at the chip edges. Pekelharing and Schuermann (1955) suggested that the grooves on the secondary cutting edge result from the abrasive action of the cusps on the machined surface. Albrecht (1956) has suggested the lateral escape of built-up edge material as the cause. Leyensetter (1956) believes that embedded particles of tool material removed on the previous revolution act as cutting edges to produce the grooves. Lambert (1961) has suggested an increase in hardness of the edge of the chip due to oxidation and Solaja (1958) has suggested that the groove might result from thermal fatigue. Solaja also suggests that several of these causes might act simultaneously. As discussed in Chapter 11, Shaw, Thurman, and Ahlgren (1966) have suggested a difference between plane stress and plane strain as a contributing cause.

Groove formation has been studied by Pekelharing (1960) and Lambert (1961) at the Delft Technical University where it has been found that groove formation occurs at a decreasing rate (with respect to helical distance cut) at cutting speeds below 200 m min^{-1} (600 f.p.m.). At higher speeds (temperatures) the rate appears to be constant. This suggests that for rapid groove formation to occur, the metal in the chip must be above the strain recrystallization temperature separating the hot-working region from the cold-working region. It also suggests that the groove is not primarily due to the side flow of BUE material, since we should then expect the rate of groove formation to increase with decrease in speed due to the increasing tendency toward BUE formation with decrease in speed.

Side Flow

The metal left behind on the secondary cutting edge is subjected to sufficiently high pressure to cause the metal to flow to the side as shown in Fig. 17.35. Figure 17.35a shows the profile left behind in the absence of side flow, or 'squeezing' as Pekelharing calls this action; while Fig. 17.35b shows the profile with side flow. The peak-to-valley distance is seen to be larger when side flow is present.

Sata (1964) has studied the influence of side flow on finish and has found that this component of roughness will be zero for a brittle material such as free cutting brass but may contribute up to 6 μm (240 μ in) to the roughness (R_t) when an alloy steel is machined. Sata suggests that the side flow contribution to roughness decreases with an increase in cutting speed.

Tool Life

Criteria for tool life in finish-machining have been discussed by Bickel (1954) and Ciragan (1955) who have shown that the arithmetic average roughness (R_a) may be used in finish machining in the same way as the size of the wear land on the clearance face (w) is used in conventional tool-life studies. The R_a vs. cutting time (T_c) curve was found to have the same general shape as the conventional wear land (w vs T_c) curve starting with a rapid increase in R_a for the first three to ten minutes and followed by a more gradual increase

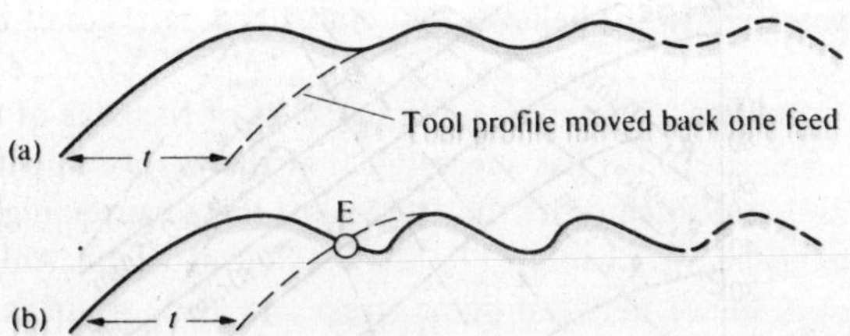

FIG. 17.35. Profile generated by tool point. (a) In absence of side flow. (b) With side flow. (After Pekelharing 1960.)

in R_a with T_c. The change in workpiece diameter with time was not found to be a sensitive measure of tool wear.

Semmler (1962) has presented tool life results based on both wear land (w) and peak-to-valley roughness (R_t) criteria. Typical results for a tool cutting AISI 1045 steel are presented in Fig. 17.36 for a feed of 0.04 mm/rev (0.0016 i.p.r.). The speeds corresponding to maximum tool life are seen to be much lower when wear land (w) is the criterion than when surface roughness (R_t) is the criterion and the tool-life curves based on roughness are seen to be much steeper than those based on land wear. This is probably due to the fact that the BUE has a greater influence on surface roughness that it does on land wear.

The controlling criterion will be the one giving the lower tool life at the operating speed V. The cutting speed should be chosen such that the total machining cost per part is a minimum (Chapter 19).

Recommended Practice

Recommended conditions for finish machining depend upon the following conditions:

1. Rigidity and stability of tool–work–machine system.
2. Limitations on speed, feed, or depth of cut.
3. Whether a new or worn tool is considered.

If an extremely stiff system is to be used then the contribution to roughness due to vibration will be small and a set of conditions that is independent of the tendency to cause chatter may be used. Otherwise, the cutting conditions may have to be shifted to decrease the tendency to chatter.

In finish machining it is not only important that the cutting temperature be sufficiently high that the BUE be eliminated, but that the layer of material adjacent to the tool face be thermally softened to provide low friction and a high shear angle. The cutting temperature will depend upon the hardness and ductility of the work material, and the cutting speed, feed, and depth of cut. Finishing cuts are usually of such small feed (0.05 mm/rev or 0.002 i.p.r.)

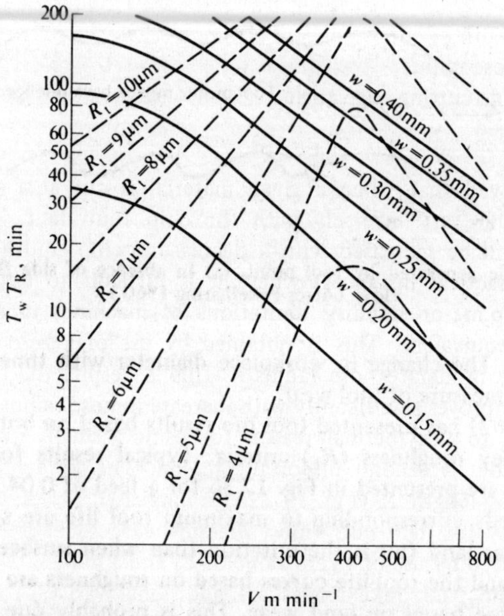

FIG. 17.36. Tool-life curves based on wear-land criterion (solid curves) and surface-roughness criterion (dashed lines). Cutting conditions: work material, AISI 1045 steel; tool material, tungsten carbide; feed, 0.04 mm/rev (0.0016 i.p.r.); depth of cut, 0.3 mm (0.120 in); side rake angle, 5°; back rake angle, 0°; clearance angle, 5°; side cutting edge angle, 30°; nose radius, 0.6 mm (0.024 in); cutting fluid, none. (After Semmler 1962.)

and small depth of cut (0.75 mm or 0.03 in) that a material such as normalized AISI 1045 steel must be machined at a very high speed in order to develop a sufficiently high cutting temperature. Figure 17.36 reveals that cost-optimum speeds are in excess of 300 m min^{-1} (1000 f.p.m.), when AISI 1045 steel is cut at a feed of 0.04 mm/rev (0.0016 i.p.r.). Such speeds are difficult to obtain when bars of ordinary size (1 in or 25 mm dia.) are cut.

The chip equivalent concept first introduced by Woxén (1931) is a convenient means for determining the influence tool geometry has on the relative cutting temperature. The chip equivalent is

$$CE = (L/bt) \tag{17.12}$$

where L is the active length of the cutting edge [see L in Fig. 17.32b], b is the depth of cut, and t is feed. It may be shown that

$$L = \frac{b - r(1 - \sin C_s)}{\cos C_s} + \frac{90 - C_s}{180}\pi r + \frac{t}{2} \tag{17.13}$$

where b = depth of cut
r = nose radius
C_s = side cutting edge angle
t = feed.

Woxén showed that when a given material is cut at a specific speed the temperature will vary inversely with the chip equivalent. Thus, the cutting temperature will be increased with a decrease in chip equivalent, and in those cases (most practical instances) where it is difficult to cut at a sufficiently high speed (r.p.m. or stability limitations of machine) it is advisable to use a short chip equivalent. This is obtained by use of a zero side cutting edge angle C_s and a moderate nose radius r. This is despite the fact that the theoretical finish improves significantly with increase in nose radius.

In addition to the thermal reason for using a small chip equivalent, tool stability provides a second important reason. The tendency toward chatter will increase as length L (Fig. 17.32b) is increased. Thus, a system that has limited stability for light cuts (most practical cases) will give far better results when L is reduced by use of a zero value of C_s and a moderate nose radius.

The side and back rake angles are additional variables that influence the cutting temperature since they influence the effective rake angle (α_e) measured in the plane of chip flow. The cutting temperature will increase when α_e is decreased by a decrease of side rake angle or as the back rake angle approaches zero. Thus, in the interest of a higher cutting temperature a side rake angle of zero degrees will usually give better results in finish-machining than the more common rake angle of $+10^\circ$ normally used for medium-sized cuts. A back rake angle of zero degrees is also found to be optimum in the finish-machining of steel.

A zero degree end cutting edge angle (C_e) is best geometrically as provided in the Kolosov design in Russia, where the tool is supplied with a secondary cutting edge that is parallel to the workpiece axis and slightly greater in length than the feed. In practice the Kolosov tool fails to give satisfactory finish due to the increased tendency toward chatter and the rapid development of grooves on the secondary cutting edge. When chatter and tool wear are taken into account the best end-cutting edge angle (C_e) is found to be about 20°. The optimum clearance angle is governed by similar groove formation considerations and should be about 10° instead of the 6° angle normally used when taking heavier cuts.

The hardness of the workpiece plays an important role in finish-machining. If the work is too soft, side flow will be too great and it will take too high a speed to get rid of the BUE and to produce a thermally softened layer on the tool face. The finish-machining performance of low-carbon steels can be improved by cold drawing and the performance of higher-carbon steels (AISI 1045 for example) may be improved by heat treatment. Semmler (1962) suggests an ultimate strength of 60 kg mm^{-2} (85 000 p.s.i.) as being optimum for AISI 1045 steel.

The ductility of the work material is also important in finish-machining. Free-machining additives such as sulfur and lead decrease chip ductility and consequently greatly improve finish by providing a smaller BUE and a smaller tendency toward side flow.

The sharpness of the cutting edge is important in finish-machining when the tool is new and it also influences the way in which the tool subsequently wears. Carbide tools that are ground and lapped with very fine diamond wheels give up to 20 percent better finish than those subjected to ordinary grinding conditions (Ciragan 1955). While there is no initial difference in the finish-machining performance between carbide and ceramic tools, it is found that worn carbide tools give better finish than worn ceramic tools. This is because the same initial cutting edge radius ends up as a much larger radius in a worn ceramic tool, particularly in the case of tools of coarser grain size.

Schuler (1957) has shown that a chemically active cutting fluid can reduce surface roughness by a factor of 4 or more at low speeds. However, the effectiveness of a fluid falls off rapidly at proper cutting speeds and in addition may cause an acceleration of tool wear due to the formation of a series of cracks that run perpendicular to the cutting edge (so called 'comb' cracks).

Example

For cutting a steel such as AISI 1045, the following recommendations may be made for best finish and reasonable life:

Cutting conditions

Speed	1000 f.p.m. (300 m min^{-1})
Feed	0.002 in (0.05 mm/rev)
Depth of cut	0.030 in (0.75 mm)

Tool geometry

Side rake angle	0° (increase for lower speed)
Back rake angle	0°
End cutting edge angle	20°
Side cutting edge angle	0°
End relief	10°
Side relief	10°
Nose radius	0.02 in (0.5 mm)
Edge radius	0.0001 in (0.003 mm) (diamond-lapped)

Work material—cold drawn with sulfur.

Fluid—none.

This should produce a finish of approximately $R_a = 32$ μin (0.8 μm) or $R_t = 4$ μm (160 μm) if the system is reasonably stable.

System Stability

The machining system can deviate from desired geometry in two general ways—static deflection leading to inaccuracy of the machined part or dynamic instability leading to periodic errors such as waviness and roughness.

Static Deflection

Static deflection may arise due to lack of stiffness of the machine structures including bearings, deflection of work or tool, differential thermal expansion, or dimensional instability of materials due to changes in residual stresses or due to a density change accompanying a structural transformation. Static deflection due to bearing elasticity may be reduced by use of preloaded nonlinear rolling contact bearings (see for example, Shaw and Macks 1949) or may be completely eliminated by use of externally pressurized bearings having variable external resistances with feedback. Such active externally pressurized bearings were first introduced in 1963 (Mayer and Shaw 1963).

Dimensional accuracy of machine tools may be improved by increasing the precision of components or by correcting for residual inaccuracies that are not convenient to eliminate. For example, lead-screw precision may be improved by use of an error-compensating nut. Similarly, cumulative errors involved in the ruling of large spectroscopic gratings cannot be reduced to a satisfactory level due to differential thermal expansion even when the machine is housed in an air-conditioned room having a temperatue variation of only a few degrees and the machine is completely immersed in an oil bath with temperature deviation of only a small fraction of a dgree during the ruling process. To counter this difficulty, the spacing of each pair of lines ruled is measured and any deviation from the required value immediately compensated.

Practically all metals contain balanced (equal plus-minus) residual elastic stresses which will cause a change in external dimensions if relaxed. An interesting example of the potential difficulty was experienced by a manufacturer of long cylindrical rolls used in the paper industry. It was found that these rolls which were very straight when leaving the plant were unacceptable relative to straightness after being transported by truck over relatively rough roads. This was traced to the relaxation of residual elastic stresses during shipment. A useful solution was to provide a mechanical stress relieving treatment before the final cylindrical grinding operation. This was done by supporting the rolls at the nodal points for the first vibrational mode and then inducing a forced vibration of sufficient amplitude to bring the maximum stress induced just to the yield-point. The cyclic force was provided by a small unbalanced mass attached to the shaft of a small electric motor clamped to the center of the freely supported beam (Fig. 17.37). When the roll was shipped after the final finish-grinding operation, it was supported at the same points as when it was subjected to the mechanical stress-relieving treatment. The roll then arrived at its destination as straight as it had been on leaving the plant since the few additional cycles it experienced during the shipment caused negligible additional relaxation of elastic residual stresses.

It is reported that large cast iron casting may be similarly 'seasoned' by storing them outdoors for a long time (a year or more) where their temperature changes cyclically. This similarly reduces the level of residual stress before final machining due to cyclic differential thermal expansion.

Dynamic Instability

Dynamic instability in which there is cyclic relative motion between tool and work is of two types—forced vibration and self-excited vibration.

A forced vibration results when a cyclically varying external load has a frequency that is close to one of the natural frequencies of the machining system (tool-work-machine). When this is the case, slightly more energy is absorbed per vibrational cycle than is given back and the net energy per cycle is available to increase the vibrational amplitude. The vibrational amplitude will increase until the energy dissipated per cycle (as heat = damping) just equals the net energy absorbed by the system per cycle. If damping were constant with amplitude, the amplitude would grow without limit. However, in practical situations, damping increases with amplitude and a steady-state vibrational amplitude is quickly achieved.

Vibrational amplitude due to a forced vibration may be reduced by

1. Elimination of the cyclic exciting forces.
2. Avoiding the coincidence of frequency of exciting force and natural frequency.
3. Increased stiffness.
4. Increased damping.

The role of the first three of the above items is obvious. In modern times, the fourth item has received considerable attention in machine tool design through modal analysis in which the natural frequencies of a complex structure may be accurately determined relative to different modes of vibrational motion. Analog and digital computers have made such analysis possible.

Self-Excited Vibrations

A self-excited vibration results when the internal energy of the system varies in such a way during a vibrational cycle that more energy is stored than is released during a single cycle.

The first source of self-excited vibration to be identified was that due to the decrease of the main (power) component of cutting force (F_P) with cutting

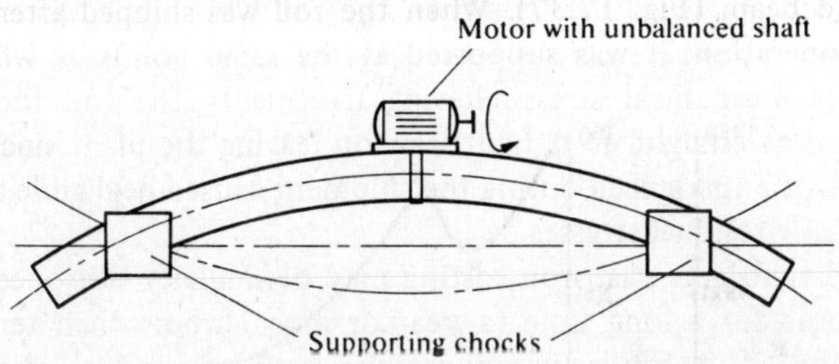

FIG. 17.37. Method of removing residual stress in bar by repeated flexing.

speed (V) at high values of V (Arnold 1946). When a tool vibrates in the (F_P) direction, the relative velocity between chip and tool will then vary in such a way that a net amount of energy will be stored in the tool per cycle. However, since F_P decreases with V only at relatively high values of V, this mechanism of self-excitation gives rise to high vibrational frequencies such as those associated with a tool (~ 2000 Hz). To induce such a self-excited vibration, the tool need only be bumped or encounter a hard spot in the work material so that it executes a single cycle. The amplitude will then grow until the damping energy per cycle becomes equal to the net energy stored per cycle. Figure 17.38 shows a typical F_P vs V curve capable of inducing self-excited vibration.

In addition to the source of self-excited vibration identified by Arnold there are undoubtedly several others that pertain at lower vibrational frequencies. One of these discussed by Shaw and Sanghani (1962) is associated with the variation in cutting energy with variable undeformed chip thickness t.

Variable Undeformed Chip Thickness

Figure 17.39a shows orthogonal cutting with constant undeformed chip thickness t while Fig. 17.39b is for an orthogonal cutting situation with cyclically varying t. The case of Fig. 17.39b is capable of causing a self-excited vibration in the F_Q direction since more energy is stored during the part of the cycle when F_Q is increasing than is released by the system when F_Q is decreasing (Shaw and Hölken 1957). The amplitude of the resulting vibration will depend upon the stiffness of the system and the degree of damping.

Figure 17.40 shows the situation for a cut where t is constant (solid lines) as well as one for which the free surface is inclined at an angle (δ) to the velocity vector V (dotted lines). The importance of such a change in direction of the free surface may be visualized by considering some simple two-dimensional slip line fields.

Figure 17.41a shows a frictionless tool engaging the corner of a workpiece. The slip lines (lines of maximum shear stress) will be inclined to both the free surface and the tool face at 45°. This becomes evident when the Mohr's circle diagram (insert) is considered. Corresponding points are similarly labeled in both

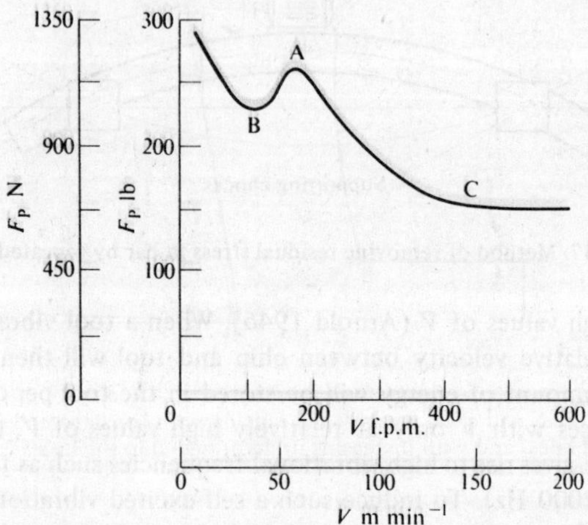

FIG. 17.38. Variation in principal cutting force F_P with cutting speed V. Work material, hot-rolled AISI 1008 steel; tool material, M-2 HSS; feed, 0.006 i.p.r. (0.15 mm/rev); rake angle, 10°; clearance angle, 5°; nose radius, 0.005 in (0.13 mm); depth of cut, 0.1 in (2.54 mm); cutting fluid, none.

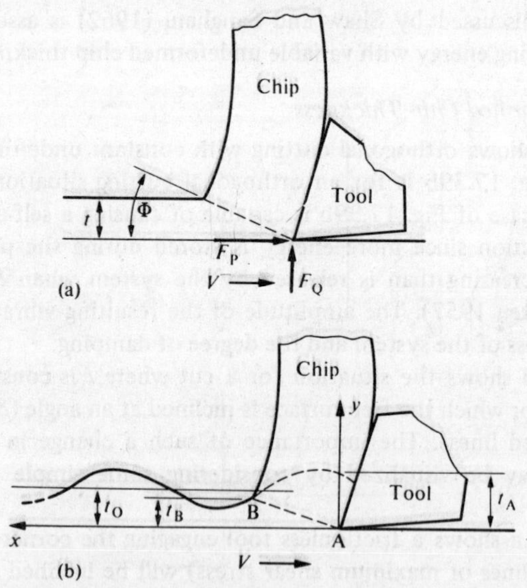

FIG. 17.39. Metal cutting. (a) With constant undeformed chip thickness t. (b) With variable undeformed chip thickness.

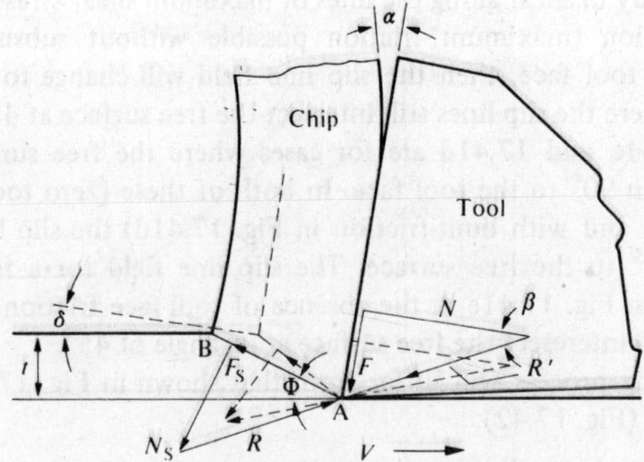

FIG. 17.40. Forces and shear angle when cutting with constant undeformed chip thickness (solid lines) and with increasing undeformed chip thickness (dotted lines).

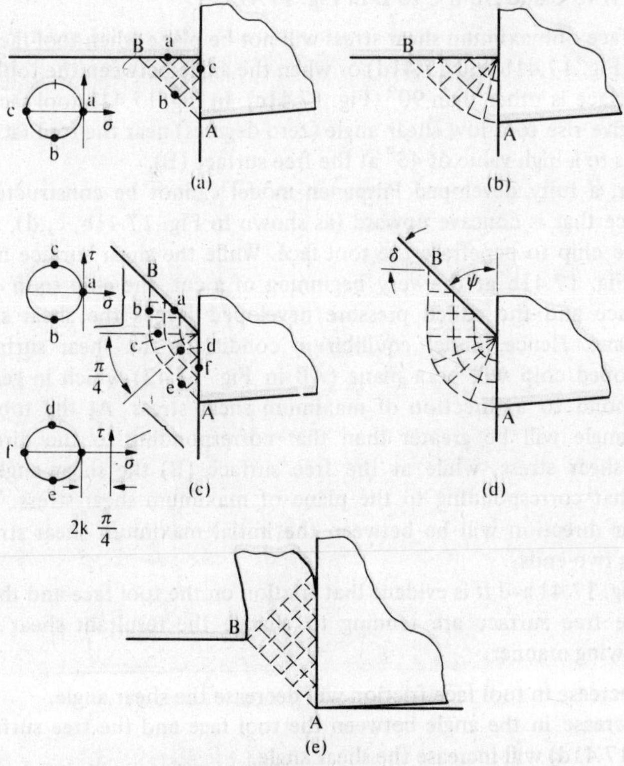

FIG. 17.41. Slip line fields. (After Shaw and Sanghani 1962.)

diagrams. If the material is homogeneous and does not strain harden, it should deform plastically in shear along the lines of maximum shear stress.

If limit-friction (maximum friction possible without subsurface flow) is present on the tool face, then the slip line field will change to that shown in Fig. 17.41b, where the slip lines still intersect the free surface at 45°.

Figures 17.41c and 17.41d are for cases where the free surface makes an angle other than 90° to the tool face. In both of these (zero tool face friction for Fig. 17.41c and with limit-friction in Fig. 17.41d) the slip lines still make an angle of 45° to the free surface. The slip line field for a fully developed chip is shown in Fig. 17.41e in the absence of tool-face friction. The slip lines are again seen to interesect the free surface at an angle of 45°.

A real cutting process will differ from that shown in Fig. 17.41e in several important ways (Fig. 17.42):

1. The rake angle α will generally not be zero.
2. Tool face friction will not be zero.
3. Additional plastic flow will occur along the tool face due to the closing of fracture points and additional burnishing action (as shown respectively from **A** to **C** and from **C** to **D** in Fig. 17.42).

The surface of maximum shear stress will not be plane when tool-face friction is present (Fig. 17.41b and 17.41d) or when the angle between the tool face and the free surface is other than 90° (Fig. 17.41c). In Fig. 17.41b tool face friction is seen to give rise to a low shear angle (zero degrees) near the tool face (A) but this changes to a high value of 45° at the free surface (B).

However, a fully developed Piispanen model cannot be constructed with a shear surface that is concave upward (as shown in Fig. 17.41b, c, d), since this requires the chip to penetrate the tool face. While the shear surface may be as shown in Fig. 17.41b at the very beginning of a cut, the chip soon curls into the tool face and the added pressure developed forces the shear surface to become plane. Hence, under equilibirum conditions the shear surface for a fully developed chip will be a plane (AB in Fig. 17.42) which in general will not correspond to a direction of maximum shear stress. At the tool tip (A) the shear angle will be greater than that corresponding to the direction of maximum shear stress, while at the free surface (B) the shear angle will be less than that corresponding to the plane of maximum shear stress. Thus, the actual shear direction will be between the initial maximum shear stress directions for its two ends.

From Fig. 17.41 a–d it is evident that friction on the tool face and the inclination of the free surface are tending to change the resultant shear angle (ϕ) in the following manner:

1. An increase in tool face friction will decrease the shear angle.
2. A decrease in the angle between the tool face and the free surface (ψ in Fig. 17.41d) will increase the shear angle.

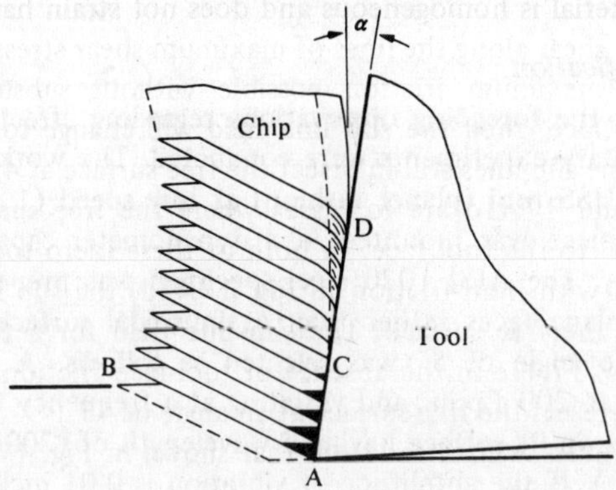

FIG. 17.42. Revised Piispanen model with secondary plastic flow along tool face. (After Shaw and Sanghani 1962.)

At the beginning of a cut, the shear direction will always intersect the free surface at the same angle (45°) regardless of the inclination of the free surface (δ) and regardless of the state of friction on the tool face. An equivalent way of expressing this is that the shear angle at the free surface at the beginning of a cut will be

$$\phi_\delta = \phi_0 + \delta \tag{17.14}$$

where ϕ_0 is the shear angle at the free surface when δ is zero (45°) and δ is positive as shown in Fig. 17.41d. The influence of tool-face friction on the direction of shear is completely independent of the influence of the direction of the free surface on the direction of shear at the beginning of a cut.

For a fully developed chip the angle between the shear surface and the free surface is forced to be other than 45° due to the compatability requirement that the shear surface be plane. Angle ϕ_0 is no longer independent of the state of friction on the tool face, and in general will not be 45°. It seems reasonable to assume that eqn 17.14 will also hold to a first order of approximation for a fully developed chip. This is equivalent to saying that the free surface and tool face friction effects on ϕ are separable and that to a first order of approximation the tool face friction contribution to ϕ will remain unchanged as δ is varied. Thus, ϕ_0 in eqn 17.14 will be considered a function of tool-face friction only and independent of changes in δ.

As in most engineering situations where it is expedient to assume an independence of variables, the error associated with the approximation should decrease as the magnitude of δ decreases.

Experimental Verification

In order to check the foregoing observations regarding effects due to changes in δ, some elementary experiments were conducted. The workpiece was moved past a stationary HSS tool (planer fashion) at low speed (1 i.p.m. = 0.03 m min^{-1}). The workpiece was mounted in a dynanometer capable of recording forces F_P and F_Q. The AISI 1020 steel specimen was prepared as shown in Fig. 17.43 with plane faces rather than a sinusoidal surface, as a matter of simplification. The angle of 5° was selected as follows. A workpiece being turned in a lathe at 200 f.p.m. and vibrating at a frequency of 200 c.p.s. will have chatter marks on its surface having a wavelength of (200 × 12/60)/200 = 0.200 inch (5 mm). If the amplitude of vibration is 0.01 inch (0.25 mm). the corresponding approximate value of δ will be:

$$\delta = \tan^{-1} \frac{0.01}{0.100} \cong 5°.$$

A representative result is shown in Fig. 17.44 where the dotted curve shows the variation in undeformed chip thickness at the tool tip (t') with time and the solid curve shows the corresponding cutting-force trace (F_P). The rake angle α was 10° and the maximum value of t was 0.077 inch (1.93 mm). Cuts were taken using CCl_4 as a cutting fluid in order to minimize effects due to a built-up edge. It is evident that the maximum force occurs long before the maximum value of t' is encountered at the tool tip. This suggests that the value of undeformed chip thickness of importance is that at the upper end of the shear plane (**B** in Fig. 17.42) rather than that at the tool tip (**A** in Fig. 17.42). This should have been recognized immediately since the length of the shear plane is the quantity of importance and this depends on t_B and not on $t_A = t'$ (Fig. 17.39b) for the situation where the tool moves in a straight line and the free surface is wavy.

It was decided to repeat the previous experiments under conditions that would enable forces to be referred to t_B rather than $t_A = t'$. At the same time

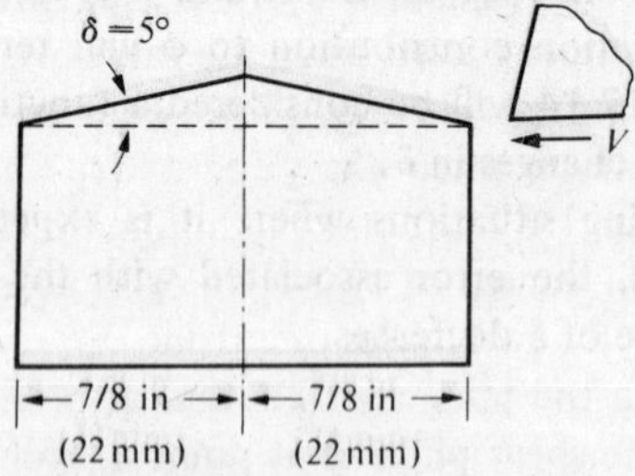

FIG. 17.43. Single-cycle specimen for tests at variable undeformed chip thickness.

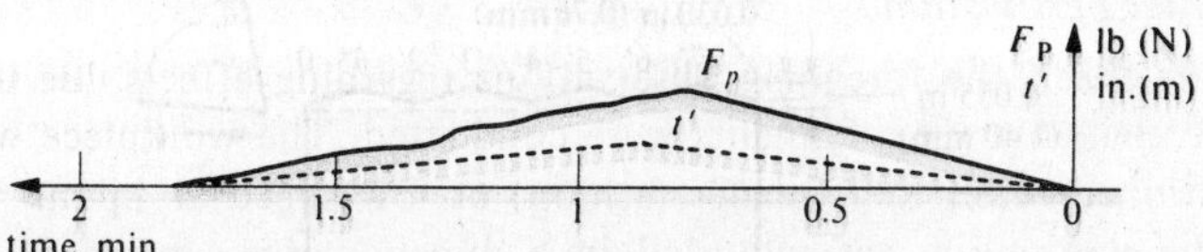

FIG. 17.44. Representative cutting-force trace for specimen of Fig. 17.43. (After Shaw and Sanghani 1962.)

it was decided to use specimens (Fig. 17.45) having several cycles present on the surface and to perform the cuts such that (t_{min}) was 0.025 inch rather than zero. The four peaks and valleys employed were numbered as shown in Fig. 17.45.

A representative force trace (F_P) is shown in Fig. 17.46. The first two cycles were usually irregular as equilibrium was established, but the last two cycles were always similar and reproducible and hence the ones considered in detail. A marking device with which the recorder was equipped enabled the point-in-time to be recorded when the upper end of the shear plane (**B** in Fig. 17.42) just reached a peak or a valley on the surface. These points are shown at the top of Fig. 17.46 and represent the locations of the tool tip when the top of the shear plane just reaches a peak or a valley.

With this information at hand it is possible to draw the surface wave to scale beneath the force plot as shown in the lower part of Fig. 17.46. From this construction we may compute the mean values of shear angle when t is increasing as well as when t is decreasing. Such calculations are given in Table 17.2 where the values in the second column were obtained by converting the measured distances between 0 and 1, 0 and 2, etc., at the top of Fig. 17.46 to equivalent distances on the specimen. The mean value of the shear angle is thus found to be 45.8° when t is increasing and 28.8° when t is decreasing.

The value of δ for the specimen of Fig. 17.45 is $\tan^{-1}(30/250) = 7°$. According to eqn 17.14 this should give rise to an increase in ϕ of approximately 7° when δ is +7° and a like decrease when δ is −7°. Half the difference between the two measured values of ϕ ($45.8 - 28.8/2 = 8.5°$) is to be compared with the value of 7° predicted by eqn 17.14. Equation 17.14 is thus found to be in reasonably good agreement with experiment, even when the value of δ is quite large.

The upper and lower portions of Fig. 17.46 when used together enable the value of F_P to be determined when the value of (t_B) has any value between the maximum (0.055 in = 1.4 mm) and the minimum (0.025 in = 0.64 mm). Values of F_P for different values of t_B, when t_B is decreasing and increasing, are recorded in Table 17.3 for two cycles of cutting.

The first value of F_P in the table (840) is that at point y, which is the value of F_P when the top of the shear plane is at point **a**. Sometimes the chip tends to break as at point x in Fig. 17.46, and this in turn will have a local influence on the force curve.

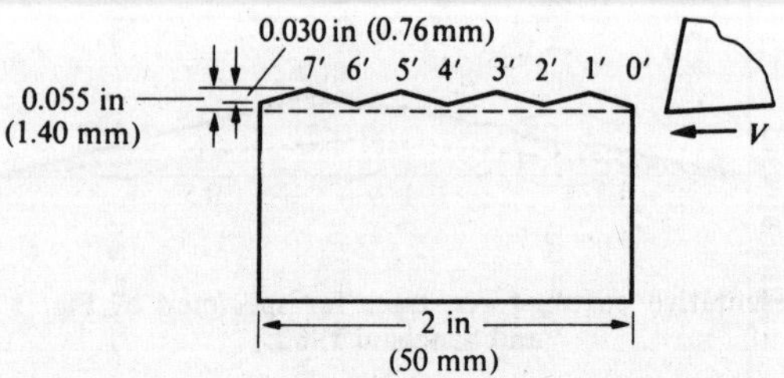

FIG. 17.45. Multi-cycle specimen for tests at variable undeformed chip thickness. (After Shaw and Sanghani 1962.)

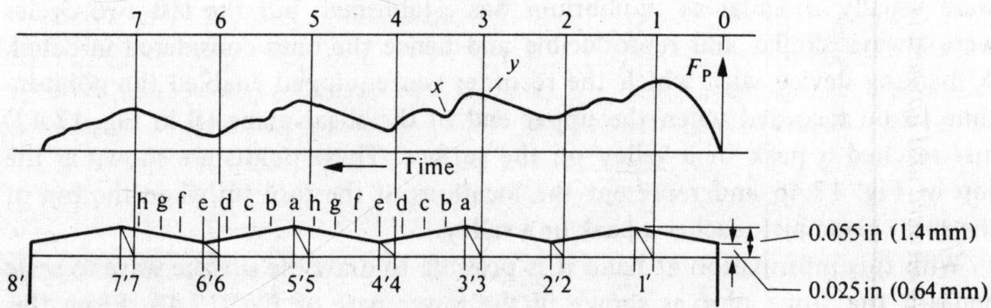

FIG. 17.46. Representative cutting-force trace for specimen of Fig. 17.45. (After Shaw and Sanghani 1962.)

It is clearly evident from Table 17.3 that F_P is less for the same value of t_B when t_B is increasing than when t_B is decreasing.

The F_Q trace paralleled that for F_P in every instance and everything that has been said regarding F_P holds for F_Q as well.

When relatively brittle materials are machined such as cast iron or beta-brass, the chips break periodically in the manner shown in Fig. 17.47, which was drawn from selected frames of a motion picture. The material is seen to fracture in shear along a line that strongly resembles slip line **AB** in Fig. 17.42.

Discontinuous Chips

Discontinuous cutting is thus seen to involve cutting with a large positive value of δ. Immediately after shear fracture has occurred, a new chip develops and as it does, δ is decreased due to a humping-up of the material. As the chip grows it rolls into the tool face and the resulting normal stress causes subsequent slip to occur on a surface that is less concave upward and farther removed from the surface of maximum shear stress. In the case of a ductile metal this action will continue until the shear plane becomes perfectly plane. With a brittle material, however, a shear stress sufficiently high to cause a shear failure will

TABLE 17.2

Mean shear angle values when cutting downward and upward (1 in = 25.4 mm)

Distance	in (× 10^3)	Distance	in (× 10^3)	Shear angle, deg: cutting downward	Shear angle, deg: cutting upward
0–1	196	1–1′	250– 196 = 54	$\phi_1 = \tan^{-1}\left(\frac{55}{54}\right) = 45.6$	
0–2	458	2–2′	500– 458 = 42		$\phi_3 = \tan^{-1}\left(\frac{25}{42}\right) = 30.8$
0–3	700	3–3′	750– 700 = 50	$\phi_3 = \tan^{-1}\left(\frac{55}{50}\right) = 47.7$	
0–4	955	4–4′	1000– 955 = 45		$\phi_4 = \tan^{-1}\left(\frac{25}{45}\right) = 29.0$
0–5	1190	5–5′	1250–1190 = 60	$\phi_5 = \tan^{-1}\left(\frac{55}{60}\right) = 42.2$	
0–6	1450	6–6′	1500–1450 = 50		$\phi_6 = \tan^{-1}\left(\frac{25}{50}\right) = 26.5$
0–7	1700	7–7′	1750–1700 = 50	$\phi_7 = \tan^{-1}\left(\frac{55}{50}\right) = 47.7$	
				Av 45.8°	Av 28.8°

TABLE 17.3.
Values of F_P for different values of t_B when t_B is decreasing and increasing (1 in = 25.4 mm, 1 lb = 4.45 N)

Position of intersection of shear plane with free surface (Fig. 17.46)	$t_B = t$ at shear plane intersection in (× 10^3)	F_P lb For cycle 3 to 5	For cycle 5 to 7	Av
a	55	840	740	790
b	47.5	800	660	730
c	40	540	480	510
d	32.5	560	440	500
e	25	400	240	320
f	32.5	380	280	330
g	40	500	400	450
h	47.5	620	520	570

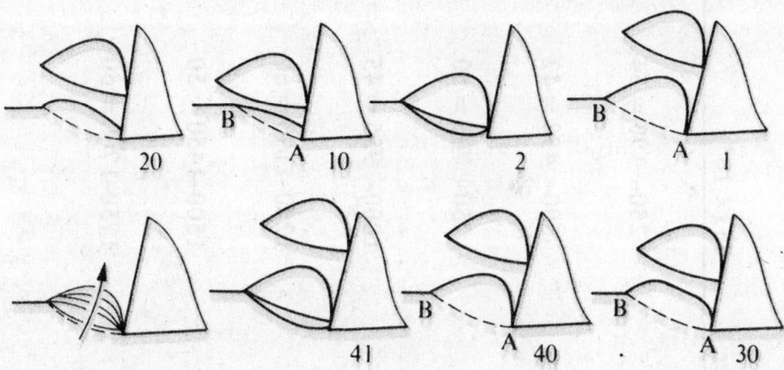

FIG. 17.47. Sketches of discontinuous cutting action for beta-brass. Numbers under sketches are movie-frame numbers. Depth of cut, 0.007 in (0.18 mm); cutting speed, 0.5 i.p.m. (13 mm min^{-1}); rake angle, 15°; width of cut, 0.161 in (4.1 mm).

be reached before the shear plane becomes plane. The metal does not fracture along the last active shear plane on which shear flow was occurring but along the surface of maximum shear stress.

It is well known that discontinuous cutting is desirable from the point of view of cutting forces. While this was originally believed due to a reduction in tool face friction associated with the periodic renewal of an adsorbed film on the face of the tool, it now appears that the high positive values of δ that are obtained in discontinuous cutting are chiefly responsible for the relatively low values of F_P in discontinuous chip formation.

Milling

Chips of variable t are characteristic of the plain milling process. In up (or conventional) milling the thickness of the chip varies from zero to a maximum value t_1 (Fig. 17.48a) while in down (or climb) milling the undeformed chip thickness goes from t_2 to zero (Fig. 17.48b).

Obvious advantages of the down-milling process are:

1. A shorter length of contact (l) for the same feed per tooth (f) and depth of cut (d).
2. Absence of rubbing at the beginning of the cut.

Because of these factors down-milling frequently gives better results than up-milling if the machine has sufficient rigidity to withstand the greater impact associated with the more abrupt tooth engagement in down-milling.

However, the extent of improvement is often not as great as expected and this could be expalined in terms of the lower forces associated with conventional milling. For the same value of t the cutting force will be greater in down-milling than in up-milling, since δ is negative for down-milling but positive for up-milling.

The approximate value of δ will be:

$$\delta = \tan^{-1}\left(\frac{t_m}{l}\right) \tag{17.15}$$

where t_m = maximum undeformed chip thickness.

If $v < V$

$$\delta = \frac{2f}{D}\cos i \tag{17.16}$$

where f = feed/tooth.

Regenerative Chatter

The foregoing discussion has been concerned with a tool moving in a straight line while cutting a wavy surface, which may be called case I (Fig. 17.49a). When a tool cuts a smooth specimen and vibrates in the y (F_Q) direction, a wavy surface will be generated as shown in Fig. 17.49b. This will be referred to as case II. The effective value of t (with regard to cutting forces) for case II is that at the tool point (t_A), since it is this that determines the length of the shear plane.

When the tool cuts from **A** to **B′** its resultant velocity (V) will be directed as shown in Fig. 17.49b and the effective rake angle (α_e) should be measured from a reference line that is perpendicular to this velocity vector. As the tool goes from **A** to **B′**, the depth of cut (t) will increase, and δ will be positive.

Thus, there are two major differences between cases I and II:

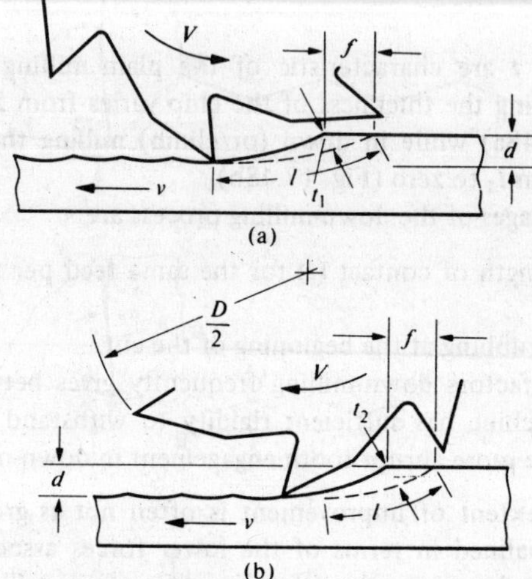

FIG. 17.48. Milling cutter action, (a) Up (conventional) milling. (b) Down (climb) milling. (f feed/tooth.)

1. For case I there is no change in rake angle but for case II the effective rake angle is increased by an amount δ when t is increasing, and vice versa.
2. The effective value of t for case I is t_B while that for case II is t_A.

Due to a change in inclination of the free surface, cutting forces will be less (and the shear angle greater) when the effective value of t is increasing for both cases I and II. In addition, for case II, the increase in rake angle that occurs when t is increasing will cause an additional decrease in cutting forces (and increase in shear angle). Similarly when t decreases, the change in rake angle for case II will cause the cutting force to increase to an even higher value than in case I.

The difference in F_Q, for a given value of t, when t is increasing and decreasing, it significant in connection with self-excited vibrations. Energy is stored elastically in a vibrating tool when the force on the tool (F_Q) is in the same direction as the motion of the tool, and vice versa. When the tool of Fig. 17.49b cuts from **A** to **B′**, energy is being released from the tool since tool-force and tool-motion are in opposite directions. Similarly as the tool goes from **B′** to **C′**, energy is stored since F_Q and tool motion are then in the same direction.

If the energy stored exceeds the energy released over a cycle, then the amplitude of vibration will increase until the damping energy per cycle just equals the net energy stored per cycle. Since force F_Q will be larger, when cutting from **B′** to **C′** (energy stored) than from **C′** to **D′** (energy released), we have

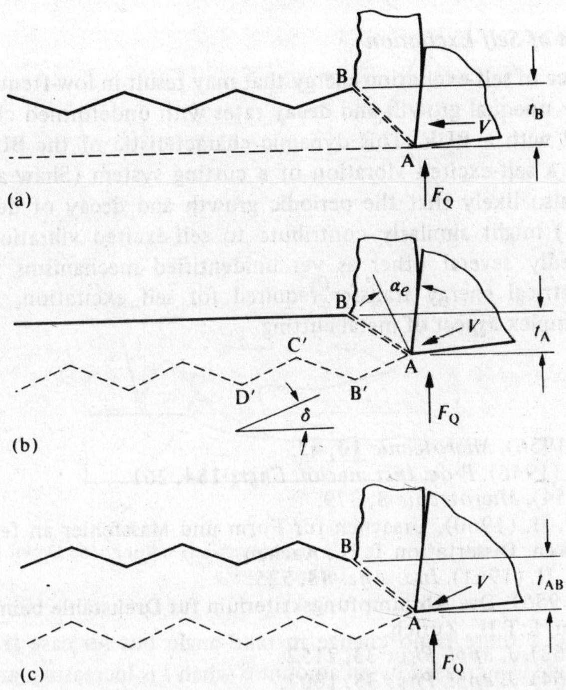

FIG. 17.49. Cutting with variable undeformed chip thickness. (a) Case I—straight tool motion and wave on free surface. (b) Case II—vibrating tool and smooth free surface. (c) Case III—vibrating tool and wave on free surface.

the condition necessary for a self-excited vibration to occur. It should be noted that the foregoing argument is independent of whether the tool moves relative to the work, the work moves relative to the tool, or both.

In a turning operation the surface is recut after each revolution and the wavy surface previously cut will act through feedback to induce additional vibration. Such a case is shown in Fig. 17.49c. The effective value of t in this instance is the vertical distance between **A** and **B** (t_{AB}). Cutting forces will be less when t_{AB} is increasing than when t_{AB} is decreasing. The net energy stored per cycle will depend upon the phase relation between the old and new waves, which will have the same frequency (the natural frequency of the tool-work system). The resulting phase angle will be that which gives the greatest stored energy per cycle (180°) and the resulting energy per cycle will be approximately double the value for the case of Fig. 17.49b. The feedback chatter just described is termed regenerative chatter.

Other Sources of Self Excitation

Another source of self-excitation energy that may result in low-frequency chatter is due to the unequal growth and decay rates with undeformed chip thickness (t) associated with a BUE. This dynamic characteristic of the BUE is capable of producing a self-excited vibration of a cutting system (Shaw and Sanghani 1963). It is also likely that the periodic growth and decay of deformed chip thickness (t_C) might similarly contribute to self-excited vibration and there are undoubtedly several other as yet unidentified mechanisms that provide the unsymmetrical energy transfer required for self excitation, thus making this a very complex aspect of metal cutting.

References

Albrecht, P. (1956). *Microtecnic* **10**, 45.

Arnold, R. N. (1946). *Proc. Inst mechn. Engrs* **154**, 261.

Bickel, E. (1954). *Microtecnic* **8**, 279.

Brammertz, P. H. (1960). Ursachen für Form und Massfehler an feinbearbeiten Werkstüchken. Dissertation T. H., Aachen.

Brammertz, P. H. (1961). *Ind. Anz.* **83**, 525.

Ciragan, O. (1955). Das Abslumpfungskriterium fur Drehstähle beim Schlichten. Dissertation, E.T.H., Zurich.

Cocks, M. (1962). *J. appl. Phys* **33**, 2152.

Cocks, M. (1964). *J. appl. Phys* **35**, 1807.

Cocks, M. (1965). *Wear* **8**, 85.

Danek, O., Polacek, M., Sparek, L., and Tlusty, J. (1962). *Selbsterregte Schwingungen an Werkzeugmaschinen*. V.E.B. Verlag Technik, East Berlin.

Doi, S. and Kato, S. (1956). *Trans. Am. Soc. mech. Engrs* **78**, 1127.

Hahn, R. S. (1953). *Trans. Am. Soc. mech. Engrs* **75**, 1073.

Ham, I., Hitomi, K., and Thuering, G. L. (1961). *Trans. Am. Soc. mech. Engrs* **83**, 142.

Ham, I., Hitomi, K., and Theuring, G. L. (1962). *Trans. Am. Soc. mech. Engrs* **83**, 282.

Ham, I., Hitomi, K., and Theuring, G. L. (1964). *Trans. Am. Soc. mech. Engrs* **86**, 141.

Hovenkamp, L. H. and Van Emden, E. (1953). *Microtecnic* **7**, 3, 117.

Lambert, H. J. (1961). *Ann CIRP* **10**, 246.

Leyensetter, W. (1956). *VDI Zeitschrift* **98**, 957.

Martellotti, M. E. (1941). *Trans. Am. Soc. mech. Engrs* **63**, 8, 677.

Mayer, J. E. and Shaw, M. C. (1963). *Trans. Am. Soc. mech. Engrs* **85**, 2.

Moll, H. (1939). Die Herstellung Hochwertiger Drehflachen. Dissertation, T. H., Aachen.

Nakayama, K. (1957). *Bull. Fac. Engng Yokohama natn* Univ. 6.

Nakayama, K., Shaw, M. C., and Brewer, R. C. (1966). *Ann. CIRP* **14**, 211.

Opitz, H. and Gappisch, M. (1962). *Int. J. Mach. Tool Des. Res.* **2**, 43.

Opitz, H. and Moll, H. (1941). *Herstellung Hochwertiger Drehflachen.* Ber Betriebswiss. Arb., VDI, Verlag, Berlin.

Ostermann, G. (1960). Neuere Erkenntnisse uber die Ursachen der Verschleisses auf der Spanflache von Hartmetalldrehwerkzeugen. Dissertation, T.H., Aachen.
Pahlitzsch, G. and Semmler, D. (1960). *Z. fur wirtschaftlich Fertigung.* **55**, 242.
Pahlitzsch, G. and Semmler, D. (1961). *Z. fur wirtschaftlich Fertigung.* **56**, 148.
Pahlitzsch, G. and Semmler, D. (1962). *Z. fur wirtschaftliche Fertigung.* **57**, 45.
Pekelharing, A. J. (1960). *Microtecnic* **14**, 2.
Pekelharing, A. J., and Schuermann, R. A. (1955). *Werkstattstech u. Machenbau*, **45**, 59.
Peters, J. (1963). In *International Research in Production Engineering, p. 486.* ASME, New York.
Sata, T. (1964). *Ann. CIRP* **12**, 190.
Schaller, E. (1964). Beitrag zur Untersunchung von Spannunger und dyannmischen Vorgangen in der Grenzschicht zwischen Werkzeug und Span bei der Stahlzerspannung mit Hartmetallwerkzeugen. Dissertation, T.H., Aachen.
Schuler, H. (1957). Untersuchungen fur einen Leistungsvergleich verschiedener Feinbearbeitungsverfahren. Dissertation, T.H., Aachen.
Semmler, D. (1962). Einfluss der Zerspannungsbedingungen auf den Verschleiss des Werkzeuges und die Gute der erzeugten Wertstuckoberflache beim aussen-Feindrehen mit keramischen Schneidwerkzeugen. Dissertation, T.H., Braunschweig.
Shaw, M. C. and Crowell, J. A. (1965). *Ann. CIRP* **13**, 5.
Shaw, M. C. and Hölken, W. (1957). *Ind. Anz.* **63**, 959.
Shaw, M. C. and Macks, E. F. (1949) *Analysis and lubrication of bearings.* McGraw-Hill, New York.
Shaw, M. C. and Sanghani, S. R. (1962). Proc. Int. Inst. Prod. Engng Res. **10, 340.**
Shaw, M. C. and Sanghani, S. R. (1963). *Ann. CIRP* **11**, 59.
Shaw, M. C., Thurman, A. L., and Ahlgren, H. J. (1966). *J. Engng Ind.* **88**, 142.
Shaw, M. C. and Usui, E. (1962). *Trans. Am. Soc. mech. Engrs* **84**, 89.
Shaw, M. C., Usui, E., and Smith, P. A. (1961). *Trans. Am. Soc. mech. Engrs* **83**, 181.
Sokolowski, A. P. (1955) *Prazision in der Metallbearbeitung.* VEB Verlag Technik, Berlin.
Solaja, V. (1958). *Wear* **2**, 1, 40.
Tobias, S. A. (1965). *Machine tool vibration.* J. Wiley and Sons, Inc., New York.
Tobias, S. A. and Fishwick, W. (1958). *Trans. Am. Soc. mech. Engrs* **80**, 1079.
Trent, E. M. (1963). *J. Iron Steel Inst.* 201.
Trigger, K. J., Zylstra, L. B., and Chao, B. T. (1952). *Trans. Am Soc. mech. Engrs* **74**, 1017.
Woxén, R. (1931). *Tekn. Tidskr* **61**, 43.

18

CHIP CONTROL

An important practical problem concerns the form of chips produced in machining since this has important implications relative to:

1. Personal safety.
2. Possible damage to equipment and product.
3. Handling and disposal of swarf after machining.
4. Cutting forces, temperatures, and tool life.

Naturally brittle materials such as gray cast iron tend to form chips that range from small segments to particles. On the other hand, ductile metals tend to form long, continuous, ribbon-shaped chips that are dangerous and difficult to handle. When machining at high speed, hot chips with sharp, jagged edges represent a hazard to the operator. Such long continuous chips tend to form tangled 'nests' that may interfere with the proper functioning of the machine. Under such conditions, it is necessary for the operator to stop the machine periodically to remove the snarled chip 'nests' with a hook. This represents a decrease in productivity. If an attempt is made to remove snarled chips while the machine is running, this represents a hazard to the operator.

A 'nest' of continuous chips has a very low apparent density making disposal difficult. Lang (1974) has introduced a chip packing ratio (R) where

$$R = \frac{\text{volume of chips}}{\text{equivalent volume of uncut metal}}.$$

In practice, R may range from 3 to 10 (or more) but satisfactory chip control calls for a value of R of about 4. To achieve this, chips must break periodically.

Chip curvature plays an important role relative to chip breaking as does the brittleness of the work material and the thickness of the chip (t_C). As chip thickness decreases, chips are more ductile and chip breaking becomes more difficult. A given work material that has been substantially strained during chip formation will have a critical strain at fracture (ϵ_f) which varies as the ratio t_C/R_C where t_C is the deformed chip thickness and R_C is the radius of curvature of the chip. If the natural chip curvature is not sufficiently small to cause fracture for a given feed rate t it becomes necessary to increase either t_c or chip curl to provide periodic chip fracture.

Chip curl may be increased by use of a so-called 'chip breaker' which is really a chip curler. There are two general types of chip breakers—an inclined obstruction clamped to the tool face (Fig. 18.1a) or a groove ground or molded into the tool face parallel to the major cutting edge (Fig. 18.1b). If a BUE is present at the tip of a tool this will cause the chip to curl away from the tool face. Nakayama (1963) has shown that the secondary shear zone which usually develops along the tool face in the absence of a BUE can also cause the chip to curl. The material in the secondary shear zone gradually increases in speed as it travels along the tool face and as a consequence of flow continuity becomes thinner. This action is equivalent to a stationary nose of decreasing thickness that plays the same role as a BUE in causing chip curl. Figure 18.2 shows a deformed set of grid lines that are rectangular before reaching the shear 'plane'. Horizontal lines become circular arcs with a common center. By continuity, the deformed, initially horizontal lines have a closer spacing as the tool face is approached and this gives a shear 'plane' that is curved slightly convex upward. Vertical lines are also convex upward. The chip surface nearest the tool face will thus be a circular arc according to Nakayama (1963). The region labelled **N** (for nose) will be either a BUE or the aforementioned equivalent secondary shear zone. Thus, for volume continuity, the chip will be 'born curled' since the shear plane will be slightly curved if a BUE or equivalent secondary shear zone is present at the tip of the tool which is practically always the case. The idea that chips are formed curved has also been suggested by several others (Ernst and Merchant 1941; Henriksen 1951; Hahn 1951; and Spaans 1971) and this appears to be the case except for the rare situation where the shear plane and tool face are perpendicular and there is no secondary shear zone or BUE.

Texture Lines

When a two-phase material is cut such as a hypoeutectoid steel having patches of pearlite which are normally approximately spherical, these patches deform into ellipsoids on crossing the shear zone. The direction of maximum deformation in the chip will thus be readily evident and make an angle ψ with the shear

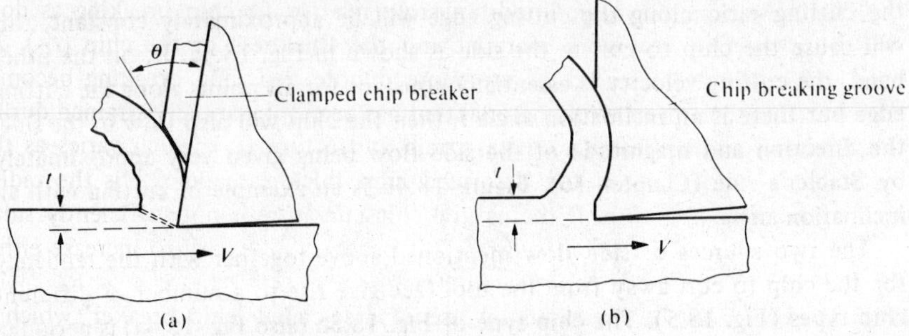

FIG. 18.1. (a) Clamped chip breaker. (b) Chip-breaking groove.

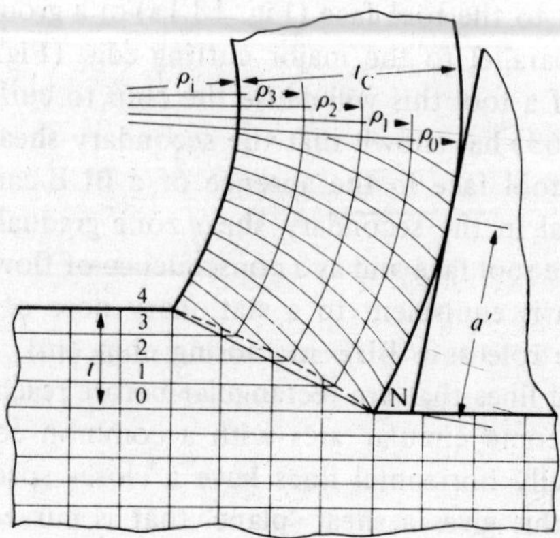

FIG. 18.2. Deformation of rectangular grid during chip formation. (After Nakayama 1963.)

'plane' (Fig. 18.3). Merchant (1945) has presented a relationship between ψ, ϕ, and α; and Towend (1947) has corrected a minor error in Merchant's analysis. However, while ψ represents a readily observable piece of data, it does not appear to lead to information concerning the cutting process that cannot be just as easily obtained via another route. The fact that the curvature of texture lines should be substantially greater than that of the shear plane does however offer a useful qualitative check on the presence of a curved shear plane.

Chip Flow Direction

In pure orthogonal cutting the chip moves across the tool face in a direction perpendicular to the cutting edge. However, there are two important situations which cause the chip flow direction to be directed otherwise. When the end of a small-diameter tube is turned (Fig. 18.4a), the cutting velocity will be substantially smaller at the inside diameter than at the outside diameter. Since the cutting ratio along the cutting edge will be approximately constant, this will cause the chip to curl to the side as shown in Fig. 18.4a. If, on the other hand, the cutting velocity is essentially constant for all points along the cutting edge but there is an inclination angle i, then the chip will also flow to the side, the direction and magnitude of the side flow being given very approximately by Stabler's rule (Chapter 16). Figure 18.4b is an example of cutting with an inclination angle.

The two sources of side flow mentioned above together with the tendency for the chip to curl away from the tool face give rise to a number of different chip types (Fig. 18.5). The chip type of Fig. 18.5c (also Fig. 18.4a) is periodically broken when the chip encounters the freshly machined surface. This may

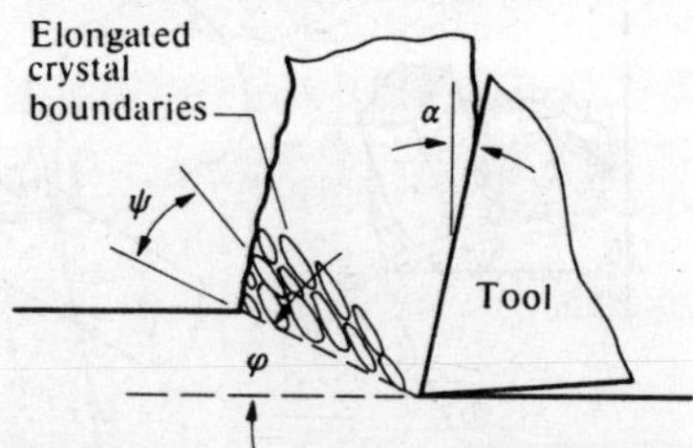

FIG. 18.3. Texture line directions in chip.

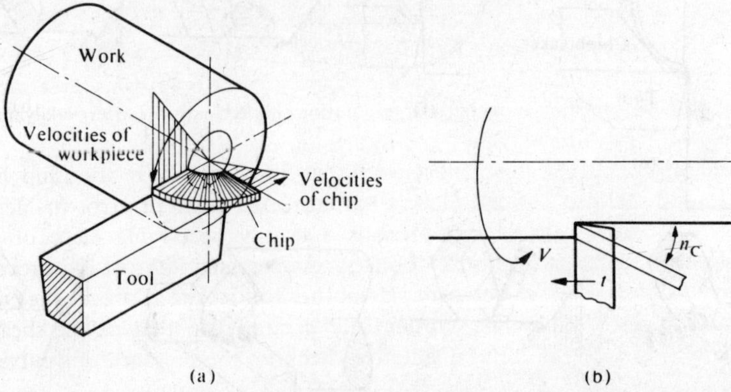

FIG. 18.4. (a) Curling of chip to side due to variation of cutting speed V along cutting edge. (After Spaans 1971.) (b) Plan view of turning operation showing alteration in chip flow direction (n_C) due to inclination angle i (zero side rake angle, positive back rake angle $\alpha_b = i$).

cause undesirable surface damage in finish machining operations. Chip types 18.5d and 18.5e are broken when an up-curling chip strikes the surface about to be machined. If side flow is sufficient, the up-curling chip may miss the old surface but encounter the clearance face of the tool giving rise to chips of the Fig. 18.5f type. If there is still greater side flow, chips of the Fig. 18.5h, i, or j type may be formed.

In addition to these types, several others are observed that represent combinations of those shown in Fig. 18.5a–j. For example, types (d), (e), (f), and (g) may form under conditions where the inclination angle i is not zero or where V is not constant across the cutting edge or both. Figure 18.5k is such an example which represents a combination of the actions of types (c) and (d). Also, chips may break periodically but every fracture may not be complete. This gives rise to connected chip segments such as the connected C-type segments shown in Fig. 18.5l. The most desirable chip types from all points of

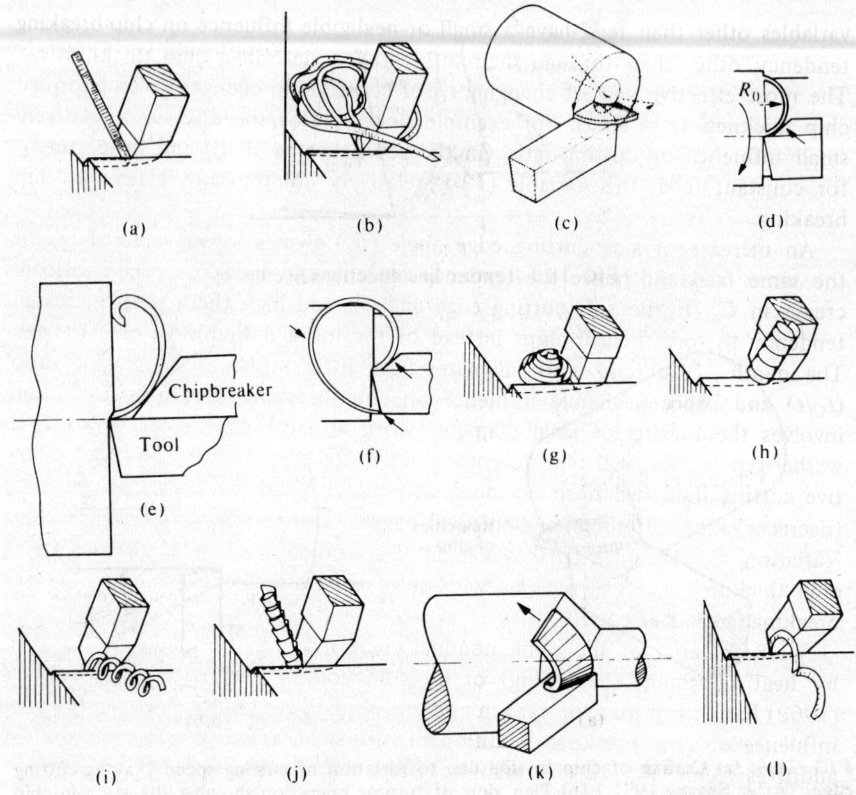

FIG. 18.5. Representative chip types (based on Spaans 1971). (a) Straight continuous ribbon ($i = 0$, $n_C = 0$). (b) Wandering continuous ribbon ($i = 0$, n_C variable). (c) Washer-type chip ($i = 0$, V = variable along cutting edge—no upward curl). (d) C-type chip ($i \cong 0$, encounters as-yet-uncut surface). (e) Short 'ear' type chip ($i \cong 0$, encounters as-yet-uncut surface). (f) Long 'ear' type chip ($i \cong 0$, encounters tool flank surface). (g) Coil-type chip ($i = 0$, V = variable along cutting edge plus upward curl). (h) Tubular helix (i positive plus curl, large depth of cut b). (i) Spring-type chip (i positive plus curl, small depth of cut b). (j) Conical helix (combination of (c) and (h)—common for drill). (k) Combination of (c) and (d) types ($i \neq 0$) (l) Connected C-type chips.

view are individual (unconnected) C- or ear-type chips: types (d), (e), (f), or (k). While type (g) chips may yield a satisfactory density ratio R, this type with two or more turns per coil will give rise to tool forces that are higher than necessary due to over curling.

Chip Breaker Performance

If chips do not break naturally then a chip curler (Fig. 18.1) must be used or the feed must be increased or the work made more brittle. Most operating variables other than feed have a small or negligible influence on chip-breaking tendency other than through their influence on deformed chip thickness t_C. The most effective way of changing t_C of course is to change the undeformed chip thickness (t = feed). For example since cutting speed V has a relatively small influence on cutting ratio (in the absence of a BUE) and hence on t_C for constant feed, this variable (V) is relatively unimportant relative to chip breaking.

An increase of side cutting edge angle (C_s) gives a lower value of t_C for the same feed and hence the tendency for chips to break decreases with increase in C_s. If the side cutting edge angle is too high there is an increased tendency to form helical chips instead of the more desirable C- or ear-types. The depth of cut and work diameter have little influence on cutting ratio (t_C/t) and hence negligible influence on chip breaking. The only exception involves the turning of small-diameter work at large depths of cut where a washer-type action will tend to enter the picture (Fig. 18.5c). Use of an effective cutting fluid will normally decrease contact length between chip and tool (decrease chip curl radius) and this will have a positive effect on chip breaking. Variation of rake angle (α) has a small effect on chip breaking. The deformed chip thickness (t_C) will increase with decrease in α and this will promote chip breaking (a small effect).

Work material of the same nominal chemistry may be made more brittle by heat treatment (hardening) or by cold-working the material. Nakayama (1962) has shown that the carbon content of plain carbon steels has a negligible influence on chip breaking. Addition of very small amounts of free-machining additives (MnS, Pb, Te, Bi, graphite, etc.) will usually improve a material's tendency to form broken chips.

Nakayama (1962) has determined the influence of a limited range of work-material chemistry on chip-breaking tendency using a clamped-type chip breaker (Fig. 18.1a) with $\theta = 45°$. The order of decreasing ease of chip breaking was found to be as follows:

1. Cr–Mo steel.
2. Carbon steel (all carbon contents).
3. Stainless steel.

Rational Approach to Chip Control

One approach to chip-breaker design is the data bank approach where a predetermined chart gives empirical chip-breaking results for a given work material. The first charts of this sort were proposed by Henriksen (1951). Nakayama (1962) has suggested a chart of the type shown in Fig. 18.6. This gives combinations of speed (V) and feed (t) that will give ear-type chips for a given work material (AISI 1045 steel) for points that plot above the curve and continuous (unbroken) chips for combinations of V and T that plot below the curve. Kaldor,

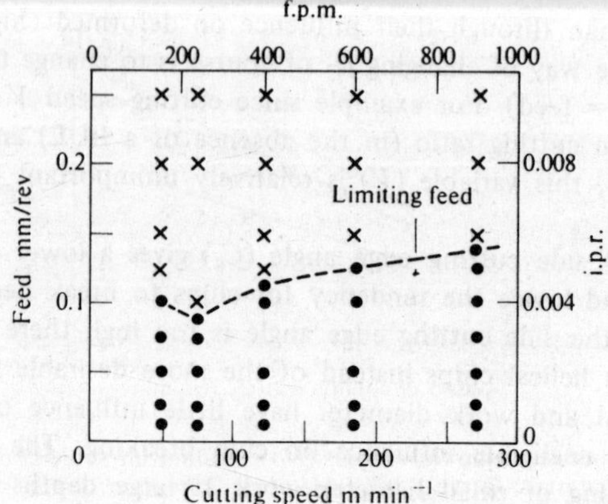

FIG. 18.6. Effects of speed and feed on chip breaking of AISI 1045 steel using clamped carbide chip breaker. $\theta = 45°$; carbide tool geometry: 0, 0, 7, 7, 8, 0, 0.5 mm; depth of cut, 2.0 mm (0.08 in); cutting fluid, none; x = broken chip, • = continuous chip. (After Nakayama 1962.)

Ber, and Lenz (1979) have suggested a map indicating different chip-types when depth of cut (b) is plotted against feed (t) for a given machining system. A number of additional chip-breaking studies are available in the literature (Henriksen 1954a, 1954b; TenHorn and Schürmann 1954; Reinhold 1959, 1962; Okushima, Hoshi and Fujinawa 1960; Trim and Boothroyd 1968; Jones 1973; Reinhart and Boothroyd 1973; Kahng and Koegler 1977; Gertler 1978, to mention but a few).

While empirical charts are useful, it often occurs that results for a given work material are not available and some experimentation is required. Both Nakayama (1962) and Spaans (1971) have suggested rational procedures to be followed in such situations.

Nakayama (1962) has suggested the following procedure for a clamped-type chip-breaker. A trial cut is first made under the desired machining conditions and the value of deformed chip thickness (t_C) determined. The limiting value of chip curl radius R_C is then obtained from the appropriate empirical curve of Fig. 18.7. This value is reduced by 20 percent to ensure chip breaking. To limit shock on contact, angle θ on the front face of the chip-breaker should not be too steep; and $\theta = 45°$ is recommended. The value of offset distance b may be estimated from the following equation which comes from Fig. 18.8:

$$b = \frac{R_C}{\cot \theta/2}. \tag{18.1}$$

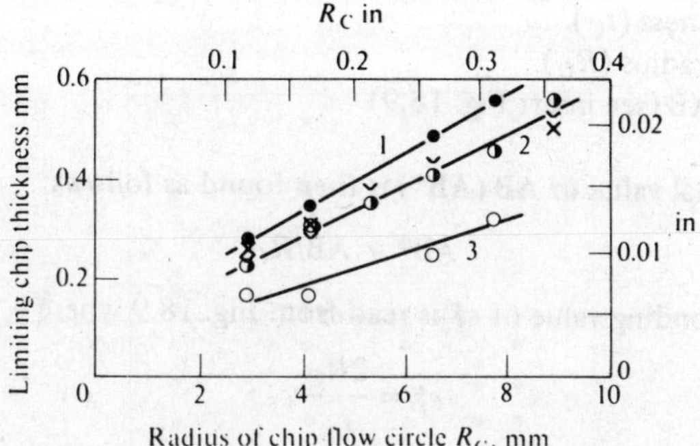

FIG. 18.7. Relation between chip flow radius (R_C) and limiting value of chip thickness (t_C) for different work materials. Curve 1 = Cr–Mo steel, 2 = plain carbon steels of all C contents, 3 = stainless steel. Machining conditions same as for Fig. 18.6. (After Nakayama 1962.)

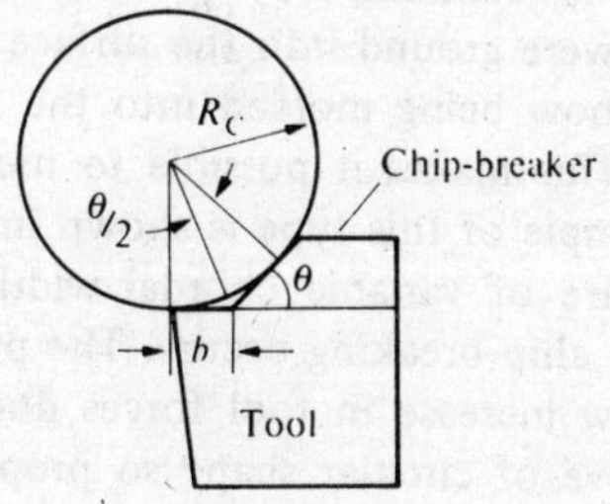

FIG. 18.8. Relation between chip flow radius (R_C) and off-set (b) of clamped chip-breaker.

For a value of θ of 45°, and R_C from Fig. 18.7, the required offset distance b is found to be

$$b = \frac{0.8R_C}{\cot(22.5°)} = 0.33R_C.$$

If the work material of interest is not given in Fig. 18.7 as will usually be the case, an estimate may be had by comparing the properties of the new material with the few that are given and interpolating, if necessary.

Spaans (1971) has suggested a useful means of rating the 'breakability' of a work material. He reasoned that values of the strain at fracture (ϵ_f) required to produce desirable ear-type chips for several materials would correlate with the ease of chip fracture. The resulting ordering of materials based on the strain at fracture (ϵ_f) he calls 'breakability'. After the chip breaker has been adjusted to give the desired 'ear' type chip, the following quantities are measured on a representative chip:

1. Chip thickness (t_C).
2. Chip curl radius (R_C).
3. Distance AB (see insert Fig. 18.9).

A nondimensional value of AB (AB*) is then found as follows:

$$AB^* = AB/R_C \tag{18.2}$$

and the corresponding value of ϵ_f^* is read from Fig. 18.9 where

$$\epsilon_f^* = \frac{2R_C}{t_C}\,\epsilon_f. \tag{18.3}$$

From this latter equation ϵ_f may be estimated.

Chip Breaking Grooves

As already mentioned, chip breaking may be induced by use of chip-curling grooves. Originally these were ground into the surface of brazed carbide tools. Chip-curling grooves are now being molded into the rake surfaces of carbide inserts before sintering. This makes it possible to use grooves that would be difficult to grind. An example of this type is shown in Fig. 18.10. The grooves in this triangular insert are of variable chordal width. This is to extend the range of feed over which chip breaking occurs. The proportions of the groove were chosen to give a low increase in tool forces due to chip breaking. This turned out to be a groove of circular shape so proportioned that the depth of the groove was about 10 percent of the chordal width.

The insert shown in Fig. 18.10 produces a chip of variable cross section (Fig. 18.11) since metal is forced into the groove parallel to the secondary cutting edge. This is important in preventing partially broken chips held together along the edge farthest from the workpiece axis. Due to the complex state of stress at the nose radius, chips tend to fracture on this edge first. This relieves the stress in the chip and prevents rupture from occurring all the way across the chip. Chip segments are thus joined together by unbroken material on the edge farthest from the nose of the tool (Fig. 18.12). This limits the range of feed over which a chip breaker will give completely broken chips. By thickening the edge of the chip adjacent to the nose of the tool, fracture is postponed and more energy is stored in the chip to cause complete chip fracture once a crack does appear. Thus, the groove segment parallel to the minor cutting edge is important as well as that parallel to the major cutting edge. The role of the latter is to promote upward-chip-curl while that of the former is to prevent premature chip-fracture leading to connected chip segments.

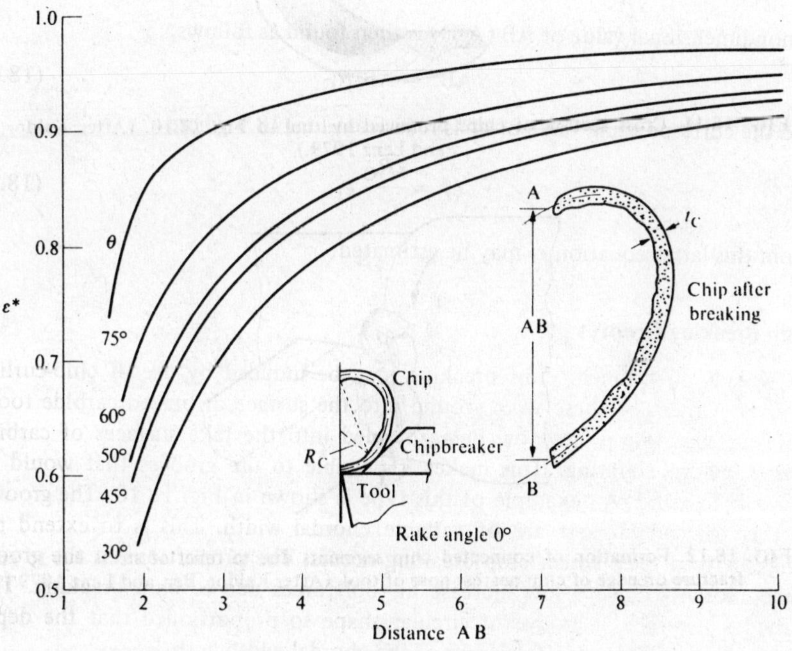

FIG. 18.9. Variation of nondimensional strain in 'ear' type chip at fracture (ϵ_f^*) with nondimensional chip size AB*. (After Spaans 1971.)

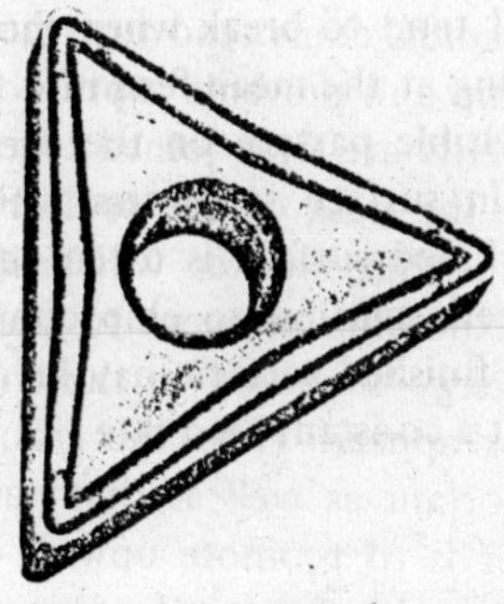

FIG. 18.10. Triangular carbide tool insert with chip-breaking grooves molded into the rake face. (After Kaldor, Ber, and Lenz 1979.)

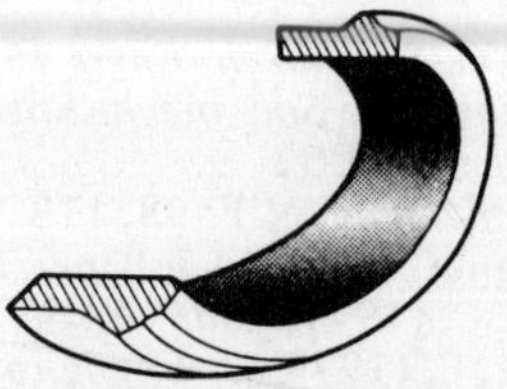

FIG. 18.11. Cross section of chips produced by tool of Fig. 18.10. (After Kaldor, Ber, and Lenz 1979.)

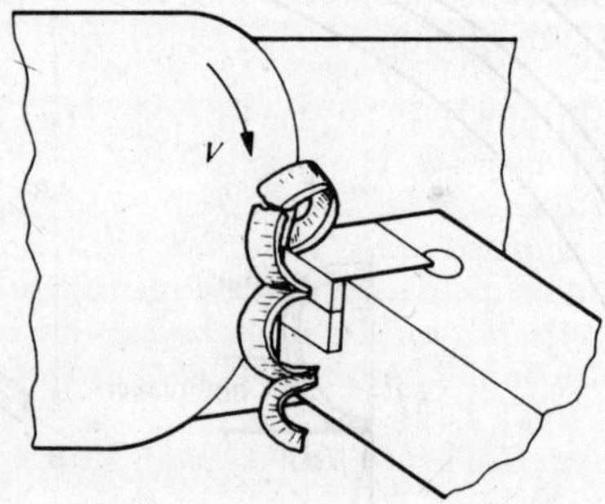

FIG. 18.12. Formation of connected chip segments due to relief of stress due to critical fracture on edge of chip nearest nose of tool. (After Kaldor, Ber, and Lenz 1979.)

Variable Feed

Since the magnitude of t_C and hence of undeformed chip thickness (t = feed rate) plays such an important role in chip-breaking, it is not surprising that chip control may be improved by use of a periodically varying feed rate. When this is done, the chip will tend to break when the feed is high even though it may not break when cutting at the mean feed rate. This method of chip control will of course lead to a visible pattern on the surface of the work that is far greater than the change in surface roughness would suggest. However, since the appearance of a machined surface is often equated to surface quality in the mind, the variable feed solution to chip control is often not acceptable even though the resulting finished surface may be equal or even superior functionally to one produced at a constant feed rate.

References

Ernst, H. and Merchant, M. E. (1941). In *Surface treatment of metals*, p. 299–378. Am. Soc. for Metals, Metals Park, Cleveland, Ohio.

Gertler, S. (1978). *Cut Tool Eng* **29**, 11.

Hahn, R. S. (1953). *Trans. Am. Soc. mech. Engrs* **75**, 581.
Henriksen, E. K. (1951). *Trans. Am. Soc. mech. Engers* **73**, 461.
Henriksen, E. K. (1954a). *Steel* **134**, 93.
Henriksen, E. K. (1954b). *Am. Mach, N.Y.* **98**, 118.
Jones, D. G. (1973). Soc. Manuf. Engrs Tech Paper 73–215.
Kahng, C. H. and Koegler W. C. (1977). *Man. Engng Trans.*, 6
Kaldor, S., Ber, A., and Lenz, E. (1979). *J. Engng Ind.* **101**, *241.*
Lang, M. (1974). Spannformer und Ablaufstufen zur Erzeugung Kurzstuckiger Spane beim Drehen von Metallen. Dissertation, T.H., Munchen.
Merchant, M. E. (1945). *J. appl. Phys.* **16**, 269.
Nakayama, K. (1962). *Bull. Jap. Soc. mech. Engrs* **5**, 192.
Nakayama, K. (1963). *Bull. Jap. Soc. precis. Eng* **1**, 25.
Okushima, K., Hoshi, T., and Fujinawa, T. (1960). *Bull Jap. Soc. mech. Engrs* **3**, 199.
Reinhart, L. E. and Boothroyd, G. (1973). In *Proc. First N. Am. Metalworking Res. Conf.* (McMaster University, Hamilton, Ont.), p. 13.
Reinhold, R. (1959). *Fertigungstechnik u. Betreib* **9**, 161.
Reinhold, R. (1962). *Fertigungstechnik u. Betrieb* **12**, 475.
Spaans, C. (1971). The fundamentals of three-dimensional chip curl, chip breaking and chip control. Dissertation, Delft Technical University, Netherlands.
TenHorn, B. L. and Schürmann, R. A. (1954). *Tool Engineer* **32**, 37.
Towend, G. H. (1947). *J. appl. Phys.* **18**, 784.
Trim, A. R. and Boothroyd, G. (1968). *Int. J. Prod. Res.* **6**, 227.

19

OPTIMIZATION

There are generally a number of ways of producing a specific part, some of which are more efficient than others. When a new design appears, the initial methods of producing the desired parts are usually not the most efficient. With experience, the processing operations are optimized or more efficiently carried out. In any optimization procedure, it is important to identify the output of chief importance and to establish a figure of merit against which different processing paths may be rated. In the area of manufacturing engineering, the most common optimization variables are specific cost (cost per part) and production rate (parts produced per unit time). The latter should normally be expected to pertain only in a time of national emergency or if a production bottleneck develops due to lack of machine capacity. However, in practice, maximum production rate often takes precedence over minimum cost per part as the optimization objective.

Levels of Optimization

There are several levels of process optimization including the following:

1. Improved details relative to present methods of production.
2. New conventional production paths.
3. Adoption of a radically new production path.

Generally, optimization levels 2 and 3 are progressively less feasible. If the present equipment available in a plant is reasonably efficient and can meet production requirements, it often cannot be justified economically to replace the old equipment even though the new equipment is more efficient. This, of course, is nature's way of softening the blow associated with loss of production capabilities due to war or natural disaster. Very often, the new equipment has a much larger production capacity than the old and a new machine may have more production capacity than a potential user requires. There is then the prospect of having the new machine lie idle part of the time or to sell excess time on the machine or excess products produced by the machine in order that the cost of meeting primary production requirements not be excessive. Another constraint associated with the adoption of new, more efficient machinery is the probability the new machinery is more sophisticated to the point it may not be consistent with the levels of skill of available operators or maintenance personnel. Another important deterrent to the adoption of new, more efficient manufacturing facilities is the fact that a new generation of machines generally involves a larger capital investment and if, as is usual, there is a shortage of capital available, a manufacturer may be prevented from replacing old machinery by new even though the new machinery can be justified on the basis of a reduced production cost per part including the true machine cost (taking into account possible unused capacity, greater maintenance cost, and the greater unit cost of a more sophisticated operator or programmer).

Level 2 optimization is best pursued by periodically 'brainstorming' major production operations with input from internal or external consultants or both. The decision to try a new conventional processing route should be based upon a sound economic analysis of alternatives with current production costs as a basis for comparison.

The situation associated with level 3 optimization is still more difficult. It will usually take a minimum of five years to bring a new manufacturing concept through the many development steps associated with its progression from a laboratory feasibility study through the pilot plant stage and finally into full-scale production. Each of these steps involves unforeseen hazards and problems that require new thinking and frequently innovative solutions. All of this takes time, careful planning, and a well-coordinated team effort. Therefore, level 3 process optimization involves important constraints of time, availability of development personnel, and risk, in addition to the constraints associated with level 2 optimization. This does not mean that optimization activity at levels 2 and 3 are less important than level 1. The wise manufacturer will periodically review the manufacturing program to be sure a new approach is not being overlooked. Such periodic review should involve a quantitative consideration of alternatives. It is important to emphasize the need for periodic review (say annually) since the options available to the manufacturing engineer are constantly changing at an ever-increasing rate.

Since the situation most frequently encountered involves optimization at level 1, the remainder of the discussion here will be concerned with optimization at this level. However, the basic approach to be used at all levels is the same:

1. Identify the appropriate optimization objective.
2. Quantitatively consider processing options available within existing constraints.
3. Adjust processing options to correspond as closely as possible to optimum values.

Constraints

Just as there are several constraints for process optimization at level 2 (capital available, unit machine size and production capacity, level of skill of personnel, etc.) and level 3 (time, risk, etc.) there are a number of constraints for level 1 optimization. These include:

1. Tool material strength and wear resistance.
2. Range of speeds, power, and stability of machines available.
3. Finish, surface integrity, and dimensional accuracy required, etc.

The problem is to adjust the machine consistent with constraints so that production is carried out as close to optimum conditions as possible.

Machining Economics

Some of the most important problems of the workshop involve the choice of cutting speeds and feeds, tool geometry, tool and work materials, cutting fluids, and machine tools themselves. There appear to be only three basic considerations associated with all of these decisions. First of all, the chosen cutting conditions must be capable of producing parts that meet the required specifications of size, shape, and finish. Secondly, the required production schedules must be met; and thirdly, the parts should be produced at the lowest possible cost. In the large-scale mass production of parts, sufficient equipment can often be procured to meet production requirements under a wide range of conditions, and the important problem might then become to choose the operating conditions so that quality requirements are just met and the total cost per part is a minimum.

Since the cost of the tools that are used on a given operation represents a major component of the total machining cost, tool life plays a major role in the choice of optimum machining conditions. The several different criteria for tool life which must be adopted under different practical situations have been discussed in Chapter 11 and it has been shown that if the range of cutting speed V or feed t is not great, the tool life (T, min) defined in any of the several possible ways may be represented as follows:

$$\text{For constant feed: } VT^n = C \qquad (19.1)$$

where n and c are constants depending on tool and work material, tool geometry, etc.

In several sections that follow, answers to the following questions will be discussed.

1. What is the most satisfactory machinability index for use in rating the ease of machining a number of different workpiece materials, and what is a reasonable approximation to this quantity?

2. On what basis should a water-base cutting fluid be chosen from a group of fluids of different cost?
3. What is the amount one is justified in paying for a stepless speed changer on a lathe?
4. Which of two tools which cost different amounts and have different wear characteristics should be chosen?
5. Which of two work materials having different machinability characteristics and cost should be used in a given instance?
6. Is the clearance which gives maximum tool life always the one which should be used in practice?

In discussing these questions, it will be assumed in the interest of brevity that the feed is fixed by considerations of finish, machine tool power, or by maximum allowable cutting force for the tool holder, and only the choice of the otpimum cutting speed will be considered. The analyses might however be readily extended to include the choice of optimum feed rate.

Machining Costs

Gilbert (1950) first approached machining economics in a rational way and the following discussion of principles involved is based on his approach.

If cutting is done at very high speed, the machine and labor cost will be low but the tool cost will be high. The opposite is true if cutting is done at very low speed. There is an optimum cutting speed for any machining operation where the total machining cost per part is a minimum.

The costs to be included in such an analysis are:

1. Direct labor and machine cost (xT_c) where x is the value of the machine and operator with overhead (cents/min) and T_c is the machining time per part.

$$T_c = \frac{\pi d l}{12 V t} = \frac{L}{V} \tag{19.2}$$

where d = work diameter, inches (mm)
l = axial length of work to be cut, inches (mm)
t = feed, i.p.r. (mm/rev)
V = cutting speed f.p.m. (m min^{-1})
L = helical length machined per part, ft (m).

When SI units are used, the 12 in eqn (19.2) is replaced by 1000.

2. Tool changing cost per part $= \left(x T_d \dfrac{T_c}{T} \right)$

where T_d is the 'down' time in minutes to change a tool.

3. Tool cost per part $= \left(y\dfrac{T_c}{T}\right)$

where y is the mean value of a single cutting edge.

The total cost per part (¢) is then

$$¢ = xT_c + xT_d\frac{T_c}{T} + y\frac{T_c}{T} \tag{19.3}$$

(work changing cost is not included since it is independent of V).

The quantity ¢ is a minimum when $\partial ¢/\partial V$ is zero, and this may be found after T_c and T are expressed as functions of V using eqns (19.1) and (19.2). In this way, it is found that ¢ is a minimum when tool life (T) has a particular value which we will call $T^{\#}$

$$T^{\#} = \left(\frac{xT_d + y}{x}\right)\left(\frac{1}{n} - 1\right) = R\left(\frac{1}{n} - 1\right) \tag{19.4}$$

where R (minutes) is called the cost ratio and n is the exponent of the tool-life equation (eqn 19.1). The corresponding cutting speed is then

$$V^{\#} = \frac{C}{(T^{\#})^n} \tag{19.5}$$

If this value is put back into the cost equation (19.3), we obtain the minimum machining cost per part

$$¢^{\#} = \frac{x}{1-n}T_c^{\#} \tag{19.6}$$

Representative tool life-speed data are shown in Fig. 19.1 for a low-carbon steel machined with a carbide insert. Two commonly used tool-life criteria are considered here.

1. Total destruction of the tool (heavy line).
2. Lines of constant wear land on the clearance face of the tool (w).

The lines of constant R shown in this figure are a convenient addition to such a plot for they immediately enable the cost optimum values ($T^{\#}$ and $V^{\#}$) to be determined. The line corresponding to $R = 10$ minutes was obtained by measuring the value of n (0.48 in the case of Fig. 19.1) and substituting the proper values into eqn (19.4) to obtain

$$T^{\#} = 10\left(\frac{1}{0.48} - 1\right) = 11 \text{ min.}$$

The line corresponding to $R = 10$ is thus drawn opposite $T = 11$ min. The corresponding value of $V^{\#}$ (speed for lowest total machining cost per part)

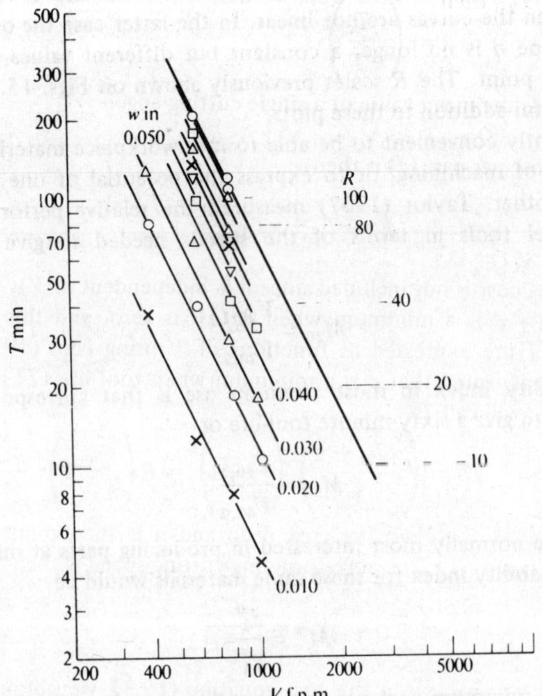

FIG. 19.1. Tool life–speed curves for hot-rolled AISI 1020 steel cut using a carbide tool having the following American Standards Association (ASA) geometry: −7, −7, 7, 7, 15, 15, $\frac{1}{32}$ in (0.75 mm). Depth of cut, 0.100 in. (2.5 mm); feed, 0.0104 i.p.r. (0.26 mm/rev); fluid, tap water applied at $1\frac{1}{2}$ g.p.m. at 60 °F (5.5 l min^{-1} at 17 °C).

for a tool operated to total destruction is seen to be over 2000 f.p.m. (610 m min^{-1}).

The cost values for the lathe and throw-away tool used in obtaining the data of Fig. 19.1 were as follows:

x = cost of operator, lathe, and overhead = 10¢/min
t_d = tool changing and adjusting time = 3 min
y = value of single cutting edge = 60 cents.

The corresponding value of R is 9 min and hence, for minimum cost per part, this tool should be operated at a speed considerably in excess of 2000 f.p.m. to give a tool life of about 9 min. This speed is beyond the range of most currently available lathes and indicates that the maximum available speeds for lathes are presently too low, even for throw-away carbide tools, not to mention ceramic tools.

Lines of constant R may be readily applied to tool life T versus speed V plots even when the curves are not linear. In the latter case the only difference is that the slope n is no longer a constant but different values must be used from point to point. The R scales previously shown on Figs. 15.19 and 15.20 represent a useful addition to these plots.

It is frequently convenient to be able to rate workpiece materials in relative order of ease of machining, or to express the potential of one tool material relative to another. Taylor (1907) measured the relative performance of his high-speed steel tools in terms of the speeds needed to give a 20-minute tool-life.

$$M_{20} = \frac{V_{20,1}}{V_{20,2}}. \qquad (19.7)$$

The machinability index in most common use is that corresponding to the ratio of speeds to give a sixty-minute tool-life or

$$M_{60} = \frac{V_{60,1}}{V_{60,2}} \qquad (19.8)$$

Since we are normally most interested in producing parts at minimum cost, a better machinability index for these same materials would be

$$M^{\#} = \frac{¢_2^{\#}}{¢_1^{\#}} \qquad (19.9)$$

where $¢_1^{\#}$ is the minimum cost per part for a given tool, work material, or fluid designated (1), and $¢_2^{\#}$ is the corresponding minimum cost per part for the second tool, work material, or cutting fluid, etc. It is of interest to know the difference which exists between these machinability indices.

From eqn (19.6)

$$M^{\#} = \frac{(xT_c^{\#}/1-n)_2}{(xT_c^{\#}/1-n)_1} \qquad (19.10)$$

but, since

$$\frac{(T_c^{\#})_2}{(T_c^{\#})_1} = \frac{V_1^{\#}}{V_2^{\#}}$$

and $x_2 = x_1$

$$M^{\#} = \frac{C_1(T_2^{\#})^{n_2}(1-n_1)}{C_2(T_1^{\#})^{n_1}(1-n_2)}. \qquad (19.11)$$

Similarly, the machinability index based on a given tool life (M_T) may be written

$$M_T = \frac{V_{T,1}}{V_{T,2}} = \frac{C_1}{C_2}\, T^{n_2-n_1}. \qquad (19.12)$$

If the value of T in eqn (19.12) is taken as the mean between $T_1^\#$ and $T_2^\#$, then

$$T = \frac{T_1^\# + T_2^\#}{2} = \frac{R}{2}\left[\frac{1-n_1}{n_1} + \frac{1-n_2}{n_2}\right] \tag{19.13}$$

since $R_1 = R_2$. The value of the ratio between the commonly used index based on speeds (M_T) and that based on minimum cost per part ($M^\#$) will be

$$\frac{M_T}{M^\#} = \left(\frac{1-n_1}{n_1} + \frac{1-n_2}{n_2}\right)^{(n_2-n_1)} \frac{[2(1-n_1/n_1)]^{n_1}}{[2(1-n_2/n_2)]^{n_2}} \left(\frac{1-n_2}{1-n_1}\right) \tag{19.14}$$

This ratio is equal to one whenever n_1 is equal to n_2. The ratio will be close to one whenever n_1 and n_2 do not differ by more than 0.2. However, whenever n_1 and n_2 differ widely, then ($M_T/M^\#$) is appreciably different from 1 and the two machinability criteria are not equivalent.

From the foregoing discussion, it is clear that there is little difference between the easily computed machinability index based on speed and the less easily computed (although more important) index based on costs, provided the slopes of the tool life curves (n) for the conditions compared do not differ very much. Since the tool life T on which the speed index M_T is based should correspond to a good average of economic tool life values $T^\#$ for a variety of applications, it would appear that $T = 60$ minutes in eqn (19.12) is a reasonable value and perhaps a better choice for present-day conditions than Taylor's original value of 20 minutes. It is surprising to find that the form of Taylor's original machinability index gives values that are so very close to those obtained by use of the more fundamental machinability index based on cost ($M^\#$).

Water-Base Cutting Fluids

The many fluids available may be divided into two general classes—oil-base fluids and water-base fluids. The oil-base fluids are generally acknowledged to be the best boundary lubricants, while the water-base fluids are capable of the greatest cooling action. At low cutting speeds, the production of good finish is the chief problem, and boundary lubrication of the tool–chip interface is of major importance; whereas, at high cutting speeds, tool life is more important and cooling the cutting tool becomes more necessary.

The objective of this discussion is to present a rational method for studying the performance of a variety of water-base cutting fluids based upon machining costs. Answers are to be sought to such questions as: Is the cost of a cutting fluid of importance if it really provides superior performance? And how much additional cost for a cutting fluid is justified in a given instance?

After a number of preliminary tests, a disposable type carbide tool was found which gave very good results, and this tool was adopted for this study. The characteristics of this tool and the cutting conditions used follow:

Carbide type: Steel cutting grade C–5

Carbide size: $\frac{3}{8}$ in × $\frac{3}{8}$ in × $1\frac{1}{2}$ in (9.53 × 9.53 × 38 mm) rectangular parallelopiped

Tool Angles:

Back-rake angle, −7°

Side-rake angle, −7°

End-relief angle, 7°

Side-relief angle, 7°

End-cutting edge angle, 15°

Side-cutting edge angle, 15°

Nose radius, $\frac{1}{32}$ inch (0.79 mm)

Feed, 0.0104 i.p.r. (0.26 mm/rev)

Depth of cut, 0.100 inch (2.5 mm)

Cutting speed, variable

Rate of fluid flow, 1.50 g.p.m. (5.7 l min^{-1})

Each tool insert had eight cutting edges and when these had been used the carbide insert was discarded. The work material used was annealed AISI 4340 steel (Brinell hardness, 185). It was thought best to run all tools to total destruction so that a comparison could be made between curves of constant wear land and total destruction. This required that a large amount of material be cut, and some tests involved as many as 16 hours of cutting time.

Tool-life results for several water-base cutting fluids are given in Fig. 19.2. In these tests the data have been confined to the practical turning region where built-up edge (BUE) is negligible. Both lines of constant wear land and total destruction are included. In each case lines of constant R have also been given and the value of R corresponding to the tool used in this study ($R = 11.5$ min) is shown by a dotted line.

Three examples will be used to compare the performance of these fluids. First, it will be assumed that we are interested in a roughing operation for which it is permissible to use tools to total destruction. The optimum cutting speed $V^{\#}$ and tool life $T^{\#}$ in this case corresponds to the point where $R = 11.5$ crosses the total destruction line for each fluid (Fig. 19.2). The values of $T^{\#}$, $V^{\#}$, and n for each of the fluids tested are given in Table 19.1.

The minimum cost per unit part $¢^{\#}$ is given by eqn (19.6) from which it is evident that (see eqn 19.2)

$$¢^{\#} = (xL)\frac{1}{V^{\#}(1-n)} \tag{19.15}$$

Since the quantity xL is constant for any operation, the optimum cost per part $¢^{\#}$ will be proportional to $(1/V^{\#}(1-n) = ¢^{\#}/xl)$ for any given cutting fluid, and values of this quantity are included in Table 19.1. The values of relative cost given in the last column of Table 19.1 are based on a cost per part in dry cutting of unity.

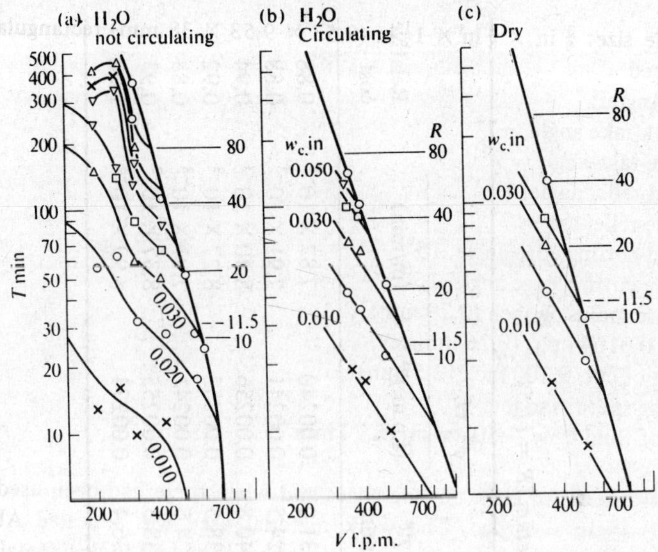

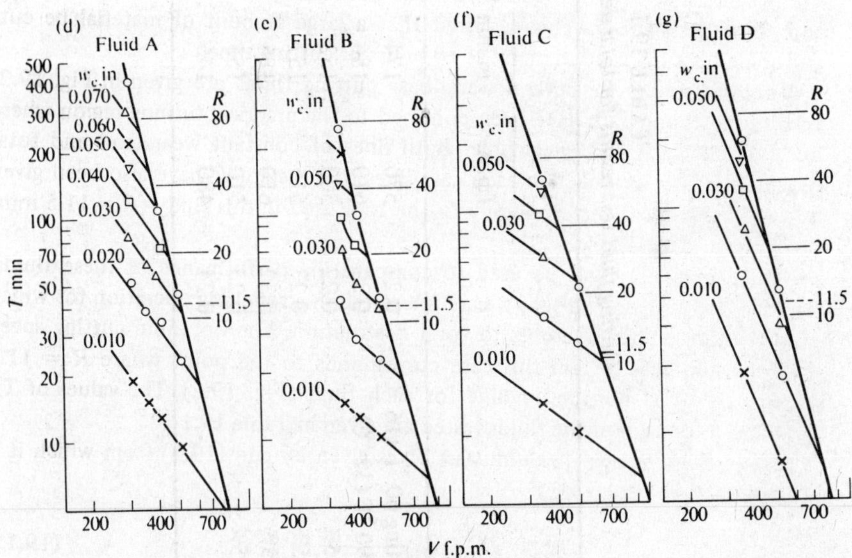

FIG. 19.2. Tool life–speed curves for different cutting fluids used in machining AISI 4340 steel with a carbide tool with the following ASA geometry: −7, −7, 7, 7, 15, 1/32 in (0.79 mm). Depth of cut, 0.100 in (2.5 mm); feed, 0.0104 i.p.r. (0.26 mm/rev). Rate of fluid flow, 1½ g.p.m. (5.7 l min⁻¹). (After Shaw 1959.) (a) Non-circulating water at 60 °F (16 °C). (b) Circulating water at 90 °F (32 °C). (c) Dry tool. (d) Cutting fluid A, 2.5% in water. (e) Cutting fluid B, 2.5% in water. (f) Cutting fluid C, 2.5% in water. (g) Cutting fluid D, 2.5% in water.

TABLE 19.1

Cutting fluid test results for total destruction ($R = 11.5$)

Fluid	n	$V^{\#}$		$T^{\#}$	$¢^{\#}/xL$		Relative
		f.p.m.	m min^{-1}	min	(f.p.m.)$^{-1}$	(m/min)$^{-1}$	cost
Water, non-circulating 60 °F (16 °C)	0.27	570	174	31.0	0.00240	7.87×10^{-3}	0.88
Water, circulating 90 °F (32 °C)	0.32	610	186	24.5	0.00241	7.91×10^{-3}	0.88
Cutting fluid A, 2.5%	0.22	500	152	40.8	0.00256	8.40×10^{-3}	0.94
Cutting fluid B, 2.5%	0.23	515	157	38.5	0.00252	8.27×10^{-3}	0.92
Cutting fluid C, 2.5%	0.32	610	186	24.5	0.00241	7.91×10^{-3}	0.88
Cutting fluid D, 2.5%	0.23	510	155	38.5	0.00255	8.37×10^{-3}	0.93
Dry tool	0.24	480	146	36.4	0.00274	8.99×10^{-3}	1.00

From this study it is evident that all water-base cutting fluids give better results than dry cutting. The poorest fluid showed a decrease in cost per part over dry cutting of about 6 percent while the best fluid showed a decrease in cost per part of 12 percent based on dry cutting. The fact that none of the commercial water-base cutting fluids was better than water indicates that the major role of the fluid in this speed range (500–600 f.p.m. or 152–183 m min^{-1}) is one of cooling. Water should be expected to be the best coolant and the performance of the water-base fluids should be expected to decrease from that of water as their tendency to form a thick oily deposit on metal surfaces increases. Such a thick oily film will tend to decrease the heat transfer coefficient between metal and fluid. The four water-base cutting fluids tested might be rated in the following order of decreasing effectiveness for use in the speed range 500–600 f.p.m. (152 to 183 m min^{-1}).

Fluid	% improvement over dry cutting
C	12 (circulating water 12%)
B	8
D	7
A	6

A second comparison may be made for the case where a wear land of 0.040 inch (1 mm) cannot be exceeded for reasons of, let us say, surface finish. The best operating point in this case ($V^{\#}$ and $T^{\#}$) will be the point where the $w = 0.040$ inch (1 mm) curve intersects the curve of total destruction. Such values of $T^{\#}$ and $V^{\#}$ are given in Table 19.2.

The equation which in this case corresponds to eqn (19.15) may be readily shown to be

$$¢^{\#} = (xL)\frac{T^{\#} + R}{T^{\#} V^{\#}} \tag{19.16}$$

where the quantity xL is a constant for any given operation. Hence, the minimum cost per part ($¢^{\#}$) with any cutting fluid will now be directly proportional to the quantity

$$\left(\frac{T^{\#} + R}{V^{\#} T^{\#}} = \frac{¢^{\#}}{xL}\right)$$

given in Table 19.2 for a value of $R = 11.5$.

The optimum speed is now seen to range from 420 to 460 f.p.m. (128 to 140 m min^{-1}), and for this speed range water and water-like fluids are the least effective. The relative rating of fluids for the 0.040 inch (1 mm) wear-land condition are now as follows in decreasing order of effectiveness.

Fluid	% improvement over dry cutting
A	9
B	8

TABLE 19.2

Values of $T^{\#}$ and $V^{\#}$ for constant wear land of 0.040 inch (1 mm) (R = 11.5)

Fluid	$V^{\#}$ f.p.m.	$V^{\#}$ m min^{-1}	$T^{\#}$ min	$¢^{\#}/xL$ (f.p.m.)$^{-1}$	$¢^{\#}/xL$ (m/min)$^{-1}$	Relative cost
Water, non-circulating 60 °F (16 °C)	525	160	40	0.00246	8.07×10^{-3}	0.87
Water, circulating 90 °F (32 °C)	410	125	88	0.00276	9.06×10^{-3}	0.97
Cutting fluid A, 2.5%	460	140	60	0.00259	8.50×10^{-3}	0.91
Cutting fluid B, 2.5%	450	137	65	0.00262	8.60×10^{-3}	0.92
Cutting fluid C, 2.5%	420	128	80	0.00272	8.92×10^{-3}	0.96
Cutting fluid D, 2.5%	440	134	72	0.00264	8.66×10^{-3}	0.93
Dry tool	420	128	60	0.00284	9.32×10^{-3}	1.00

D 7
C 4 (circulating water = 3%)

Apparently lubrication begins to be important in this speed range and fluid A that was poorest at the high-speed range of 500-600 f.p.m. (157-183 m min^{-1}) (Table 19.1), because of its poorer heat transfer, now appears to be the most effective fluid for the 420-460 f.p.m. (128-140 m min^{-1}) speed range. The range of effectiveness of the commercial fluids is still seen to be about 2:1 but the relative order of effectiveness is entirely changed.

Still another comparison of the effectiveness of these cutting fluids can be made for a case where the tooling is relatively expensive. Let us consider the specific case where tools are run to just short of total destruction and the value of R has been increased by a factor of 4 (i.e. to a value of 46). The quantities of interest for this example are given in Table 19.3.

The optimum speed is now still lower, ranging from 345 to 390 f.p.m. (105 to 110 m min^{-1}). The relative ratings of the fluids are as follows:

Fluid	% improvement over dry cutting
A	9
B	9
D	8
C	1 (circulating water = 2%)

Now the water and water-like fluid (C) are seen to be not only the poorest fluids of the group but to be practically without effect.

The foregoing analysis of water-base cutting fluids has clearly shown that two distinct actions are present which are not completely compatible—a lubrication action and a cooling action. For the AISI 4340 steel–carbide-tool combination used in these experiments, cooling seems to be of negligible importance for speeds of 300 f.p.m. (91 m min^{-1}) and below, while lubrication assumes an increasing role of importance. On the other hand, at speeds of 600 f.p.m. (183 m min^{-1}) and above, only cooling seems to be beneficial and any lubricating properties a fluid may have appear to be detrimental. Of the fluids tested it would appear that fluids A and B have the best lubricating characteristics, while water and fluid C have about the same cooling characteristics. For cutting at speeds from 300 to 500 f.p.m. (91 to 152 m min^{-1}), fluids A or B should be used, but for speeds above 500 f.p.m. (152 m min^{-1}), fluid C would be distinctly better.

The quantity $\cent^{\#}/xL$ is not only useful in deciding which fluid of a group is best, but it can also be used to determine how much one is justified in spending for a particular cutting fluid. This is most effectively illustrated by an example. Assume that a part is to be produced under the following conditions:

x = value of machine and labor ≐ 10 cents/min ($6 per hour)

TABLE 19.3

Optimum cost values for expensive tool for total destruction (R = 46)

Fluid	n	$V^{\#}$ f.p.m.	$V^{\#}$ m min^{-1}	$T^{\#}$ min	$¢^{\#}/xl$ (f.p.m.)$^{-1}$	$¢^{\#}/xl$ (m/min)$^{-1}$	Relative cost
Water, non-circulating, 60 °F (16 °C)	0.27	390	119	124	0.00352	11.55×10^{-3}	0.92
Water, circulating, 90 °F (32 °C)	0.32	390	119	98	0.00376	12.34×10^{-3}	0.98
Cutting fluid A, 2.5%	0.22	370	113	163	0.00346	11.35×10^{-3}	0.91
Cutting fluid B, 2.5%	0.23	375	114	154	0.00347	11.38×10^{-3}	0.91
Cutting fluid C, 2.5%	0.32	390	119	98	0.00377	13.37×10^{-3}	0.99
Cutting fluid D, 2.5%	0.23	370	113	154	0.00351	11.52×10^{-3}	0.92
Dry tool	0.24	345	105	146	0.00382	12.53×10^{-3}	1.00

d = diameter of part = 2 in (50 mm)
l = axial length of part = 2 in (50 mm)
t = feed = 0.010 i.p.r. (0.25 mm/rev)

The quantity in parentheses in eqn (19.15) is then

$$xL = \frac{x\pi dl}{12t} = \frac{(10)\pi(2)(2)}{12(0.010)} = 1050.$$

From Table 19.1 the value of cutting fluid A to this operation (assuming tools are used to total destruction) is clearly

$$¢^{\#}_{\text{dry}} - ¢^{\#}_{\text{A}} = (1050)\left(\frac{¢^{\#}}{xL_{\text{dry}}}\right) - \left(\frac{¢^{\#}}{xL_{\text{A}}}\right) = 0.19 \text{ cents per part}$$

If this fluid is used as a $2\frac{1}{2}$ percent solution and the concentrate costs g cents per gallon, then the cost of the solution will be (0.025)g cents/gallon. If N parts can be made per gallon of solution (including drag-out with chips, evaporation, replacement, etc.), then the cost of fluid per part will be (0.025 g/N) cents. The use of cutting fluid A is justified on a cost basis if

$$¢^{\#}_{\text{dry}} - ¢^{\#}_{\text{A}} > \frac{0.025g}{N}.$$

In the above example fluid A costs \$2.00 per gallon and hence, for the fluid to be justified on a cost basis

$$N > \frac{¢^{\#}_{\text{dry}} - ¢^{\#}_{\text{A}}}{(0.025)(200)} = \frac{0.19}{(0.025)(200)} = 0.04$$

The number of parts that may be produced per gallon (3.78 l) of solution will greatly exceed 0.04 and hence, it would appear that the fluid is certainly justified on a cost basis.

The optimum total cost per part including machining costs and fluid costs will be

$$\left(¢^{\#} + \frac{cg}{N}\right)$$

in any particular case where c is the concentration of the active ingredient that is used. The fluid of a group for which this quantity is a minimum is the most economical one to use and hence normally should be the one used if other conditions are equivalent. The cost of the fluid per part (cg/N) will normally be completely negligible compared with the machining cost per part and hence, that fluid which gives the smallest machining cost per part is normally the most economical regardless of the cost of the fluid concentrate. Thus, it is clear that if a particular fluid really does a superior job and the machining cost per part ($¢^{\#}$) is reduced any measurable amount by using it, this water-base fluid should be used regardless of whether the concentrate costs one, ten, or twenty dollars per gallon.

Stepless Speed Change

Most lathes are constructed with a number of finite speeds which differ one from the next by a constant ratio $(1 + K)$, where K is usually from 0.2 to 0.4. A stepless speed control can be supplied at a greater machine tool cost and it is of interest to know how much additional cost is justified for this feature.

If, in a given instance, the speed corresponding to minimum cost per part $(¢^{\#})$ is $(V^{\#})$, then the maximum discrepancy in speed due to the lack of a stepless speed changer will be $(\pm KV^{\#}/2)$, while the average discrepancy will be $(\pm KV^{\#}/4)$. The + or − sign pertains to whether the actual speed employed is greater or less than $V^{\#}$. The ideal cost per part when $V^{\#}$ can be realized will be

$$¢^{\#} = xT_c^{\#}\left(1 + \frac{R}{T^{\#}}\right). \tag{19.17}$$

The cost per part with the average discrepancy present will be

$$¢_1 = \frac{xT_c^{\#}}{1 \pm K/4}\left[1 + \frac{R}{T^{\#}}\left(1 \pm \frac{K}{4}\right)^{1/n}\right]. \tag{19.18}$$

The average additional cost per part due to the lack of a stepless speed changer will be

$$¢_1 - ¢^{\#} = xT_c^{\#}\left\{\frac{\mp K/4}{1 \pm K/4} + \frac{R}{T^{\#}}\left[\left(1 \pm \frac{K}{4}\right)^{(1/n-1)} - 1\right]\right\} \tag{19.19}$$

If Z parts are to be made during the entire life of the machine, then the total saving associated with the stepless speed control is

$$S = Z(¢_1 - ¢^{\#}) = ZxT_c^{\#}\left\{\frac{\mp K/4}{1 \pm K/4} + \left(\frac{n}{1-n}\right)\left[\left(1 \pm \frac{K}{4}\right)^{(1/n-1)} - 1\right]\right\}. \tag{19.20}$$

The quantity (ZT_c) is the total time the machine will be making useful parts. If the use-factor for the machine is F (i.e. F is the percentage of the nominal time the machine is in use that it will be cutting) and its total useful life is M minutes then (ZT_c) may be replaced by (FM).

In evaluating eqn (19.20), the average between the solutions with the plus and minus signs should be taken and hence

$$\bar{S} = \frac{FMx}{Z}\left\{\frac{2(K/4)^2}{1-(K/4)^2} + \left(\frac{n}{1-n}\right)\left[\left(1 + \frac{K}{4}\right)^{(1/n-1)} + \left(1 - \frac{K}{4}\right)^{(1/n-1)} - 2\right]\right\}$$

$$= FMx\left\{\left(\frac{K}{2}\right)^2\left[\frac{1}{1-(K/4)^2}+\frac{1-2n}{2n}\right]\right\} = FMx\psi(K, n) \qquad (19.21)$$

which is obtained by expanding the

$$\left(1+\frac{K}{4}\right)^{(1/n-1)} \text{ and } \left(1-\frac{K}{4}\right)^{(1/n-1)}$$

terms into a binomial series and noting that $(K/4)^2 \ll 1$.

Use factor $F = 0.8$
Life of machine (M) = 2 shifts for five days per week for 5 years = 1.2×16^6 min
Value of machine, labor, and overhead $(x) = 10$ cents/min
$K = 0.3$
$n = 0.1, 0.2, 0.3, 0.5.$

The results of the calculation for the total savings associated with the stepless speed change $(\bar{S})$ are given in Table 19.4.

The amount one is justified in paying for a stepless speed control on a representative lathe is seen to be about $1500 when no interest is charged on the investment and nothing is allowed for maintenance. Let the worth of the invested capital be taken at 10 percent per year and let 5 percent of the initial investment be allowed for maintenance per year. If the amount we are justified in spending for the speed changer at the beginning is x dollars, then the net investment at the end of the first year will be

$$x(1.1) + 0.05\,x - 300$$

where 300 is the amount of benefit derived from the speed changer during the first year. At the end of the second year the net investment in the speed changer is

$$[x\,(1.1) + 0.05\,x - 300]\,1.1 + 0.05\,x - 300 = (1.1)^2 x + (0.05\,x - 300)\,(1.1 + 1)$$

At the end of the fifth year the net investment is:

$$(1.1)^5\,x + (0.05\,x - 300)\,[(1.1)^4 + (1.1)^3 + (1.1)^2 + 1.1 + 1]$$

which will be zero for the correct value of x.

Solving for x we obtain

$$x = \$953$$

Thus, for this example, the justified expenditure for the speed changer at the time of purchase is $953. This value depends on the expected return on the investment which may be three or more times the value (10 percent) assumed in the above example. Further discussion of the economics of stepless machine tool drives, particularly for facing operations is to be found in Shaw (1960).

TABLE 19.4
Savings with stepless speed changer

n	$\psi(K, n)$	$\bar{S}$, \$
0.1	0.0281	2700
0.2	0.0140	1350
0.3	0.0094	900
0.5	0.0056	540

Tool and Work Materials

In the past the improvement associated with changes in tool and work materials have usually been studied in terms of tool life. While tool life is the proper criterion in such cases only when the costs of the two processes are the same, a comparison based on cost per part is always proper. A cost analysis for two ceramic tools has been presented by Shaw (1957) in which it was found that the tool which gave the poorest life but which cost the least was the one to be used since it resulted in the lowest cost per part. There are many problems of this sort that arise from day to day in the workshop but only one is considered here to illustrate rational procedures which can be followed in such cases.

A problem that frequently arises is that of deciding which of two steels, that are equivalent with regard to end-product, should be used, when the machinability and the cost of the two steels differ. The two steels are designated A and B. The cost per pound for these steels is M_A and M_B respectively (where $M_B > M_A$), while the machinability equations (eqn 19.1) for the steels are

$$VT^n = C_A$$

$$VT^n = C_B$$

where $C_B > C_A$ and the values of n are assumed to be approximately the same.

From eqn (19.6), the difference in minimum cost per part will be

$$¢_A^{\#} - ¢_B^{\#} = \frac{x}{1-n}\,[T_{c,A}^{\#} - T_{c,B}^{\#}]. \qquad (19.22)$$

By use of eqn (19.2)

$$¢_A^{\#} - ¢_B^{\#} = \frac{\pi dlx}{12t(1-n)}\left(\frac{1}{V_A^{\#}} - \frac{1}{V_B^{\#}}\right). \qquad (19.23)$$

The difference in cost of the workpiece material will be $W(M_B - M_A)$ where W is the weight of the workpiece before machining. In order that use of the new work material be justified on a cost basis,

$$W(M_B - M_A) < ¢_A^{\#} - ¢_B^{\#} \qquad (19.24)$$

or

$$M_B - M_A < \frac{\pi dlx}{12t(1-n)W}\left(\frac{1}{V_A^\#} - \frac{1}{V_B^\#}\right). \tag{19.25}$$

As an example consider the following case:

Diameter to be cut (d) = 2 in (50 mm)
Axial length of cut (l) = 4 in (100 mm)
Value of machine, operator, and overhead, (x) = 10 cents/min.
Weight of workpiece (W) = 4 lb (1.82 kg)
$R = 20$ min
$n = 0.2$
$t = 0.01$ i.p.r. (0.25 mm/rev)

If the present material A costs 15 cents per pound while the new material B costs 17 cents per pound, then

$$\frac{1}{V_A^\#} - \frac{1}{V_B^\#} > \frac{12(0.01)(0.8)4}{\pi(2)(4)(10)}(2) = 0.00304.$$

This means that if the cost-optimum cutting speed for the present material ($V_A^\#$) is 100 f.p.m. (30.5 m min^{-1}) then

$$V_B^\# > \frac{1}{1/100 - 0.00304} = 144$$

in order that the proposed substitution be justified on a cost basis.

Relief Angle

There are two criteria on which the relief angle for a tool may be chosen:

1. Minimum cost per part.
2. Close dimensional tolerance.

The volume rate of wear on the clearance face of a tool is experimentally found to increase linearly with cutting time or the helical length of the surface machined. If the tool life T corresponds to a certain value of wear land, then it follows that

$$T \sim \tan\theta \tag{19.26}$$

where θ is the relief angle (see Fig. 19.3). This means that the volume available for wear to give a certain wear land w should be as large as possible. However, when the relief angle is increased there is less metal at the cutting edge to support the load and to conduct heat away. In practice the maximum tool life based on a 0.03 inch (0.75 mm) wear land is found to correspond to a relief angle of about 12°.

However, this does not mean that such a large clearance angle should always

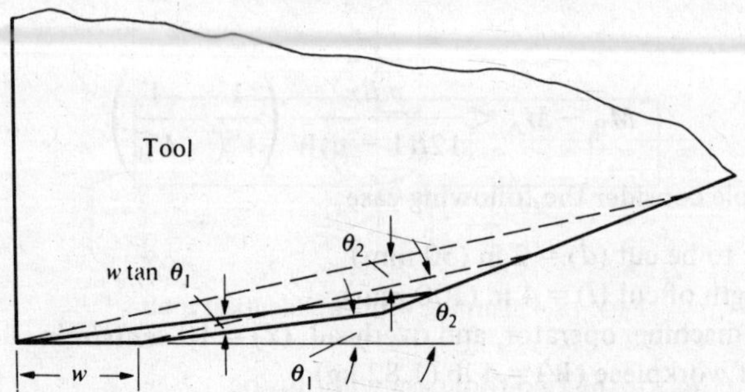

FIG. 19.3. Material to be removed when resharpening tools of different relief angle θ.

be used in practice, and it becomes evident why this is so when costs are considered. The mean cost of a tool and the cost of reconditioning a tool often represent an appreciable portion of the total cost for machining a part. The amount of metal that must be removed when a tool is reconditioned is (Fig. 19.3)

$$x = w \tan \theta. \tag{19.27}$$

The larger the layer (x) that must be removed to recondition the tool, the more grinding time will be required and the fewer the number of regrinds that will be possible before the tool must be replaced. It is thus evident that, except in the case of the throw-away tool, the relief angle giving minimum cost does not correspond to that giving maximum tool life, but will be a somewhat smaller relief angle. This explains why tool life can usually be increased by increasing the relief angles beyond values in general use in workshops but why it is frequently not wise to increase these relief angles even though an increase in tool life results.

From Fig. 19.3, it is evident that whenever dimensional accuracy is of major concern, both w and θ must be as small as possible. Thus, broaches (where dimensional accuracy is extremely important) are fitted with tools of very small relief angle (1° to 3°) and are resharpened when the wear land is but a few thousandths-of-an-inch. On the other hand, rough-turning tools (where minimum cost is of major interest) are fitted with relief angles of from 5° to 10° and the wear land is usually measured in hundredths-of-an-inch.

Tools Cutting Simultaneously

When two identical tools take identical cuts at the same time on the same machine involving a single operator, the entire operation will be completed in half the time (Fig. 19.4) and this alters the economic balance. The cost of machine, operator, and overhead per minute (x) will be the same as when a single tool is used but the value of the down-time to change tools (T_d) and the cost of the new tools ($2y$) will be double the value for the single-tool

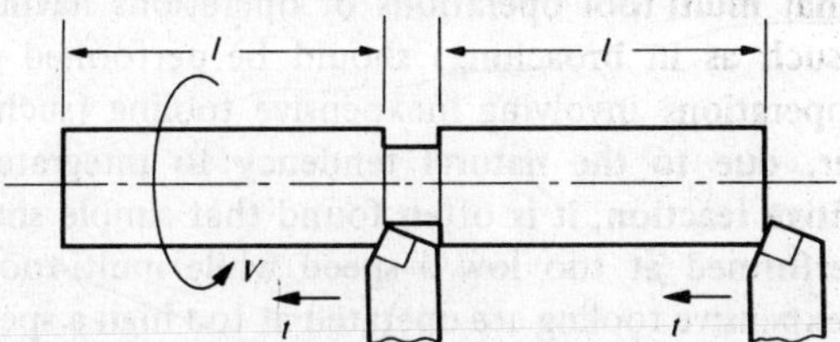

FIG. 19.4. Turning with two identical tools

case. The cost ratio for the two-tool case will thus be

$$R_2 = \frac{xT_d + y}{x} = 2R_1 \tag{19.28}$$

where R_1 is the value of R for the single-tool case. From eqn (19.4), the cost optimum tool life (min) for the two-tool case will be

$$T_2^{\#} = \left(\frac{1}{n} - 1\right) R_2 = 2T_1^{\#} \tag{19.29}$$

where $T_1^{\#}$ is the cost-optimum tool-life for the single-tool case.

It is thus evident that as the total tool and tool changing cost increases relative to machine and labor cost per minute (x), the optimum tool-life increases and the machine should be operated at a lower speed.

An extreme multi-tool situation is the multispindle automatic screw machine. In this case several tools operate simultaneously on each of several spindles. After the slowest tool has completed its cut, the spindles are indexed and a completed part is obtained after a given spindle has been indexed through all stations.

All spindles operate at the same r.p.m. and an important problem is to determine the cost optimum r.p.m. There is one tool that requires the greatest time to complete its cut and this tool determines the cycle time (time to complete a single part = cutting time for this critical tool times the number of spindles). If all tools are changed at the same time (this need not be the case), then a good approximation to this complex optimization problem is to compute the R value for each tool and then determine the optimum tool life ($T^{\#}$) as though the critical tool (one determining cycle time) is acting alone with an R value equal to the sum of the R values of all tools. The cost optimum tool life ($T^{\#}$) may thus be estimated by substituting the sum of the R values for all tools for R in eqn (19.4) and using the value of n for the critical tool.

In actual practice, tools are normally changed in groups rather than all at the same time but this does not change the fact that a multispindle automatic screw machine is one having an unusually high R value and it is therefore to be expected that the cost-optimum operating speed for the critical tool will be significantly lower than for an equivalent single-tool situation. It is therefore to be expected that multi-tool operations or operations having unusually expensive tooling (such as in broaching) should be performed at lower speeds than single-tool operations involving inexpensive tooling (such as single-point turning). However, due to the natural tendency to integrate all experience into a single intuitive reaction, it is often found that simple single-tool turning operations are performed at too low a speed while multi-tool operations involving relatively expensive tooling are operated at too high a speed.

Optimum Production Rate

An alternative optimization procedure involves the selection of cutting speed for maximum rate of production. This involves writing an expression for the total machining time including cutting time and tool and work changing times, and then differentiating this with respect to cutting speed and equating to zero.

The total machining time will be

$$\Sigma T = T_c + T_d \frac{T_c}{T} + T_w \tag{19.30}$$

where T_c = cutting time given by eqn (19.2)
T_d = down-time to change and reset tool
T = tool-life, min
T_w = work-changing time (assumed independent of V).

Substituting for T_c and T from eqns (19.1) and (19.2), taking the partial derivative relative to V (t constant), and equating to zero gives the following value for production-rate-optimum tool-life

$$T^{\#\#} = T_d\left(\frac{1}{n} - 1\right). \tag{19.31}$$

When this equation is compared with eqn (19.4):

$$\frac{T^{\#\#}}{T^{\#}} = \frac{T_d}{R} = \frac{T_d}{T_d + y/x} \tag{19.32}$$

it is evident that the optimum tool-life based on rate of production will be less than that based on minimum cost per part.

The cutting speed corresponding to maximum production rate will be

$$V^{\#\#} = \frac{C}{(T^{\#\#})^n}. \tag{19.33}$$

Figure 19.5 shows diagrammatically the relationship between the optimum cutting speeds for minimum cost ($V^{\#}$) and maximum production rate ($V^{\#\#}$).

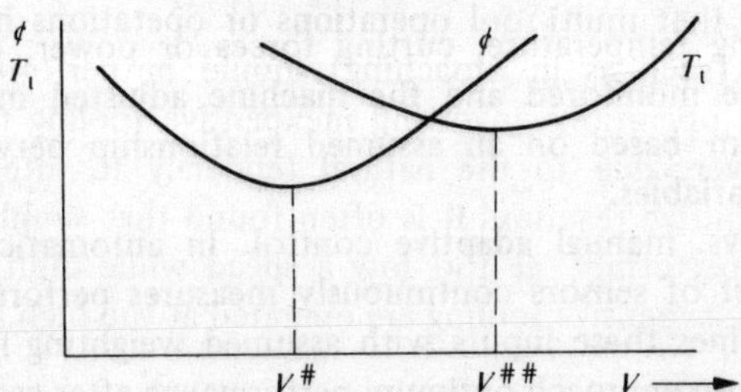

FIG. 19.5. Relation of cost-optimum cutting speed ($V^{\#}$) and production-rate-optimum cutting speed ($V^{\#\#}$).

Adaptive Control

Data Banks vs. Adaptive Control

Level 1 optimization is most widely carried out in industry by use of a data-bank approach. Past experience of the company is stored in a computer and used to estimate the unit production cost when using a variety of tool materials, tool geometries, feeds, speeds, etc. By comparing the unit costs for each combination, it is possible for the production engineer to estimate the optimum set of conditions for a job and to specify these conditions on the production order going to the operator with the parts to be processed and the expendible tools to be used. The difficulty with this approach is that the number of variables that must be specified to characterize a job completely is too large to be used in practice and when only the most important variables are considered, the result is too approximate. Also, the inputs to the process such as workpiece hardness, workshop temperature, or even operator fatigue are not static but vary with time. The problem of predicting the optimum set of conditions for a given machining operation is similar to the problem of predicting the weather. For any one month, the average of all weather predictions is apt to be fairly accurate even though the day-to-day predictions are relatively inaccurate. In the case of data-bank-based predictions, the mean performance for a number of jobs will similarly be far closer to what is predicted than the performance for individual jobs will be a data-bank approach, although approximate, usually represents a good starting point. To do better, some form of adaptive control must be used in which performance is monitored and the machine adjusted according to the results obtained.

Adaptive control may be classified as follows:

1. Primary adaptive control vs. secondary. In primary adaptive control, the actual optimization variable is measured or calculated and the machine adjusted stepwise to approach optimization.
2. In secondary adaptive control, secondary variables such as rate of tool

wear, cutting temperature, cutting forces or power, chatter, finish, or accuracy are monitored and the machine adjusted in the direction of the optimum based on an assumed relationship between primary and secondary variables.

3. Automatic vs. manual adaptive control. In automatic adaptive control (AAC), a set of sensors continuously measures performance and a computer combines these inputs with assumed weighting factors and adjusts the machine to approach optimum performance after each set of inputs.

   In manual adaptive control (MAC), the operator is part of the control loop. He replaces the required group of sensors and also serves as the interface between an inexpensive programmable computer and the machine tool. Manual adaptive control is a particularly attractive alternative when capital funds are in short supply.

Automatic Adaptive Control (AAC)

Practically all applications of AAC are of the secondary type. For example it is well established that in a turning operation, the feed should be as high as possible consistent with the maximum force on the cutting tool to cause tool fracture and the surface finish required. In some instances, the hardness of the workpiece will vary significantly from point to point on a given workpiece and from one workpiece to another. When this is the case the cost per part and the production rate can be improved by holding the cutting speed constant and adjusting the feed to the highest value consistent with a maximum allowable cutting force that is safe relative to tool fracture and a maximum allowable amplitude of chatter vibration relative to the finish required. If either the allowable force or amplitude of chatter is exceeded, the feed is automatically decreased until either the force or vibrational amplitude decreases to what is thought to be the acceptable level. In this case, both cutting force and vibrational amplitude are secondary variables related in a complex way to the items of primary interest—tool fracture and surface finish. Since there is no known way of sensing when a tool is about to fail in brittle fracture, one must be content with monitoring a secondary variable—cutting force. Likewise, tool chatter is only one component of surface roughness and since there are no workshop-proven in-process sensors for surface roughness, one must be content to measure a secondary variable (vibrational amplitude) that is related in a complex way to the main variable of interest (surface roughness).

The automatic adaptive control of this example is useful even though far from perfect. It is useful in that the rate of material removal will be greater than if the feed rate were to be set at a level such that constant values of tool force or vibrational amplitude not be exceeded regardless of the local workpiece hardness, presence of surface scale, or workpiece-to-workpiece variation in hardness or structure. It is less than perfect in that there is a wide dispersion of tool-to-tool fracture strength and the fact than an interaction exists between tool-fracture tendency and surface finish produced due to a change in workpiece hardness. When workpiece hardness increases, the cutting force increases, calling for a decrease in feed. However, an increase in workpiece hardness leads to a better finish (aside from chatter-induced roughness) calling for an increase in feed.

The adaptive control system just described automatically senses secondary variables: cutting force and vibrational amplitude, related in a complex way to the primary variables of interest (tool-facture tendency and surface finish). It may therefore be classified as an example of secondary automatic adaptive control (SAAC).

The only known example of primary automatic adaptive control (PAAC) involves the control of an abrasive cutoff machine (Gall 1969). Figure 19.6 shows schematically the abrasive cut-off operation. A high-speed grinding wheel (V = 12 000 f.p.m. or 3600 m min^{-1}) is fed vertically downward at a feed rate d. The variable of interest in this case will usually be the total cost per cut which involves the following components:

1. Cost of machine, labor, and overhead (cents/cut).
2. Value of abrasive consumed per cut (cents/cut).
3. Work-changing cost per cut (cents/cut).

The optimum (minimum) cost per cut (¢#) is obtained when operating at the highest wheel-speed (V) possible consistent with safety and machine capability. However, an optimum downfeed (d#) exists that usually lies in the practical range of downfeeds available. This optimum value of d depends on workpiece size, shape, and properties, several grinding-wheel characteristics as well as environmental and operator influences in such a complex way that it is very difficult to predict the cost-optimum value of d, based on the general accumulated past experience.

A more practical approach is to sense the change in wheel-diameter after each cut and introduce this into a computer along with the downfeed pertaining and several fixed quantities (wheel speed, mean value of grinding wheel [cents/vol], value of machine, operator, and overhead [cents/min], and certain workpiece and grinding-wheel dimensions). The computer calculates the cost per cut, takes the previous value from storage and automatically changes the downfeed by a small increment (Δd) upward or downward depending on the sign of Δd and whether the new cost per cut is greater or less than the previous one. In this way the machine may be adjusted to begin operation at the best estimate of d based on previous experience but will automatically adjust d to correspond to the cost-optimum value d# after a few trials. By monitoring each cut, the machine will seek a new cost-optimum downfeed d# as soon as any one of a large number of variables changes sufficiently to require a new optimum. This includes such obvious variables as workpiece hardness and grinding-wheel homogeneity as well as less obvious ones such as workshop

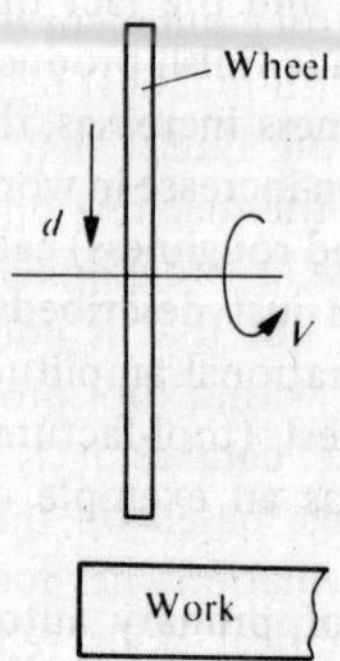

FIG. 19.6. Abrasive cut-off operation.

temperature and operator fatigue. If the machine is manually serviced, the work-changing time should be included in the time per cut. The optimization will then include any increase in work-changing cost associated with an increase in rate of production and operator fatigue.

The above is an example of primary automatic adaptive control (PAAC). While the procedure may appear complicated, the hardware involved is relatively inexpensive since it employs readily available analog elements assembled into an inexpensive dedicated computer. The only sensor required (measurement of wheel diameter) is relatively simple in this case. Versatility is provided by allowing quantities such as cost of machine, and operator (x, cents/min) and mean value of consumable wheel volume (y, cents/vol) to be dialed in at the beginning of each job. Details on this development are to be found in Gall (1969).

Manual Adaptive Control

The difficult elements of automatic adaptive control are the sensors, interface between sensors and computer, and the cost of the computer. The appearance of small, inexpensive, programmable computers makes it possible to expand the concept of an incremental approach to optimum operation of a machine tool by having the machine operator perform the function of the required group of sensors as well as the machine–computer interface. This type of adaptive control is termed manual adaptive control (MAC) to distinguish it from the more sophisticated automatic adaptive control (AAC) and has been the subject of several articles (Shaw and Komanduri 1977; Shaw and Avery 1979).

A trend in industry is to reduce the input of the operator by removing him from the decision-making process and by depending less on his operating skills. This has tended to reduce the role of the operator to that of a watchman leading to boredom and aggravating the job-satisfaction problem.

The MAC approach to adaptor control appears to be useful until sensor technology and the cost of control equipment drops to the point where primary automatic adaptive control can be a practical reality.

It should be emphasized at the outset that in manual adaptive control, the operator uses the calculator at the machine on the shop floor. Of course, it is possible for an engineer to use a programmable calculator to help analyze data collected in the workshop but that is not what MAC is about. The basic idea is to provide the operator with a tool programmed by an engineer which enables the operator to analyze data as it is generated by the machine and to make wise decisions-for-change based on this analysis. The operator is thus enabled to make important calculations he does not have time to make even when he possesses the mathematical skills required. The system is designed to work as though the engineer were continuously at the side of the operator to analyze results coming from the machine and to make the changes in operating conditions necessary to optimize the process. The programmable calculator merely enables the engineer to perform these activities indirectly *in absentia*.

In order to focus attention on the principles involved, a very simple example will be considered—that of turning a workpiece in a lathe by a single-point tool. However, detailed programs have been prepared and applied in industry to manufacturing operations involving multispindle automatic screw machines, boring, milling, sawing, and a variety of grinding situations. For turning on a lathe, the quantity to be optimized will usually be cost per part (to be minimized) or production rate in parts per hour (to be maximized). In both cases, the feed should be as large as possible consistent with constraints (power available, tool fracture, finish, etc.). In this case there is an optimum cutting speed that will usually be in the practical range of values and the problem is to operate as close to this speed as possible.

When MAC is applied to a straight single-tool turning situation, the operator introduces the program prepared previously by an engineer by running the small magnetic strip containing the program through a hand-held calculator. This program contains approximate values of machine, operator, and overhead cost (x) and the mean cost of a cutting edge (y) for the corresponding machine operator and tool to be used. The program (magnetic strip) should be delivered to the operator with the shop order, tools, and parts to be machined.

The operator decides when the tool should be changed which may be because of any one of a combination of reasons (tool breakage, tool wear, excessive cutting temperature, chatter, poor finish, lack of dimensional accuracy, etc.). After each tool is replaced, the operator keys in two numbers:

1. Time elapsed from last tool change.
2. Number of parts machined.

A key is then pressed that causes the calculator to determine the mean cost per part and subract this value from the corresponding value taken from storage for the previous tool. If this difference is positive, the operator increases the spindle speed by a small increment (say 10 percent or the next available speed on the machine if the speed is not continuously variable) and repeats the procedure. If the number that comes up is negative, the operator decreases the speed by a corresponding amount. In this way, the machine is adjusted to reach the optimum cutting speed ($V\#$) in a few trials. If workpiece hardness changes, or any other of a large number of items that will affect tool life, the machine will automatically be guided to a new optimum value of cutting speed.

In each application of MAC, the operation must be analyzed to determine whether one or more operating variables may be optimized and then a program written and tested in the workshop. This requires a certain amount of skill and has proven to be an excellent vehicle for training students to approach manufacturing problems in a rational way.

MAC was successfully applied to a centerless grinding machine used to grind cylindrical track-pins in a tracked vehicle manufacturing plant. A preliminary consideration of the process indicated that there should be an optimum production rate for minimum cost. At a high rate of production, the cost of wheel-dressing (including wheel and diamond cost and downtime during dressing) is excessive while at a lower rate of production, labor cost is too high. The program was written to provide total cost per part when the time between dressing operations and the number of parts produced were entered into the calculator. Production rate was changed by changing the r.p.m. of the regulating wheel. The initial setting for the first part tested was ground at the feed rate given by the data bank programmed into the company's central computer. This was based on actual performance values for the machine in use that had been accumulated over a fairly long period of time. After the first wheel-dressing cycle, the calculator called for an incremental increase in feed rate. This was repeated again and again until the optimum value was reached. At this point the total cost for this particular grinding operation had been cut approximately in half. The next job that was put on this machine was similarly treated. This time the calculator led the operator downward in speed and finally the optimum cost was appreciably less than the starting value suggested by the data bank.

These results are typical of those to be expected from a data-bank approach. While the average production rate will be about right for a variety of jobs, the predicted value for each individual case may be appreciably in error—some values being too low and some too high. More details concerning this case may be found in Stelson, Komanduri, and Shaw (1977).

The MAC system is also of value when used in connection with a preprogrammed numerically controlled machine. In one example, a large sophisticated NC machining center was programmed to produce parts based on previous experience. However, this machine was provided with an override capability that enabled the operator to change the production rate without changing the proportional time involved with each detail. When MAC was used to adjust the cycle time for minimum cost per part, an appreciable saving in total cost per part was realized. This illustrates that MAC can play a valuable role when used to fine-tune and monitor a preprogrammed sophisticated NC machine.

The MAC system has also proven to be of use in evaluating different tool materials on a given job. In such a case, it is important that conditions be optimized for each material before comparing the resulting cost per part. Details on how this may be done and an example of where production rate is optimized instead of cost per part is to be found in Fucich, Obermiller, Shuster, and Shaw (1978).

Role of Capital in Productivity

It should be noted that throughout the discussion of optimization presented here, both labor and capital cost were included. This is consistent with the concept that productivity is not alone a matter of the output per man hour but should take into account the capital involved in increasing output per hour. As the cost of more sophisticated machinery increases, as interest rates rise, and as capital available to industry for investment decreases, it becomes ever more important to be sure that available capital is wisely used.

The big problem relative to process-optimization is not in knowing what to do but in applying what is already known. There are many reasons why it is difficult to implement in the workshop concepts that are fully developed and available in the literature. The single most important reason is a lack of broadly trained, innovative, manufacturing engineers.

References

Fucich, L., Obermiller, K., Shuster, A., and Shaw, M.C. (1978). *Proc. Sixth N. Am. Metal Working Res. Conf.* p. 372. Soc. Manuf. Engrs, Dearborn, MI.

Gall, D. (1969). *Ann. CIRP* **17**, 359.

Gilbert, W. W. (1950). *Machining theory and practice*, p. 465. Am. Soc. for Metals, Metals Park, Cleveland, OH.

Shaw, M. C. (1957). *Ind. Anz.* **79**, 847.

Shaw, M. C. (1959). *Microtecnic* **13**, 103.

Shaw, M. C. (1960). *Int. J. mech. Sci* **1**, 89.

Shaw, M. C. and Avery, J. P. (1979). *Proc. Fifth World Congress on Theory of Machines and Mechanisms* **1**, 777. Concordia University, Montreal.

Shaw, M. C. and Komanduri, R. (1977). *American Machinist* **121**, 81.

Stelson, T. S., Komanduri, R., Shaw, M.C. (1977). In *Proc. Fourth Int. Prod. Eng. Conf.* (Tokyo). Taylor and Francis, London.

Taylor, F. W. (1907). *Trans. Am. Soc. mech. Engrs* **28**, 33.

AUTHOR INDEX

SUBJECT INDEX

19